*Instant Notes*

# BIOCHEMISTRY

**Second Edition**

**The INSTANT NOTES series**

Series editor
**B.D. Hames**
*School of Biochemistry and Molecular Biology, University of Leeds, Leeds, UK*

---

Animal Biology
Ecology
Genetics
Microbiology
Chemistry for Biologists
Immunology
Biochemistry 2nd edition
Molecular Biology 2nd edition
Neuroscience
Psychology
Developmental Biology

*Forthcoming titles*
Plant Biology
Bioinformatics

**The INSTANT NOTES Chemistry Series**
*Consulting editor: Howard Stanbury*

Organic Chemistry
Inorganic Chemistry
Physical Chemistry

*Forthcoming titles*
Analytical Chemistry
Medicinal Chemistry

# *Instant Notes*

# BIOCHEMISTRY

## Second Edition

*B.D. Hames & N.M. Hooper*
School of Biochemistry and Molecular Biology,
University of Leeds, Leeds, UK

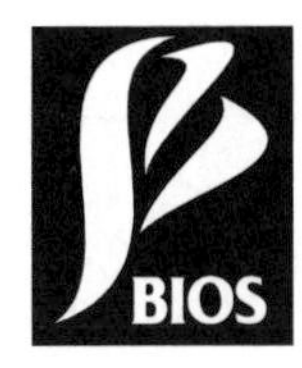

First published 1997
Second edition published 2000
Reprinted 2002



A CIP catalogue record for this book is available from the British Library.

ISBN 1 85996 142 8

**BIOS Scientific Publishers Ltd**
**9 Newtec Place, Magdalen Road, Oxford OX4 1RE, UK**
**Tel. +44 (0)1865 726286. Fax +44 (0)1865 246823**
**World Wide Web home page: http://www.bio.co.uk/**

Published in the United States of America, its dependent territories and Canada by Springer-Verlag New York Inc., 175 Fifth Avenue, New York, NY 10010-7858, in association with BIOS Scientific Publishers Ltd

Published in Hong Kong, Taiwan, Singapore, Thailand, Cambodia, Korea, The Philippines, Indonesia, The People's Republic of China, Brunei, Laos, Malaysia, Macau and Vietnam by Springer-Verlag Singapore Pte Ltd, 1 Tannery Road, Singapore 347719, in association with BIOS Scientific Publishers Ltd.

Production Editor: Fran Kingston
Typeset by illustrated by J&L Composition Ltd, Filey, UK
Printed by Biddles Ltd, Guildford, UK

**Cover image:** molecular surface rendering of HIV-1 reverse transcriptase complexed with a highly specific RNA pseudoknot inhibitor (J. Jaeger *et al. EMBO J.* **17**(15): 4535–42, 1988). The inhibitor was generated by SELEX method and has a sub-nanamolar binding constant. Image courtesy of Dr. J. Jaeger, Astbury Centre for Structural Molecular Biology, University of Leeds.

# Contents

# ABBREVIATIONS

| | |
|---|---|
| A | adenine |
| ACAT | acyl-CoA cholesterol acyltransferase |
| ACP | acyl carrier protein |
| ADP | adenosine diphosphate |
| AIDS | acquired immune deficiency syndrome |
| Ala | alanine |
| ALA | aminolaevulinic acid |
| AMP | adenosine monophosphate |
| Arg | arginine |
| Asn | asparagine |
| Asp | aspartic acid |
| ATCase | aspartate transcarbamoylase |
| ATP | adenosine 5′-triphosphate |
| ATPase | adenosine triphosphatase |
| bp | base pairs |
| C | cytosine |
| cAMP | 3′, 5′ cyclic AMP |
| CAP | catabolite activator protein |
| cDNA | complementary DNA |
| CDP | cytidine diphosphate |
| cGMP | cyclic GMP |
| CM | carboxymethyl |
| CMP | cytidine monophosphate |
| CNBr | cyanogen bromide |
| CoA | coenzyme A |
| CoQ | coenzyme Q (ubiquinone) |
| $CoQH_2$ | reduced coenzyme Q (ubiquinol) |
| CRP | cAMP receptor protein |
| CTL | cytotoxic T lymphocyte |
| CTP | cytosine triphosphate |
| Cys | cysteine |
| $\Delta E_0'$ | change in redox potential under standard conditions |
| $\Delta G$ | Gibbs free energy |
| $\Delta G^{\ddagger}$ | Gibbs free energy of activation |
| $\Delta G^{0\prime}$ | Gibbs free energy under standard conditions |
| DAG | 1,2-diacylglycerol |
| dATP | deoxyadenosine 5′-triphosphate |
| dCTP | deoxycytidine 5′-triphosphate |
| ddNTP | dideoxynucleoside triphosphate |
| DEAE | diethylaminoethyl |
| dGTP | deoxyguanosine 5′-triphosphate |
| DIPF | diisopropylfluorophosphate |
| DNA | deoxyribonucleic acid |
| DNase | deoxyribonuclease |
| DNP | 2,4-dinitrophenol |
| dTTP | deoxythymidine 5′-triphosphate |
| $E$ | redox potential |
| EC | Enzyme Commission |
| EF | elongation factor |
| eIF | eukaryotic initiation factor |
| ELISA | enzyme-linked immunosorbent assay |
| ER | endoplasmic reticulum |
| ETS | external transcribed spacer |
| F-2,6-BP | fructose 2,6-bisphosphate |
| FAB-MS | fast atom bombardment mass spectrometry |
| FACS | fluorescence-activated cell sorter |
| FAD | flavin adenine dinucleotide (oxidized) |
| $FADH_2$ | flavin adenine dinucleotide (reduced) |
| FBPase | fructose bisphosphatase |
| *N*-fMet | *N*-formylmethionine |
| $FMNH_2$ | flavin mononucleotide (reduced) |
| FMN | flavin mononucleotide (oxidized) |
| GalNAc | *N*-acetylgalactosamine |
| GDP | guanosine diphosphate |
| GlcNAc | *N*-acetylglucosamine |
| Gln | glutamine |
| Glu | glutamic acid |
| Gly | glycine |
| GMP | guanosine monophosphate |
| GPI | glycosyl phosphatidylinositol |
| GTP | guanosine 5′-triphosphate |
| Hb | hemoglobin |
| HbA | adult hemoglobin |
| HbF | fetal hemoglobin |
| HbS | sickle cell hemoglobin |
| HDL | high density lipoprotein |
| His | histidine |
| HIV | human immunodeficiency virus |
| HMG | 3-hydroxy-3-methylglutaryl |
| HMM | heavy meromyosin |
| hnRNA | heterogeneous nuclear RNA |
| hnRNP | heterogeneous nuclear ribonucleoprotein |
| HPLC | high-performance liquid chromatography |
| hsp | heat shock protein |
| Hyl | 5-hydroxylysine |
| Hyp | 4-hydroxyproline |
| IDL | intermediate density lipoprotein |
| IF | initiation factor |
| Ig | immunoglobulin |

| | |
|---|---|
| IgG | immunoglobulin G |
| Ile | isoleucine |
| $IP_3$ | inositol 1,4,5-trisphosphate |
| IPTG | isopropyl-β-D-thiogalactopyranoside |
| IRES | internal ribosome entry sites |
| ITS | internal transcribed spacer |
| K | equilibrium constant |
| $K_m$ | Michaelis constant |
| LCAT | lecithin–cholesterol acyltransferase |
| LDH | lactate dehydrogenase |
| LDL | low density lipoprotein |
| Leu | leucine |
| LMM | light meromyosin |
| Lys | lysine |
| Met | methionine |
| MS | mass spectrometry |
| mV | millivolt |
| mRNA | messenger RNA |
| $NAD^+$ | nicotinamide adenine dinucleotide (oxidized) |
| NADH | nicotinamide adenine dinucleotide (reduced) |
| $NADP^+$ | nicotinamide adenine dinucleotide phosphate (oxidized) |
| NADPH | nicotinamide adenine dinucleotide phosphate (reduced) |
| NAM | *N*-acetylmuramic acid |
| NHP | nonhistone protein |
| NMR | nuclear magnetic resonance |
| ORF | open reading frame |
| PAGE | polyacrylamide gel electrophoresis |
| PC | plastocyanin |
| PCR | polymerase chain reaction |
| PEP | phosphoenolpyruvate |
| PFK | phosphofructokinase |
| Phe | phenylalanine |
| $P_i$ | inorganic phosphate |
| pI | isoelectric point |
| pK | dissociation constant |
| PKA | protein kinase A |
| $PP_i$ | inorganic pyrophosphate |
| Pro | proline |
| PQ | plastoquinone |
| PSI | photosystem I |
| PSII | photosystem II |
| PTH | phenylthiohydantoin |
| Q | ubiquinone (coenzyme Q) |
| $QH_2$ | ubiquinol ($CoQH_2$) |
| RER | rough endoplasmic reticulum |
| RF | release factor |
| RFLP | restriction fragment length polymorphism |
| RNA | ribonucleic acid |
| RNase | ribonuclease |
| rRNA | ribosomal RNA |
| rubisco | ribulose bisphosphate carboxylase |
| SDS | sodium dodecyl sulfate |
| Ser | serine |
| SER | smooth endoplasmic reticulum |
| snoRNA | small nucleolar RNA |
| snoRNP | small nucleolar ribonucleoprotein |
| snRNA | small nuclear RNA |
| snRNP | small nuclear ribonucleoprotein |
| SRP | signal recognition particle |
| SSB | single-stranded DNA-binding (protein) |
| TBP | TATA box-binding protein |
| TFII | transcription factor for RNA polymerase II |
| TFIIIA | transcription factor IIIA |
| Thr | threonine |
| $T_m$ | melting point |
| Tris | Tris(hydroxymethyl)aminomethane |
| tRNA | transfer RNA |
| Trp | tryptophan |
| Tyr | tyrosine |
| UDP | uridine diphosphate |
| UMP | uridine monophosphate |
| URE | upstream regulatory element |
| UTP | uridine 5′-triphosphate |
| UV | ultraviolet |
| Val | valine |
| $V_0$ | initial rate of reaction |
| VLDL | very low density lipoprotein |
| $V_{max}$ | maximum rate of reaction |

# PREFACE

Three years ago, the sight of first-year students wading through acres of fine print in enormous biochemistry textbooks led us to believe that there must be a better way; a book that presented the core information in a much more accessible format. Hence *Instant Notes in Biochemistry* was born. The tremendous success of this book has proved the concept. However, not surprisingly, we did not get everything right at the first attempt. Student readers and lecturing staff told us about the relatively scant coverage of gene expression, for example, plus a host of other more minor, but significant points. We have addressed all of these issues in this new edition. There is a major expansion of coverage of gene transcription and its regulation in both prokaryotes and eukaryotes, as well as RNA processing and protein synthesis (sections G and H). Many other topics have been added or rewritten in the light of comments, including acids and bases, pH, ionization of amino acids, thermodynamics, protein stability, protein folding, protein structure determination, flow cytometry, and peptide synthesis. Whilst writing the new edition, we have also looked at each illustration again and made modifications as necessary to make these even clearer for the student reader. Many new illustrations have also been included. Naturally, all of this has led to a substantial lengthening of the book. However, in every case, whether considering the text or the illustrations, we have been at pains to include only the information that we believe is essential for a good student understanding of the subject. The key features of this new book therefore remain the same as for the first edition: to present the core information on biochemistry in an easily accessible format that is ideally suited to student understanding – and to revision when the dreaded examinations come! We have been told by students that the first edition did just that. We have great hopes that the same will hold true for this new update.

*David Hames*
*Nigel Hooper*

# A1 PROKARYOTES

## Key Notes

**Prokaryotes**

Prokaryotes (bacteria and blue-green algae) are the most abundant organisms on earth. A prokaryotic cell does not contain a membrane-bound nucleus. Bacteria are either cocci, bacilli or spirilla in shape, and fall into two groups, the eubacteria and the archaebacteria.

**Cell structure**

Each prokaryotic cell is surrounded by a plasma membrane. The cell has no subcellular organelles, only infoldings of the plasma membrane called mesosomes. The deoxyribonucleic acid (DNA) is condensed within the cytosol to form the nucleoid. Some prokaryotes have tail-like flagella.

**Bacterial cell walls**

The peptidoglycan (protein and oligosaccharide) cell wall protects the prokaryotic cell from mechanical and osmotic pressure. A Gram-positive bacterium has a thick cell wall surrounding the plasma membrane, whereas Gram-negative bacteria have a thinner cell wall and an outer membrane, between which is the periplasmic space.

**Related topics**

Eukaryotes (A2)
Amino acids (B1)
Membrane lipids (E1)
Chromosomes (F2)
Cilia and flagella (N2)

## Prokaryotes

Prokaryotes are the most numerous and widespread organisms on earth, and are so classified because they have no defined membrane-bound nucleus. Prokaryotes range in size from 0.1 to 10 μm, and have one of three basic shapes: spherical (**cocci**), rodlike (**bacilli**) or helically coiled (**spirilla**). They can be divided into two separate groups: the **eubacteria** and the **archaebacteria**. The eubacteria are the commonly encountered bacteria in soil, water and living in or on larger organisms, and include the Gram-positive and Gram-negative bacteria, and cyanobacteria (photosynthetic blue-green algae). The archaebacteria grow in unusual environments such as salt brines, hot acid springs and in the ocean depths, and include the sulfur bacteria and the methanogens.

## Cell structure

Like all cells, a prokaryotic cell is bounded by a **plasma membrane** that completely encloses the cytosol and separates the cell from the external environment. The plasma membrane, which is about 8 nm thick, consists of a lipid bilayer containing proteins (see Topic E1). Although prokaryotes lack the membranous subcellular organelles characteristic of eukaryotes (see Topic A2), their plasma membrane may be infolded to form **mesosomes** (*Fig. 1*). The mesosomes may be the sites of deoxyribonucleic acid (DNA) replication and other specialized enzymatic reactions. In photosynthetic bacteria, the mesosomes contain the proteins and pigments that trap light and generate adenosine triphosphate (ATP). The aqueous cytosol contains the macromolecules [enzymes, messenger ribonucleic acid (mRNA), transfer RNA (tRNA) and ribosomes], organic compounds and

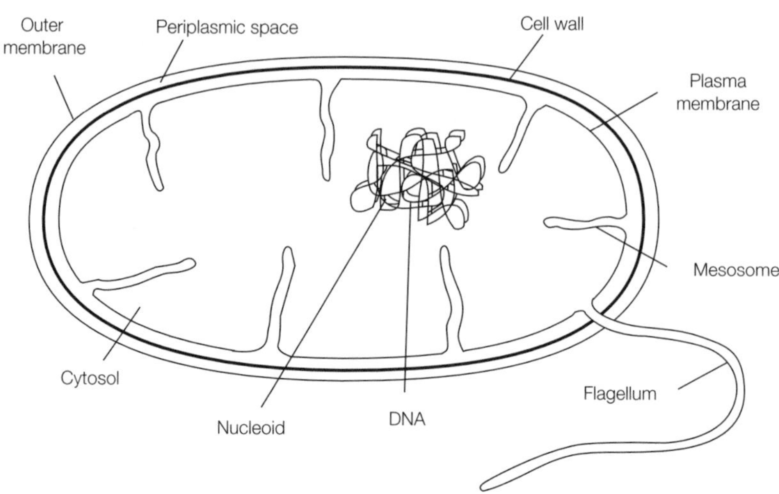


*Fig. 1. Prokaryote cell structure.*

ions needed for cellular metabolism. Also within the cytosol is the prokaryotic 'chromosome' consisting of a single circular molecule of DNA which is condensed to form a body known as the **nucleoid** (*Fig. 1*) (see Topic F2). Many bacterial cells have one or more tail-like appendages known as **flagella** which are used to move the cell through its environment (see Topic N2).

## Bacterial cell walls

To protect the cell from mechanical injury and osmotic pressure, most prokaryotes are surrounded by a rigid 3–25 nm thick **cell wall** (*Fig. 1*). The cell wall is composed of **peptidoglycan**, a complex of **oligosaccharides** and **proteins**. The oligosaccharide component consists of linear chains of alternating *N*-acetylglucosamine (GlcNAc) and *N*-acetylmuramic acid (NAM) linked β(1–4) (see Topic J1). Attached via an amide bond to the lactic acid group on NAM is a **D-amino acid**-containing tetrapeptide. Adjacent parallel peptidoglycan chains are covalently cross-linked through the tetrapeptide side-chains by other short peptides. The extensive cross-linking in the peptidoglycan cell wall gives it its strength and rigidity. The presence of D-amino acids in the peptidoglycan renders the cell wall resistant to the action of **proteases** which act on the more commonly occurring L-amino acids (see Topic B1), but provides a unique target for the action of certain **antibiotics** such as **penicillin**. Penicillin acts by inhibiting the enzyme that forms the covalent cross-links in the peptidoglycan, thereby weakening the cell wall. The β(1–4) glycosidic linkage between NAM and GlcNAc is susceptible to hydrolysis by the enzyme **lysozyme** which is present in tears, mucus and other body secretions.

Bacteria can be classified as either **Gram-positive** or **Gram-negative** depending on whether or not they take up the **Gram stain**. Gram-positive bacteria (e.g. *Bacillus polymyxa*) have a thick (25 nm) cell wall surrounding their plasma membrane, whereas Gram-negative bacteria (e.g. *Escherichia coli*) have a thinner (3 nm) cell wall and a second **outer membrane** (*Fig. 2*). In contrast with the plasma membrane (see Topic E3), this outer membrane is very permeable to the passage of relatively large molecules (molecular weight > 1000 Da) due to **porin proteins** which form pores in the lipid bilayer. Between the outer membrane and the cell wall is the **periplasm**, a space occupied by proteins secreted from the cell.

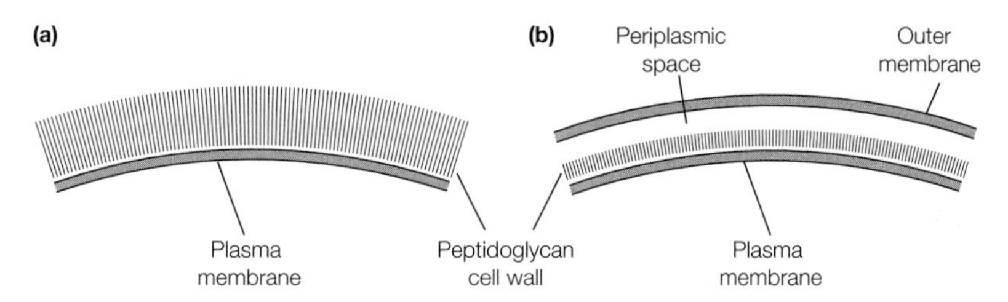


*Fig. 2. Cell wall structure of (a) Gram-positive and (b) Gram-negative bacteria.*

# A2 EUKARYOTES

## Key Notes

**Eukaryotes**

Eukaryotic cells have a membrane-bound nucleus and a number of other membrane-bound subcellular (internal) organelles, each of which has a specific function.

**Plasma membrane**

The plasma membrane surrounds the cell, separating it from the external environment. The plasma membrane is a selectively permeable barrier due to the presence of specific transport proteins. It is also involved in receiving information when ligands bind to receptor proteins on its surface, and in the processes of exocytosis and endocytosis.

**Nucleus**

The nucleus stores the cell's genetic information as DNA in chromosomes. It is bounded by a double membrane but pores in this membrane allow molecules to move in and out of the nucleus. The nucleolus within the nucleus is the site of ribosomal ribonucleic acid (rRNA) synthesis.

**Endoplasmic reticulum**

This interconnected network of membrane vesicles is divided into two distinct parts. The rough endoplasmic reticulum (RER), which is studded with ribosomes, is the site of membrane and secretory protein biosynthesis and their post-translational modification. The smooth endoplasmic reticulum (SER) is involved in phospholipid biosynthesis and in the detoxification of toxic compounds.

**Golgi apparatus**

The Golgi apparatus, a system of flattened membrane-bound sacs, is the sorting and packaging center of the cell. It receives membrane vesicles from the RER, further modifies the proteins within them, and then packages the modified proteins in other vesicles which eventually fuse with the plasma membrane or other subcellular organelles.

**Mitochondria**

Mitochondria have an inner and an outer membrane separated by the intermembrane space. The outer membrane is more permeable than the inner membrane due to the presence of porin proteins. The inner membrane, which is folded to form cristae, is the site of oxidative phosphorylation, which produces ATP. The central matrix is the site of fatty acid degradation and the citric acid cycle.

**Chloroplasts**

Chloroplasts in plant cells are surrounded by a double membrane and have an internal membrane system of thylakoid vesicles that are stacked up to form grana. The thylakoid vesicles contain chlorophyll and are the site of photosynthesis. Carbon dioxide ($CO_2$) fixation takes place in the stroma, the soluble matter around the thylakoid vesicles.

**Lysosomes**

Lysosomes in animal cells are bounded by a single membrane. They have an acidic internal pH (pH 4–5), maintained by proteins in the membrane that pump in $H^+$ ions. Within the lysosomes are acid hydrolases; enzymes involved in the degradation of macromolecules, including those internalized by endocytosis.

**Peroxisomes**

Peroxisomes contain enzymes involved in the breakdown of amino acids and fatty acids, a byproduct of which is hydrogen peroxide. This toxic compound is rapidly degraded by the enzyme catalase, also found within the peroxisomes.

**Cytosol**

The cytosol is the soluble part of the cytoplasm where a large number of metabolic reactions take place. Within the cytosol is the cytoskeleton, a network of fibers (microtubules, intermediate filaments and microfilaments) that maintain the shape of the cell.

**Cytoskeleton**

Eukaryotic cells have an internal scaffold, the cytoskeleton, that controls the shape and movement of the cell. The cytoskeleton is made up of actin microfilaments, intermediate filaments and microtubules.

**Microtubules**

Microtubule filaments are hollow cylinders made of the protein tubulin. The wall of the microtubule is made up of a helical array of alternating α- and β-tubulin subunits. The mitotic spindle involved in separating the chromosomes during cell division is made of microtubules. Colchicine inhibits microtubule formation, whereas the anticancer agent, taxol, stabilizes microtubules and interferes with mitosis.

**Plant cell wall**

The cell wall surrounding a plant cell is made up of the polysaccharide cellulose. In woody plants, the phenolic polymer called lignin gives the cell wall additional strength and rigidity.

**Plant cell vacuole**

The membrane-bound vacuole is used to store nutrients and waste products, has an acidic pH and, due to the influx of water, creates turgor pressure inside the cell as it pushes out against the cell wall.

**Related topics**

Microscopy (A3)
Membrane transport: macromolecules (E4)
Signal transduction (E5)
Chromosomes (F2)
Protein targeting (H4)
Electron transport and oxidative phosphorylation (L2)
Photosynthesis (L3)
Cilia and flagella (N2)

## Eukaryotes

A eukaryotic cell is surrounded by a **plasma membrane**, has a membrane-bound nucleus and contains a number of other distinct **subcellular organelles** (*Fig. 1*). These organelles are membrane-bounded structures, each having a unique role and each containing a specific complement of proteins and other molecules. Animal and plant cells have the same basic structure, although some organelles and structures are found in one and not the other (e.g. chloroplasts, vacuoles and cell wall in plant cells, lysosomes in animal cells).

## Plasma membrane

The plasma membrane envelops the cell, separating it from the external environment and maintaining the correct ionic composition and osmotic pressure

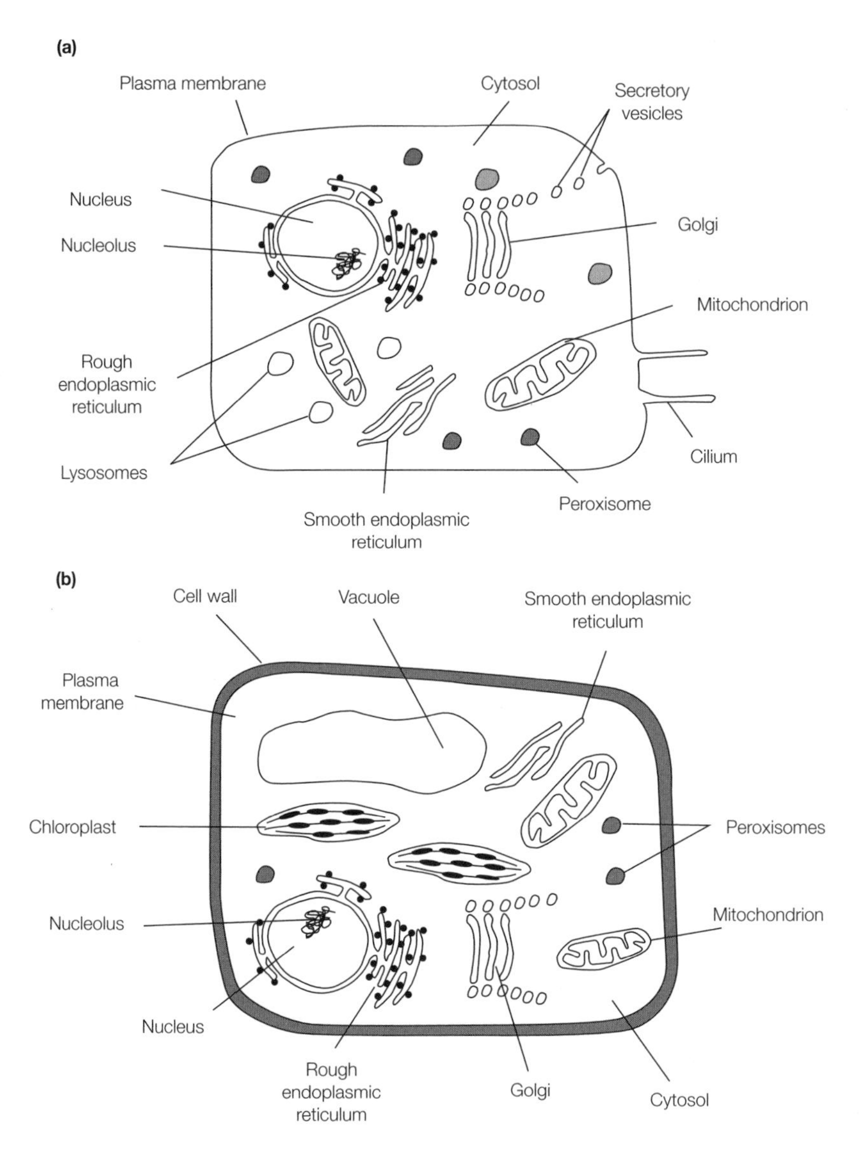


*Fig. 1. Eukaryote cell structure. (a) Structure of a typical animal cell, (b) structure of a typical plant cell.*

of the cytosol. The plasma membrane, like all membranes, is impermeable to most substances but the presence of specific proteins in the membrane allows certain molecules to pass through, therefore making it **selectively permeable** (see Topic E3). The plasma membrane is also involved in communicating with other cells, in particular through the binding of ligands (small molecules such as hormones, neurotransmitters, etc.) to **receptor proteins** on its surface (see Topic E5). The plasma membrane is also involved in the **exocytosis** (secretion) and **endocytosis** (internalization) of macromolecules (see Topic E4).

## Nucleus

The nucleus is bounded by two membranes, the **inner and outer nuclear membranes**. These two membranes fuse together at the **nuclear pores** through

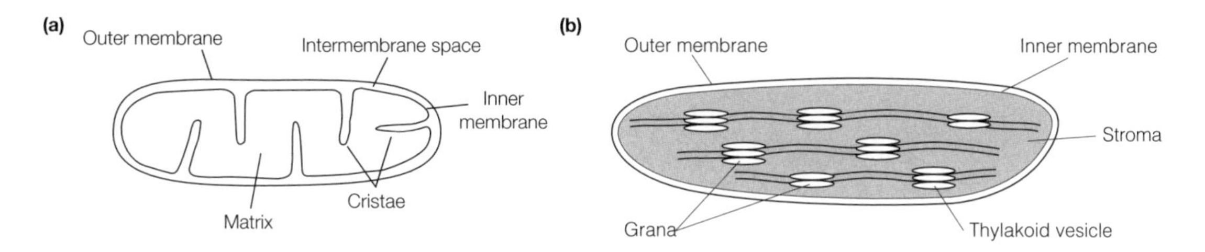


*Fig. 2. Structure of (a) a mitochondrion and (b) a chloroplast.*

which molecules [messenger ribonucleic acid (mRNA), proteins, ribosomes, etc.] can move between the nucleus and the cytosol. Other proteins, for example those involved in regulating gene expression, can pass through the pores from the cytosol to the nucleus. The outer nuclear membrane is often continuous with the rough endoplasmic reticulum (RER). Within the nucleus the **DNA** is tightly coiled around **histone proteins** and organized into complexes called **chromosomes** (see Topic F2). Visible under the light microscope (see Topic A3) is the **nucleolus**, a subregion of the nucleus which is the site of ribosomal ribonucleic acid (rRNA) synthesis.

## Endoplasmic reticulum

The endoplasmic reticulum (ER) is an interconnected network of membrane vesicles. The **rough endoplasmic reticulum (RER)** is studded on the cytosolic face with **ribosomes**, the sites of **membrane and secretory protein biosynthesis** (see Topic H3). Within the lumen of the RER are enzymes involved in the **post-translational modification** (glycosylation, proteolysis, etc.) of membrane and secretory proteins (see Topic H5). The **smooth endoplasmic reticulum (SER)**, which is not studded with ribosomes, is the site of **phospholipid biosynthesis**, and is where a number of **detoxification reactions** take place.

## Golgi apparatus

The Golgi apparatus, a system of flattened membrane-bound sacs, is the **sorting center** of the cell. Membrane vesicles from the RER, containing membrane and secretory proteins, fuse with the Golgi apparatus and release their contents into it. On transit through the Golgi apparatus, further **post-translational modifications** to these proteins take place and they are then sorted and packaged into different vesicles (see Topic H5). These vesicles bud off from the Golgi and are transported through the cytosol, eventually fusing either with the plasma membrane to release their contents into the extracellular space (a process known as **exocytosis**; see Topic E4) or with other internal organelles (lysosomes, peroxisomes, etc.).

## Mitochondria

A mitochondrion has an **inner and an outer membrane** between which is the **intermembrane space** (*Fig. 2a*). The outer membrane contains **porin proteins** which make it permeable to molecules of up to 10 kDa. The inner membrane, which is considerably less permeable, has large infoldings called **cristae** which protrude into the **central matrix**. The inner membrane is the site of oxidative phosphorylation and electron transport involved in ATP production (see Topic L2). The central matrix is the site of numerous metabolic reactions including the citric acid cycle (see Topic L1) and fatty acid breakdown (see Topic K2). Also within the matrix is found the mitochondrial DNA which encodes some of the mitochondrial proteins.

**Chloroplasts**

Chloroplasts also have **inner and outer membranes**. In addition, there is an extensive internal membrane system made up of **thylakoid vesicles** (interconnected vesicles flattened to form discs) stacked upon each other to form **grana** (*Fig. 2b*). Within the thylakoid vesicles is the green pigment **chlorophyll** (see Topic M4), along with the enzymes that trap light energy and convert it into chemical energy in the form of ATP (see Topic L3). The **stroma**, the space surrounding the thylakoid vesicles, is the site of carbon dioxide ($CO_2$) fixation – the conversion of $CO_2$ into organic compounds. Chloroplasts, like mitochondria, contain DNA which encodes some of the chloroplast proteins.

**Lysosomes**

Lysosomes, which are found only in animal cells, have a single boundary membrane. The internal pH of these organelles is **mildly acidic** (pH 4–5), and is maintained by integral membrane proteins which pump $H^+$ ions into them (see Topic E3). The lysosomes contain a range of hydrolases that are optimally active at this acidic pH (and hence are termed **acid hydrolases**) but which are inactive at the neutral pH of the cytosol and extracellular fluid. These enzymes are involved in the degradation of host and foreign macromolecules into their monomeric subunits; **proteases** degrade proteins, **lipases** degrade lipids, **phosphatases** remove phosphate groups from nucleotides and phospholipids, and **nucleases** degrade DNA and RNA. Lysosomes are involved in the degradation of extracellular macromolecules that have been brought into the cell by **endocytosis** (see Topic E4).

**Peroxisomes**

These organelles have a single boundary membrane and contain enzymes that degrade fatty acids and amino acids. A byproduct of these reactions is **hydrogen peroxide**, which is toxic to the cell. The presence of large amounts of the enzyme **catalase** in the peroxisomes rapidly converts the toxic hydrogen peroxide into harmless $H_2O$ and $O_2$:

$$2H_2O_2 \xrightarrow{\textit{Catalase}} 2H_2O + O_2$$

**Cytosol**

The cytosol is that part of the **cytoplasm** not included within any of the subcellular organelles, and is a major site of cellular metabolism. It contains a large number of different enzymes and other proteins. The cytosol is not a homogenous 'soup' but has within it the **cytoskeleton**, a network of fibers criss-crossing through the cell that helps to maintain the shape of the cell. The cytoskeletal fibers include **microtubules** (25 nm in diameter), **intermediate filaments** (10 nm in diameter) and **microfilaments** (8 nm in diameter) (see Topic N2). Also found within the cytosol of many cells are **inclusion bodies** (granules of material that are not membrane-bounded) such as glycogen granules in liver and muscle cells, and droplets of triacylglycerol in the fat cells of adipose tissue.

**Cytoskeleton**

In the cytosol of eukaryotic cells is an **internal scaffold**, the cytoskeleton (see Topic E2). The cytoskeleton is important in maintaining and altering the **shape of the cell**, in enabling the cell to **move** from one place to another, and in **transporting intracellular vesicles**. Three types of **filaments** make up the cytoskeleton: microfilaments, intermediate filaments and microtubules. The **microfilaments**, diameter approximately 7 nm, are made of **actin** and have a mechanically supportive function. Through their interaction with myosin (see Topic N1), the microfilaments form contractile assemblies that are involved in various intracellular movements such as cytoplasmic streaming and the formation of membrane invaginations (see Topic E4). The **intermediate filaments** (7–11 nm in diameter) are probably involved in a load-bearing function within the cell. For example, the skin in higher animals contains an extensive network of intermediate filaments made up of the protein **keratin** that has a two-stranded α-helical coiled-coil structure.

**Microtubules**

The third type of cytoskeletal filaments, the **microtubules**, are hollow cylindrical structures with an outer diameter of 30 nm that are built from the protein **tubulin**. The rigid wall of a microtubule is made up of a helical array of alternating α- and β-tubulin subunits, each of 50 kDa. A cross-section through a microtubule reveals that there are 13 tubulin subunits per turn of the filament. Microtubules in cells are formed by the addition of α- and β-tubulin molecules to pre-existing filaments or nucleation centers. The microtubules form a supportive framework that guides the movement of subcellular organelles within the cell. For example, the **mitotic spindle** involved in separating the replicated chromosomes during mitosis is an assembly of microtubules. The drug **colchicine** inhibits the polymerization of microtubules, thus blocking cell processes such as cell division that depend on functioning microtubules. Another compound, **taxol**, stabilizes tubulin in microtubules and promotes polymerization. It is being used as an anticancer drug since it blocks the proliferation of rapidly dividing cells by interfering with the mitotic spindle.

**Plant cell wall**

Surrounding the plasma membrane of a plant cell is the cell wall, which imparts strength and rigidity to the cell. This is built primarily of **cellulose**, a rod-like **polysaccharide** of repeating glucose units linked β(1–4) (see Topic J1). These cellulose molecules are aggregated together by hydrogen bonding into bundles of fibers, and the fibers in turn are cross-linked together by other polysaccharides. In woody plants another compound, **lignin**, imparts added strength and rigidity to the cell wall. Lignin is a complex water-insoluble phenolic polymer.

**Plant cell vacuole**

Plant cells usually contain one or more **membrane-bounded vacuoles**. These are used to store nutrients (e.g. sucrose), water, ions and waste products (especially excess nitrogen-containing compounds). Like lysosomes in animal cells, vacuoles have an **acidic pH** maintained by $H^+$ pumps in the membrane and contain a variety of **degradative enzymes**. Entry of water into the vacuole causes it to expand, creating hydrostatic pressure (**turgor**) inside the cell which is balanced by the mechanical resistance of the cell wall.

# A3 Microscopy

## Key Notes

**Light microscopy**

In light microscopy, a beam of light is focused through a microscope using glass lenses to produce an enlarged image of the specimen.

**Standard light microscopy**

The specimen to be viewed is first fixed with alcohol or formaldehyde, embedded in wax and then cut into thin sections. A section is illuminated from below with the beam of light being focused on to it by the condenser lens. The incident light that passes through the specimen is then focused by the objective lens on to its focal plane, creating a magnified image.

**Staining**

Subcellular organelles cannot readily be distinguished under the light microscope without first staining the specimen with a chemical. Proteins can be stained with eosin or methylene blue, DNA with fuchsin. The location of an enzyme in a specimen can be revealed by cytochemical staining using a substrate which is converted into a colored product by the enzyme.

**Dark-field microscopy**

In dark-field microscopy, light from the condenser lens is directed at an angle on to the specimen such that only light which has been refracted or diffracted by the specimen enters the objective lens and forms an image.

**Phase-contrast microscopy**

In phase-contrast microscopy, the light microscope is adapted to alter the phase of the light waves to produce an image in which the degree of brightness of a region of the specimen depends on its refractive index.

**Immunofluorescence microscopy**

In immunofluorescence microscopy, fluorescent compounds (which absorb light at the exciting wavelength and then emit it at the emission wavelength) are attached to an antibody specific for the subcellular structure under investigation. The antibody is then added to the specimen and allowed to bind. Unbound antibody is removed and the specimen is illuminated at the exciting wavelength, to visualize where the antibody has bound.

**Confocal scanning microscopy**

This variation of immunofluorescence microscopy uses a laser to focus light of the exciting wavelength on to the specimen so that only a thin section of it is illuminated. The laser beam is moved through the sample, producing a series of images which are then reassembled by a computer to produce a three-dimensional picture of the specimen.

**Electron microscopy**

In electron microscopy, a beam of electrons is focused using electromagnetic lenses. The specimen is mounted within a vacuum so that the electrons are not absorbed by atoms in the air.

**Transmission electron microscopy**

In transmission electron microscopy, the beam of electrons is passed through a thin section of the specimen that has been stained with heavy metals. The electron-dense metals scatter the incident electrons, thereby producing an image of the specimen.

**Scanning electron microscopy**

In scanning electron microscopy, the surface of a whole specimen is coated with a layer of heavy metal and then scanned with an electron beam. Excited molecules in the specimen release secondary electrons which are focused to produce a three-dimensional image of the specimen.

**Related topics**

Eukaryotes (A2)
Antibodies as tools (D5)
Membrane protein and carbohydrate (E2)

## Light microscopy

In 1835 Schleiden and Schwann used a primitive light microscope to look at and identify individual cells for the first time. From these studies they proposed their **cell theory** "that the nucleated cell is the unit of structure and function in plants and animals". In light microscopy, **glass lenses** are used to focus a beam of light on to the **specimen** under investigation. The light passing through the specimen is then focused by other lenses to produce a **magnified image**. Technological advances since 1835 have resulted in the manufacture of much more powerful and sophisticated instruments, which have enabled detailed studies of the structure and function of cells to take place.

## Standard light microscopy

Standard (bright-field) light microscopy is the most common microscopy technique in use today and uses a **compound microscope**. The specimen to be examined is first fixed with a solution containing alcohol or formaldehyde. These compounds denature proteins and, in the case of formaldehyde, introduce covalent cross-links between amino groups on adjacent molecules which stabilize protein–protein and protein–nucleic acid interactions. The **fixed specimen** is then embedded in paraffin wax and cut into **thin sections** (approximately 1 μm thick). Each section is mounted on a glass slide and then positioned on the movable specimen stage of the microscope. The specimen is illuminated from underneath by a lamp in the base of the microscope (*Fig. 1a*), with the light being focused on to the plane of the specimen by a **condenser**

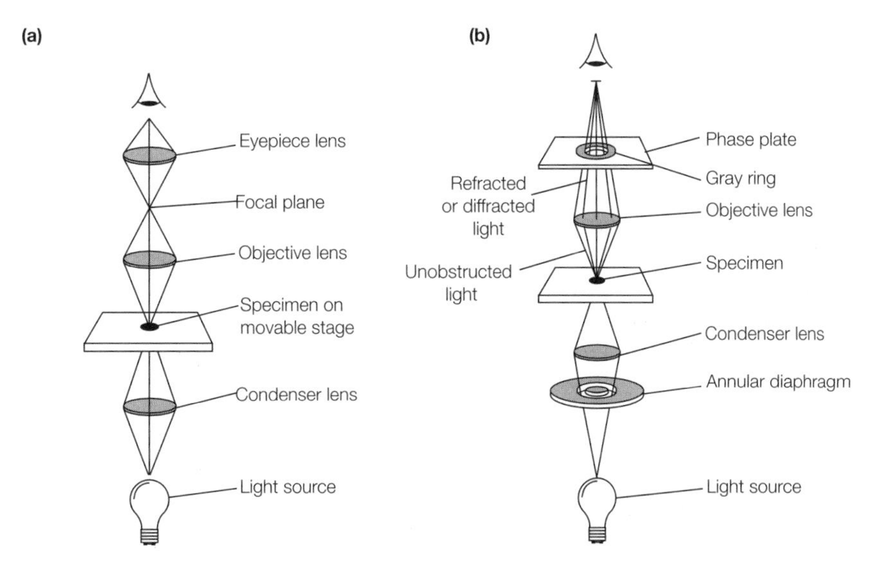


*Fig. 1. Optical pathway of (a) a compound microscope and (b) a phase-contrast microscope.*

**lens.** Incident light coming through the specimen is picked up by the **objective lens** and focused on to its focal plane, creating a magnified image. This image is further magnified by the eyepiece, with the total magnification achieved being the sum of the magnifications of the individual lenses. In order to increase the resolution achieved by a compound microscope, the specimen is often overlaid with **immersion oil** into which the objective lens is placed. The limit of resolution of the light microscope using visible light is approximately 0.2 μm.

### Staining

The various subcellular constituents (nucleus, mitochondria, cytosol, etc.) absorb about the same degree of visible light, making it difficult to distinguish them under the light microscope without first staining the specimen. Many **chemical stains** bind to biological molecules; for example, **eosin** and **methylene blue** bind to proteins, and **fuchsin** binds to DNA. Another useful way of visualizing specific structures within cells is **cytochemical staining** in which an enzyme reaction catalyzes the production of a colored precipitate from a colorless precursor. The colored precipitate can then be seen in the light microscope wherever the enzyme is present. For example, peroxisomes can be visualized by using a cytochemical stain for catalase (see Topic A2).

### Dark-field microscopy

In dark-field microscopy, light is directed from the condenser lens at an angle so that none of the incident light enters the objective lens; only light **refracted** (bent) or **diffracted** (scattered) by the specimen can enter the lens. The resolution of dark-field microscopy is not particularly good, but this method does allow small objects that refract a large proportion of the incident light to appear as bright particles, and so it is widely used in microbiology to detect bacteria.

### Phase-contrast microscopy

In phase-contrast microscopy, a glass **phase plate** between the specimen and the observer further increases the difference in contrast. The incident light is passed through an **annular diaphragm** which focuses a circular ring of light on the specimen (*Fig. 1b*). Light that passes unobstructed through the specimen is focused by the objective lens on to the gray ring in the phase plate which absorbs some of it and alters its phase. Light refracted or diffracted by the specimen will have its phase altered and will pass through the clear region of the phase plate. The refracted and diffracted light waves then recombine with the unrefracted light waves, producing an image in which the degree of brightness or darkness of a region of the specimen depends on the **refractive index** of that region. Phase-contrast microscopy is useful for examining the structure and movement of larger organelles (nucleus, mitochondria, etc.) in living cells but is suitable only for single cells or thin cell layers.

### Immunofluorescence microscopy

In immunofluorescence microscopy, the light microscope is adapted to detect the light emitted by a **fluorescent compound**, that is a compound which absorbs light at one wavelength (the **excitation wavelength**) and then emits light at a longer wavelength (the **emission wavelength**). Two commonly used compounds in fluorescent microscopy are **rhodamine**, which emits red light, and **fluorescein**, which emits green light. First, the fluorescent compound is chemically coupled to an **antibody** specific for a particular protein or other macromolecule in the cell under investigation (see Topic D5). Then the fluorescently tagged antibody is added to the tissue section or permeabilized cell, and the specimen is illuminated with light at the exciting wavelength. The structures in the specimen to which the antibody has bound can then be visualized. Fluorescence microscopy can also be applied to living cells, which allows the movement of the cells and structures within them to be followed with time (see Topic E2 for an example of this).

## Confocal scanning microscopy

Confocal scanning microscopy is a refinement of normal immunofluorescence microscopy which produces clearer images of whole cells or larger specimens. In normal immunofluorescence microscopy, the fluorescent light emitted by the compound comes from molecules above and below the **plane of focus**, blurring the image and making it difficult to determine the actual three-dimensional molecular arrangement. With the confocal scanning microscope, only molecules in the plane of focus fluoresce due to the use of a focused **laser beam** at the exciting wavelength. The laser beam is moved to different parts of the specimen, allowing a series of images to be taken at different depths through the sample. The images are then combined by a computer to provide the complete three-dimensional image.

## Electron microscopy

In contrast with light microscopy where optical lenses focus a beam of light, in electron microscopy **electromagnetic lenses** focus a beam of **electrons**. Because electrons are absorbed by atoms in the air, the specimen has to be mounted in a **vacuum** within an evacuated tube. The resolution of the electron microscope with biological materials is at best 0.10 nm.

## Transmission electron microscopy

In transmission electron microscopy, a beam of electrons is directed through the specimen and electromagnetic lenses are used to focus the **transmitted electrons** to produce an image either on a viewing screen or on photographic film (*Fig. 2a*). As in standard light microscopy, thin sections of the specimen are viewed. However, for transmission electron microscopy the sections must be much thinner (50–100 nm thick). Since electrons pass uniformly through biological material, unstained specimens give very poor images. Therefore, the specimen must routinely be stained in order to scatter some of the incident electrons which are then not focused by the electromagnetic lenses and so do

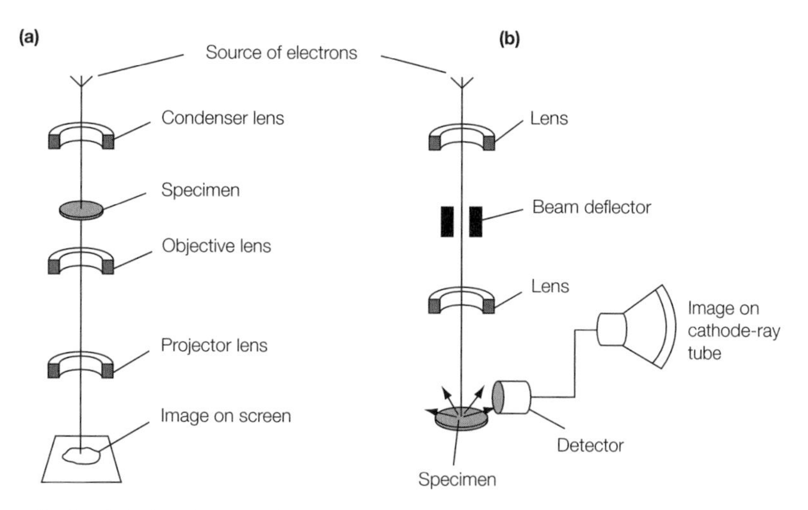


*Fig. 2. Principal features of (a) a transmission electron microscope and (b) a scanning electron microscope.*

not form the image. **Heavy metals** such as gold and osmium are often used to stain biological materials. In particular **osmium tetroxide** preferentially stains certain cellular components, such as membranes, which appear black in the image. The transmission electron microscope has sufficiently high resolution that it can be used to obtain information about the shapes of purified proteins, viruses and subcellular organelles.

**Antibodies** can be tagged with electron-dense gold particles in a similar way to being tagged with a fluorescent compound in immunofluorescence microscopy, and then bound to specific target proteins in the thin sections of the specimen. When viewed in the electron microscope, small dark spots due to the gold particles are seen in the image wherever an antibody molecule has bound to its antigen (see Topic D5) and so the technique can be used to localize specific antigens.

## Scanning electron microscopy

In scanning electron microscopy, an (unsectioned) specimen is fixed and then coated with a thin layer of a **heavy metal** such as **platinum**. An electron beam then **scans** over the specimen, exciting molecules within it that release secondary electrons. These secondary electrons are focused on to a scintillation detector and the resulting image displayed on a cathode-ray tube (*Fig. 2b*). The scanning electron microscope produces a **three-dimensional image** because the number of secondary electrons produced by any one point on the specimen depends on the angle of the electron beam in relation to the surface of the specimen. The resolution of the scanning electron microscope is 10 nm, some 100-fold less than that of the transmission electron microscope.

# A4 CELLULAR FRACTIONATION

## Key Notes

**Subcellular fractionation: overview**

Subcellular fractionation is the breaking open of a cell (e.g. by homogenization) and the separation of the various organelles from one another by centrifugation.

**Differential velocity centrifugation**

Differential velocity centrifugation separates the subcellular organelles on the basis of their size. A centrifuge is used to generate powerful forces to separate the various organelles which pellet to the bottom of the centrifuge tube. At lower forces, nuclei, mitochondria, chloroplasts and lysosomes pellet, whereas higher forces are needed to pellet the endoplasmic reticulum, Golgi apparatus and plasma membrane.

**Equilibrium density-gradient centrifugation**

This procedure uses a gradient of a dense solution (e.g. sucrose solution) to separate out subcellular organelles on the basis of their density. An ultracentrifuge is used to sediment the organelles to an equilibrium position in the gradient where their density is equal to that of the sucrose.

**Rate-zonal centrifugation**

In rate-zonal centrifugation, the sample is centrifuged through a weak sucrose solution until the organelles have separated from each other. Separation is on the basis of size. If centrifuged for too long all the organelles will end up in the pellet at the bottom of the centrifuge tube.

**Marker enzymes**

A convenient way of determining the purity of an organelle preparation is to measure the activity of a marker enzyme in the various subcellular fractions. A marker enzyme is one that is found within only one particular compartment of the cell.

**Flow cytometry**

Individual cells can be identified using a flow cytometer. Antibodies, coupled to fluorescent compounds, that bind to molecules on the surface of particular types of cells can be used to separate cells from each other in a fluorescent-activated cell sorter.

**Related topics**

Eukaryotes (A2)
Microscopy (A3)
Protein purification (B6)
Introduction to enzymes (C1)

### Subcellular fractionation: overview

In order to study macromolecules and metabolic processes within cells it is often helpful to isolate one type of **subcellular organelle** (see Topic A2) from the rest of the cell contents by subcellular fractionation. Initially, the plasma membrane (and cell wall if present) has to be ruptured. To do this, the tissue or cell sample is suspended in an isotonic sucrose solution (0.25–0.32 M) buffered at the appropriate pH, and the cells are then broken open by **homogenization** in a blender or homogenizer, by **sonication** or by subjecting them to high pressures (French press or nitrogen bomb). The initial homogenization, and the following subcellular fractionation, are usually carried out at 4°C in order to minimize enzymic degradation of the cell's constituents. The sample of broken cells is often strained through muslin or other fine gauze to remove larger lumps of material before proceeding further.

**Differential velocity centrifugation**

In differential velocity centrifugation, the various subcellular organelles are separated from one another on the basis of their **size**. A **centrifuge** is used to generate powerful forces; up to 100 000 times the force of gravity (*g*). The homogenized sample is placed in an appropriate centrifuge tube which is then loaded in the **rotor** of the centrifuge and subjected to centrifugation (*Fig. 1a*). At first relatively low *g* forces are used for short periods of time but then increasingly higher *g* forces are used for longer time periods. For example, centrifugation at 600 *g* for 3 min would pellet the **nuclei**, the largest organelles (*Fig. 1b*). The supernatant from this step is removed to a fresh tube and then centrifuged at 6000 *g* for 8 min to pellet out **mitochondria**, **peroxisomes** and, if present, **lysosomes** or **chloroplasts**. Centrifugation of this next supernatant at 40 000 *g* for 30 min will pellet out the **plasma membrane**, and fragments of the **endoplasmic reticulum** and **Golgi apparatus**. A final centrifugation at 100 000 *g* for 90 min would result in a **ribosomal pellet** and a supernatant that is essentially free of particulate matter and is considered to be the true soluble **cytosolic fraction**. However, the fractions isolated by differential velocity centrifugation are not usually entirely free of other subcellular organelles and so may need to be purified further. For separations at low *g* forces, a preparative centrifuge is used which has a rotor spinning in air at ambient pressure. However, an ultracentrifuge is required for separations at higher *g* forces. The chamber of the ultracentrifuge is kept in a high vacuum to reduce friction, and subsequent heating, which would otherwise occur between the spinning rotor and air.

**Equilibrium density-gradient centrifugation**

Equilibrium density-gradient centrifugation is often used to further purify organelles following their partial separation by differential velocity centrifugation. In this procedure the organelles are separated on the basis of their **density**, not their size. The impure organelle fraction is loaded at the top of a centrifuge tube that contains a gradient of a **dense solution** (e.g. a sucrose solution; *Fig. 2*).

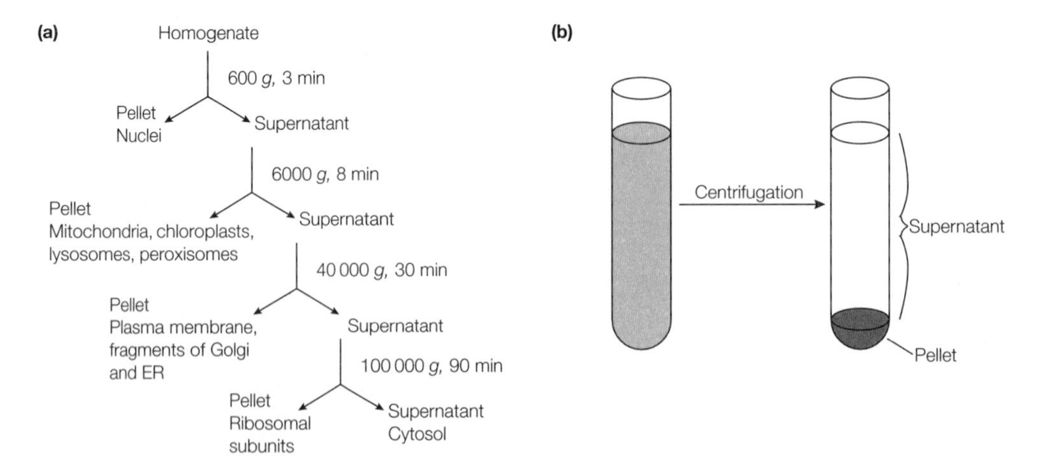


*Fig. 1. Cell fractionation by differential velocity centrifugation. (a) Scheme for subcellular fractionation of a tissue sample, (b) appearance of a sample in the centrifuge tube before and after centrifugation.*

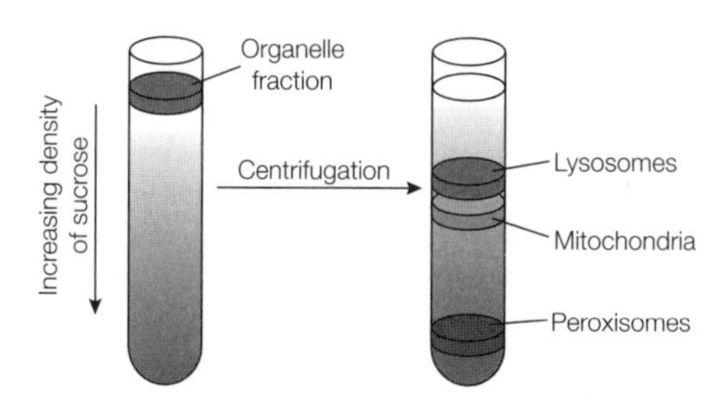


*Fig. 2. Separation of organelles by equilibrium density-gradient centrifugation.*

The sucrose solution is most concentrated (dense) at the bottom of the tube, and decreases in concentration (and density) towards the top of the tube. During centrifugation (e.g. 160 000 *g* for 3 h) the various organelles move down the tube to an **equilibrium position** where their density is equal to that of the sucrose at that position. The **forces of sedimentation** tend to make the organelles move further down the tube but, if they do so, they enter a region of higher density than the organelle density and so they float back to their previous position. Mitochondria, lysosomes and peroxisomes all differ in density and so can be effectively separated from one another by density-gradient centrifugation (*Fig. 2*). Similarly, the rough endoplasmic reticulum, Golgi apparatus and plasma membrane can be separated using a gradient of lower density. The more dense **cesium chloride** is used to make the density gradient for the separation of denser particles such as DNA, RNA and proteins by equilibrium centrifugation.

## Rate-zonal centrifugation

In rate-zonal centrifugation the sample is layered at the top of a centrifuge tube that contains a sucrose solution of low concentration. In this case, the sucrose is not being used to separate samples by density but simply serves to prevent convection mixing, and is subjected to centrifugation. The organelles move down the tube at a rate determined by the centrifugal force, their mass, the difference between their density and that of the surrounding solution, and the friction between them and the surrounding solution. On completion of centrifugation, different sized organelles are found in different zones of the centrifuge tube. The sample has to be centrifuged for just long enough to separate the organelles of interest; if centrifuged for too long, all of the organelles will end up in the pellet at the bottom of the tube.

## Marker enzymes

When the cell homogenate has been fractionated, the purity of the different organelle preparations needs to be assessed. One way in which this can be done is by assessing **morphology** in the electron microscope (see Topic A3). A more readily available alternative though is to measure the activity of (to assay for) a particular **enzyme** which is characteristic of that organelle and is not found elsewhere in the cell (see Topic C1). For example, **catalase** is a good marker enzyme for peroxisomes, **succinate dehydrogenase** for mitochondria, **cathepsin C** or **acid phosphatase** for lysosomes, and **alkaline phosphatase** for the plasma membrane. A good indication of the **purity**/degree of contamination of an organelle preparation can be ascertained by measuring the activity of such enzymes in the various isolated fractions.

## Flow cytometry

Different cells can be identified by measuring the light they scatter, or the fluorescence they emit, as they pass a laser beam in a **flow cytometer**. In a **fluorescence-activated cell sorter** or **FACS** (*Fig. 3*), an instrument based on flow cytometry, cells can be identified and separated from each other. The cells of interest are first labeled with an **antibody** which is specific for a particular cell-surface molecule. The antibody is coupled to a fluorescent dye (see Topic A3), such that when the individual cells pass a laser beam in single file in a narrow stream, the fluorescence of each cell is measured. A vibrating nozzle then forms tiny droplets each containing a single cell which are given a positive or negative charge depending on whether the cell they contain is fluorescing. A strong electric field deflects the different charged droplets into separate containers so that each container eventually has a homogenous population of cells with respect to the cell-surface molecule tagged with fluorescent antibody. These homogenous populations can then be used for biochemical analysis or grown in culture. The DNA and RNA content of a cell can also be measured by flow cytometry.

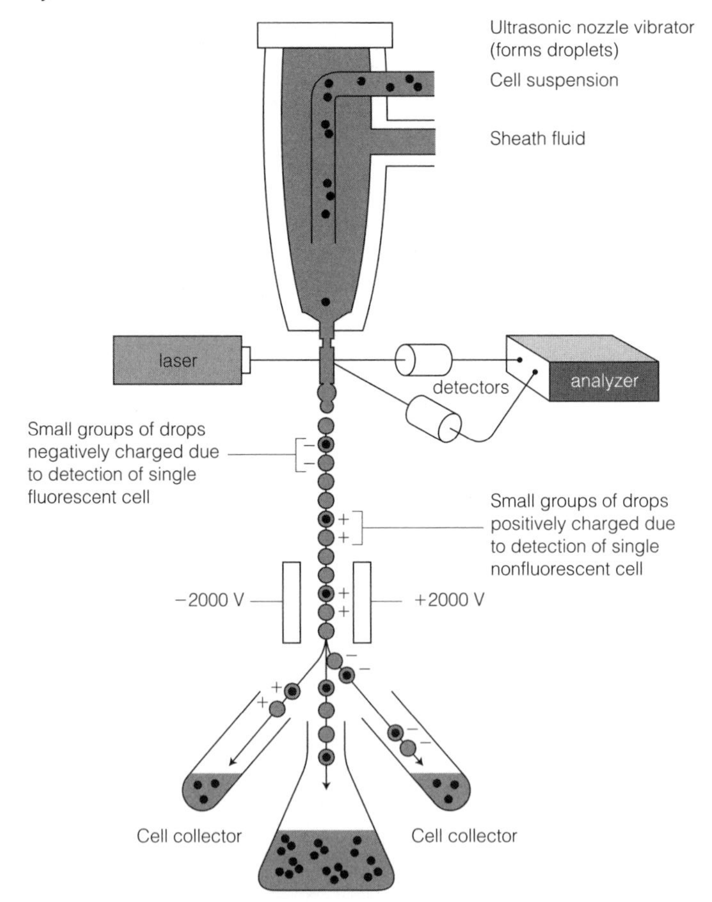


*Fig. 3. A fluorescence-activated cell sorter. When a cell passes through the laser beam it is monitored for fluorescence. Droplets containing single cells are given a positive or negative charge, depending on whether the cell has bound the fluorescently tagged antibody or not. The droplets containing a single cell are then deflected by an electric field into collection tubes according to their charge.*

# B1 Amino acids

## Key Notes

**Amino acids**

All proteins are made up from the same set of 20 standard amino acids. A typical amino acid has a primary amino group, a carboxyl group, a hydrogen atom and a side-chain (R group) attached to a central α-carbon atom ($C_\alpha$). Proline is the exception to the rule in that it has a secondary amino group.

**Enantiomers**

All of the 20 standard amino acids, except for glycine, have four different groups arranged tetrahedrally around the $C_\alpha$ atom and thus can exist in either the D or L configuration. These two enantiomers are nonsuperimposable mirror images that can only be distinguished on the basis of their different rotation of plane-polarized light. Only the L isomer is found in proteins.

**The 20 standard amino acids**

The standard set of 20 amino acids have different side-chains or R groups and display different physicochemical properties (polarity, acidity, basicity, aromaticity, bulkiness, conformational inflexibility, ability to form hydrogen bonds, ability to cross-link and chemical reactivity). Glycine (Gly, G) has a hydrogen atom as its R group. Alanine (Ala, A), valine (Val, V), leucine (Leu, L), isoleucine (Ile, I) and methionine (Met, M) have aliphatic side-chains of differing structures that are hydrophobic and chemically inert. The aromatic side-chains of phenylalanine (Phe, F), tyrosine (Tyr, Y) and tryptophan (Trp, W) are also hydrophobic in nature. The conformationally rigid proline (Pro, P) has its aliphatic side-chain bonded back on to the amino group and thus is really an imino acid. The hydrophobic, sulfur-containing side-chain of cysteine (Cys, C) is highly reactive and can form a disulfide bond with another cysteine residue. The basic amino acids arginine (Arg, R) and lysine (Lys, K) have positively charged side-chains, whilst the side-chain of histidine (His, H) can be either positively charged or uncharged at neutral pH. The side-chains of the acidic amino acids aspartic acid (Asp, D) and glutamic acid (Glu, E) are negatively charged at neutral pH. The amide side-chains of asparagine (Asn, N) and glutamine (Gln, Q), and the hydroxyl side-chains of serine (Ser, S) and threonine (Thr, T) are uncharged and polar, and can form hydrogen bonds.

**Related topics** Acids and bases (B2) Protein structure (B3)

## Amino acids

Amino acids are the building blocks of **proteins** (see Topic B3). Proteins of all species, from bacteria to humans, are made up from the same set of **20 standard amino acids**. Nineteen of these are α-amino acids with a **primary amino**

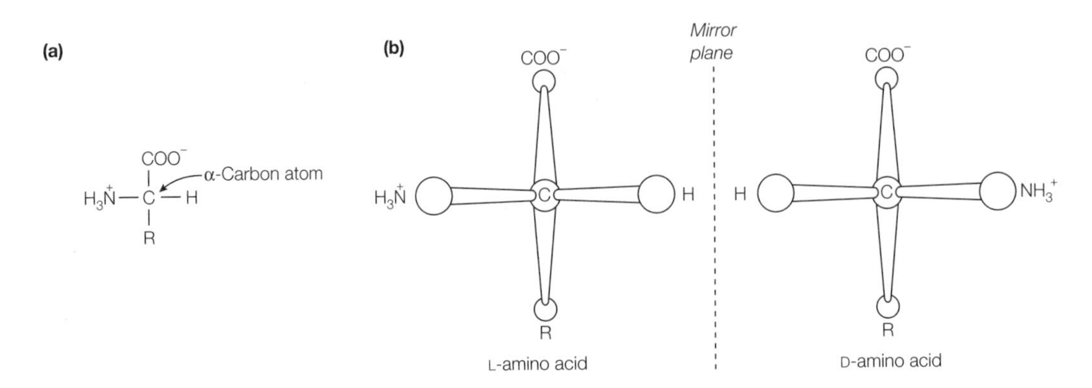


*Fig. 1. (a) Basic structure of an amino acid showing the four different groups around the central α-carbon atom, (b) the two enantiomers of an amino acid.*

**group** ($-NH_3^+$) and a **carboxylic acid** (carboxyl; –COOH) group attached to a central carbon atom, which is called the **α-carbon atom** ($C_\alpha$) because it is adjacent to the carboxyl group (*Fig. 1a*). Also attached to the $C_\alpha$ atom is a hydrogen atom and a **variable side-chain** or 'R' group. The one exception to this general structure is proline, which has a secondary amino group and is really an **α-imino acid**. The names of the amino acids are often abbreviated, either to three letters or to a single letter. Thus, for example, proline is abbreviated to Pro or P (see *Fig. 2*).

## Enantiomers

All of the amino acids, except for glycine (Gly or G; see *Fig. 2*), have four different groups arranged **tetrahedrally** around the central $C_\alpha$ atom which is thus known as an **asymmetric center** or **chiral center** and has the property of **chirality** (Greek; *cheir*, hand) (*Fig. 1b*). The two **nonsuperimposable, mirror images** are termed **enantiomers**. Enantiomers are physically and chemically indistinguishable by most techniques, but can be distinguished on the basis of their different optical rotation of plane-polarized light. Molecules are classified as dextrorotatory (D; Greek *'dextro'* = right) or levorotatory (L; Greek *'levo'* = left) depending on whether they rotate the plane of plane-polarized light clockwise or anticlockwise. Only the **L-amino acids** are found in proteins. **D-Amino acids** rarely occur in nature, but are found in bacterial cell walls (see Topic A1) and certain antibiotics.

## The 20 standard amino acids

The standard 20 amino acids differ only in the structure of the **side-chain** or 'R' group (*Figs 2* and *3*). They can be subdivided into smaller groupings on the basis of similarities in the properties of their side-chains. They display different **physicochemical properties** depending on the nature of their side-chain. Some are acidic, others are basic. Some have small side-chains, others large, bulky side-chains. Some have aromatic side-chains, others are polar. Some confer conformational inflexibility, others can participate either in hydrogen bonding or covalent bonding. Some are chemically reactive.

### *Hydrophobic, aliphatic amino acids*

**Glycine** (Gly or G) (*Fig. 2a*), the smallest amino acid with the simplest structure, has a hydrogen atom in the side-chain position, and thus does not exist as a pair

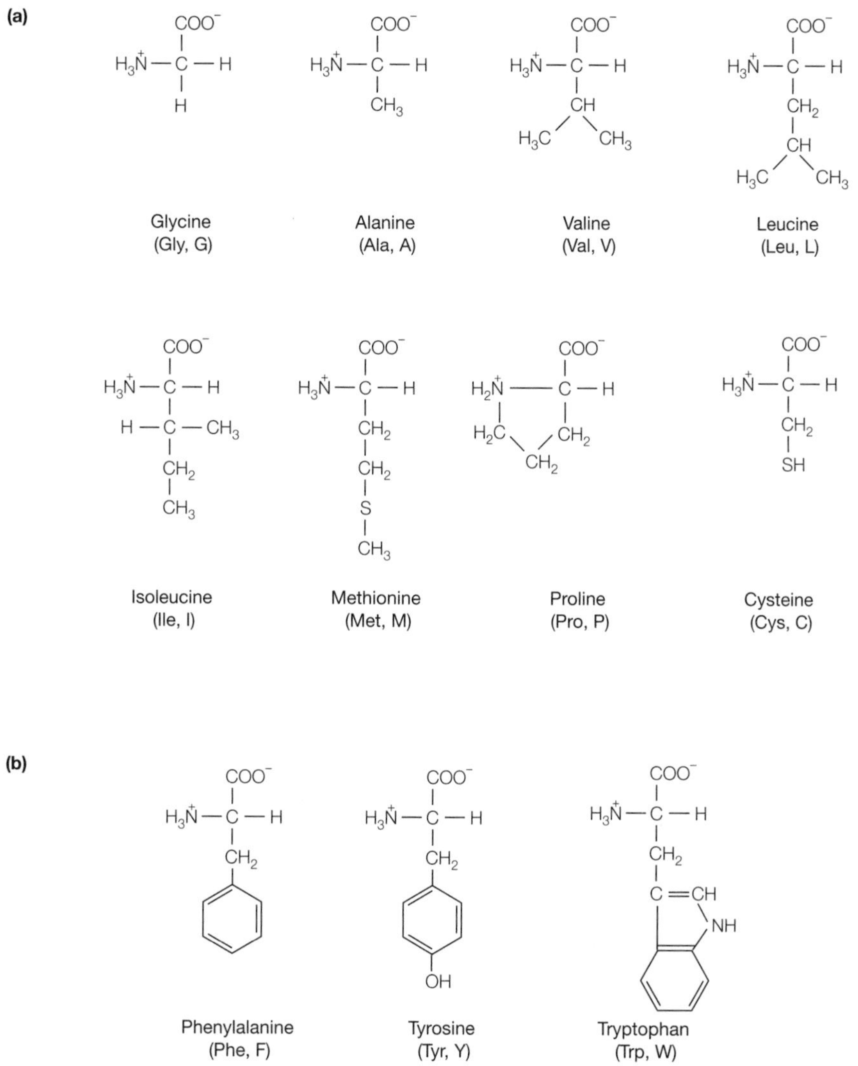

(a)

Glycine (Gly, G) Alanine (Ala, A) Valine (Val, V) Leucine (Leu, L)

Isoleucine (Ile, I) Methionine (Met, M) Proline (Pro, P) Cysteine (Cys, C)

(b)

Phenylalanine (Phe, F) Tyrosine (Tyr, Y) Tryptophan (Trp, W)

*Fig. 2. The standard amino acids. (a) Hydrophobic, aliphatic R groups, (b) hydrophobic, aromatic R groups. The molecular weights of the amino acids are given in Topic B2, Table 1.*

of stereoisomers since there are two identical groups (hydrogen atoms) attached to the $C_\alpha$ atom. The aliphatic side-chains of **alanine** (Ala or A), **valine** (Val or V), **leucine** (Leu or L), **isoleucine** (Ile or I) and **methionine** (Met or M) (*Fig. 2a*) are chemically unreactive, but hydrophobic in nature. **Proline** (Pro or P) (*Fig. 2a*) is also hydrophobic but, with its aliphatic side-chain bonded back on to the amino group, it is conformationally rigid. The sulfur-containing side-chain of **cysteine** (Cys or C) (*Fig. 2a*) is also hydrophobic and is highly reactive, capable of reacting with another cysteine to form a disulfide bond (see Topic B3).

### *Hydrophobic, aromatic amino acids*

**Phenylalanine** (Phe or F), **tyrosine** (Tyr or Y) and **tryptophan** (Trp or W) (*Fig. 2b*) are hydrophobic by virtue of their aromatic rings.

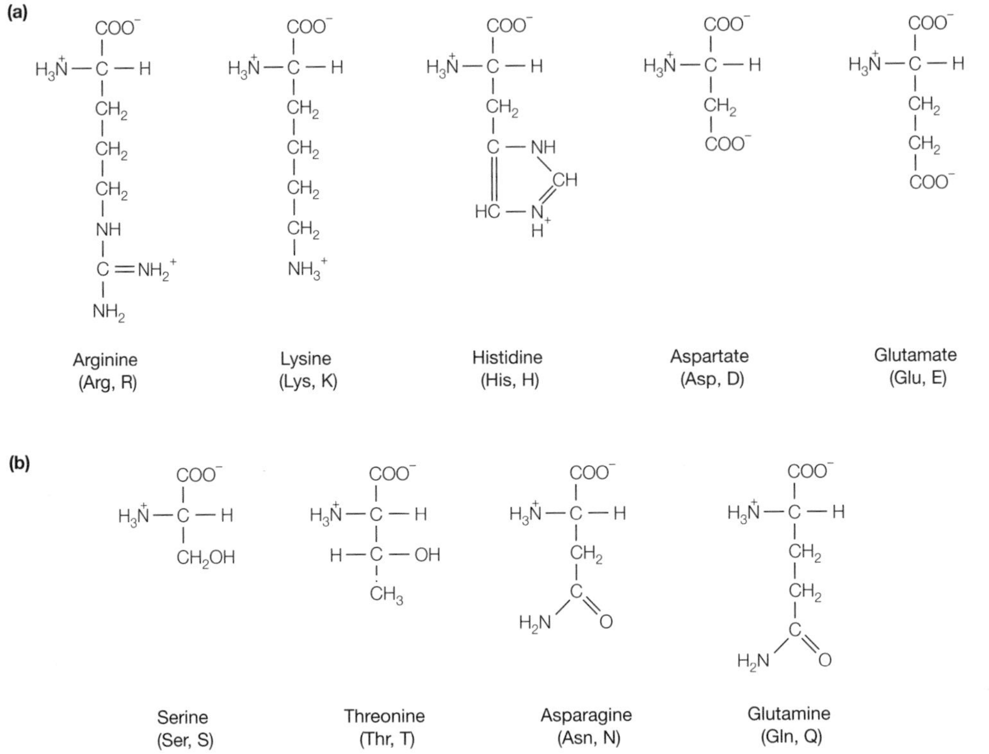

(a)

Arginine (Arg, R) | Lysine (Lys, K) | Histidine (His, H) | Aspartate (Asp, D) | Glutamate (Glu, E)

(b)

Serine (Ser, S) | Threonine (Thr, T) | Asparagine (Asn, N) | Glutamine (Gln, Q)

*Fig. 3. The standard amino acids. (a) Polar, charged R groups, (b) polar, uncharged R groups. The molecular weights of the amino acids are given in Topic B2, Table 1.*

### *Polar, charged amino acids*

The remaining amino acids all have polar, hydrophilic side-chains, some of which are charged at neutral pH. The amino groups on the side-chains of the basic amino acids **arginine** (Arg or R) and **lysine** (Lys or K) (*Fig. 3a*) are protonated and thus positively charged at neutral pH. The side-chain of **histidine** (His or H) (*Fig. 3a*) can be either positively charged or uncharged at neutral pH. In contrast, at neutral pH the carboxyl groups on the side-chains of the acidic amino acids **aspartic acid** (aspartate; Asp or D) and **glutamic acid** (glutamate; Glu or E) (*Fig. 3a*) are de-protonated and possess a negative charge.

### *Polar, uncharged amino acids*

The side-chains of **asparagine** (Asn or N) and **glutamine** (Gln or Q) (*Fig. 3b*), the amide derivatives of Asp and Glu, respectively, are uncharged but can participate in hydrogen bonding. **Serine** (Ser or S) and **threonine** (Thr or T) (*Fig. 3b*) are polar amino acids due to the reactive hydroxyl group in the side-chain, and can also participate in hydrogen bonding (as can the hydroxyl group of the aromatic amino acid Tyr).

# B2 Acids and bases

## Key Notes

**Acids, bases and pH**

pH is a measure of the concentration of $H^+$ in a solution. An acid is a proton donor, a base is a proton acceptor. Ionization of an acid yields its conjugate base, and the two are termed a conjugate acid–base pair, for example acetic acid ($CH_3COOH$) and acetate ($CH_3COO^-$). The p*K* of an acid is the pH at which it is half dissociated. The Henderson–Hasselbach equation expresses the relationship between pH, p*K* and the ratio of acid to base, and can be used to calculate these values.

**Buffers**

An acid–base conjugate pair can act as a buffer, resisting changes in pH. From a titration curve of an acid the inflexion point indicates the p*K* value. The buffering capacity of the acid–base pair is the p*K* ± 1 pH unit. In biological fluids the phosphate and carbonate ions act as buffers. Amino acids, proteins, nucleic acids and lipids also have some buffering capacity. In the laboratory other compounds, such as TRIS, are used to buffer solutions at the appropriate pH.

**Ionization of amino acids**

The α-amino and α-carboxyl groups on amino acids act as acid–base groups, donating or accepting a proton as the pH is altered. At low pH, both groups are fully protonated, but as the pH is increased first the carboxyl group and then the amino group loses a hydrogen ion. For the standard 20 amino acids, the p*K* is in the range 1.8–2.9 for the α-carboxyl group and 8.8–10.8 for the α-amino group. Those amino acids with an ionizable side-chain have an additional acid–base group with a distinctive p*K*.

**Related topic**

Amino acids (B1)

## Acids, bases and pH

The **pH** of a solution is a measure of its **concentration of protons** ($H^+$), and pH is defined as:

$$pH = \log_{10} \frac{1}{H^+} = -\log_{10} [H^+]$$

in which the square brackets denote a **molar concentration**.

An **acid** can be defined as a proton donor and a **base** as a proton acceptor:

$$\text{Acid} \rightleftharpoons H^+ + \text{base.}$$

For example;

$$\underset{\text{Acetic acid}}{CH_3COOH} \rightleftharpoons H^+ + \underset{\text{Acetate}}{CH_3COO^-}$$

$$\underset{\text{Ammonium ion}}{NH_4^+} \rightleftharpoons H^+ + \underset{\text{Ammonia}}{NH_3}$$

The species formed by the **ionization** of an acid is its conjugate base. Conversely, protonation of a base yields its conjugate acid. So, for example, acetic acid and acetate are a **conjugate acid–base pair**.

The ionization of a weak acid is given by:

$$HA \rightleftharpoons H^+ + A^-$$

The apparent **equilibrium constant** ($K$) for this ionization is defined as:

$$K = \frac{[H^+][A^-]}{[HA]} \qquad \text{(Equation 1)}$$

The **p*K*** of an acid is defined as:

$$pK = -\log K = \log \frac{1}{K}$$

The p$K$ of an acid is the pH at which it is half dissociated, i.e. when $[A^-] = [HA]$.

The **Henderson–Hasselbach equation** expresses the relationship between pH and the ratio of acid to base. It is derived as follows. Rearrangement of Equation 1 gives:

$$\frac{1}{[H^+]} = \frac{1}{K} \times \frac{[A^-]}{[HA]}$$

Taking the logarithm of both sides of this equation gives:

$$\log \frac{1}{[H^+]} = \log \frac{1}{K} + \log \frac{[A^-]}{[HA]}$$

Substituting pH for $\log 1/[H^+]$ and p$K$ for $\log 1/K$ gives:

$$pH = pK + \log \frac{[A^-]}{[HA]}$$

which is the Henderson–Hasselbach equation. This equation indicates that the p$K$ of an acid is numerically equal to the pH of the solution when the molar concentration of the acid is equal to that of its conjugate base. The pH of a solution can be calculated from the Henderson–Hasselbach equation if the molar concentrations of $A^-$ and HA, and the p$K$ of HA are known. Similarly, the p$K$ of an acid can be calculated if the molar concentrations of $A^-$ and HA, and the pH of the solution are known.

## Buffers

An acid–base conjugate pair, such as acetic acid and acetate, is able to resist changes in the pH of a solution. That is, it can act as a **buffer**. On addition of hydroxide ($OH^-$) to a solution of acetic acid the following happens:

$$CH_3COOH + OH^- \rightleftharpoons CH_3COO^- + H_2O$$

A plot of the dependence of the pH of this solution on the amount of $OH^-$ added is called a **titration curve** (*Fig. 1*). There is an inflection point in the curve at pH 4.8 which is the p$K$ of acetic acid. In the vicinity of this pH, a relatively large amount of $OH^-$ (or $H^+$) produces little change in pH as the added $OH^-$ (or $H^+$) reacts with $CH_3COOH$ (or $CH_3COO^-$), respectively. Weak acids are most effective in buffering against changes in pH within 1 pH unit of the p$K$ (see *Fig. 1*), often referred to as p$K \pm 1$, the **buffering capacity**.

Biological fluids, including the cytosol and extracellular fluids such as blood, are buffered. For example, in healthy individuals the pH of the blood is carefully controlled at pH 7.4. The major buffering components in most biological fluids

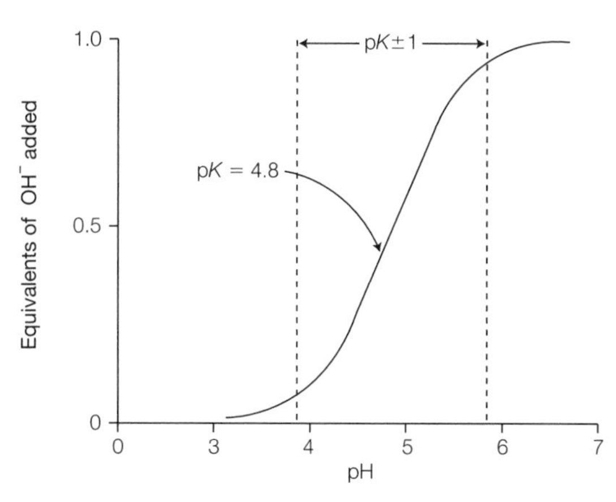


*Fig. 1. Titration curve of acetic acid.*

are the phosphate ion ($H_2PO_4^-$, p*K* 6.82) and the carbonate ion ($H_2CO_3$, p*K* 6.35) because they have p*K* values in this range. However, many biological molecules, including amino acids, proteins, nucleic acids and lipids, have multiple acid–base groups that are effective at buffering in the physiological pH range (pH 6–8).

When working with enzymes, proteins and other biological molecules it is often crucial to buffer the pH of the solution in order to avoid **denaturation** (loss of activity) of the component of interest (see Topic C3). Numerous buffers are used in laboratories for this purpose. One of the commonest is tris(hydroxymethyl)aminomethane or **TRIS** which has a p*K* of 8.08.

## Ionization of amino acids

The 20 standard amino acids have two **acid–base groups**: the α-amino and α-carboxyl groups attached to the $C_\alpha$ atom. Those amino acids with an **ionizable side-chain** (Asp, Glu, Arg, Lys, His, Cys, Tyr) have an additional acid–base group. The **titration curve** of Gly is shown in *Fig. 2a*. At low pH (i.e. high hydrogen ion concentration) both the amino group and the carboxyl group are fully protonated so that the amino acid is in the cationic form $H_3N^+CH_2COOH$ (*Fig. 2b*). As the amino acid in solution is titrated with increasing amounts of a strong base (e.g. NaOH), it loses two protons, first from the carboxyl group which has the lower **p*K*** value (p*K* = 2.3) and then from the amino group which has the higher p*K* value (p*K* = 9.6). The pH at which Gly has no net charge is termed its **isoelectric point, pI**. The α-carboxyl groups of all the 20 standard amino acids have p*K* values in the range 1.8–2.9, whilst their α-amino groups have p*K* values in the range 8.8–10.8 (*Table 1*). The side-chains of the acidic amino acids Asp and Glu have p*K* values of 3.9 and 4.1, respectively, whereas those of the basic amino acids Arg and Lys, have p*K* values of 12.5 and 10.8, respectively. Only the side-chain of His, with a p*K* value of 6.0, is ionized within the **physiological pH range** (pH 6–8). It should be borne in mind that when the amino acids are linked together in proteins, only the side-chain groups and the terminal α-amino and α-carboxyl groups are free to ionize.

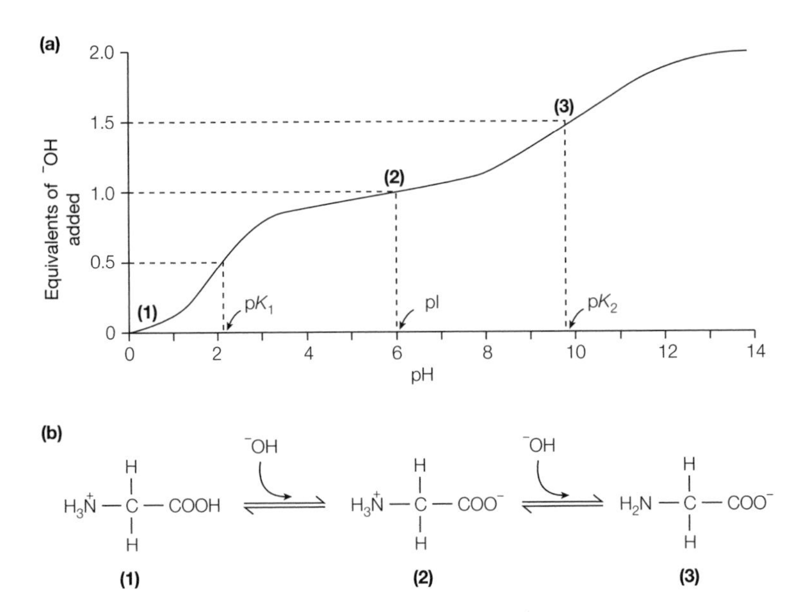


*Fig. 2. Ionization of glycine. (a) Titration curve of glycine, (b) dissociation of glycine. Numbers in bold in parentheses in (a) correspond to the structures in (b).*

*Table 1. pK values and molecular weights of the 20 standard amino acids*

| Amino acid | Mol. Wt | $pK$ α-COOH | $pK$ α-$NH_3^+$ | $pK$ side chain |
|---|---|---|---|---|
| Alanine | 89.1 | 2.35 | 9.87 | |
| Arginine | 174.2 | 1.82 | 8.99 | 12.48 |
| Asparagine | 132.1 | 2.14 | 8.72 | |
| Aspartic acid | 133.1 | 1.99 | 9.90 | 3.90 |
| Cysteine | 121.2 | 1.92 | 10.70 | 8.37 |
| Glutamic acid | 147.1 | 2.10 | 9.47 | 4.07 |
| Glutamine | 146.2 | 2.17 | 9.13 | |
| Glycine | 75.1 | 2.35 | 9.78 | |
| Histidine | 155.2 | 1.80 | 9.33 | 6.04 |
| Isoleucine | 131.2 | 2.32 | 9.76 | |
| Leucine | 131.2 | 2.33 | 9.74 | |
| Lysine | 146.2 | 2.16 | 9.06 | 10.54 |
| Methionine | 149.2 | 2.13 | 9.28 | |
| Phenylalanine | 165.2 | 2.20 | 9.31 | |
| Proline | 115.1 | 1.95 | 10.64 | |
| Serine | 105.1 | 2.19 | 9.21 | |
| Threonine | 119.1 | 2.09 | 9.10 | |
| Tryptophan | 204.2 | 2.46 | 9.41 | |
| Tyrosine | 181.2 | 2.20 | 9.21 | 10.46 |
| Valine | 117.1 | 2.29 | 9.74 | |

# B3 PROTEIN STRUCTURE

## Key Notes

**Peptide bond**

A protein is a linear sequence of amino acids linked together by peptide bonds. The peptide bond is a covalent bond between the α-amino group of one amino acid and the α-carboxyl group of another. The peptide bond has partial double bond character and is nearly always in the *trans* configuration. The backbone conformation of a polypeptide is specified by the rotation angles about the $C_\alpha$–N bond (*phi*, $\phi$) and $C_\alpha$–C bond (*psi*, $\psi$) of each of its amino acid residues. The sterically allowed values of $\phi$ and $\psi$ are visualized in a Ramachandran plot. When two amino acids are joined by a peptide bond they form a dipeptide. Addition of further amino acids results in long chains called oligopeptides and polypeptides.

**Primary structure**

The linear sequence of amino acids joined together by peptide bonds is termed the primary structure of the protein. The position of covalent disulfide bonds between cysteine residues is also included in the primary structure.

**Secondary structure**

Secondary structure in a protein refers to the regular folding of regions of the polypeptide chain. The two most common types of secondary structure are the α-helix and the β-pleated sheet. The α-helix is a cylindrical, rod-like helical arrangement of the amino acids in the polypeptide chain which is maintained by hydrogen bonds parallel to the helix axis. In a β-pleated sheet, hydrogen bonds form between adjacent sections of polypeptides that are either running in the same direction (parallel β-pleated sheet) or in the opposite direction (antiparallel β-pleated sheet). β-Turns reverse the direction of the polypeptide chain and are often found connecting the ends of antiparallel β-pleated sheets.

**Tertiary structure**

Tertiary structure in a protein refers to the three-dimensional arrangement of all the amino acids in the polypeptide chain. This biologically active, native conformation is maintained by multiple noncovalent bonds.

**Quaternary structure**

If a protein is made up of more than one polypeptide chain it is said to have quaternary structure. This refers to the spatial arrangement of the polypeptide subunits and the nature of the interactions between them.

**Protein stability**

In addition to the peptide bonds between individual amino acid residues, the three-dimensional structure of a protein is maintained by a combination of noncovalent interactions (electrostatic forces, van der Waals forces, hydrogen bonds, hydrophobic forces) and covalent interactions (disulfide bonds).

**Protein folding**

Proteins spontaneously fold into their native conformation, with the primary structure of the protein dictating its three-dimensional structure. Protein folding is driven primarily by hydrophobic forces and proceeds through an ordered set of pathways. Accessory proteins, including protein disulfide isomerases, peptidyl prolyl *cis–trans* isomerases, and molecular chaperones, assist proteins to fold correctly in the cell.

**Protein structure determination**

X-Ray crystallography and nuclear magnetic resonance (NMR) spectroscopy can be used to determine the three-dimensional structure of a protein.

**Related topics**

Amino acids (B1)
Myoglobin and hemoglobin (B4)
Collagen (B5)
The genetic code (H1)

## Peptide bond

Proteins are linear sequences of amino acids linked together by peptide bonds. The peptide bond is a chemical, covalent bond formed between the α-amino group of one amino acid and the α-carboxyl group of another (*Fig. 1a*) (see Topic B1). Once two amino acids are joined together via a peptide bond to form a dipeptide, there is still a free amino group at one end and a free carboxyl group at the other, each of which can in turn be linked to further amino acids. Thus, long, unbranched chains of amino acids can be linked together by peptide bonds to form oligopeptides (up to 25 amino acid residues) and polypeptides (> 25 amino acid residues). Note that the polypeptide still has a free α-amino group and a free α-carboxyl group. Convention has it that peptide chains are written down with the free α-amino group on the left, the free α-carboxyl group on the right and a hyphen between the amino acids to indicate the peptide bonds. Thus, the tripeptide $^{+}H_3N$-serine–leucine–phenylalanine-$COO^-$ would be written simply as Ser-Leu-Phe or S-L-F.

The peptide bond between the carbon and nitrogen exhibits **partial double-bond character** due to the closeness of the carbonyl carbon–oxygen double-bond allowing the **resonance structures** in *Fig. 1b* to exist. Because of this, the C–N

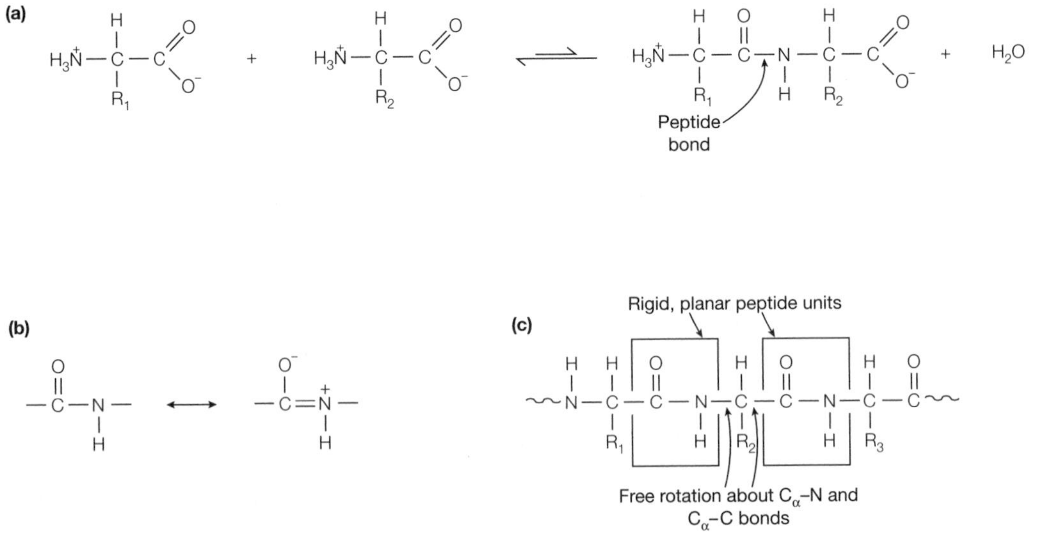

*Fig. 1. (a) Formation of a peptide bond, (b) resonance structures of the peptide bond, (c) peptide units within a polypeptide.*

bond length is also shorter than normal C–N single bonds. The **peptide unit** which is made up of the CO–NH atoms is thus relatively rigid and planar, although free rotation can take place about the $C_\alpha$–N and $C_\alpha$–C bonds (the bonds either side of the peptide bond), permitting adjacent peptide units to be at different angles (*Fig. 1c*). The hydrogen of the amino group is nearly always on the opposite side (*trans*) of the double bond to the oxygen of the carbonyl group, rather than on the same side (*cis*).

The backbone of a protein is a linked sequence of rigid planar peptide groups. The backbone conformation of a polypeptide is specified by the **rotation angles** or **torsion angles** about the $C_\alpha$–N bond (*phi*, $\phi$) and $C_\alpha$–C bond (*psi*, $\psi$) of each of its amino acid residues. When the polypeptide chain is in its planar, fully extended (all-*trans*) conformation the $\phi$ and $\psi$ angles are both defined as 180°, and increase for a clockwise rotation when viewed from $C_\alpha$ (*Fig. 2*). The **conformational range** of the torsion angles, $\phi$ and $\psi$, in a polypeptide backbone are restricted by steric hindrance. The sterically allowed values of $\phi$ and $\psi$ can be determined by calculating the distances between the atoms of a tripeptide at all values of $\phi$ and $\psi$ for the central peptide unit. These values are visualized in a steric contour diagram, otherwise known as a conformation map or **Ramachandran plot** (*Fig. 3*). From *Fig. 3* it can be seen that most areas of the Ramachandran plot (most combinations of $\phi$ and $\psi$) are conformationally inaccessible to a polypeptide chain. Only three small regions of the conformation map are physically accessible to a polypeptide chain, and within these regions are the $\phi$–$\psi$ values that produce the right-handed α-helix, the parallel and antiparallel β-pleated sheets and the collagen helix (see below and Topic B5).

The polypeptide chain folds up to form a specific shape (**conformation**) in the protein. This conformation is the **three-dimensional arrangement** of atoms in the structure and is determined by the amino acid sequence. There are four levels of structure in proteins: **primary**, **secondary**, **tertiary** and, sometimes but not always, **quaternary**.

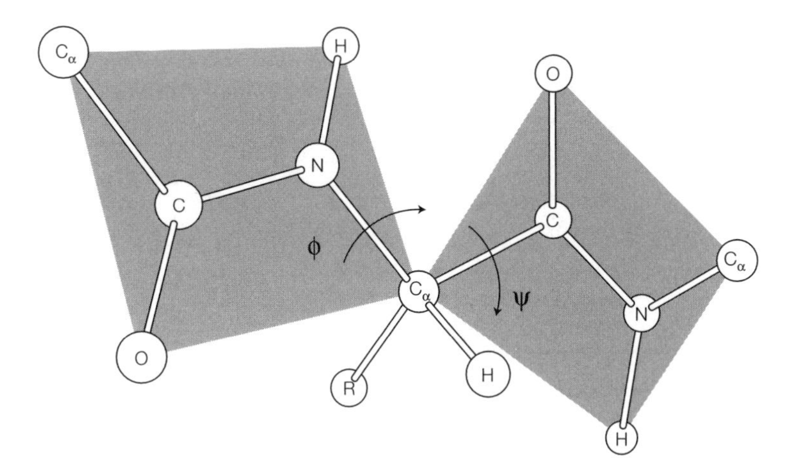


*Fig. 2. A segment of a polypeptide chain showing the torsion angles about the $C_\alpha$–N bond ($\phi$) and $C_\alpha$–C bond ($\psi$).*

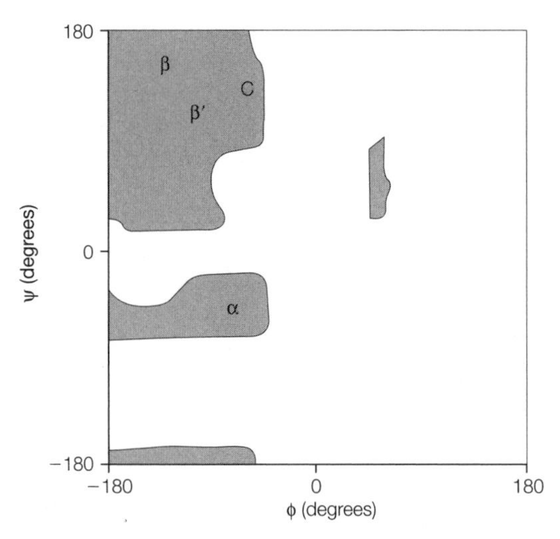


*Fig. 3. Ramachandran plot showing the allowed angles for poly-L-alanine (grey regions). α, φ–ψ values that produce the right-handed α-helix; β, the antiparallel β-pleated sheet; β′, the parallel β-pleated sheet; C, the collagen helix.*

## Primary structure

The primary level of structure in a protein is the **linear sequence of amino acids** as joined together by peptide bonds. This sequence is determined by the sequence of nucleotide bases in the gene encoding the protein (see Topic H1). Also included under primary structure is the location of any other **covalent bonds**. These are primarily **disulfide bonds** between cysteine residues that are adjacent in space but not in the linear amino acid sequence. These covalent cross-links between separate polypeptide chains or between different parts of the same chain are formed by the oxidation of the SH groups on cysteine residues that are juxtaposed in space (*Fig. 4*). The resulting disulfide is called a **cystine** residue. Disulfide bonds are often present in extracellular proteins, but are rarely found in intracellular proteins. Some proteins, such as collagen, have covalent cross-links formed between the side-chains of Lys residues (see Topic B5).

## Secondary structure

The secondary level of structure in a protein is the regular folding of regions of the polypeptide chain. The two most common types of protein fold are the **α-helix** and the **β-pleated sheet**. In the rod-like α-helix, the amino acids arrange

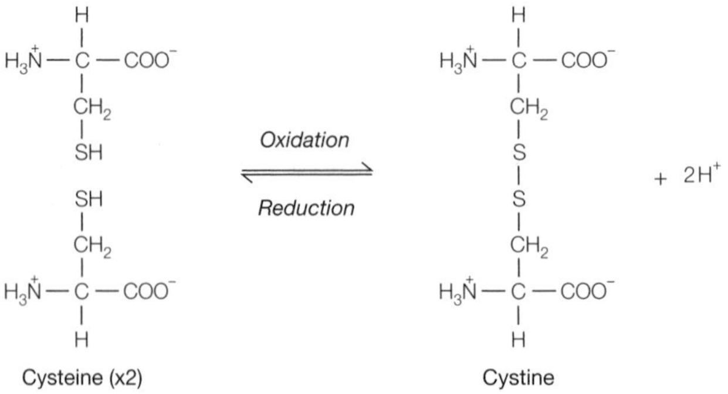

$$2\ H_3\overset{+}{N}-CH(COO^-)-CH_2-SH \underset{\text{Reduction}}{\overset{\text{Oxidation}}{\rightleftharpoons}} H_3\overset{+}{N}-CH(COO^-)-CH_2-S-S-CH_2-CH(COO^-)-\overset{+}{N}H_3 + 2H^+$$

Cysteine (x2) — Cystine

*Fig. 4. Formation of a disulfide bond between two cysteine residues, generating a cystine residue.*

themselves in a regular helical conformation (*Fig. 5a*). The carbonyl oxygen of each peptide bond is **hydrogen bonded** to the hydrogen on the amino group of the fourth amino acid away (*Fig. 5b*), with the hydrogen bonds running nearly parallel to the axis of the helix. In an α-helix there are 3.6 amino acids per turn of the helix covering a distance of 0.54 nm, and each amino acid residue represents an advance of 0.15 nm along the axis of the helix (*Fig. 5a*). The side-chains of the amino acids are all positioned along the outside of the cylindrical helix (*Fig. 5c*). Certain amino acids are more often found in α-helices than others. In particular, Pro is rarely found in α-helical regions as it cannot form the correct pattern of hydrogen bonds due to the lack of a hydrogen atom on its nitrogen atom. For this reason, Pro is often found at the end of an α-helix, where it alters the direction of the polypeptide chain and terminates the helix. Different proteins have a different amount of the polypeptide chain folded up into α-helices. For example, the single polypeptide chain of myoglobin has eight α-helices (see Topic B4).

In the **β-pleated sheet** hydrogen bonds form between the peptide bonds either in different polypeptide chains or in different sections of the same polypeptide chain (*Fig. 6a*). The planarity of the peptide bond forces the polypeptide to be pleated with the side-chains of the amino acids protruding above and below the sheet (*Fig. 6b*). Adjacent polypeptide chains in β-pleated sheets can be either **parallel** or **antiparallel** depending on whether they run in the same direction or in opposite directions, respectively (*Fig. 6c*). The polypeptide chain within a β-pleated sheet is fully extended, such that there is a distance of 0.35 nm from

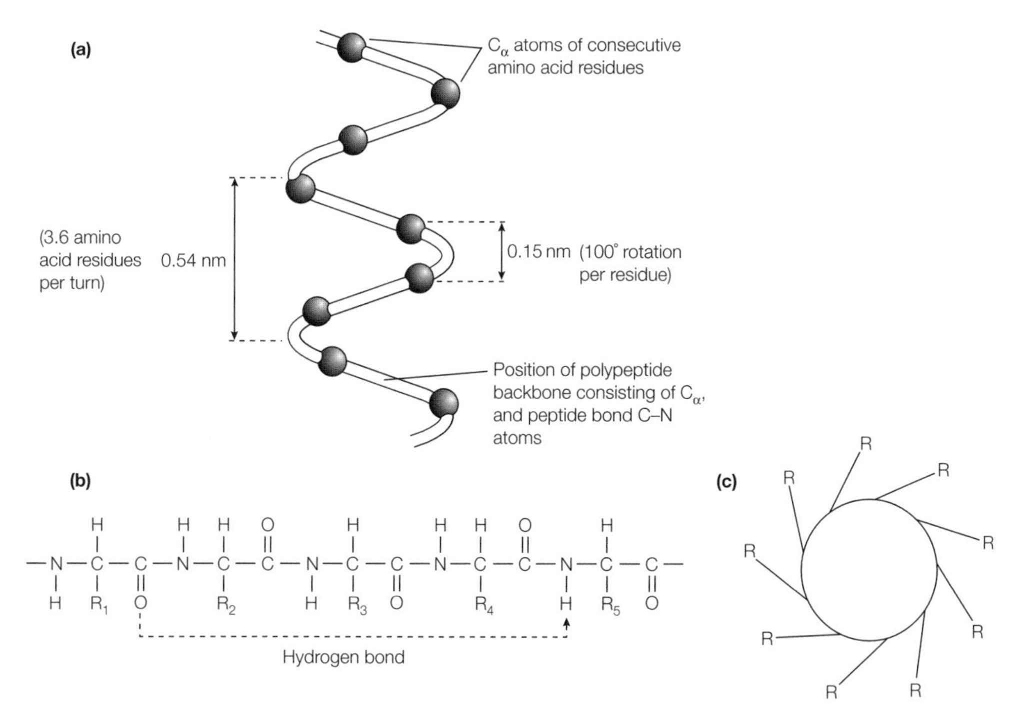


*Fig. 5. The folding of the polypeptide chain into an α-helix. (a) Model of an α-helix with only the $C_\alpha$ atoms along the backbone shown; (b) in the α-helix the CO group of residue* n *is hydrogen bonded to the NH group on residue (*n *+ 4); (c) cross-sectional view of an α-helix showing the positions of the side-chains (R groups) of the amino acids on the outside of the helix.*

one $C_\alpha$ atom to the next. β-Pleated sheets are always slightly curved and, if several polypeptides are involved, the sheet can close up to form a **β-barrel**. Multiple β-pleated sheets provide strength and rigidity in many structural proteins, such as silk fibroin, which consists almost entirely of stacks of antiparallel β-pleated sheets.

In order to fold tightly into the compact shape of a globular protein, the polypeptide chain often reverses direction, making a hairpin or **β-turn**. In these β-turns the carbonyl oxygen of one amino acid is hydrogen bonded to the hydrogen on the amino group of the fourth amino acid along (*Fig. 7*). β-Turns are often found connecting the ends of antiparallel β-pleated sheets. Regions of the polypeptide chain that are not in a regular secondary structure are said to have a **coil** or **loop conformation**. About half the polypeptide chain of a typical globular protein will be in such a conformation.

(a)

Hydrogen bonds

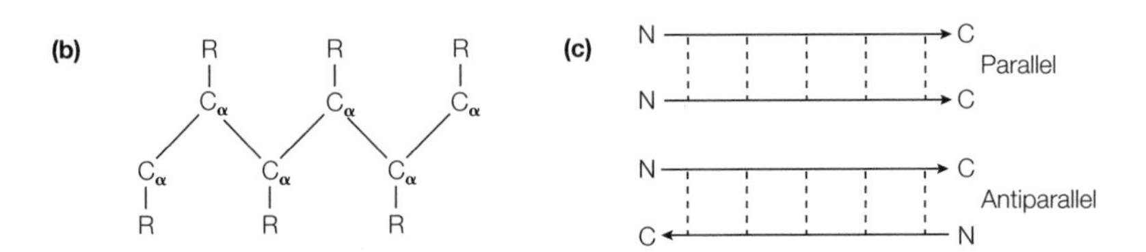


*Fig. 6. The folding of the polypeptide chain in a β-pleated sheet. (a) Hydrogen bonding between two sections of a polypeptide chain forming a β-pleated sheet; (b) a side-view of one of the polypeptide chains in a β-pleated sheet showing the side-chains (R groups) attached to the $C_\alpha$ atoms protruding above and below the sheet; (c) because the polypeptide chain has polarity, either parallel or antiparallel β-pleated sheets can form.*

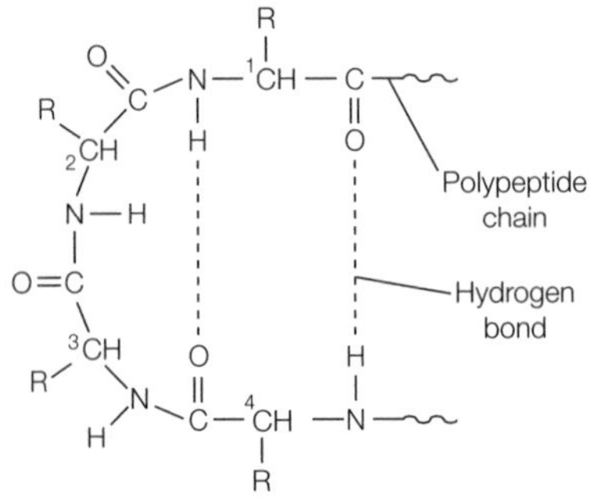

Polypeptide chain

Hydrogen bond

*Fig. 7. The folding of the polypeptide chain in a β-turn.*

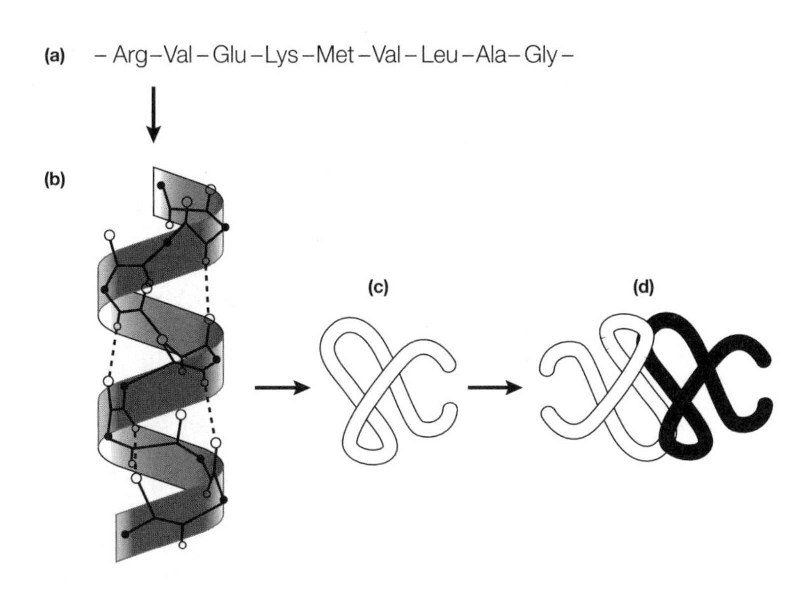


*Fig. 8.* *The four levels of structure in proteins. (a) Primary structure (amino acid sequence), (b) secondary structure (α-helix), (c) tertiary structure, (d) quaternary structure.*

## Tertiary structure

The third level of structure found in proteins, tertiary structure, refers to the spatial arrangement of amino acids that are far apart in the linear sequence as well as those residues that are adjacent. Again, it is the sequence of amino acids that specifies this final **three-dimensional structure** (*Fig. 8*). In water-soluble globular proteins such as myoglobin (see Topic B4), the main driving force behind the folding of the polypeptide chain is the energetic requirement to bury the nonpolar amino acids in the hydrophobic interior away from the surrounding aqueous, hydrophilic medium. The polypeptide chain folds spontaneously so that the majority of its hydrophobic side-chains are buried in the interior, and the majority of its polar, charged side-chains are on the surface. Once folded, the **three-dimensional, biologically active (native) conformation** of the protein is maintained not only by hydrophobic interactions, but also by electrostatic forces, hydrogen bonding and, if present, the covalent disulfide bonds. The electrostatic forces include salt bridges between oppositely charged groups and the multiple weak van der Waals interactions between the tightly packed aliphatic side-chains in the interior of the protein.

## Quaternary structure

Proteins containing more than one polypeptide chain, such as hemoglobin (see Topic B4), exhibit a fourth level of protein structure called quaternary structure (*Fig. 8*). This level of structure refers to the spatial arrangement of the polypeptide **subunits** and the nature of the interactions between them. These interactions may be covalent links (e.g. disulfide bonds) or noncovalent interactions (electrostatic forces, hydrogen bonding, hydrophobic interactions).

## Protein stability

The native three-dimensional conformation of a protein is maintained by a range of noncovalent interactions (electrostatic forces, hydrogen bonds, hydrophobic forces) and covalent interactions (disulfide bonds), in addition to the peptide bonds between individual amino acids.

- **Electrostatic forces**: these include the interactions between two ionic groups of opposite charge, for example the ammonium group of Lys and the carboxyl group of Asp, often referred to as an **ion pair** or **salt bridge**. In addition, the noncovalent associations between electrically neutral molecules, collectively referred to as **van der Waals forces**, arise from electrostatic interactions between permanent and/or induced dipoles, such as the carbonyl group in peptide bonds.
- **Hydrogen bonds**: these are predominantly electrostatic interactions between a weakly acidic donor group and an acceptor atom that bears a lone pair of electrons, which thus has a partial negative charge that attracts the hydrogen atom. In biological systems the donor group is an oxygen or nitrogen atom that has a covalently attached hydrogen atom, and the acceptor is either oxygen or nitrogen (*Fig. 9*). Hydrogen bonds are normally in the range 0.27–0.31 nm and are highly directional, i.e. the donor, hydrogen and acceptor atoms are colinear. Hydrogen bonds are stronger than van der Waals forces but much weaker than covalent bonds. Hydrogen bonds not only play an important role in protein structure, but also in the structure of other biological macromolecules such as the DNA double helix (see Topic F1) and lipid bilayers (see Topic E1). In addition, hydrogen bonds are critical to both the properties of water and to its role as a biochemical solvent.
- **Hydrophobic forces**: The **hydrophobic effect** is the name given to those forces that cause nonpolar molecules to minimize their contact with water. This is clearly seen with amphipathic molecules such as lipids and detergents which form micelles in aqueous solution (see Topic E1). Proteins, too, find a conformation in which their nonpolar side chains are largely out of contact with the aqueous solvent, and thus hydrophobic forces are an important determinant of protein structure, folding and stability. In proteins, the effects of hydrophobic forces are often termed **hydrophobic bonding**, to indicate the specific nature of protein folding under the influence of the hydrophobic effect.
- **Disulfide bonds**: These covalent bonds form between Cys residues that are close together in the final conformation of the protein (see *Fig. 4*) and function to stabilize its three-dimensional structure. Disulfide bonds are really only formed in the oxidizing environment of the endoplasmic reticulum (see Topic A2), and thus are found primarily in extracellular and secreted proteins.

## Protein folding

Under appropriate physiological conditions, proteins **spontaneously fold** into their native conformation. As there is no need for external templates, this implies

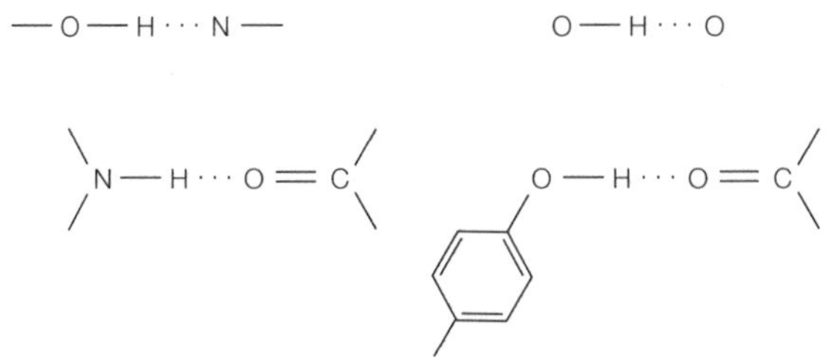

—O—H···N—    O—H···O

N—H···O=C    O—H···O=C

*Fig. 9. Examples of hydrogen bonds (shown as dotted lines).*

that the primary structure of the protein dictates its three-dimensional structure. From experiments with the protein **RNase A** it has been observed that it is mainly the internal residues of a protein that direct its folding to the native conformation. Alteration of surface residues by mutation is less likely to affect the folding than changes to internal residues. It has also been observed that protein folding is driven primarily by **hydrophobic forces**. Proteins fold into their native conformation through an **ordered set of pathways** rather than by a random exploration of all the possible conformations until the correct one is stumbled upon.

Although proteins can fold *in vitro* (in the laboratory) without the presence of accessory proteins, this process can take minutes to days. *In vivo* (in the cell) this process requires only a few minutes because the cells contain **accessory proteins** which assist the polypeptides to fold to their native conformation. There are three main classes of protein folding accessory proteins:

- **protein disulfide isomerases** catalyze disulfide interchange reactions, thereby facilitating the shuffling of the disulfide bonds in a protein until they achieve their correct pairing.
- **peptidyl prolyl *cis–trans* isomerases** catalyze the otherwise slow interconversion of Xaa–Pro peptide bonds between their *cis* and *trans* conformations, thereby accelerating the folding of Pro-containing polypeptides. One of the classes of peptidyl prolyl *cis–trans* isomerases is inhibited by the **immunosuppressive drug** cyclosporin A.
- **molecular chaperones**, which include proteins such as the heat shock proteins 70 (Hsp 70), the chaperonins and the nucleoplasmins. These prevent the improper folding and aggregation of proteins that may otherwise occur as internal hydrophobic regions are exposed to one another.

## Protein structure determination

The three-dimensional structure of a protein can be determined almost to the atomic level by the techniques of **X-ray crystallography** and **nuclear magnetic resonance (NMR) spectroscopy**. In X-ray crystallography a crystal of the protein to be visualized is exposed to a beam of X-rays and the resulting diffraction pattern caused as the X-rays encounter the protein crystal is recorded on photographic film. The intensities of the diffraction maxima (the darkness of the spots on the film) are then used to mathematically construct the three-dimensional image of the protein crystal. NMR spectroscopy can be used to determine the three-dimensional structures of small (up to approximately 30 kDa) proteins in aqueous solution.

# B4 MYOGLOBIN AND HEMOGLOBIN

## Key Notes

**Oxygen-binding proteins**

Hemoglobin and myoglobin are the two oxygen-binding proteins present in large multicellular organisms. Hemoglobin transports oxygen in the blood and is located in the erythrocytes; myoglobin stores the oxygen in the muscles.

**Myoglobin**

Myoglobin was the first protein to have its three-dimensional structure solved by X-ray crystallography. It is a globular protein made up of a single polypeptide chain of 153 amino acid residues that is folded into eight α-helices. The heme prosthetic group is located within a hydrophobic cleft of the folded polypeptide chain.

**Hemoglobin**

Hemoglobin has quaternary structure as it is made up of four polypeptide chains; two α-chains and two β-chains ($\alpha_2\beta_2$), each with a heme prosthetic group. Despite little similarity in their primary sequences, the individual polypeptides of hemoglobin have a three-dimensional structure almost identical to the polypeptide chain of myoglobin.

**Binding of oxygen to heme**

The heme prosthetic group consists of a protoporphyrin IX ring and a central $Fe^{2+}$ atom which forms four bonds with the porphyrin ring. In addition, on one side of the porphyrin ring the $Fe^{2+}$ forms a bond with the proximal histidine (His F8); a residue eight amino acids along the F-helix of hemoglobin. The sixth bond from the $Fe^{2+}$ is to a molecule of $O_2$. Close to where the $O_2$ binds is another histidine residue, the distal histidine (His E7), which prevents carbon monoxide binding most efficiently.

**Allostery**

Hemoglobin is an allosteric protein. The binding of $O_2$ is cooperative; the binding of $O_2$ to one subunit increases the ease of binding of further $O_2$ molecules to the other subunits. The oxygen dissociation curve for hemoglobin is sigmoidal whereas that for myoglobin is hyperbolic. Myoglobin has a greater affinity for $O_2$ than does hemoglobin.

**Mechanism of the allosteric change**

Oxyhemoglobin has a different quaternary structure from deoxyhemoglobin. As $O_2$ binds to the $Fe^{2+}$ it distorts the heme group and moves the proximal histidine. This in turn moves helix F and alters the interactions between the four subunits.

**The Bohr effect**

$H^+$, $CO_2$ and 2,3-bisphosphoglycerate are allosteric effectors, promoting the release of $O_2$ from hemoglobin. $H^+$ and $CO_2$ bind to different parts of the polypeptide chains, while 2,3-bisphosphoglycerate binds in the central cavity between the four subunits.

**Fetal hemoglobin**

Hemoglobin F (HbF) which consists of two α-chains and two γ-chains ($\alpha_2\gamma_2$) is present in the fetus. HbF binds 2,3-bisphosphoglycerate less strongly than adult hemoglobin (HbA) and thus has a higher affinity for $O_2$ which promotes the transfer of $O_2$ from the maternal to the fetal circulation.

**Hemoglobinopathies**

Comparison of hemoglobin sequences from different species reveals that only nine amino acid residues are invariant. Some residues are subject to conservative substitution of one residue by another with similar properties, others to nonconservative substitution where one amino acid residue is replaced by another with different properties. Hemoglobinopathies are diseases caused by abnormal hemoglobins. The best characterized of these is the genetically transmitted, hemolytic disease sickle-cell anemia. This is caused by the nonconservative substitution of a glutamate by a valine, resulting in the appearance of a hydrophobic sticky patch on the surface of the protein. This allows long aggregated fibers of hemoglobin molecules to form which distort the shape of the red blood cells. Heterozygotes carrying only one copy of the sickle-cell gene are more resistant to malaria than those homozygous for the normal gene.

**Related topics** Protein structure (B3) Hemes and chlorophylls (M4)

## Oxygen-binding proteins

**Hemoglobin** is one of two **oxygen-binding proteins** found in vertebrates. The function of hemoglobin is to carry $O_2$ in the blood from the lungs to the other tissues in the body, in order to supply the cells with the $O_2$ required by them for the oxidative phosphorylation of foodstuffs (see Topic L2). The hemoglobin is found in the blood within the **erythrocytes** (red blood cells). These cells essentially act, amongst other things, as a sack for carrying hemoglobin, since mature erythrocytes lack any internal organelles (nucleus, mitochondria, etc.). The other $O_2$-binding protein is **myoglobin**, which stores the oxygen in the tissues of the body ready for when the cells require it. The highest concentrations of myoglobin are found in skeletal and cardiac **muscle** which require large amounts of $O_2$ because of their need for large amounts of energy during contraction (see Topic N1).

## Myoglobin

Myoglobin is a relatively small protein of mass 17.8 kDa made up of 153 amino acids in a single polypeptide chain. It was the first protein to have its **three-dimensional structure** determined by **X-ray crystallography** by John Kendrew in 1957. Myoglobin is a typical **globular protein** in that it is a highly folded compact structure with most of the hydrophobic amino acid residues buried in the interior and many of the polar residues on the surface. X-ray crystallography revealed that the single polypeptide chain of myoglobin consists entirely of **α-helical secondary structure** (see Topic B3). In fact there are eight α-helices (labeled A–H) in myoglobin (*Fig. 1a*). Within a hydrophobic crevice formed by the folding of the polypeptide chain is the **heme prosthetic group** (*Fig. 1a*). This nonpolypeptide unit is noncovalently bound to myoglobin and is essential for the biological activity of the protein (i.e. the binding of $O_2$).

## Hemoglobin

The three-dimensional structure of hemoglobin was solved using **X-ray crystallography** in 1959 by Max Perutz. This revealed that hemoglobin is made up of four polypeptide chains, each of which has a very similar three-dimensional structure to the single polypeptide chain in myoglobin (*Fig. 1b*) despite the fact that their amino acid sequences differ at 83% of the residues. This highlights a relatively common theme in protein structure:

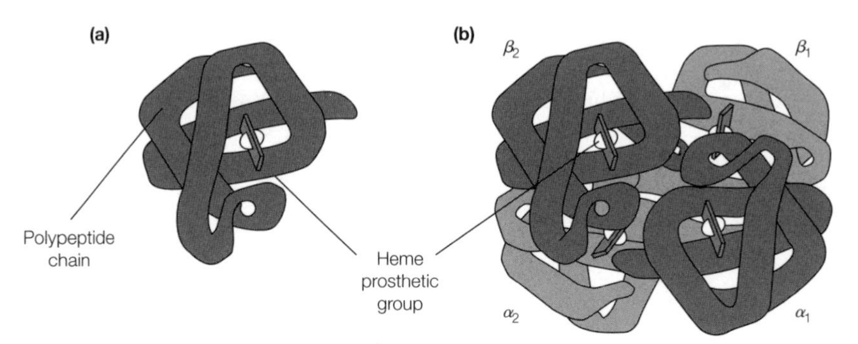


*Fig. 1. Structure of (a) myoglobin and (b) hemoglobin, showing the α and β polypeptide chains.*

that very different primary sequences can specify very similar three-dimensional structures. The major type of hemoglobin found in adults (HbA) is made up of two different polypeptide chains: the **α-chain** that consists of 141 amino acid residues, and the **β-chain** of 146 residues ($\alpha_2\beta_2$; *Fig. 1b*). Each chain, like that in myoglobin, consists of eight α-helices and each contains a heme prosthetic group (*Fig. 1b*). Therefore, hemoglobin can bind four molecules of $O_2$. The four polypeptide chains, two α and two β, are packed tightly together in a tetrahedral array to form an overall spherically shaped molecule that is held together by multiple noncovalent interactions.

## Binding of oxygen to heme

The **heme prosthetic group** in myoglobin and hemoglobin is made up of a **protoporphyrin IX** ring structure with an **iron atom** in the ferrous ($Fe^{2+}$) oxidation state (see Topic M4; *Fig. 2*). This $Fe^{2+}$ bonds with four nitrogen atoms in the

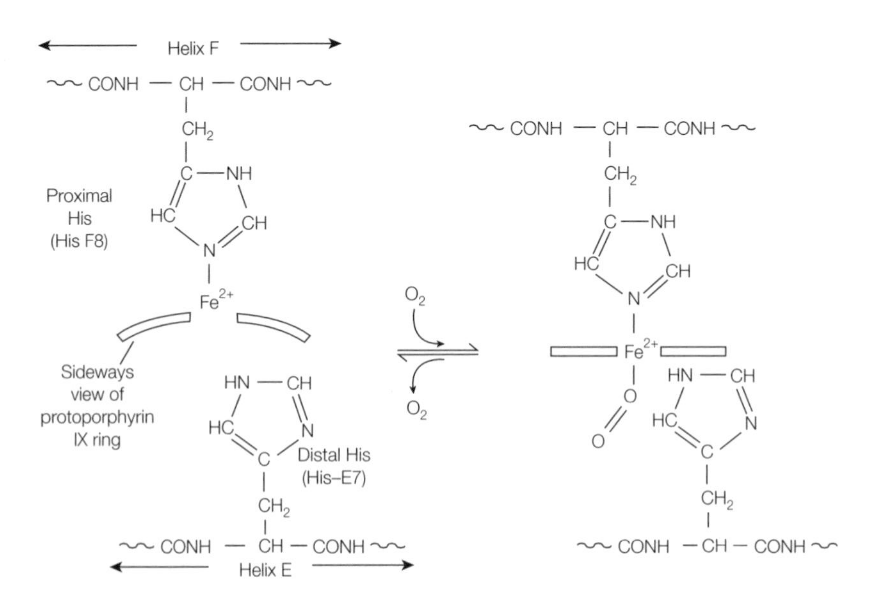


*Fig. 2. Binding of $O_2$ to heme. The $Fe^{2+}$ of the protoporphyrin ring is bonded to His F8 but not to His E7 which is located nearby. As the heme $Fe^{2+}$ binds $O_2$, helix F moves closer to helix E (see the text for details).*

center of the protoporphyrin ring and forms two additional bonds on either side of the plane of the protoporphyrin ring. One of these is to a histidine residue which lies eight residues along helix F of hemoglobin, the **proximal histidine** (His F8) (*Fig.* 2). The sixth bond is to one of the oxygen atoms in a molecule of $\mathbf{O_2}$ (*Fig.* 2). Near to where the $O_2$ binds to the heme group is another histidine residue, the **distal histidine** (His E7) (*Fig.* 2). This serves two very important functions. First, it prevents heme groups on neighboring hemoglobin molecules coming into contact with one another and oxidizing to the $Fe^{3+}$ state in which they can no longer bind $O_2$. Second, it prevents **carbon monoxide** (CO) binding with the most favorable configuration to the $Fe^{2+}$, thereby lowering the affinity of heme for CO. This is important because once CO has bound irreversibly to the heme, the protein can no longer bind $O_2$. Thus, although the oxygen binding site in hemoglobin and myoglobin is only a small part of the whole protein, the polypeptide chain modulates the function of the heme prosthetic group.

## Allostery

Hemoglobin is an **allosteric protein**. This means that the binding of $O_2$ to one of the subunits is affected by its interactions with the other subunits. In fact the binding of $O_2$ to one hemoglobin subunit induces conformational changes (see below and *Fig.* 2) that are relayed to the other subunits, making them more able to also bind $O_2$ by raising their affinity for this molecule. Thus binding of $O_2$ to hemoglobin is said to be **cooperative**. In contrast, the binding of $O_2$ to the single polypeptide unit of myoglobin is **noncooperative**. This is clearly apparent from the **oxygen dissociation curves** for the two proteins: that for hemoglobin is **sigmoidal**, reflecting this cooperative binding, whereas that for myoglobin is **hyperbolic** (*Fig.* 3). From the $O_2$ dissociation curve it can also be seen that for any particular oxygen pressure the degree of saturation of myoglobin is higher than that for hemoglobin. In other words, myoglobin has a higher affinity for $O_2$ than does hemoglobin. This means that in the blood capillaries in the muscle, for example, hemoglobin will release its $O_2$ to myoglobin for storage there.

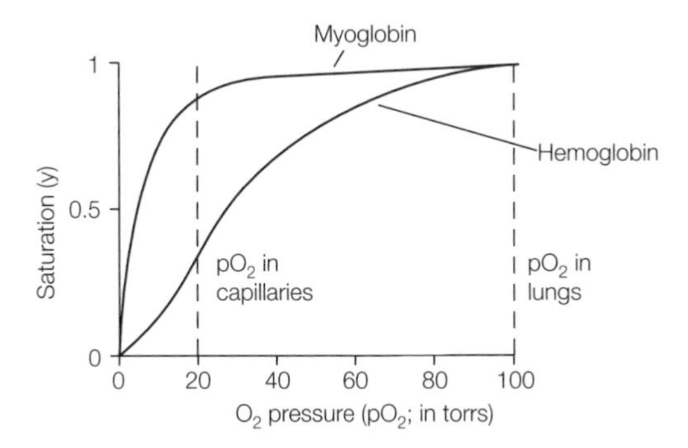


*Fig. 3. Oxygen dissociation curves for hemoglobin and myoglobin.*

## Mechanism of the allosteric change

X-ray crystallography revealed that **oxyhemoglobin**, the form that has four $O_2$ molecules bound, differs markedly in its **quaternary structure** from **deoxy-hemoglobin**, the form with no $O_2$ bound. In the absence of bound $O_2$, the $Fe^{2+}$ lies slightly to one side of the porphyrin ring, which itself is slightly curved (*Fig.* 2). As a molecule of $O_2$ binds to the heme prosthetic group it pulls the $Fe^{2+}$ into the plane of the porphyrin ring (*Fig.* 2), flattening out the ring in the process. Movement of the $Fe^{2+}$ causes the **proximal histidine** to move also. This, in turn, shifts the position of helix F and regions of the polypeptide chain at either end of the helix. Thus, movement in the center of the subunit is transmitted to the surfaces, where it causes the ionic interactions holding the four subunits together to be broken and to reform in a different position, thereby altering the quaternary structure, leading to the cooperative binding of $O_2$ to Hb.

## The Bohr effect

The binding of $O_2$ to hemoglobin is affected by the concentration of **$H^+$ ions** and **$CO_2$** in the surrounding tissue; the Bohr effect. In actively metabolizing tissue, such as muscle, the concentrations of these two substances are relatively high. This effectively causes a shift of the $O_2$ dissociation curve for hemoglobin to the right, promoting the release of $O_2$. This comes about because there are $H^+$ binding sites, primarily His146 in the β-chain, which have a higher affinity for binding $H^+$ in deoxyhemoglobin than in oxyhemoglobin. An increase in $CO_2$ also causes an increase in $H^+$ due to the action of the enzyme **carbonic anhydrase** which catalyzes the reaction:

$$CO_2 + H_2O \rightleftharpoons HCO_3^- + H^+$$

In addition, $CO_2$ can react with the primary amino groups in the polypeptide chain to form a negatively charged carbamate. Again, this change from a positive to a negative charge favors the conformation of deoxyhemoglobin. On returning in the blood to the lungs, the concentrations of $H^+$ and $CO_2$ are relatively lower and that of $O_2$ higher, so that the process is reversed and $O_2$ binds to hemoglobin. Thus, it can be seen that not only does hemoglobin carry $O_2$ but it also carries $CO_2$ back to the lungs where it is expelled.

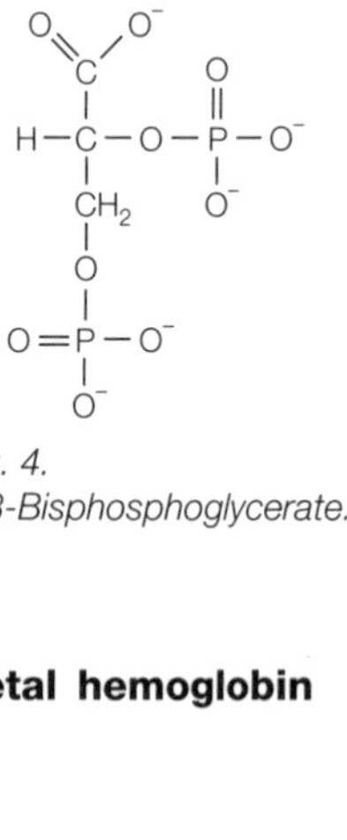

**2,3-Bisphosphoglycerate** is a highly anionic organic phosphate molecule (*Fig. 4*) that is present in erythrocytes along with the hemoglobin. This molecule promotes the release of $O_2$ from hemoglobin by lowering the affinity of the protein for $O_2$. 2,3-Bisphosphoglycerate binds in the small cavity in the center of the four subunits. In oxyhemoglobin this cavity is too small for it, whereas in deoxyhemoglobin it is large enough to accommodate a single molecule of 2,3-bisphosphoglycerate. On binding in the central cavity of deoxyhemoglobin it forms ionic bonds with the positively charged amino acid side-chains in the β-subunits, stabilizing the quaternary structure. $H^+$, $CO_2$ and 2,3-bisphosphoglycerate are all **allosteric effectors** as they favor the conformation of deoxyhemoglobin and therefore promote the release of $O_2$. Because these three molecules act at different sites, their effects are additive.

*Fig. 4. 2,3-Bisphosphoglycerate.*

## Fetal hemoglobin

In the fetus there is a different kind of hemoglobin, **hemoglobin F** (HbF) which consists of two α-chains and two γ-chains ($\alpha_2\gamma_2$), in contrast to adult hemoglobin (HbA, $\alpha_2\beta_2$). HbF has a **higher affinity** for $O_2$ under physiological conditions than HbA, which optimizes the transfer of oxygen from the maternal to the fetal circulation across the placenta. The molecular basis for this difference in $O_2$ affinity is that HbF binds 2,3-bisphosphoglycerate less strongly than does HbA. Near birth the synthesis of the γ-chain is switched off, and that of the β-chain (which is present in HbA) is switched on (*Fig. 5*).

## Hemoglobinopathies

Comparison of the primary sequences of hemoglobin chains from more than 60 different species reveals that only nine residues in the polypeptide chain are **invariant** (i.e. the same) between all of the species. These nine residues include the **proximal and distal histidines** which are essential for the correct functioning of the protein. Many of the other residues are replaced from one species

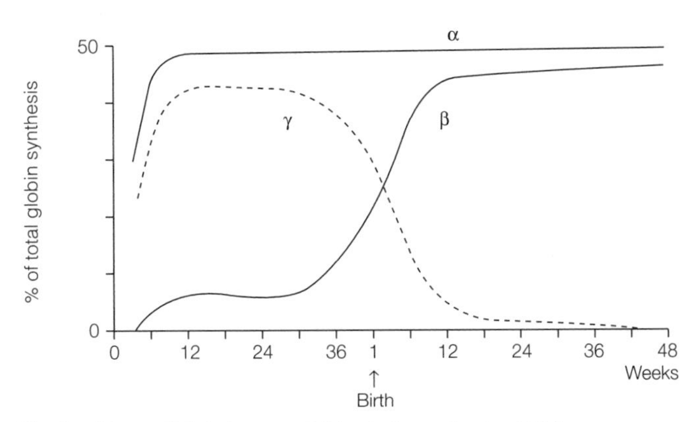


*Fig. 5. The switch in human globin chain synthesis at birth.*

to another by residues with similar properties (e.g. the hydrophobic valine is replaced with the hydrophobic isoleucine, or the polar serine is replaced with the polar asparagine), so-called **conservative substitutions**. In contrast, only a few residues have changed between species to a completely different residue (e.g. a hydrophobic leucine to a positively charged lysine or a negatively charged glutamate to a positively charged arginine), so-called **nonconservative substitutions**, since this type of change could have a major effect on the structure and function of the protein.

Several hundred **abnormal hemoglobins** have been characterized, giving rise to the so-called **hemoglobinopathies**. Probably the best characterized hemoglobinopathy is **sickle-cell anemia** (sickle-cell hemoglobin; HbS). This disease is characterized by the patient's erythrocytes having a characteristic sickle or crescent shape. The molecular basis for this disease is the change of a glutamic acid residue for a valine at position 6 of the β-chain, resulting in the substitution of a polar residue by a hydrophobic one. This **nonconservative substitution** of valine for glutamate gives HbS a **sticky hydrophobic patch** on the outside of each of its β-chains. In the corner between helices E and F of the β-chain of deoxy-HbS is a hydrophobic site that is complementary to the sticky patch (*Fig. 6*). Thus the complementary site on one deoxy-HbS molecule can bind to the sticky patch on another deoxy-HbS molecule, resulting in the formation of **long fibers** of hemoglobin molecules that distort the erythrocyte. In oxy-HbS

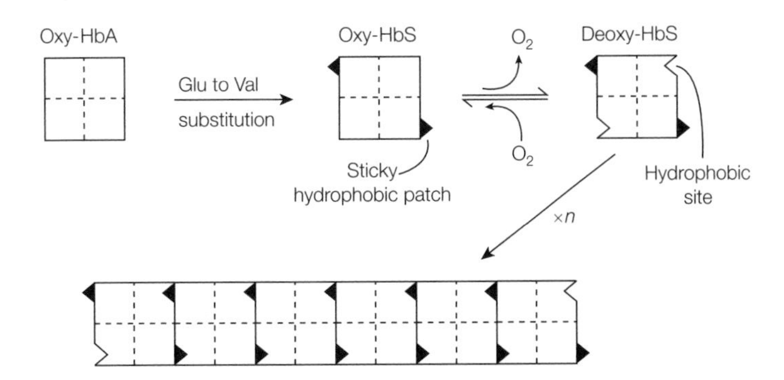


*Fig. 6. Molecular basis for the aggregation of deoxyhemoglobin molecules in sickle-cell anemia.*

the complementary site is masked, so the formation of the long fibers occurs only when there is a high concentration of the deoxygenated form of HbS.

Sickle-cell anemia is a **genetically transmitted**, hemolytic disease. The sickled cells are more fragile than normal erythrocytes, lysing more easily and having a shorter half-life, which leads to severe anemia. As sickle-cell anemia is genetically transmitted, **homozygotes** have two copies of the abnormal gene whereas **heterozygotes** have one abnormal and one normal copy. Homozygotes often have a reduced life-span as a result of infection, renal failure, cardiac failure or thrombosis, due to the sickled cells becoming trapped in small blood vessels leading to tissue damage. In contrast, heterozygotes are usually not symptomatic as only approximately 1% of their erythrocytes are sickled, compared with approximately 50% in a homozygote. The frequency of the sickle gene is relatively high in certain parts of Africa and correlates with the incidence of **malaria**. The reason for this is that heterozygotes are protected against the most lethal form of malaria, whereas normal homozygotes are more vulnerable to the disease. Inheritance of the abnormal hemoglobin gene can now be monitored by recombinant DNA techniques (see Topic I1).

# B5 COLLAGEN

## Key Notes

**Function and diversity**

Collagen is the name given to a family of structurally related proteins that form strong insoluble fibers. Collagens consist of three polypeptide chains, the identity and distribution of which vary between collagen types. The different types of collagen are found in different locations in the body.

**Biosynthesis: overview**

The collagen polypeptides are post-translationally modified by hydroxylation and glycosylation on transport through the rough endoplasmic reticulum and Golgi. The three polypeptides form the triple-helical procollagen which is secreted out of the cell. The extension peptides are removed to form tropocollagen which then aggregates into a microfibril and is covalently cross-linked to form the mature collagen fiber.

**Composition and post-translational modifications**

One-third of the amino acid residues in collagen are Gly, while another quarter are Pro. The hydroxylated amino acids 4-hydroxyproline (Hyp) and 5-hydroxylysine (Hyl) are formed post-translationally by the action of proline hydroxylase and lysine hydroxylase. These $Fe^{2+}$-containing enzymes require ascorbic acid (vitamin C) for activity. In the vitamin C deficiency disease scurvy, collagen does not form correctly due to the inability to hydroxylate Pro and Lys. Hyl residues are often post-translationally modified with carbohydrate.

**Structure**

Collagen contains a repeating tripeptide sequence of Gly–X–Y, where X is often Pro and Y is often Hyp. Each polypeptide in collagen folds into a helix with 3.3 residues per turn. Three polypeptide chains then come together to form a triple-helical cable that is held together by hydrogen bonds between the chains. Every third residue passes through the center of the triple helix, which is so crowded that only Gly is small enough to fit. One form of osteogenesis imperfecta (brittle bones) is caused by the mutation of a Gly residue to another amino acid, which prevents the triple-helical cable folding correctly and results in defective collagen.

**Secretion and aggregation**

The extension peptides on both the N and C termini of the polypeptide chains direct the formation of the triple-helical cable and prevent the premature aggregation of the procollagen molecules within the cell. Following secretion out of the cell, the extension peptides are cleaved off by peptidases, and the resulting tropocollagen molecules aggregate together in a staggered array.

**Cross-links**

Covalent cross-links both between and within the tropocollagen molecules confer strength and rigidity on the collagen fiber. These cross-links are formed between Lys and its aldehyde derivative allysine. Allysine is derived from Lys by the action of the copper-containing lysyl oxidase which requires pyridoxal phosphate for activity. The disease lathyrism is caused by the inhibition of lysyl oxidase by the chemical β-aminopropionitrile in sweet pea seeds, and results in defective collagen due to the lack of cross-links.

**Bone formation** Hydroxyapatite (calcium phosphate) is deposited in nucleation sites between the ends of tropocollagen molecules as the first step in bone formation.

**Related topics** Protein structure (B3) Myoglobin and hemoglobin (B4) Translation in eukaryotes (H3)

## Function and diversity

**Collagen**, which is present in all multicellular organisms, is not one protein but a family of structurally related proteins. It is the most abundant protein in mammals and is present in most organs of the body, where it serves to hold cells together in discrete units. It is also the major **fibrous element** of skin, bones, tendons, cartilage, blood vessels and teeth. The different collagen proteins have very diverse functions. The extremely hard structures of bones and teeth contain collagen and a calcium phosphate polymer. In tendons, collagen forms rope-like fibers of high tensile strength, while in the skin collagen forms loosely woven fibers that can expand in all directions. The different types of collagen are characterized by different polypeptide compositions (*Table 1*). Each collagen is composed of three polypeptide chains, which may be all identical (as in types II and III) or may be of two different chains (types I, IV and V). A single molecule of type I collagen has a molecular mass of 285 kDa, a width of 1.5 nm and a length of 300 nm.

*Table 1. Types of collagen*

| Type | Polypeptide composition | Distribution |
|---|---|---|
| I | $[\alpha 1(I)]_2\ \alpha 2(I)$ | Skin, bone, tendon, cornea, blood vessels |
| II | $[\alpha 1(II)]_3$ | Cartilage, intervertebral disk |
| III | $[\alpha 1(III)]_3$ | Fetal skin, blood vessels |
| IV | $[\alpha 1(IV)]_2\ \alpha 2(IV)$ | Basement membrane |
| V | $[\alpha 1(V)]_2\ \alpha 2(V)$ | Placenta, skin |

## Biosynthesis: overview

Like other secreted proteins, collagen polypeptides are synthesized by ribosomes on the rough endoplasmic reticulum (RER; see Topic H3). The polypeptide chain then passes through the RER and Golgi apparatus before being secreted. Along the way it is **post-translationally modified**: Pro and Lys residues are hydroxylated and carbohydrate is added (*Fig. 1*). Before secretion, three polypeptide chains come together to form a triple-helical structure known as **procollagen**. The procollagen is then secreted into the extracellular spaces of the connective tissue where extensions of the polypeptide chains at both the N and C termini (**extension peptides**) are removed by peptidases to form **tropocollagen** (*Fig. 1*). The tropocollagen molecules aggregate into a **microfibril** and are extensively **cross-linked** to produce the mature **collagen fiber** (*Fig. 1*).

## Composition and post-translational modifications

The amino acid composition of collagen is quite distinctive. Nearly *one-third* of its residues are **Gly**, while another *one-quarter* are **Pro**, significantly higher proportions than are found in other proteins. The hydroxylated amino acids **4-hydroxyproline (Hyp)** and **5-hydroxylysine (Hyl)** (*Fig. 2*) are found exclusively in collagen. These hydroxylated amino acids are formed from the parent amino acid by the action of **proline hydroxylase** and **lysine hydroxylase**, respectively

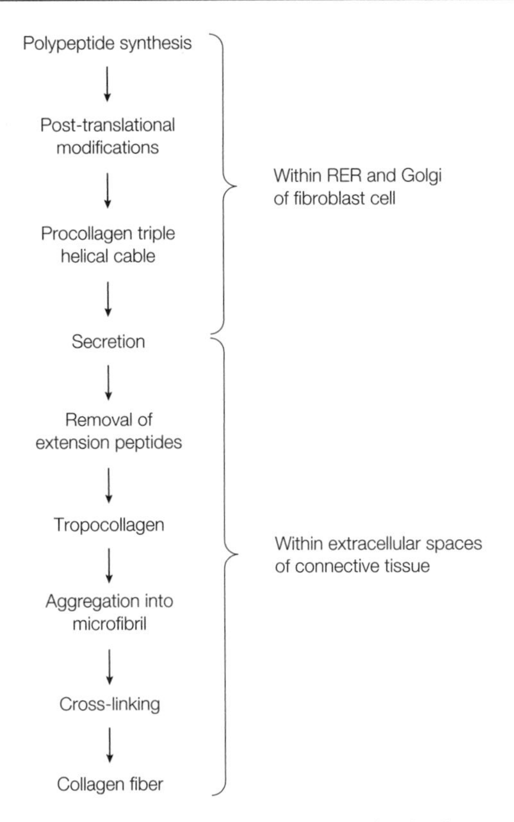


*Fig. 1. Overview of the biosynthesis of collagen.*

(*Fig. 2*). These enzymes have an $Fe^{2+}$ ion at their active site and require **ascorbic acid (vitamin C)** for activity. The ascorbic acid acts as an antioxidant, keeping the $Fe^{2+}$ ion in its reduced state. Proline hydroxylase and lysine hydroxylase are dioxygenases, using a molecule of $O_2$. α-Ketoglutarate, the citric acid cycle intermediate (see Topic L1), is an obligatory substrate and is converted into succinate during the reaction (*Fig. 2*). Both enzymes will hydroxylate only Pro and Lys residues that are incorporated in a polypeptide chain, and then only when the residue is on the N-terminal side of Gly. Hyp is important in stabilizing the structure of collagen through hydrogen bond formation (see below). In vitamin C deficiency, Hyp (and Hyl) are not synthesized, resulting in the weakening of the collagen fibers. This leads to the skin lesions, fragile blood vessels and poor wound healing that are characteristic of the disease **scurvy**.

The other post-translational modification that occurs to collagen is **glycosylation**. In this case the sugar residues, usually only glucose, galactose and their disaccharides, are attached to the hydroxyl group in the newly formed Hyl residues, rather than to Asn or Ser/Thr residues as occurs in the more widespread N- and O-linked glycosylation (see Topic H5). The amount of attached carbohydrate in collagen varies from 0.4 to 12% by weight depending on the tissue in which it is synthesized.

## Structure

The **primary structure** of each polypeptide in collagen is characterized by a repeating **tripeptide** sequence of **Gly–X–Y** where X is often, but not exclusively,

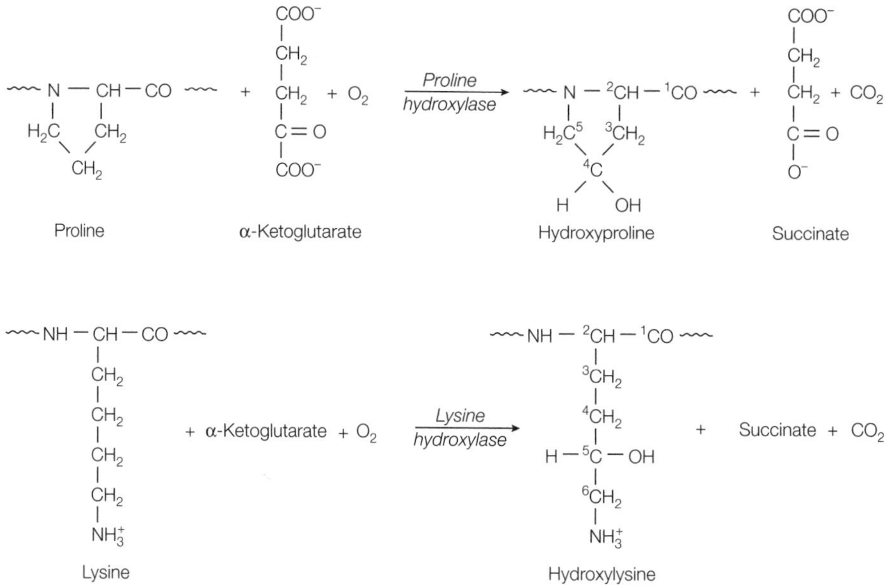

Proline + α-Ketoglutarate + $O_2$ —(Proline hydroxylase)→ Hydroxyproline + Succinate + $CO_2$

Lysine + α-Ketoglutarate + $O_2$ —(Lysine hydroxylase)→ Hydroxylysine + Succinate + $CO_2$

*Fig. 2. Formation of hydroxyproline and hydroxylysine.*

Pro and Y is often Hyp. Each of the three polypeptide chains in collagen is some 1000 residues long and they each fold up into a **helix** that has only 3.3 residues per turn, rather than the 3.6 residues per turn of an α-helix (see Topic B3). This **secondary structure** is unique to collagen and is often called the **collagen helix**. The three polypeptide chains lie parallel and wind round one another with a slight right-handed, rope-like twist to form a **triple-helical cable** (*Fig. 3*). Every third residue of each polypeptide passes through the center of the triple helix, which is so crowded that only the small side chain of Gly can fit in. This explains the absolute requirement for Gly at every third residue. The residues in the X and Y positions are located on the outside of the triple-helical cable, where there is room for the bulky side-chains of Pro and other residues. The three polypeptide chains are also staggered so that the Gly residue in one chain is aligned with the X residue in the second and the Y residue in the third. The triple helix is held together by an extensive network of **hydrogen bonds**, in particular between the primary amino group of Gly in one helix and the primary carboxyl group of Pro in position X of one of the other helices. In addition, the hydroxyl groups of Hyp residues participate in stabilizing the structure. The relatively inflexible Pro and Hyp also confer rigidity on the collagen structure.

The importance of Gly at every third residue is seen when a **mutation** in the DNA leads to the incorporation of a different amino acid at just one position in the 1000 residue polypeptide chain. For example, if a mutation leads to the incorporation of Cys instead of Gly, the triple helix is disrupted as the -$CH_2$-SH side-chain of Cys is too large to fit in the interior of the triple helix. This leads to a partly unfolded structure that is susceptible to excessive hydroxylation and glycosylation and is not efficiently secreted by the fibroblast cells. This, in turn, results in a defective collagen structure that can give rise to **brittle bones** and **skeletal deformities**. A whole spectrum of such mutations

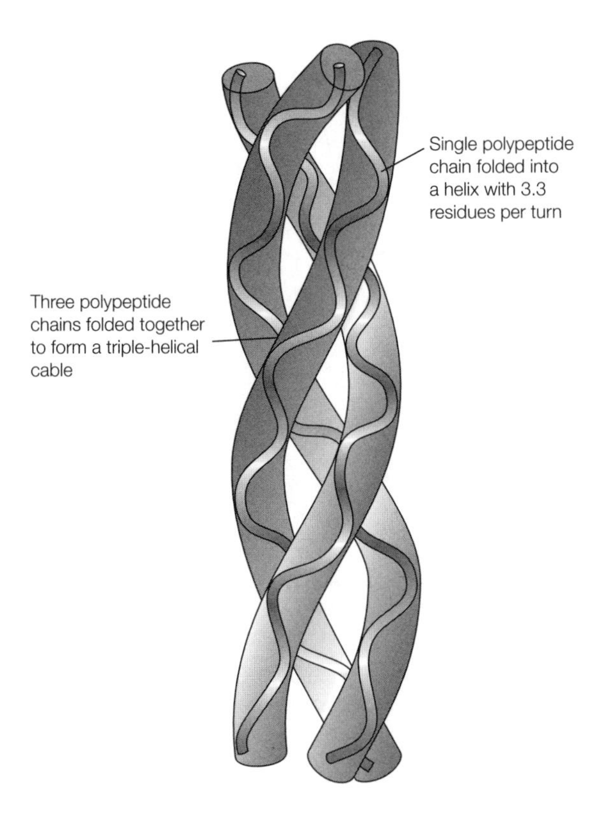


*Fig. 3. Arrangement of the three polypeptide chains in collagen.*

are known which cause the production of defective collagen and result in **osteogenesis imperfecta** (brittle bones).

## Secretion and aggregation

When the collagen polypeptides are synthesized they have additional amino acid residues (100–300) on both their N and C termini that are absent in the mature collagen fiber (*Fig. 4*). These **extension peptides** often contain Cys residues, which are usually absent from the remainder of the polypeptide chain. The extension peptides help to correctly align the three polypeptides as they come together in the triple helix, a process that may be aided by the formation of disulfide bonds between extension peptides on neighboring polypeptide chains. The extension peptides also prevent the premature aggregation of the procollagen triple helices within the cell. On **secretion** out of the fibroblast the extension peptides are removed by the action of extracellular **peptidases** (*Fig. 4*). The resulting tropocollagen molecules then **aggregate** together in a staggered head-to-tail arrangement in the collagen fiber (*Fig. 4*).

## Cross-links

The strength and rigidity of a collagen fiber is imparted by **covalent cross-links** both between and within the tropocollagen molecules. As there are few, if any, Cys residues in the final mature collagen, these covalent cross-links are not disulfide bonds as commonly found in proteins, but rather are unique cross-links formed between **Lys** and its aldehyde derivative **allysine**. Allysine residues are formed from Lys by the action of the monooxygenase **lysyl oxidase**

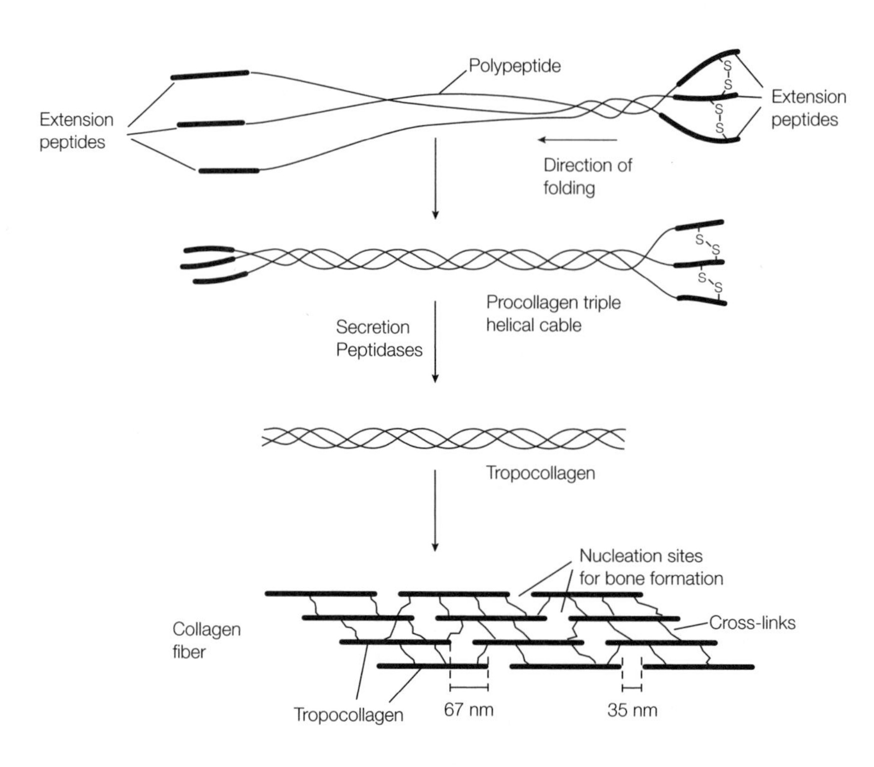


*Fig. 4. Role of the extension peptides in the folding and secretion of procollagen. Once secreted out of the cell, the extension peptides are removed and the resulting tropocollagen molecules aggregate and are cross-linked to form a microfibril.*

(*Fig. 5*). This **copper**-containing enzyme requires the coenzyme **pyridoxal phosphate**, derived from vitamin $B_6$ (see Topic M2), for activity. The aldehyde group on allysine then reacts spontaneously with either the side-chain amino group of Lys or with other allysine residues on other polypeptide chains to form covalent interchain bonds.

$$\text{\textasciitilde NH–CH(}(CH_2)_4\text{–}\overset{+}{N}H_3\text{)–CO\textasciitilde} + O_2 \xrightarrow{\text{Lysyl oxidase}} \text{\textasciitilde NH–CH(}(CH_2)_3\text{–CHO)–CO\textasciitilde} + \overset{+}{N}H_4 + H_2O$$

Lysine → Allysine

*Fig. 5. Conversion of lysine to allysine by lysyl oxidase.*

The importance of cross-linking to the normal functioning of collagen is demonstrated by the disease **lathyrism**. This occurs in humans and other animals through the ingestion of sweet pea (*Lathyrus odoratus*) seeds which contain the chemical **β-aminopropionitrile**. This compound irreversibly inhibits lysyl oxidase, thereby preventing the cross-linking of the tropocollagen molecules, resulting in serious abnormalities of the bones, joints and large blood vessels due to the fragile collagen. One collagen deficiency disease, the

**Ehlers–Danlos syndrome type V**, is due to a deficiency in lysyl oxidase and results in hypermobile joints and hyperextensibility of the skin. The 'India-rubber man' found in circuses probably had such a collagen deficiency.

## Bone formation

The spaces between the ends of the tropocollagen molecules in a collagen fiber (see *Fig. 4*) are the **nucleation sites** for the deposition of a form of **calcium phosphate, hydroxyapatite**, in bone formation. Further hydroxyapatite is added until the nucleation sites grow and join with one another to form the mature bone structure.

# B6 PROTEIN PURIFICATION

## Key Notes

**Principles of protein purification**

The aim of protein purification is to isolate one particular protein from all the others in the starting material. A combination of fractionation techniques is used that exploits the solubility, size, charge or/and specific binding affinity of the protein of interest.

**Selection of a protein source**

Because proteins have different distributions in biological materials, it is important to make the right choice of starting material from which to purify the protein. This will usually be a source that is relatively rich in the protein of interest and which is readily available.

**Homogenization and solubilization**

The protein has to be obtained in solution prior to its purification. Thus tissues and cells must be disrupted by homogenization or osmotic lysis and then subjected to differential centrifugation to isolate the subcellular fraction in which the protein is located. For membrane-bound proteins, the membrane structure has to be solubilized with a detergent to liberate the protein.

**Stabilization of proteins**

Certain precautions have to be taken in order to prevent proteins being denatured or inactivated during purification by physical or biological factors. These include buffering the pH of the solutions, undertaking the procedures at a low temperature and including protease inhibitors to prevent unwanted proteolysis.

**Assay of proteins**

In order to monitor the progress of the purification of a protein it is necessary to have an assay for it. Depending on the protein, the assay may involve measuring the enzyme activity or ligand-binding properties, or may quantify the protein present using antibodies directed against it.

**Ammonium sulfate precipitation**

The solubility of proteins decreases as the concentration of ammonium sulfate in the solution is increased. The concentration of ammonium sulfate at which a particular protein comes out of solution and precipitates may be sufficiently different from other proteins in the mixture to effect a separation.

**Dialysis**

Proteins can be separated from small molecules by dialysis through a semi-permeable membrane which has pores that allow small molecules to pass through but not proteins.

**Ultracentrifugation**

Proteins with large differences in molecular mass can be separated by rate-zonal centrifugation using a gradient of a dense material such as sucrose.

**Related topics**

Eukaryotes (A2)
Cellular fractionation (A4)
Chromatography of proteins (B7)
Electrophoresis of proteins (B8)
Introduction to enzymes (C1)
Antibodies as tools (D5)
Membrane protein and carbohydrate (E2)

**Principles of protein purification**

The basic aim in protein purification is to isolate one particular protein of interest from other contaminating proteins so that its structure and/or other properties can be studied. Once a suitable cellular **source** of the protein has been identified, the protein is liberated into solution and then separated from contaminating material by sequential use of a series of different **fractionation techniques** or **separations**. These separations exploit one or more of the following basic properties of the protein: its **solubility**, its **size**, its **charge** or its **specific binding affinity**. These separations may be **chromatographic techniques** such as **ion exchange**, **gel filtration** or **affinity chromatography** (see Topic B7), hydrophobic interaction chromatography, in which the protein binds to a hydrophobic material, or **electrophoretic techniques** such as **isoelectric focusing** (see Topic B8). Other electrophoretic procedures, mainly **sodium dodecyl sulfate (SDS) polyacrylamide gel electrophoresis (PAGE)** but also **native PAGE** (see Topic B8), are used to monitor the extent of purification and to determine the molecular mass and subunit composition of the purified protein.

**Selection of a protein source**

Before attempting to purify a protein, the first thing to consider is the source of **starting material**. Proteins differ in their cellular and tissue distribution, and thus if a protein is known to be abundant in one particular tissue (e.g. kidney) it makes sense to start the purification from this source. Also, some sources are more readily available than others and this should be taken into account too. Nowadays, with the use of **recombinant DNA techniques** (see Topics I1 and I6), even scarce proteins can be expressed in bacteria or eukaryotic cells and relatively large amounts of the protein subsequently obtained.

**Homogenization and solubilization**

Once a suitable source has been identified, the next step is to obtain the protein in solution. For proteins in biological fluids, such as blood serum, this is already the case, but for the majority of proteins the tissues and cells need to be disrupted and broken open (lysed). **Homogenization** and subsequent **differential centrifugation** of biological samples is detailed in Topic A4. In addition to the procedures described there, another simple way of breaking open cells that do not have a rigid cell wall to release the cytosolic contents is **osmotic lysis**. When animal cells are placed in a hypotonic solution (such as water or a buffered solution without added sucrose), the water in the surrounding solution diffuses into the more concentrated cytosol, causing the cell to swell and burst. Differential centrifugation is then employed to remove contaminating subcellular organelles (see Topic A4). Those proteins that are bound to membranes require a further solubilization step. After isolation by differential centrifugation, the appropriate membrane is treated with a **detergent** such as Triton X-100 to disrupt the lipid bilayer and to release the integral membrane proteins into solution (see Topic E2 for more details).

**Stabilization of proteins**

Throughout the purification procedure, steps have to be taken to ensure that the protein of interest is not **destroyed (inactivated** or **denatured)** either by physical or biological factors. The pH of the solutions used needs to be carefully **buffered** at a pH in which the protein is stable, usually around pH 7. The temperature often needs to be maintained below 25°C (usually around 4°C) to avoid **thermal denaturation** and to minimize the activity of **proteases**. Upon homogenization, **proteases** within the source material that are normally in a different subcellular compartment will be liberated into solution and come into contact with the protein of interest and may degrade it. For example, the acid hydrolases in lysosomes (see Topic A2) could be liberated into solution and rapidly degrade the cytosolic protein of interest. Thus, as well as carrying out the procedures at low temperature, **protease inhibitors** are often included in the buffer used in the early stages of the isolation procedure in order to minimize unwanted proteolysis (see Topic C4).

## Assay of proteins

A suitable means of detecting (assaying) the protein must be available to monitor the success of each stage in the purification procedure. The most straightforward **assays** are those for enzymes that catalyze reactions with readily detectable products (see Topic C1). Proteins which are not enzymes may be assayed through the observation of their biological effects. For example, a receptor can be assayed by measuring its ability to bind its specific ligand. Immunological techniques are often used to assay for the protein of interest using antibodies that specifically recognize it [e.g. radioimmunoassay, enzyme-linked immunosorbent assay (ELISA), or Western blot analysis (see Topics B8 and D5)].

## Ammonium sulfate precipitation

A commonly employed first separation step is **ammonium sulfate precipitation**. This technique exploits the fact that the **solubility** of most proteins is lowered at high salt concentrations. As the salt concentration is increased, a point is reached where the protein comes out of solution and precipitates. The concentration of salt required for this **salting out effect** varies from protein to protein, and thus this procedure can be used to fractionate a mixture of proteins. For example, 0.8 M ammonium sulfate precipitates out the clotting protein fibrinogen from blood serum, whereas 2.4 M ammonium sulfate is required to precipitate albumin. Salting out is also sometimes used at later stages in a purification procedure to **concentrate** a dilute solution of the protein since the protein precipitates and can then be redissolved in a smaller volume of buffer.

## Dialysis

Proteins can be separated from small molecules by dialysis through a **semi-permeable membrane** such as cellophane (cellulose acetate). **Pores** in the membrane allow molecules up to approximately 10 kDa to pass through, whereas larger molecules are retained inside the dialysis bag (*Fig. 1*). As most proteins have molecular masses greater than 10 kDa, this technique is not suitable for fractionating proteins, but is often used to remove small molecules such

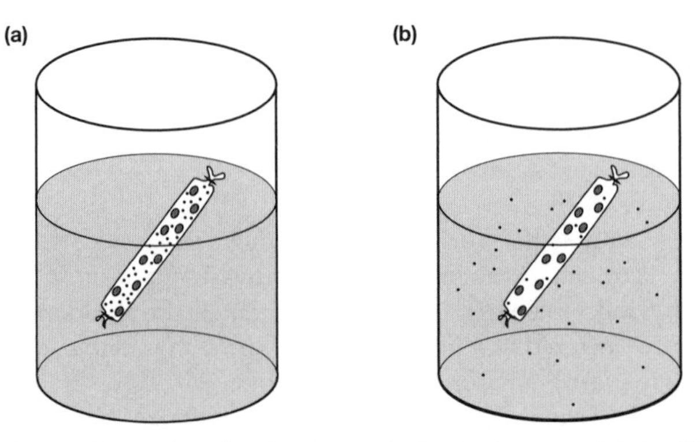


*Fig. 1. Separation of molecules on the basis of size by dialysis. (a) Starting point, (b) at equilibrium.*

as salts from a protein solution. It should be noted that at equilibrium, the concentration of small molecules inside a dialysis bag will be equal to that outside (*Fig. 1b*), and so several changes of the surrounding solution are often required to lower the concentration of the small molecule in the protein solution sufficiently.

**Ultracentrifugation**

Although ultracentrifugation was used extensively in the past to isolate proteins, advances in other separation techniques have generally superseded this method. Proteins which have large differences in molecular mass can be separated by **rate-zonal centrifugation** using a gradient of a dense material such as sucrose (see Topic A4). Most proteins, though, have similar densities and, therefore, cannot readily be separated by density gradient centrifugation (Topic A4).

# B7 CHROMATOGRAPHY OF PROTEINS

## Key Notes

**Gel filtration chromatography**

Gel filtration chromatography separates proteins on the basis of their size and shape using porous beads packed in a column. Large or elongated proteins cannot enter the pores in the beads and elute from the bottom of the column first, whereas smaller proteins can enter the beads, have a larger volume of liquid accessible to them and move through the column more slowly, eluting later. Gel filtration chromatography can be used to de-salt a protein mixture and to estimate the molecular mass of a protein.

**Ion exchange chromatography**

In ion exchange chromatography, proteins are separated on the basis of their net charge. In anion exchange chromatography a column containing positively charged beads is used to which proteins with a net negative charge will bind, whereas in cation exchange chromatography negatively charged beads are used to which proteins with a net positive charge will bind. The bound proteins are then eluted by adding a solution of sodium chloride or by altering the pH of the buffer.

**Affinity chromatography**

Affinity chromatography exploits the specific binding of a protein for another molecule, its ligand (e.g. an enzyme for its inhibitor). The ligand is immobilized on an insoluble support and packed in a column. On adding a mixture of proteins, only the protein of interest binds to the ligand. All other proteins pass straight through. The bound protein is then eluted from the immobilized ligand in a highly purified form.

**Related topics** Protein purification (B6) Antibodies as tools (D5)

## Gel filtration chromatography

In gel filtration chromatography (**size exclusion chromatography** or **molecular sieve chromatography**), molecules are separated on the basis of their **size and shape**. The protein sample in a small volume is applied to the top of a column of **porous beads** (diameter 0.1 mm) that are made of an insoluble but highly hydrated polymer such as polyacrylamide (Bio-Gel) or the carbohydrates dextran (Sephadex) or agarose (Sepharose) (*Fig. 1a*). Small molecules can enter the **pores** in the beads whereas larger or more elongated molecules cannot. The smaller molecules therefore have a larger volume of liquid accessible to them: both the liquid surrounding the porous beads and that inside the beads. In contrast, the larger molecules have only the liquid surrounding the beads accessible to them, and thus move through the column faster, emerging out of the bottom (**eluting**) first (*Fig. 1a* and *b*). The smaller molecules move more slowly through the column and elute later. Beads of differing pore sizes are available, allowing proteins of different sizes to be effectively separated. Gel filtration chromatography is often used to de-salt a protein sample (for example to remove the ammonium sulfate after ammonium sulfate precipitation; Topic B6),

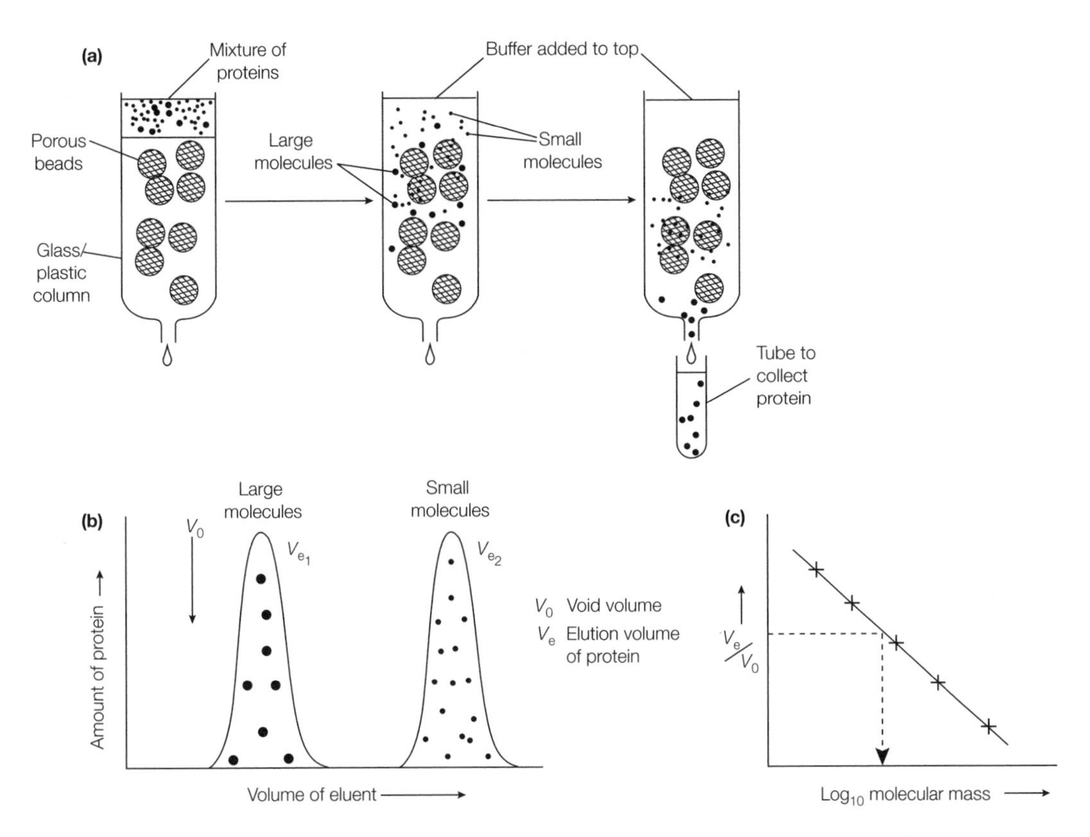


*Fig. 1. Gel filtration chromatography. (a) Schematic illustration of gel filtration chromatography; (b) elution diagram indicating the separation; (c) a plot of relative elution volume versus the logarithm of molecular mass for known proteins, indicating how the molecular mass of an unknown can be read off when its relative elution volume is known.*

since the salt enters the porous beads and is eluted late, whereas the protein does not enter the beads and is eluted early. Gel filtration chromatography can also be used to estimate the **molecular mass** of a protein. There is a linear relationship between the relative elution volume of a protein ($V_e/V_o$ where $V_e$ is the elution volume of a given protein and $V_o$ is the void volume of the column, that is the volume of the solvent space surrounding the beads; *Fig. 1b*) and the logarithm of its molecular mass. Thus a 'standard' curve of $V_e/V_o$ against $\log_{10}$ molecular mass can be estimated for the column using proteins of known mass. The elution volume of any sample protein then allows its molecular mass to be estimated by reference to its position on the standard curve (*Fig. 1c*).

## Ion exchange chromatography

In ion exchange chromatography, proteins are separated on the basis of their **overall (net) charge**. If a protein has a net negative charge at pH 7, it will bind to a column containing positively charged beads, whereas a protein with no charge or a net positive charge will not bind (*Fig. 2a*). The negatively charged proteins bound to such a column can then be eluted by washing the column with an increasing gradient (increasing concentration) of a solution of **sodium chloride** ($Na^+$ $Cl^-$ ions) at the appropriate pH. The $Cl^-$ ions compete with the protein for the positively charged groups on the column. Proteins having a low density of negative charge elute first, followed by those with a higher density of negative charge (*Fig. 2b*). Columns containing positively charged diethylaminoethyl

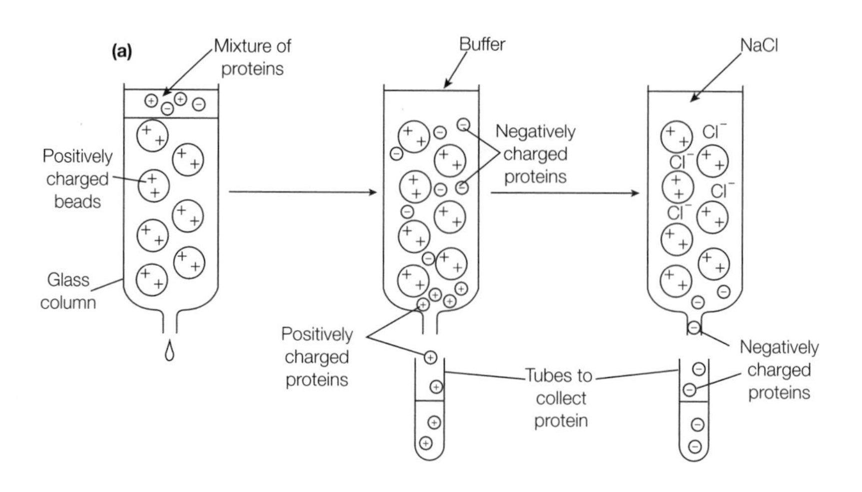


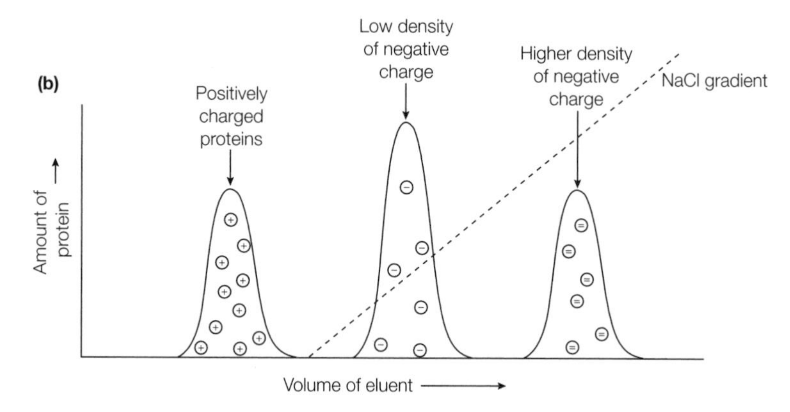


*Fig. 2. Ion exchange chromatography. (a) Schematic illustration of ion exchange chromatography; (b) elution diagram indicating the separation of a protein of net positive charge that does not bind to the positively charged beads and passes straight through the column, and of two proteins with different net negative charges that bind to the positively charged beads and are eluted on increasing the concentration of NaCl applied to the column. The protein with the lower density of negative charge elutes earlier than the protein with the higher density of negative charge.*

(DEAE) groups (such as DEAE-cellulose or DEAE-Sephadex) are used for separation of negatively charged proteins (anionic proteins). This is called **anion exchange chromatography**. Columns containing negatively charged carboxymethyl (CM) groups (such as CM-cellulose or CM-Sephadex) are used for the separation of positively charged proteins (cationic proteins). This is called **cation exchange chromatography**. As an alternative to elution with a gradient of NaCl, proteins can be eluted from anion exchange columns by decreasing the pH of the buffer, and from cation exchange columns by increasing the pH of the buffer, thus altering the ionization state of the amino acid side-chains (see Topic B2) and hence the net charge on the protein.

## Affinity chromatography

Affinity chromatography exploits the specific, high affinity, noncovalent binding of a protein to another molecule, the **ligand**. First, the ligand is covalently attached to an inert and porous matrix (such as Sepharose). The protein mixture is then

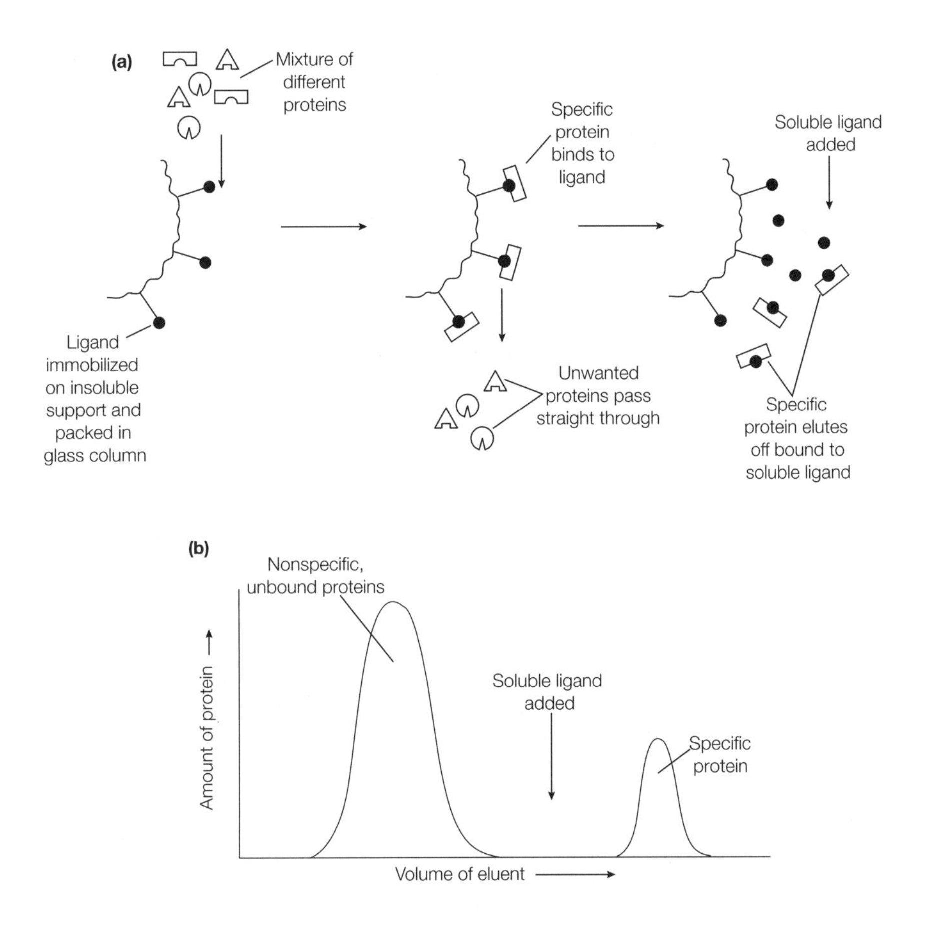


*Fig. 3. Affinity chromatography. (a) Schematic diagram of affinity chromatography; (b) elution diagram indicating that nonspecific proteins that do not bind to the immobilized ligand pass straight through the column, while the specific protein binds to the immobilized ligand and is eluted from the column only on addition of soluble ligand.*

passed down a column containing the **immobilized ligand**. The protein of interest will bind to the ligand, whereas all other proteins pass straight through (*Fig. 3*). After extensive washing of the column with buffer to remove nonspecifically bound proteins, the bound protein is released from the immobilized ligand either by adding soluble ligand which competes with the immobilized ligand for the protein, or by altering the properties of the buffer (changing the pH or salt concentration). If soluble ligand is used to elute the protein from the column, extensive **dialysis** often then has to be used to remove the small ligand from the larger protein (see Topic B6). Because this technique exploits the specific, often unique, binding properties of the protein, it is often possible to separate the protein from a mixture of hundreds of other proteins in a single chromatographic step. Commonly employed combinations of immobilized ligand and protein to be purified used in affinity chromatographic systems include an **inhibitor** to purify an **enzyme** (see Topic C4), an **antibody** to purify its **antigen** (see Topic D5), a **hormone** (e.g. insulin) to purify its **receptor** (see Topic E5), and a **lectin** (e.g. concanavalin A) to purify a **glycoprotein** (see Topics E2 and H5).

# B8 Electrophoresis of proteins

## Key Notes

**Native PAGE**

In native polyacrylamide gel electrophoresis (PAGE) proteins are applied to a porous polyacrylamide gel and separated in an electric field on the basis of their net negative charge and their size. Small/more negatively charged proteins migrate further through the gel than larger/less negatively charged proteins.

**SDS-PAGE**

In SDS-PAGE, the protein sample is treated with a reducing agent to break disulfide bonds and then with the anionic detergent sodium dodecyl sulfate (SDS) which denatures the proteins and covers them with an overall negative charge. The sample is then fractionated by electrophoresis through a polyacrylamide gel. As all the proteins now have an identical charge to mass ratio, they are separated on the basis of their mass. The smallest proteins move farthest. SDS-PAGE can be used to determine the degree of purity of a protein sample, the molecular mass of a protein and the number of polypeptide subunits in a protein.

**Isoelectric focusing**

In isoelectric focusing, proteins are separated by electrophoresis in a pH gradient in a gel. They separate on the basis of their relative content of positively and negatively charged residues. Each protein migrates through the gel until it reaches the point where it has no net charge, its isoelectric point (pI).

**Visualization of proteins in gels**

Proteins can be visualized directly in gels by staining them with the dye Coomassie brilliant blue or with a silver stain. Radioactively labeled proteins can be detected by overlaying the gel with X-ray film and observing the darkened areas on the developed autoradiograph that correspond to the radiolabeled proteins. A specific protein of interest can be detected by immunoblot (Western blot) following its transfer from the gel to nitrocellulose using an antibody that specifically recognizes it. This primary antibody is then detected with either a radiolabeled or enzyme-linked secondary antibody.

**Related topics**

Protein structure (B3)
Protein purification (B6)
Antibodies as tools (D5)

## Native PAGE

When placed in an **electric field,** molecules with a net charge, such as proteins, will move towards one electrode or the other, a phenomenon known as **electrophoresis**. The greater the net charge the faster the molecule will move. In native polyacrylamide gel electrophoresis (PAGE) the molecular separation is based on the **size** of the protein as well as its net **charge** since the electrophoretic separation is carried out in a gel which serves as a molecular sieve. Small molecules move readily through the pores in the gel, whereas larger molecules are retarded. The gels are commonly made of **polyacrylamide** which is chemically inert and which is readily formed by the polymerization of acrylamide. The **pore sizes** in the gel can be controlled by choosing appropriate concentrations of acrylamide and the cross-linking reagent, methylene bisacrylamide. The higher the concentration of acrylamide used, the smaller the pore size in the final gel. The gel is usually cast between two glass plates of 7–20 $cm^2$ separated by a distance of 0.5–1.0 mm. The protein sample is added to wells in the top of the gel, which are formed by placing a plastic comb in the gel solution before it sets (*Fig. 1*). A blue dye (bromophenol blue) is mixed with the protein sample to aid its loading on to the gel. Because bromophenol blue is a small molecule, it also migrates quickly through the gel during electrophoresis and so indicates the progress of electrophoresis. In the simplest form of native PAGE, the buffer, which is the same in both the upper and lower reservoirs and in the gel, has a pH of approximately 9, such that most proteins have **net negative charges** and will migrate towards the anode in the lower reservoir. An electric current (approximately 300 V) is applied across the gel from top to bottom for 30–90 min in order to move the proteins through the gel. The gel is then removed from the electrophoresis apparatus, and the proteins within it visualized.

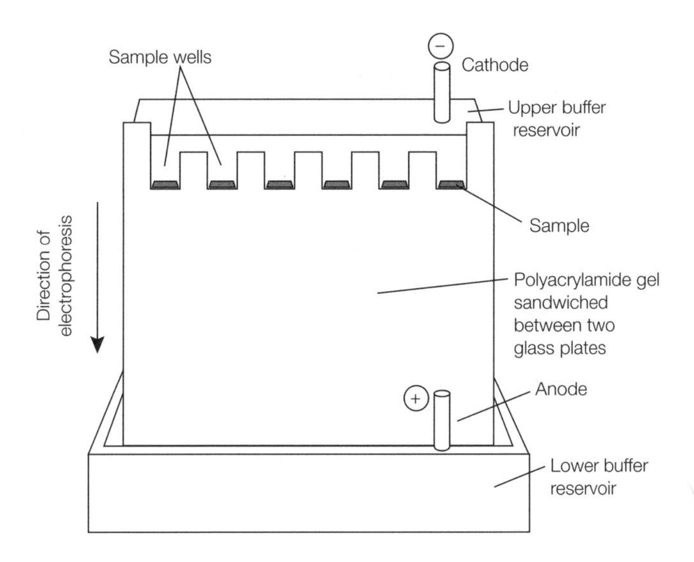


*Fig. 1. Native polyacrylamide gel electrophoresis. The protein samples are loaded into the sample wells formed in the top of the gel. An electric field is applied across the gel from top to bottom and the proteins migrate down through the gel. The smaller the protein and the greater its net negative charge, the further it will migrate.*

## SDS-PAGE

In **SDS-PAGE**, the proteins are **denatured** and coated with an **overall negative charge** [due to bound sodium dodecyl sulfate (SDS) molecules] and thus the basis for their separation is only their **mass**. The protein mixture is first treated with a **reducing agent** such as 2-mercaptoethanol or dithiothreitol to break all the disulfide bonds (see Topic B3). The strong **anionic detergent SDS** is then added which disrupts nearly all the noncovalent interactions in the protein, unfolding the polypeptide chain. Approximately one molecule of SDS binds via its hydrophobic alkyl chain to the polypeptide backbone for every two amino acid residues, which gives the denatured protein a large net negative charge that is proportional to its mass. The SDS/protein mixture is then applied to sample

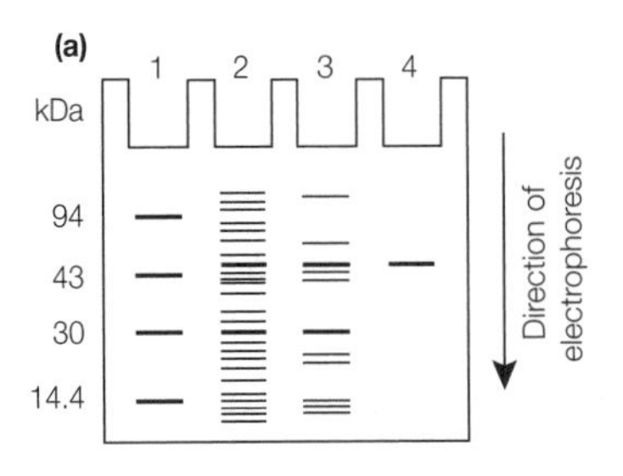


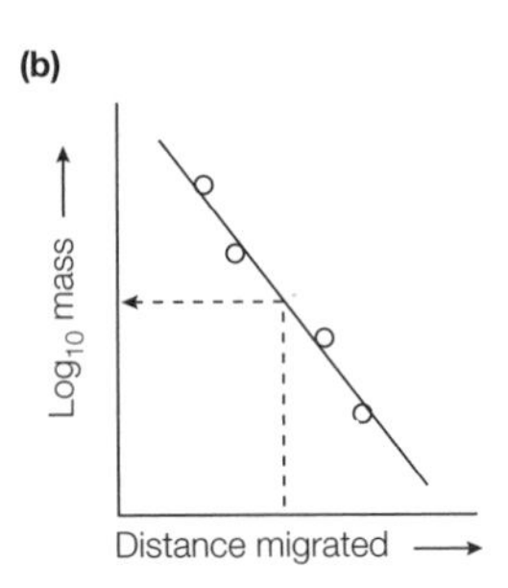


*Fig. 2. SDS-PAGE. (a) Appearance of proteins after electrophoresis on an SDS polyacrylamide gel. Lane 1, proteins (markers) of known molecular mass; lane 2, unpurified mixture of proteins; lane 3, partially purified protein; lane 4, protein purified to apparent homogeneity; (b) determination of the molecular mass of an unknown protein by comparison of its electrophoretic mobility (distance migrated) with those of proteins (markers) of known molecular mass.*

wells in the top of a **polyacrylamide gel** as in native PAGE (see *Fig. 1*). After carrying out electrophoresis, the gel is removed from the apparatus and the proteins visualized (*Fig. 2a*). Small proteins move furthest through the gel, whereas large ones move more slowly as they are held back by the cross-linking in the gel and remain near the top. Under these conditions, the mobility of most polypeptide chains is linearly proportional to the logarithm of their mass. Thus, if proteins of known molecular mass are electrophoresed alongside the samples, the mass of the unknown proteins can be determined (*Fig. 2b*). Proteins that differ in mass by about 2% (e.g. 40 and 41 kDa; a difference of approximately 10 amino acid residues) can usually be distinguished. SDS-PAGE is a rapid, sensitive and widely used technique from which can be determined the degree of purity of a protein sample, the molecular mass of an unknown protein and the number of polypeptide subunits within a protein (see Topic B3).

## Isoelectric focusing

Isoelectric focusing electrophoretically separates proteins on the basis of their relative content of positively and negatively charged groups. When a protein is at its **pI** (see Topic B2), its **net charge is zero** and hence it will not move in an electric field.

In isoelectric focusing, a polyacrylamide gel is used which has large pores (so as not to impede protein migration) and contains a mixture of **polyampholytes** (small multi-charged polymers that have many pI values). If an electric field is applied to the gel, the polyampholytes migrate and produce a **pH gradient**. To separate proteins by isoelectric focusing, they are electrophoresed through such a gel. Each protein will migrate through the gel until it reaches a position at which the pH is equal to its pI (*Fig. 3*). If a protein diffuses away from this position, its net charge will change as it moves into a region of different pH and the resulting electrophoretic forces will move it back to its isoelectric position. In this way each protein is focused into a narrow band (as thin as 0.01 pH unit) about its pI.

Isoelectric focusing can be combined with SDS-PAGE to obtain very high resolution separations in a procedure known as **two-dimensional gel electrophoresis**. The protein sample is first subjected to isoelectric focusing in a narrow strip of gel containing polyampholytes. This gel strip is then placed on top of an SDS-polyacrylamide gel and electrophoresed to produce a two-dimensional pattern of spots in which the proteins have been separated in the horizontal direction on the basis of their pI, and in the vertical direction on the basis of their mass (*Fig. 4*). The overall result is that proteins are separated both on the basis of their size and their

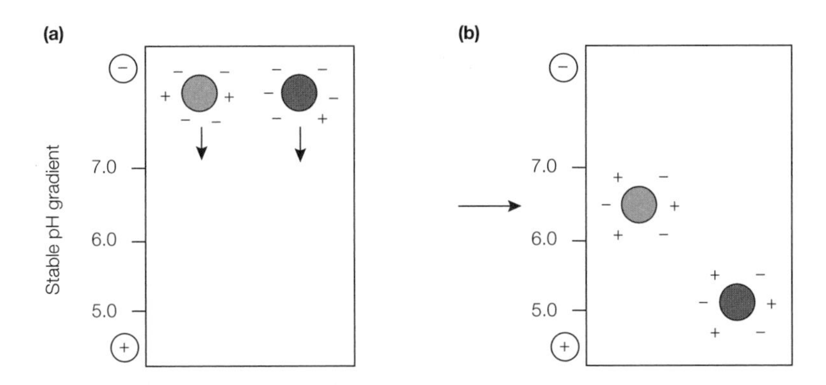


*Fig. 3. Isoelectric focusing. (a) Before applying an electric current. (b) after applying an electric current the proteins migrate to a position at which their net charge is zero (isoelectric point, pI).*

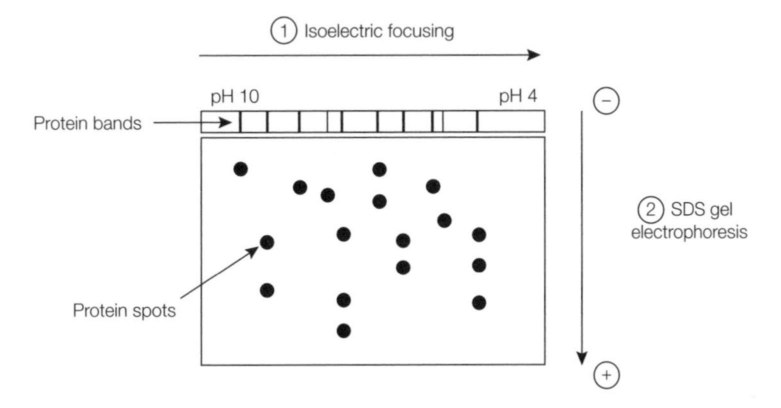


*Fig. 4. Two-dimensional gel electrophoresis. The protein sample is first subjected to isoelectric focusing in one dimension and then to SDS-PAGE in the second dimension.*

charge. Thus two proteins that have very similar or identical pIs, and produce a single band by isoelectric focusing, will produce two spots by two-dimensional gel electrophoresis (see *Fig. 4*). Similarly, proteins with similar or identical molecular masses, which would produce a single band by SDS-PAGE, also produce two spots because of the initial separation by isoelectric focusing.

## Visualization of proteins in gels

As most proteins are not directly visible on gels to the naked eye, a method has to be employed in order to visualize them following electrophoresis. The most commonly used protein stain is the dye **Coomassie brilliant blue**. After electrophoresis, the gel containing the separated proteins is immersed in an acidic alcoholic solution of the dye. This **denatures** the proteins, **fixes** them in the gel so that they do not wash out, and allows the dye to bind to them. After washing away excess dye, the proteins are visible as discrete blue bands (see *Fig. 2a*). As little as 0.1–1.0 μg of a protein in a gel can be visualized using Coomassie brilliant blue. A more sensitive general protein stain involves soaking the gel in a **silver** salt solution. However, this technique is rather more difficult to apply. If the protein sample is **radioactive** the proteins can be visualized indirectly by overlaying the gel with a sheet of **X-ray film**. With time (hours to weeks depending on the radioactivity of the sample proteins), the radiation emitted will cause a darkening of the film. Upon development of the film the resulting **autoradiograph** will have darkened areas corresponding to the positions of the radiolabeled proteins.

Another way of visualizing the protein of interest is to use an **antibody** against the protein in an **immunoblot (Western blot)** (see Topic D5). For this technique, the proteins have to be transferred out of the gel on to a sheet of **nitrocellulose** or nylon membrane. This is accomplished by overlaying the gel with the nitrocellulose and **blotting** the protein on to it by applying an electric current. The nitrocellulose then has an exact image of the pattern of proteins that was in the gel. The excess binding sites on the nitrocellulose are then blocked with a nonspecific protein solution such as milk powder, before placing the nitrocellulose in a solution containing the antibody that recognizes the protein of interest (the **primary antibody**). After removing excess unbound antibody, the primary antibody that is now specifically bound to the protein of interest is detected with either a **radiolabeled** or **enzyme-coupled secondary antibody**. Finally, the secondary antibody is detected either by placing the nitrocellulose against a sheet of X-ray film (if a radiolabeled secondary antibody has been used), or by adding to the nitrocellose a solution of a substrate that is converted into a colored insoluble product by the enzyme that is coupled to the secondary antibody.

# B9 PROTEIN SEQUENCING AND PEPTIDE SYNTHESIS

## Key Notes

**Amino acid composition analysis**

The number of each type of amino acid in a protein can be determined by acid hydrolysis and separation of the individual amino acids by ion exchange chromatography. The amino acids are detected by colorimetric reaction with, for example, ninhydrin or fluorescamine.

**Edman degradation**

The N-terminal amino acid of a protein can be determined by reacting the protein with dansyl chloride or fluorodinitrobenzene prior to acid hydrolysis. The amino acid sequence of a protein can be determined by Edman degradation which sequentially removes one residue at a time from the N terminus. This uses phenyl isothiocyanate to label the N-terminal amino acid prior to its release from the protein as a cyclic phenylthiohydantoin amino acid.

**Sequencing strategy**

In order to sequence an entire protein, the polypeptide chain has to be broken down into smaller fragments using either chemicals (e.g. cyanogen bromide) or enzymes (e.g. chymotrypsin and trypsin). The resulting smaller fragments are then sequenced by Edman degradation. The complete sequence is assembled by analyzing overlapping fragments generated by cleaving the polypeptide with different reagents. Aminopeptidase and carboxypeptidase release the N- and C-terminal amino acids from a protein, respectively. The polypeptides in a multi-subunit protein have to be dissociated and separated prior to sequencing using urea or guanidine hydrochloride which disrupt noncovalent interactions, and 2-mercaptoethanol or dithiothreitol that break disulfide bonds.

**Peptide sequencing by mass spectrometry**

Short polypeptides can be sequenced rapidly by fast atom bombardment mass spectrometry (FAB-MS). This technique not only provides the amino acid sequence of the peptide but also information on post-translational modifications.

**Recombinant DNA technology**

The sequence of a protein can be determined using recombinant DNA technology to identify and sequence the piece of DNA encoding the protein. The amino acid sequence of the protein can then be deduced from its DNA sequence using the genetic code.

**Information derived from protein sequences**

The amino acid sequence of a protein not only reveals the primary structure of the protein but also information on possible protein families or groups and evolutionary relationships, potential gene duplication(s) and possible post-translational modifications. In addition, a knowledge of the amino acid sequence can be used to generate specific antibodies and DNA probes.

**Peptide synthesis**

In solid phase peptide synthesis, polypeptides are chemically synthesized by addition of free amino acids to a tethered peptide. To prevent unwanted reactions, the α-amino group and reactive side chain groups of the free amino acids are chemically protected or blocked, and then deprotected or deblocked once the amino acid is attached to the growing polypeptide chain.

**Related topics**

Chromatography of proteins (B7)
Electrophoresis of proteins (B8)
The genetic code (H1)

## Amino acid composition analysis

The number of each type of **amino acid** in a protein sample can be determined by amino acid composition analysis. The purified protein sample is hydrolyzed into its constituent amino acids by heating it in 6 M HCl at 110°C for 24 h in an evacuated and sealed tube. The resulting mixture (**hydrolysate**) of amino acids is subjected to **ion exchange chromatography** (see Topic B7) on a column of sulfonated polystyrene to separate out the **20 standard amino acids** (see Topic B1). The separated amino acids are then detected and quantified by reacting them with **ninhydrin**. The α-amino acids produce a blue color, whereas the imino acid proline produces a yellow color. The amount of each amino acid in an unknown sample can be determined by comparison of the optical absorbance with a known amount of each of the individual amino acids in a standard sample. With ninhydrin, as little as 10 nmol of an amino acid can be detected. A more sensitive detection system (detecting down to 10 pmol of an amino acid) uses **fluorescamine** to react with the α-amino group to form a fluorescent product. Amino acid composition analysis indicates the number of each amino acid residue in a peptide, but it does not provide information on the sequence of the amino acids. For example, the amino acid composition of the oligopeptide:

Val-Phe-Asp-Lys-Gly-Phe-Val-Glu-Arg

would be:

(Arg, Asp, Glu, Gly, Leu, Lys, $Phe_2$, $Val_2$)

where the parentheses and the commas between each amino acid denote that this is the amino acid composition, not the sequence.

## Edman degradation

The **amino-terminal** (N-terminal) residue of a protein can be identified by reacting the protein with a compound that forms a stable covalent link with the free α-amino group, prior to hydrolysis with 6 M HCl. The labeled N-terminal amino acid can then be identified by comparison of its chromatographic properties with standard amino acid derivatives. Commonly used reagents for N-terminal analysis are **fluorodinitrobenzene** and **dansyl chloride**. If this technique was applied to the oligopeptide above, the N-terminal residue would be identified as Val, but the remainder of the sequence would still be unknown. Further reaction with dansyl chloride would not reveal the next residue in the sequence since the peptide is totally degraded in the acid hydrolysis step.

This problem was overcome by Pehr Edman who devised a method for labeling the N-terminal residue and then cleaving it from the rest of the peptide without breaking the peptide bonds between the other amino acids. In so-called

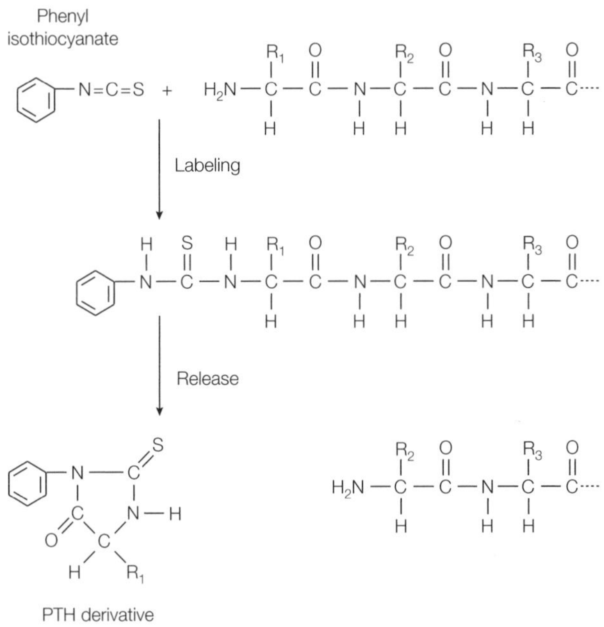

Phenyl isothiocyanate

Labeling

Release

PTH derivative

*Fig. 1. Edman degradation. The N-terminal amino acid is labeled with phenyl isothiocyanate. Upon mild acid hydrolysis this residue is released as a PTH-derivative and the peptide is shortened by one residue, ready for another round of labeling and release.*

**Edman degradation**, one residue at a time is sequentially removed from the N-terminal end of a peptide or protein and identified. The uncharged N-terminal amino group of the protein is reacted with **phenyl isothiocyanate** to form a phenylthiocarbamoyl derivative which is then released from the rest of the protein as a cyclic **phenylthiohydantoin (PTH) amino acid** under mildly acidic conditions (*Fig. 1*). This milder cleavage reaction leaves the remainder of the peptide intact, available for another round of labeling and release. The released PTH amino acid is identified by **high performance liquid chromatography** (HPLC). This sequencing technique has been **automated** and refined so that upwards of 50 residues from the N terminus of a protein can be sequenced from picomole quantities of material.

## Sequencing strategy

An 'average' sized protein of 50 kDa would contain approximately 500 amino acids. Thus, even with large amounts of highly purified material, only about the N-terminal one-tenth of the protein can be sequenced by Edman degradation. In order to sequence a larger protein, the first step is to **cleave** it into **smaller fragments** of 20–100 residues which are then separated and sequenced. Specific cleavage can be achieved by **chemical** or **enzymatic** methods. For example, the chemical **cyanogen bromide** (CNBr) cleaves polypeptide chains on the C-terminal side of Met residues, whereas the enzymes **trypsin** and **chymotrypsin** cleave on the C-terminal side of basic (Arg, Lys) and aromatic (Phe, Trp, Tyr) residues, respectively. On digestion with trypsin, a protein with six Lys and five Arg would yield 12 tryptic peptides, each of which would end with Arg or Lys, apart from the C-terminal peptide. The peptide fragments obtained by specific chemical or enzymatic cleavage are then separated by **chromatography** (e.g. ion exchange chromatography; see Topic B7) and the sequence of each in turn determined by Edman degradation.

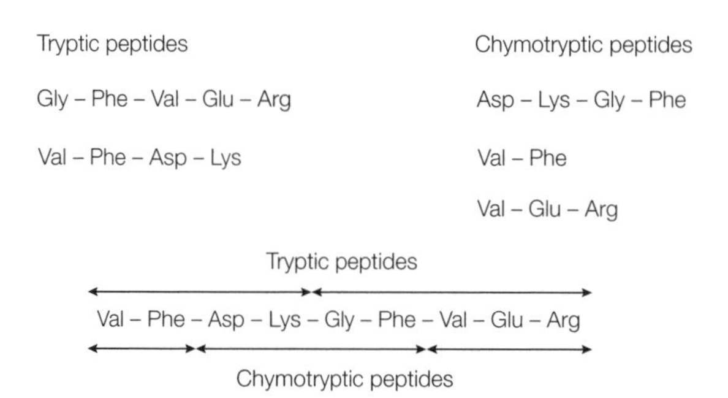


*Fig. 2. The use of overlapping fragments to determine the sequence of a peptide. The protein is first digested with trypsin and the resulting peptides separated and sequenced. The protein is separately digested with chymotrypsin and the resulting peptides again separated and sequenced. The order of the peptide fragments in the protein can be determined by comparing the sequences obtained.*

Although the sequence of each peptide fragment would now be known, the order of these fragments in the polypeptide chain would not. The next stage is to generate **overlapping fragments** by going back and cleaving the original polypeptide chain with a different chemical or enzyme (e.g. chymotrypsin), separating the fragments and then sequencing them. These **chymotryptic peptides** will overlap one or more of the **tryptic peptides**, enabling the order of the fragments to be established (*Fig. 2*). In this way, the entire length of the polypeptide chain can be sequenced.

Specific enzymes, called **exopeptidases**, that cleave one residue at a time from the end of a polypeptide chain, can be used to provide information on the **terminal residues**. **Aminopeptidases** cleave amino acids from the N terminus; **carboxypeptidases** from the C terminus. The released amino acid can then be identified as above by comparison with known standards.

To sequence the polypeptides in a **multi-subunit protein**, the individual polypeptide chains must first be dissociated by disrupting the **noncovalent interactions** with **denaturing agents** such as **urea** or **guanidine hydrochloride**. The **disulfide bonds** in the protein also have to be broken by reduction with **2-mercaptoethanol** or **dithiothreitol**. To prevent the cysteine residues recombining, **iodoacetate** is added to form stable *S*-carboxymethyl derivatives. The individual polypeptide chains then have to be separated by, for example, ion exchange chromatography (see Topic B7) before sequencing each. Nowadays, as little as picomole amounts of proteins can be sequenced following their separation by SDS-PAGE either using the polyacrylamide gel containing the protein directly, or following their transfer to nitrocellulose (see Topic B8).

## Peptide sequencing by mass spectrometry

Polypeptides of up to approximately 25 residues can be sequenced by the technique of **mass spectrometry** (MS), which involves an ionization technique called **fast atom bombardment** (FAB) in concert with a **tandem mass spectrometer** (two mass spectrometers coupled in series). The sequence of the polypeptide can be obtained from the molecular masses of the various fragments produced in the ionization stage in only a few minutes compared to the hour required for just one cycle of Edman degradation. In addition, mass spectrometry can be used to sequence several polypeptides in a mixture, alleviating the need to completely purify the sample prior to analysis. Other advantages of mass spectrometry are that it can be used to determine the sequence of peptides which have **blocked N-termini**, such as pyroglutamate, a derivative of glutamate in which the side chain carboxyl group forms an amide bond with its primary amino group (a common eukaryotic post-translational modification that prevents Edman degradation) and to characterize other **post-translational modifications** such as glycosylation and phosphorylation.

## Recombinant DNA technology

Although numerous proteins have been sequenced by Edman degradation using the above strategy, the determination of the sequences of large proteins by this method is a demanding and time-consuming process. Nowadays, **recombinant DNA technology** has enabled the sequences of even very large proteins to be determined by first sequencing the stretch of **DNA** encoding the protein and then using the **genetic code** to decipher the protein sequence (see Topic H1). Even so, some direct protein sequence data is often required to confirm that the protein sequence obtained is the correct one. Thus, currently, protein sequencing and DNA sequencing are techniques that are used together to determine the complete sequence of a protein.

## Information derived from protein sequences

The amino acid sequence can provide information over and above the **primary structure** of the protein.

1. The sequence of interest can be compared with other known sequences to see whether there are similarities. For example, the sequences of hemoglobin and myoglobin indicate that they belong to the globin group or **family of proteins** (see Topic B4).
2. The comparison of the sequences of the same protein in different species can provide information about **evolutionary relationships**.
3. The presence of repeating stretches of sequence would indicate that the protein may have arisen by **gene duplication** (e.g. in antibody molecules; see Topic D2).
4. Within the amino acid sequence there may be specific sequences which act as signals for the **post-translational processing** of the protein (e.g. glycosylation or proteolytic processing; see Topic H5).
5. The amino acid sequence data can be used to prepare **antibodies** specific for the protein of interest which can be used to study its structure and function (see Topic D5).
6. The amino acid sequence can be used for designing **DNA probes** that are specific for the gene encoding the protein (see Topics I2 and I6).

## Peptide synthesis

Polypeptides can be **chemically synthesized** by covalently linking amino acids to the end of a growing polypeptide chain. In **solid phase peptide synthesis** the growing polypeptide chain is covalently anchored at its C-terminus to an insoluble support such as polystyrene beads. The next amino acid in the sequence has to react with the free α-amino group on the tethered peptide, but it has a free α-amino group itself which will also react. To overcome this problem the free amino acid has its α-amino group **chemically protected** (blocked) so that it does not react with other molecules. Once the new amino acid is coupled, its now N-terminal α-amino group is **deprotected** (deblocked) so that the next peptide bond can be formed. Every cycle of amino acid addition therefore requires a **coupling step** and a **deblocking step**. In addition, reactive side chain groups must also be blocked to prevent unwanted reactions occurring.

# C1 Introduction to Enzymes

## Key Notes

**Enzymes as catalysts**

Enzymes are catalysts that change the rate of a reaction without being changed themselves. Enzymes are highly specific and their activity can be regulated. Virtually all enzymes are proteins, although some catalytically active RNAs have been identified.

**Active site**

The active site is the region of the enzyme that binds the substrate, to form an enzyme–substrate complex, and transforms it into product. The active site is a three-dimensional entity, often a cleft or crevice on the surface of the protein, in which the substrate is bound by multiple weak interactions. Two models have been proposed to explain how an enzyme binds its substrate: the lock-and-key model and the induced-fit model.

**Substrate specificity**

The substrate specificity of an enzyme is determined by the properties and spatial arrangement of the amino acid residues forming the active site. The serine proteases trypsin, chymotrypsin and elastase cleave peptide bonds in protein substrates on the carboxyl side of positively charged, aromatic and small side-chain amino acid residues, respectively, due to complementary residues in their active sites.

**Enzyme classification**

Enzymes are classified into six major groups on the basis of the type of reaction that they catalyze. Each enzyme has a unique four-digit classification number.

**Enzyme assays**

An enzyme assay measures the conversion of substrate to product, under conditions of cofactors, pH and temperature at which the enzyme is optimally active. High substrate concentrations are used so that the initial reaction rate is proportional to the enzyme concentration. Either the rate of appearance of product or the rate of disappearance of substrate is measured, often by following the change in absorbance using a spectrophotometer. Reduced nicotinamide adenine dinucleotide (NADH) and reduced nicotinamide adenine dinucleotide phosphate (NADPH), which absorb light at 340 nm, are often used to monitor the progress of an enzyme reaction.

**Linked enzyme assays**

If neither the substrates nor products of an enzyme-catalyzed reaction absorb light at an appropriate wavelength, the enzyme can be assayed by linking it to another enzyme-catalyzed reaction that does involve a change in absorbance. The second enzyme must be in excess, so that the rate-limiting step in the linked assay is the action of the first enzyme.

**Coenzymes and prosthetic groups**

Some enzymes require the presence of cofactors, small nonprotein units, to function. Cofactors may be inorganic ions or complex organic molecules called coenzymes. A cofactor that is covalently attached to the enzyme is called a prosthetic group. A holoenzyme is the catalytically active form of the enzyme with its cofactor, whereas an apoenzyme is the protein part on its own. Many coenzymes are derived from dietary vitamin precursors, and deficiencies in them lead to certain diseases. Nicotinamide adenine dinucleotide ($NAD^+$), nicotinamide adenine dinucleotide phosphate ($NADP^+$), flavin adenine dinucleotide (FAD) and flavin mononucleotide (FMN) are widely occurring coenzymes involved in oxidation–reduction reactions.

**Isoenzymes**

Isoenzymes are different forms of an enzyme which catalyze the same reaction, but which exhibit different physical or kinetic properties. The isoenzymes of lactate dehydrogenase (LDH) can be separated electrophoretically and can be used clinically to diagnose a myocardial infarction.

**Related topics**

Thermodynamics (C2)
Enzyme kinetics (C3)
Enzyme inhibition (C4)
Regulation of enzyme activity (C5)

## Enzymes as catalysts

Enzymes are **catalysts** that increase the rate of a chemical reaction without being changed themselves in the process. In the absence of an enzyme, the reaction may hardly proceed at all, whereas in its presence the rate can be increased up to $10^7$-fold. Enzyme catalyzed reactions usually take place under relatively mild conditions (temperatures well below 100°C, atmospheric pressure and neutral pH) as compared to the corresponding chemical reactions. Enzymes are also **highly specific** with respect to the substrates that they act on and the products that they form. In addition, enzyme activity can be **regulated**, varying in response to the concentration of substrates or other molecules (see Topic C5). Nearly all enzymes are **proteins**, although a few catalytically active **RNA molecules** have been identified.

## Active site

The **active site** of an enzyme is the region that **binds the substrate** and converts it into product. It is usually a relatively small part of the whole enzyme molecule and is a **three-dimensional entity** formed by amino acid residues that can lie far apart in the linear polypeptide chain (see Topic B3). The active site is often a cleft or crevice on the surface of the enzyme that forms a predominantly nonpolar environment which enhances the binding of the substrate. The substrate(s) is bound in the active site by **multiple weak forces** (electrostatic interactions, hydrogen bonds, van der Waals bonds, hydrophobic interactions; see Topic B3) and in some cases by reversible covalent bonds. Having bound the substrate molecule, and formed an **enzyme–substrate complex**, catalytically active residues within the active site of the enzyme act on the substrate molecule to transform it first into the transition state complex (see Topic C2) and then into product, which is released into solution. The enzyme is now free to bind another molecule of substrate and begin its catalytic cycle again.

Two models have been proposed to explain how an enzyme binds its substrate. In the **lock-and-key** model proposed by Emil Fischer in 1894, the shape of the substrate and the active site of the enzyme are thought to fit together like a key into its lock (*Fig. 1a*). The two shapes are considered as rigid and fixed, and perfectly complement each other when brought together in the right alignment. In the **induced-fit model** proposed in 1958 by Daniel E.

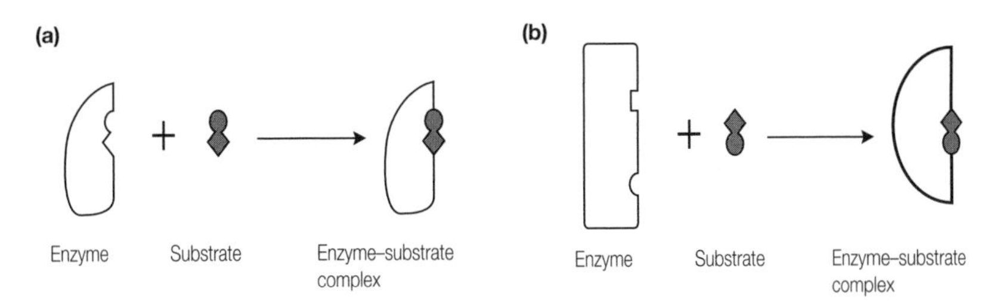


*Fig. 1. Binding of a substrate to an enzyme. (a) Lock-and-key model; (b) induced-fit model.*

Koshland, Jr., the binding of substrate **induces a conformational change** in the active site of the enzyme (*Fig. 1b*). In addition, the enzyme may distort the substrate, forcing it into a conformation similar to that of the transition state (see Topic C2). For example, the binding of **glucose** to **hexokinase** induces a conformational change in the structure of the enzyme such that the active site assumes a shape that is complementary to the substrate (glucose) only after it has bound to the enzyme. Different enzymes show features of both models, with some complementarity and some conformational change.

## Substrate specificity

The properties and spatial arrangement of the amino acid residues forming the active site of an enzyme will determine which molecules can bind and be substrates for that enzyme. **Substrate specificity** is often determined by changes in relatively few amino acids in the active site. This is clearly seen in the three digestive enzymes **trypsin**, **chymotrypsin** and **elastase** (see Topic C5). These three enzymes belong to a family of enzymes called the **serine proteases** – 'serine' because they have a serine residue in the active site that is critically involved in catalysis and 'proteases' because they catalyze the hydrolysis of peptide bonds in proteins. The three enzymes cleave peptide bonds in protein substrates on the carboxyl side of certain amino acid residues.

Trypsin cleaves on the carboxyl side of positively charged Lys or Arg residues, chymotrypsin cleaves on the carboxyl side of bulky aromatic and hydrophobic amino acid residues, and elastase cleaves on the carboxyl side of residues with small uncharged side chains. Their differing specificities are determined by the nature of the amino acid groups in their substrate binding sites which are complementary to the substrates that they act upon. Thus trypsin has a negatively charged Asp residue in its substrate binding site which interacts with the positive charge on the Lys and Arg side chains of the substrate (*Fig. 2a*). Chymotrypsin has amino acid residues with small side chains, such as Gly and Ser, in its substrate binding site that allow access of the bulky side chain of the substrate (*Fig. 2b*). In contrast, elastase has the relatively large uncharged amino

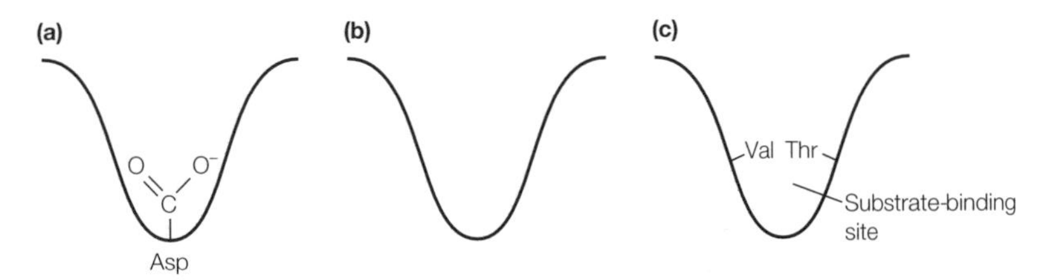


*Fig. 2. Schematic representation of the substrate-binding sites in the serine proteases (a) trypsin, (b) chymotrypsin and (c) elastase.*

acid side chains of Val and Thr protruding into its substrate binding site, preventing access of all but the small side chains on Gly and Ala (*Fig. 2c*).

## Enzyme classification

Many enzymes are named by adding the suffix '-ase' to the name of their substrate. Thus **urease** is the enzyme that catalyzes the hydrolysis of urea, and **fructose-1,6-bisphosphatase** hydrolyzes fructose-1,6-bisphosphate. However, other enzymes, such as trypsin and chymotrypsin, have names that do not denote their substrate. Some enzymes have several alternative names. To rationalize enzyme names, a system of **enzyme nomenclature** has been internationally agreed. This system places all enzymes into one of **six major classes** based on the type of reaction catalyzed (*Table 1*). Each enzyme is then uniquely identified with a four-digit classification number. Thus trypsin has the Enzyme Commission (EC) number 3.4.21.4, where the first number (3) denotes that it is a hydrolase, the second number (4) that it is a protease that hydrolyzes peptide bonds, the third number (21) that it is a serine protease with a critical serine residue at the active site, and the fourth number (4) indicates that it was the fourth enzyme to be assigned to this class. For comparison, chymotrypsin has the EC number 3.4.21.1, and elastase 3.4.21.36.

## Enzyme assays

The amount of enzyme protein present can be determined (assayed) in terms of the catalytic effect it produces, that is the conversion of substrate to product. In order to **assay** (monitor the activity of) an enzyme, the overall equation of the reaction being catalyzed must be known, and an analytical procedure must be available for determining either the disappearance of substrate or the appearance of product. In addition, one must take into account whether the enzyme requires any **cofactors**, and the **pH** and **temperature** at which the enzyme is optimally active (see Topic C3). For mammalian enzymes, this is usually in the range 25–37°C. Finally, it is essential that the rate of the reaction being assayed is a measure of the enzyme activity present and is not limited by an insufficient supply of substrate. Therefore, very high substrate concentrations are generally required so that the **initial reaction rate**, which is determined experimentally, is proportional to the enzyme concentration (see Topic C3).

An enzyme is most conveniently assayed by measuring the **rate of appearance of product** or the **rate of disappearance of substrate**. If the substrate (or product) absorbs light at a specific wavelength, then changes in the concentration of these molecules can be measured by following the **change of absorbance** at this wavelength. Typically this is carried out using a **spectrophotometer**. Since absorbance is proportional to concentration, the rate of change in

*Table 1. International classification of enzymes*

| Class | Name | Type of reaction catalyzed | | Example |
|---|---|---|---|---|
| 1 | Oxidoreductases | Transfer of electrons | $A^- + B \rightarrow A + B^-$ | Alcohol dehydrogenase |
| 2 | Transferases | Transfer of functional groups | A–B + C → A + B–C | Hexokinase |
| 3 | Hydrolases | Hydrolysis reactions | A–B + $H_2O$ → A–H + B–OH | Trypsin |
| 4 | Lyases | Cleavage of C–C, C–O, C–N and other bonds, often forming a double bond | A–B (with X, Y attached) → A=B + X–Y | Pyruvate decarboxylase |
| 5 | Isomerases | Transfer of groups within a molecule | A–B (with X, Y attached) → A–B (with Y, X attached) | Maleate isomerase |
| 6 | Ligases (or synthases) | Bond formation coupled to ATP hydrolysis | A + B → A–B | Pyruvate carboxylase |

absorbance is proportional to the rate of enzyme activity in moles of substrate used (or product formed) per unit time.

Two of the most common molecules used for absorbance measurement in enzyme assays are the coenzymes **reduced nicotinamide adenine dinucleotide (NADH)** and **reduced nicotinamide adenine dinucleotide phosphate (NADPH)** (see below) which each absorb in the ultraviolet (UV) region at 340 nm. Thus, if NADH or NADPH is produced during the course of the reaction there will be a corresponding increase in absorbance at 340 nm, whilst if the reaction involves the oxidation of NADH or NADPH to $NAD^+$ or $NADP^+$, respectively, there will be a corresponding decrease in absorbance, since these oxidized forms do not absorb at 340 nm. One example, is that the activity of **lactate dehydrogenase** with lactate as substrate can be assayed by following the increase in absorbance at 340 nm, according to the following equation:

$$\underset{\text{lactate}}{CH_3CH(OH)COO^-} + NAD^+ \rightleftharpoons \underset{\text{pyruvate}}{CH_3COCOO^-} + NADH + H^+$$

## Linked enzyme assays

Numerous reactions do not involve substrates or products that absorb light at a suitable wavelength. In this case it is often possible to assay the enzyme that catalyzes this reaction by **linking** (or **coupling**) it to a second enzyme reaction that does involve a characteristic absorbance change. For example, the action of the enzyme **glucose oxidase**, which is often used to measure the concentration of glucose in the blood of diabetic patients, does not result in a change in absorbance upon conversion of substrates to products (*Fig. 3*). However, the hydrogen peroxide produced in this reaction can be acted on by a second enzyme, peroxidase, which simultaneously converts a colorless compound into a colored one (**chromogen**) whose absorbance can be easily measured (*Fig. 3*).

If the activity of the first enzyme (glucose oxidase) is to be measured accurately, the second enzyme (peroxidase) and its co-substrates or coenzymes

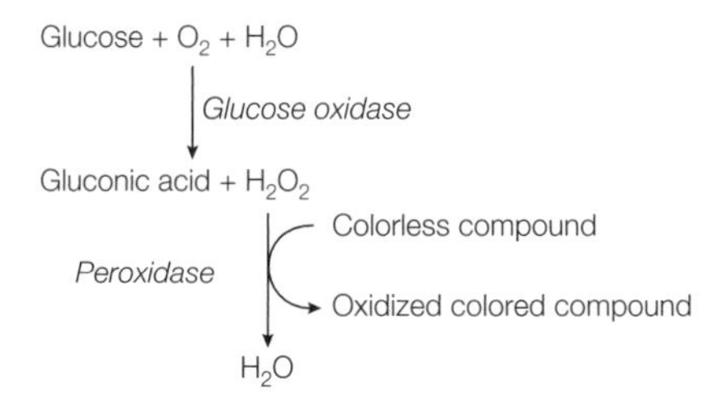


*Fig. 3. A linked enzyme assay with glucose oxidase and peroxidase can be used to measure the amount of glucose in a blood sample.*

*Table 2. Some common coenzymes, their vitamin precursors and deficiency diseases*

| Coenzyme | Precursor | Deficiency disease |
|---|---|---|
| Coenzyme A | Pantothenic acid | Dermatitis |
| FAD, FMN | Riboflavin (vitamin $B_2$) | Growth retardation |
| $NAD^+$, $NADP^+$ | Niacin | Pellagra |
| Thiamine pyrophosphate | Thiamine (vitamin $B_1$) | Beriberi |
| Tetrahydrofolate | Folic acid | Anemia |
| Deoxyadenosyl cobalamin | Cobalamin (vitamin $B_{12}$) | Pernicious anemia |
| Co-substrate in the hydroxylation of proline in collagen | Vitamin C (ascorbic acid) | Scurvy |
| Pyridoxal phosphate | Pyridoxine (vitamin $B_6$) | Dermatitis |

must be **in excess** so as not to be the **rate-limiting step** of the linked assay. This will ensure that the rate of production of the colored chromogen is proportional to the rate of production of $H_2O_2$, whose production in turn is proportional to the activity of glucose oxidase.

## Coenzymes and prosthetic groups

Many enzymes require the presence of small, nonprotein units or **cofactors** to carry out their particular reaction. Cofactors may be either one or more **inorganic ions**, such as $Zn^{2+}$ or $Fe^{2+}$, or a complex organic molecule called a **coenzyme**. A metal or coenzyme that is covalently attached to the enzyme is called a **prosthetic group** (cf. heme in hemoglobin; see Topic B4). A complete catalytically-active enzyme together with its coenzyme or metal ion is called a **holoenzyme**. The protein part of the enzyme on its own without its cofactor is termed an **apoenzyme**. Some coenzymes, such as $NAD^+$, are bound and released by the enzyme during its catalytic cycle and in effect function as cosubstrates. Many coenzymes are derived from **vitamin precursors** (*Table 2*) which are often essential components of the organism's diet, thus giving rise to **deficiency diseases** when in inadequate supply.

**Nicotinamide adenine dinucleotide** (**$NAD^+$**) and **nicotinamide adenine dinucleotide phosphate** (**$NADP^+$**) coenzymes are based on a common structure consisting of the base adenine, two ribose sugars linked by phosphate groups and a nicotinamide ring (*Fig. 4*). $NADP^+$ differs from $NAD^+$ in having an additional phosphate group attached to one of the ribose sugars (*Fig. 4*). These two coenzymes share a common function as they both act as carriers of electrons and are involved in **oxidation–reduction reactions**. $NAD^+$ is more commonly used in **catabolic** (breakdown) reactions, whilst $NADP^+$ is used in **anabolic** (biosynthetic) reactions. The reactive part of both molecules is the **nicotinamide ring** which exists in either a reduced or an oxidized form, and so acts to accept or donate electrons in an enzymic reaction. The reaction also involves the transfer of protons, according to the equation:

$$NAD^+ + H^+ + 2e^- \rightleftharpoons NADH$$

**Flavin adenine dinucleotide** (**FAD**) and **flavin mononucleotide** (**FMN**) are also carriers of electrons and have related chemical structures (*Fig. 5*). Both of these coenzymes consist of a **flavine mononucleotide unit** which contains the reactive

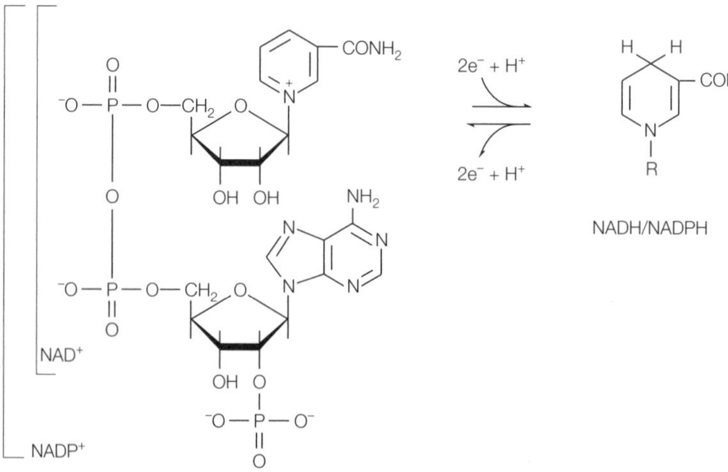

*Fig. 4. The structures of the coenzymes $NAD^+$ and $NADP^+$.*

site. FAD has an additional sugar group and an adenine base which complete its structure. FAD and FMN react with two protons, as well as two electrons, in alternating between the reduced and oxidized state:

$$FAD + 2H^+ + 2e^- \rightleftharpoons FADH_2$$

## Isoenzymes

Isoenzymes (isozymes) are different forms of an enzyme which **catalyze the same reaction**, but which exhibit **different physical or kinetic properties**, such as isoelectric point, pH optimum, substrate affinity or effect of inhibitors. Different isoenzyme forms of a given enzyme are usually derived from different genes and often occur in different tissues of the body.

An example of an enzyme which has different isoenzyme forms is **lactate dehydrogenase** (LDH) which catalyzes the reversible conversion of pyruvate into lactate in the presence of the coenzyme NADH (see above). LDH is a tetramer of two different types of **subunits**, called H and M, which have small differences in amino acid sequence. The two subunits can combine randomly with each other, forming five isoenzymes that have the compositions $H_4$, $H_3M$, $H_2M_2$, $HM_3$ and $M_4$. The five isoenzymes can be resolved electrophoretically (see Topic B8). M subunits predominate in skeletal muscle and liver, whereas H subunits predominate in the heart. $H_4$ and $H_3M$ isoenzymes are found predominantly in the heart and red blood cells; $H_2M_2$ is found predominantly in the brain and kidney; while $HM_3$ and $M_4$ are found predominantly in the liver and skeletal muscle. Thus, the isoenzyme pattern is characteristic of a particular tissue, a factor which is of immense diagnostic importance in medicine. **Myocardial infarction**, **infectious hepatitis** and **muscle diseases** involve cell death of the affected tissue, with release of the cell contents into the blood. As LDH is a soluble, cytosolic protein it is readily released in these conditions. Under normal circumstances there is little LDH in the blood. Therefore the pattern of LDH isoenzymes in the blood is indicative of the tissue that released the isoenzymes and so can be used to diagnose a condition, such as a myocardial infarction, and to monitor the progress of treatment.

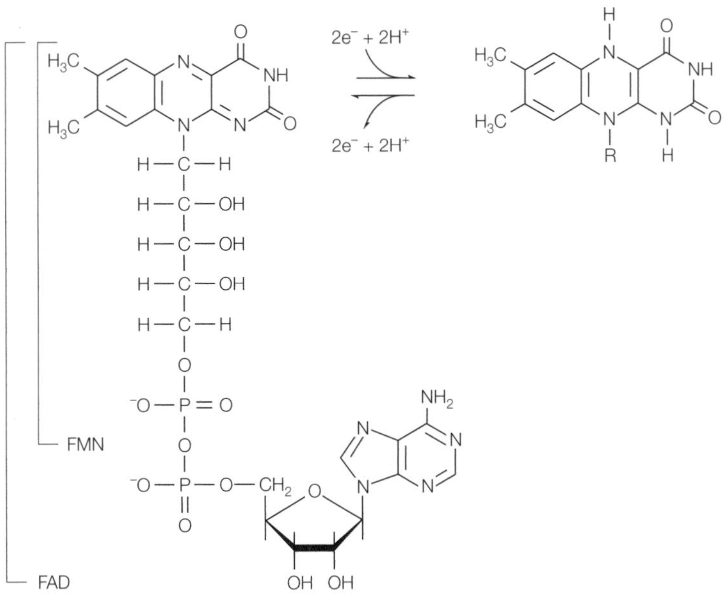

*Fig. 5. The structures of the coenzymes FAD and FMN.*

# C2 Thermodynamics

## Key Notes

**Thermodynamics**

A knowledge of thermodynamics, which is the description of the relationships among the various forms of energy and how energy affects matter, enables one to determine whether a physical process is possible. The first and second laws of thermodynamics are combined in the thermodynamic function, free energy ($G$). If the change in free energy ($\Delta G$) of a reaction is negative, that reaction can occur spontaneously. If $\Delta G$ is positive, an input of energy is required to drive the reaction. The unit of energy is the Joule (J) or the calorie (cal).

**Activation energy and transition state**

For a biochemical reaction to proceed, the energy barrier needed to transform the substrate molecules into the transition state has to be overcome. The transition state has the highest free energy in the reaction pathway. The difference in free energy between the substrate and the transition state is termed the Gibbs free energy of activation ($\Delta G‡$). An enzyme stabilizes the transition state and lowers $\Delta G‡$, thus increasing the rate at which the reaction occurs.

**Free energy change**

The difference in energy level between the substrates and products is termed the change in Gibbs free energy ($\Delta G$). A negative $\Delta G$ indicates that the reaction is thermodynamically favorable in the direction indicated, whereas a positive $\Delta G$ indicates that the reaction is not thermodynamically favorable and requires an input of energy to proceed in the direction indicated. An energetically unfavorable reaction is often driven by linking it to an energetically favorable reaction, such as the hydrolysis of ATP.

**Chemical equilibria**

A chemical reaction often exists in a state of dynamic equilibrium. The equilibrium constant ($K$) defines the ratio of the concentrations of substrates and products at equilibrium. Enzymes do not alter the equilibrium position, but do accelerate the attainment of the equilibrium position by speeding up the forward and reverse reactions.

**Related topics**

Introduction to enzymes (C1)
Enzyme kinetics (C3)
Enzyme inhibition (C4)
Regulation of enzyme activity (C5)

## Thermodynamics

A knowledge of thermodynamics enables one to determine whether a physical process is possible, and is required for understanding why proteins fold to their native conformation, why some enzyme-catalyzed reactions require an input of energy, how muscles generate mechanical force, etc. **Thermodynamics** (Greek: *therme*, heat; *dynamis*, power) is the description of the relationships among the various forms of energy and how energy affects matter on the macroscopic level. As it applies to biochemistry, thermodynamics is most often concerned with describing the conditions under which processes occur spontaneously (by themselves).

In thermodynamics, a **system** is the matter within a defined region. The matter in the rest of the universe is called the **surroundings**. **The first law of thermodynamics**, a mathematical statement of the law of conservation of energy, states that the total energy of a system and its surroundings is a constant:

$$\Delta E = E_B - E_A = Q - W$$

in which $E_A$ is the energy of the system at the start of a process and $E_B$ at the end of the process. $Q$ is the heat absorbed by the system and $W$ is the work done by the system. The change in energy of a system depends only on the initial and final states and not on how it reached that state. Processes in which the system releases heat (i.e. have a negative $Q$) are known as **exothermic** processes; those in which the system gains heat (i.e. have a positive $Q$) are known as **endothermic**. The SI **unit of energy** is the Joule (J), although the calorie (cal) is still often used (1 kcal = 4.184 kJ).

The first law of thermodynamics cannot be used to predict whether a reaction can occur spontaneously, as some spontaneous reactions have a positive $\Delta E$. Therefore a function different from $\Delta E$ is required. One such function is **entropy** ($S$), which is a measure of the degree of randomness or disorder of a system. The entropy of a system increases ($\Delta S$ is positive) when the system becomes more disordered. **The second law of thermodynamics** states that a process can occur spontaneously only if the sum of the entropies of the system and its surroundings increases (or that the universe tends towards maximum disorder), that is:

$$(\Delta S_{system} + \Delta S_{surroundings}) > 0 \text{ for a spontaneous process.}$$

However, using entropy as a criterion of whether a biochemical process can occur spontaneously is difficult, as the entropy changes of chemical reactions are not readily measured, and the entropy change of both the system and its surroundings must be known. These difficulties are overcome by using a different thermodynamic function, **free energy** ($G$), proposed by Josiah Willard Gibbs which combines the first and second laws of thermodynamics:

$$\Delta G = \Delta H - T\Delta S$$

in which $\Delta G$ is the free energy of a system undergoing a transformation at constant pressure ($P$) and temperature ($T$), $\Delta H$ is the change in **enthalpy** (heat content) of this system, and $\Delta S$ is the change in the entropy of this system. The enthalpy change is given by:

$$\Delta H = \Delta E + P\Delta V.$$

The volume change ($\Delta V$) is small for nearly all biochemical reactions, and so $\Delta H$ is nearly equal to $\Delta E$. Therefore

$$\Delta G = \Delta E - T\Delta S.$$

Thus, the $\Delta G$ of a reaction depends both on the change in internal energy and on the change in entropy of the system. The change in free energy $\Delta G$ of a reaction is a valuable criterion of whether that reaction can occur spontaneously:

- a reaction can occur spontaneously only if $\Delta G$ is negative;
- a system is at equilibrium if $\Delta G$ is zero;

- a reaction cannot occur spontaneously if $\Delta G$ is positive. An input of energy is required to drive such a reaction;
- the $\Delta G$ of a reaction is independent of the path of the transformation;
- $\Delta G$ provides no information about the rate of a reaction.

## Activation energy and transition state

The energy changes that take place during the course of a particular biochemical reaction are shown in *Fig. 1*. In all reactions there is an **energy barrier** that has to be overcome in order for the reaction to proceed. This is the energy needed to transform the substrate molecules into the **transition state** – an unstable chemical form part-way between the substrates and the products. The transition state has the **highest free energy** of any component in the reaction pathway. The **Gibbs free energy of activation** ($\Delta G^{\ddagger}$) is equal to the difference in free energy between the transition state and the substrate (*Fig. 1*). An enzyme works by **stabilizing the transition state** of a chemical reaction and decreasing $\Delta G^{\ddagger}$ (*Fig. 1*). The enzyme does not alter the energy levels of the substrates or the products. Thus an enzyme increases the rate at which the reaction occurs, but has no effect on the overall change in energy of the reaction.

## Free energy change

The change in **Gibbs free energy** ($\Delta G$) dictates whether a reaction will be energetically favorable or not. *Figure 1* shows an example where the overall energy change of the reaction makes it energetically favorable (i.e. the products are at a lower energy level than the substrates and $\Delta G$ is negative). It should be noted that $\Delta G$ is unrelated to $\Delta G^{\ddagger}$. The $\Delta G$ of a reaction is independent of the path of the reaction, and it provides no information about the rate of a reaction since the rate of the reaction is governed by $\Delta G^{\ddagger}$. A negative $\Delta G$ indicates that the reaction is thermodynamically favorable in the direction indicated (i.e. that it is likely to occur without an input of energy), whereas a positive $\Delta G$ indicates that the reaction is not thermodynamically favorable and requires an **input of energy** to proceed in the direction indicated. In biochemical systems, this input of energy is often achieved by **coupling** the energetically unfavorable reaction with a more energetically favorable one (**coupled reactions**).

It is often convenient to refer to $\Delta G$ under a standard set of conditions, defined as when the substrates and products of a reaction are all present at concentrations of 1.0 M and the reaction is taking place at a constant pH of 7.0. Under these conditions a slightly different value for $\Delta G$ is found, and this is called $\Delta G^{\circ\prime}$. An example of an energetically favorable reaction which has a large

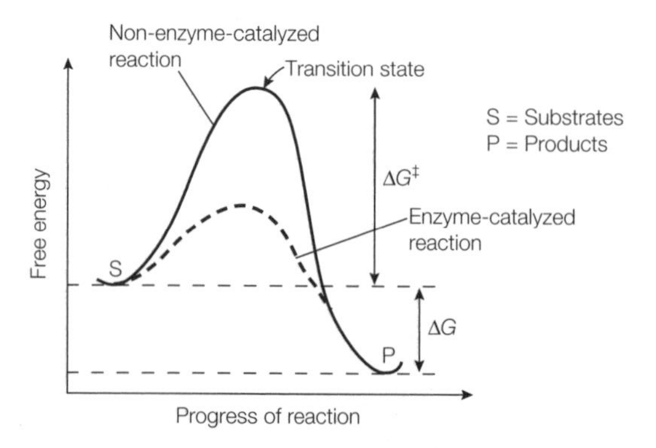


*Fig. 1. The energy changes taking place during the course of a biochemical reaction.*

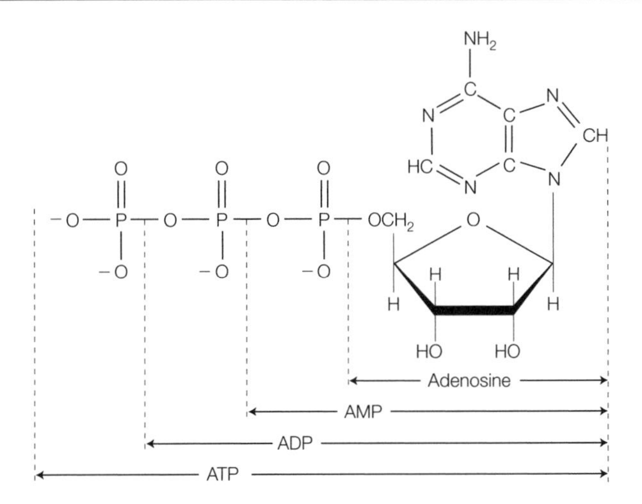


*Fig. 2. Structure of adenosine triphosphate (ATP), adenosine diphosphate (ADP), adenosine monophosphate (AMP) and adenosine.*

negative $\Delta G^{\circ\prime}$ and is commonly used to drive less energetically favorable reactions is the hydrolysis of adenosine triphosphate (ATP; *Fig. 2*) to form adenosine diphosphate (ADP) and free inorganic phosphate ($P_i$):

$$\text{ATP} + \text{H}_2\text{O} \rightarrow \text{ADP} + \text{P}_i \quad \Delta G^{\circ\prime} = -30.5 \text{ kJ mol}^{-1}$$
$$-7.3 \text{ kcal mol}^{-1}$$

## Chemical equilibria

A chemical reaction usually exists in a state of **dynamic equilibrium,** where although new molecules of substrate and product are continually being transformed and formed, the ratio of substrate to product remains at a constant value.

Consider the reaction:

$$\text{A} \underset{10^{-6}\ \text{sec}^{-1}}{\overset{10^{-4}\ \text{sec}^{-1}}{\rightleftharpoons}} \text{B}$$

where the rate of the forward reaction is $10^{-4}$ per second ($\text{sec}^{-1}$) and the rate of the reverse reaction is $10^{-6}$ $\text{sec}^{-1}$. At equilibrium the ratio of the concentrations of the substrate and product gives a constant value, known as the **equilibrium constant** (*K*). The equilibrium constant for a given reaction is defined as:

$$K = \frac{[\text{products}]_{eq}}{[\text{reactants}]_{eq}} = \frac{[\text{B}]_{eq}}{[\text{A}]_{eq}}$$

where square brackets indicate concentration. The equilibrium constant is also given by the ratio of the forward reaction rate ($k_f$) and the reverse reaction rate ($k_b$):

$$K = \frac{k_f}{k_b} = \frac{10^{-4}}{10^{-6}} = 100$$

Thus, for the above reaction at equilibrium, there is 100 times more of product B than there is of substrate A, regardless of whether there is enzyme present or not.

This is because enzymes do not alter the equilibrium position of a reaction, but accelerate the forward and reverse reactions to the same extent. In other words, **enzymes accelerate the attainment of the equilibrium position** but do not shift its position. For the hypothetical reaction shown above, in the absence of added enzyme the reaction may take over an hour to reach the equilibrium position, whereas in the presence of enzyme the equilibrium position may be reached in less than 1 sec.

# C3 Enzyme kinetics

## Key Notes

**Enzyme velocity**

Enzyme activity is commonly expressed by the initial rate ($V_0$) of the reaction being catalyzed. The units of $V_0$ are μmol min$^{-1}$, which can also be represented by the enzyme unit (U) or the katal (kat), where 1 μmol min$^{-1}$ = 1 U = 16.67 nanokat. The term activity (or total activity) refers to the total units of enzyme in a sample, whereas specific activity is the number of units per milligram of protein (units mg$^{-1}$).

**Substrate concentration**

At low substrate concentrations ([S]) a doubling of [S] leads to a doubling of $V_0$, whereas at higher [S] the enzyme becomes saturated and there is no further increase in $V_0$. A graph of $V_0$ against [S] will give a hyperbolic curve.

**Enzyme concentration**

When [S] is saturating, a doubling of the enzyme concentration leads to a doubling of $V_0$.

**Temperature**

Temperature affects the rate of an enzyme-catalyzed reaction by increasing the thermal energy of the substrate molecules. This increases the proportion of molecules with sufficient energy to overcome the activation barrier and hence increases the rate of the reaction. In addition, the thermal energy of the component molecules of the enzyme is increased, which leads to an increased rate of denaturation of the enzyme protein due to the disruption of the noncovalent interactions holding the structure together.

**pH**

Each enzyme has an optimum pH at which the rate of the reaction that it catalyzes is at its maximum. Slight deviations in the pH from the optimum lead to a decrease in the reaction rate. Larger deviations in pH lead to denaturation of the enzyme due to changes in the ionization of amino acid residues and the disruption of noncovalent interactions.

**Michaelis–Menten model**

The Michaelis–Menten model uses the following concept of enzyme catalysis:

$$E + S \underset{k_2}{\overset{k_1}{\rightleftharpoons}} ES \overset{k_3}{\rightarrow} E + P$$

where the rate constants $k_1$, $k_2$ and $k_3$ describe the rates associated with each step of the catalytic process. At low [S], $V_0$ is directly proportional to [S], while at high [S] the velocity tends towards a maximum velocity ($V_{max}$). The Michaelis–Menten equation:

$$V_0 = \frac{V_{max} \cdot [S]}{K_m + [S]}$$

describes these observations and predicts a hyperbolic curve of $V_0$ against [S]. The Michaelis constant, $K_m$, is equal to the sum of the rates of breakdown of the enzyme–substrate complex over its rate of formation, and is a measure of the affinity of an enzyme for its substrate.

**Lineweaver–Burk plot**

$V_{max}$ and $K_m$ can be determined experimentally by measuring $V_0$ at different substrate concentrations, and then plotting $1/V_0$ against 1/[S] in a double reciprocal or Lineweaver–Burk plot. The intercept on the $y$-axis is equal to $1/V_{max}$, the intercept on the $x$-axis is equal to $-1/K_m$ and the slope of the line is equal to $K_m / V_{max}$.

**Related topics**

Introduction to enzymes (C1)
Thermodynamics (C2)
Enzyme inhibition (C4)
Regulation of enzyme activity (C5)

## Enzyme velocity

The rate of an enzyme-catalyzed reaction is often called its velocity. Enzyme velocities are normally reported as values at time zero (**initial velocity**, symbol $V_0$; μmol min$^{-1}$), since the rate is fastest at the point where no product is yet present. This is because the substrate concentration is greatest before any substrate has been transformed to product, because enzymes may be subject to **feedback inhibition** by their own products and/or because with a reversible reaction the products will fuel the reverse reaction. Experimentally $V_0$ is measured before more than approximately 10% of the substrate has been converted to product in order to minimize such complicating factors. A typical **plot of product formed against time** for an enzyme-catalyzed reaction shows an initial period of rapid product formation which gives the linear portion of the plot (*Fig. 1*). This is followed by a slowing down of the enzyme rate as substrate is used up and/or as the enzyme loses activity. $V_0$ is obtained by drawing a straight line through the linear part of the curve, starting at the zero time-point (*Fig. 1*). The slope of this straight line is equal to $V_0$.

### *Enzyme units*

Enzyme activity may be expressed in a number of ways. The commonest is by the initial rate ($V_0$) of the reaction being catalyzed (e.g. μmol of substrate transformed per minute; μmol min$^{-1}$). There are also two standard units of enzyme activity, the **enzyme unit** (**U**) and the **katal** (**kat**). An enzyme unit is that amount

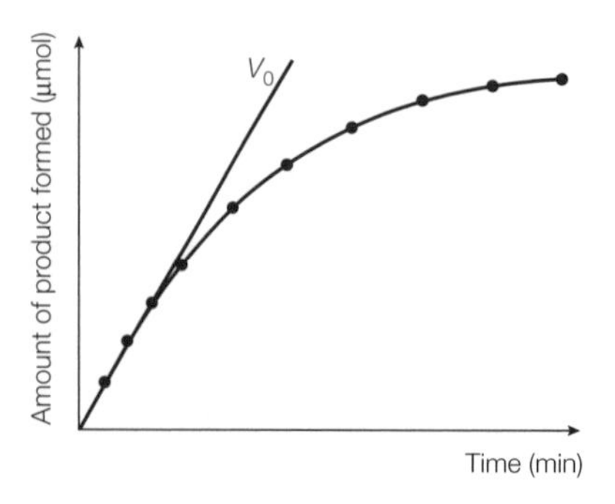


*Fig. 1. The relationship between product formation and time for an enzyme-catalyzed reaction.*

of enzyme which will catalyze the transformation of 1 μmol of substrate per minute at 25°C under optimal conditions for that enzyme. The katal is the accepted SI unit of enzyme activity and is defined as that catalytic activity which will raise the rate of a reaction by one mole per second in a specified system. It is possible to convert between these different units of activity using 1 μmol $min^{-1}$ = 1 U = 16.67 nanokat. The term **activity** (or **total activity**) refers to the total units of enzyme in the sample, whereas the **specific activity** is the number of enzyme units per milligram of protein (units $mg^{-1}$). The specific activity is a measure of the purity of an enzyme; during the purification of the enzyme its specific activity increases and becomes maximal and constant when the enzyme is pure.

## Substrate concentration

The normal pattern of dependence of enzyme rate on **substrate concentration** ([S]) is that at low substrate concentrations a doubling of [S] will lead to a doubling of the initial velocity ($V_0$). However, at higher substrate concentrations the enzyme becomes **saturated**, and further increases in [S] lead to very small changes in $V_0$. This occurs because at saturating substrate concentrations effectively all of the enzyme molecules have bound substrate. The overall enzyme rate is now dependent on the rate at which the product can dissociate from the enzyme, and adding further substrate will not affect this. The shape of the resulting graph when $V_0$ is plotted against [S] is called a **hyperbolic curve** (*Fig. 2*).

## Enzyme concentration

In situations where the substrate concentration is saturating (i.e. all the enzyme molecules are bound to substrate), a doubling of the **enzyme concentration** will lead to a doubling of $V_0$. This gives a straight line graph when $V_0$ is plotted against enzyme concentration.

## Temperature

**Temperature** affects the rate of enzyme-catalyzed reactions in two ways. First, a rise in temperature increases the **thermal energy** of the substrate molecules. This raises the proportion of substrate molecules with sufficient energy to overcome the Gibbs free energy of activation ($\Delta G‡$) (see Topic C2), and hence increases the rate of the reaction. However, a second effect comes into play at higher temperatures. Increasing the thermal energy of the molecules which make up the protein structure of the enzyme itself will increase the chances of breaking the multiple weak, noncovalent interactions (hydrogen bonds, van der Waals forces, etc.) which hold the three-dimensional structure of the enzyme

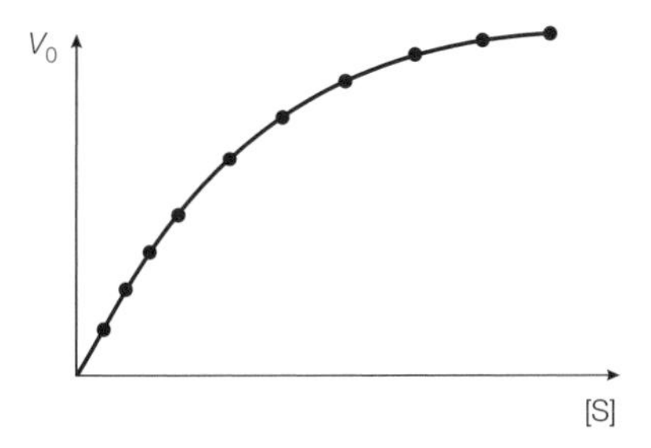


*Fig. 2. The relationship between substrate concentration [S] and initial reaction velocity ($V_0$).*

together (see Topic B3). Ultimately this will lead to the **denaturation** (unfolding) of the enzyme, but even small changes in the three-dimensional shape of the enzyme can alter the structure of the active site and lead to a decrease in catalytic activity. The overall effect of a rise in temperature on the reaction rate of the enzyme is a balance between these two opposing effects. A graph of temperature plotted against $V_0$ will therefore show a curve, with a well-defined temperature optimum (*Fig. 3a*). For many mammalian enzymes this is around 37°C, but there are also organisms which have enzymes adapted to working at considerably higher or lower temperatures. For example, ***Taq* polymerase** that is used in the polymerase chain reaction (see Topic I6), is found in a bacterium that lives at high temperatures in hot springs, and thus is adapted to work optimally at high temperatures.

## pH

Each enzyme has an **optimum pH** at which the rate of the reaction that it catalyzes is at its maximum. Small deviations in pH from the optimum value lead to decreased activity due to changes in the ionization of groups at the active site of the enzyme. Larger deviations in pH lead to the **denaturation** of the enzyme protein itself, due to interference with the many weak noncovalent bonds maintaining its three-dimensional structure. A graph of $V_0$ plotted against pH will usually give a bell shaped curve (*Fig. 3b*). Many enzymes have a pH optimum of around 6.8, but there is great diversity in the pH optima of enzymes, due to the different environments in which they are adapted to work. For example, the digestive enzyme **pepsin** is adapted to work at the acidic pH of the stomach (around pH 2.0).

## Michaelis–Menten model

The **Michaelis–Menten model** uses the following concept of enzyme catalysis:

$$E + S \underset{k_2}{\overset{k_1}{\rightleftharpoons}} ES \overset{k_3}{\rightarrow} E + P.$$

The enzyme (E), combines with its substrate (S) to form an **enzyme–substrate complex** (ES). The ES complex can dissociate again to form E + S, or can proceed chemically to form E and the product P. The **rate constants** $k_1$, $k_2$ and $k_3$ describe the rates associated with each step of the catalytic process. It is assumed that there is no significant rate for the backward reaction of enzyme and product (E + P) being converted to ES complex. [ES] remains approximately constant until nearly all the substrate is used, hence the rate of synthesis

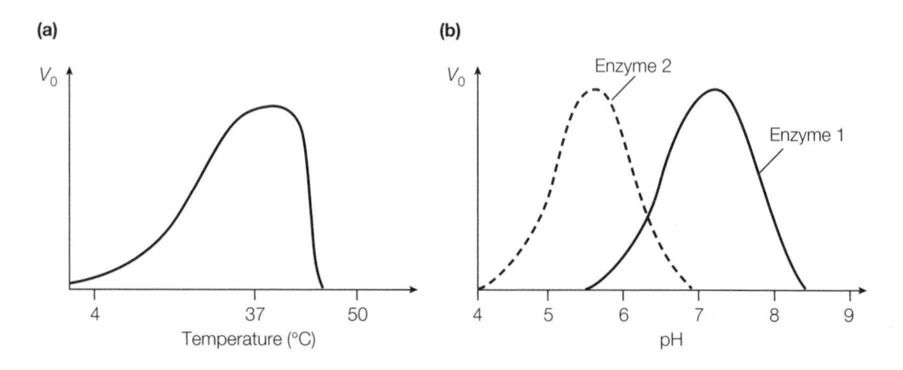


*Fig. 3. The effect of (a) temperature and (b) pH on enzyme activity.*

of ES equals its rate of consumption over most of the course of the reaction; that is, [ES] maintains a **steady state**. From the observation of the properties of many enzymes it was known that the initial velocity ($V_0$) at low substrate concentrations is directly proportional to [S], while at high substrate concentrations the velocity tends towards a maximum value, that is the rate becomes independent of [S] (*Fig. 4a*). This maximum velocity is called $V_{max}$ (μmol min$^{-1}$). The **initial velocity** ($V_0$) is the velocity measured experimentally before more than approximately 10% of the substrate has been converted to product in order to minimize such complicating factors as the effects of reversible reactions, inhibition of the enzyme by product, and progressive inactivation of the enzyme (see above).

Michaelis and Menten derived an equation to describe these observations, the **Michaelis–Menten equation**:

$$V_0 = \frac{V_{max} \cdot [S]}{K_m + [S]}$$

The equation describes a **hyperbolic curve** of the type shown for the experimental data in *Fig. 1a*. In deriving the equation, Michaelis and Menten defined a new constant, $K_m$, the **Michaelis constant** [units: Molar (i.e. per mole), M]:

$$K_m = \frac{k_2 + k_3}{k_1}$$

$K_m$ is a measure of the **stability of the ES complex**, being equal to the sum of the rates of breakdown of ES over its rate of formation. For many enzymes $k_2$ is much greater than $k_3$. Under these circumstances $K_m$ becomes a measure of the **affinity** of an enzyme for its substrate since its value depends on the relative values of $k_1$ and $k_2$ for ES formation and dissociation, respectively. A high $K_m$ indicates weak substrate binding ($k_2$ predominant over $k_1$), a low $K_m$ indicates strong substrate binding ($k_1$ predominant over $k_2$). $K_m$ can be determined experimentally by the fact that its value is equivalent to the substrate concentration at which the velocity is equal to half of $V_{max}$.

## Lineweaver–Burk plot

Because $V_{max}$ is achieved at infinite substrate concentration, it is impossible to estimate $V_{max}$ (and hence $K_m$) from a hyperbolic plot as shown in *Fig. 4a*. However, $V_{max}$ and $K_m$ can be determined experimentally by measuring $V_0$ at

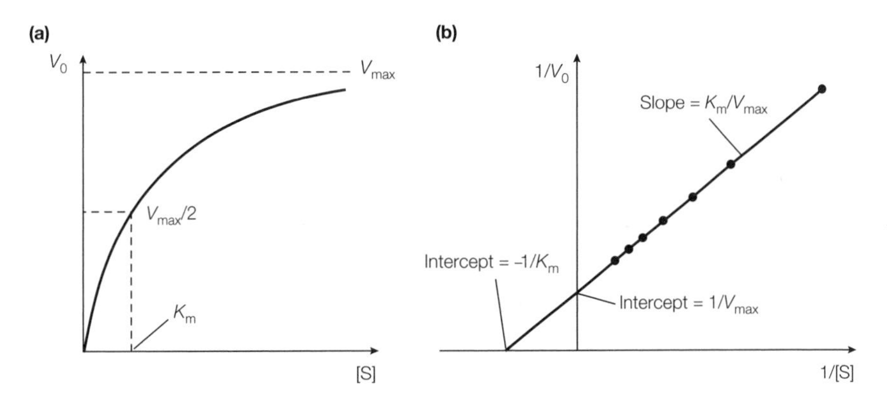


*Fig. 4. The relationship between substrate concentration [S] and initial reaction velocity ($V_0$). (a) A direct plot, (b) a Lineweaver–Burk double-reciprocal plot.*

different substrate concentrations (see *Fig. 1*). Then a **double reciprocal** or **Lineweaver–Burk** plot of $1/V_0$ against $1/[S]$ is made (*Fig. 4b*). This plot is a derivation of the Michaelis–Menten equation:

$$\frac{1}{V_0} = \frac{1}{V_{max}} + \frac{K_m}{V_{max}} \cdot \frac{1}{[S]}$$

which gives a straight line, with the **intercept on the *y*-axis equal to $1/V_{max}$**, and the **intercept on the *x*-axis equal to $-1/K_m$**. The slope of the line is equal to $K_m/V_{max}$ (*Fig. 4b*). The Lineweaver–Burk plot is also a useful way of determining how an inhibitor binds to an enzyme (see Topic C4).

Although the Michaelis–Menten model provides a very good model of the experimental data for many enzymes, a few enzymes do not conform to Michaelis–Menten kinetics. These enzymes, such as aspartate transcarbamoylase (ATCase), are called **allosteric enzymes** (see Topic C5).

# C4 Enzyme inhibition

## Key Notes

**Enzyme inhibition**

The catalytic rate of an enzyme can be lowered by inhibitor molecules. Many inhibitors exist, including normal body metabolites, foreign drugs and toxins. Enzyme inhibition can be of two main types: irreversible or reversible. Reversible inhibition can be subdivided into competitive and noncompetitive.

**Irreversible inhibition**

An irreversible inhibitor binds tightly, often covalently, to amino acid residues at the active site of the enzyme, permanently inactivating the enzyme. Examples of irreversible inhibitors are diisopropylfluorophosphate (DIPF), iodoacetamide and penicillin.

**Reversible competitive inhibition**

A competitive inhibitor competes with the substrate molecules for binding to the active site of the enzyme. At high substrate concentration, the effect of a competitive inhibitor can be overcome. On a Lineweaver–Burk plot a competitive inhibitor can be seen to increase the $K_m$ but leave $V_{max}$ unchanged.

**Reversible noncompetitive inhibition**

A noncompetitive inhibitor binds at a site other than the active site of the enzyme and decreases its catalytic rate by causing a conformational change in the three-dimensional shape of the enzyme. The effect of a noncompetitive inhibitor cannot be overcome at high substrate concentrations. On a Lineweaver–Burk plot a noncompetitive inhibitor can be seen to decrease $V_{max}$ but leave $K_m$ unchanged.

**Related topics**

Introduction to enzymes (C1)
Enzyme kinetics (C3)
Regulation of enzyme activity (C5)

## Enzyme inhibition

Many types of molecule exist which are capable of interfering with the activity of an individual enzyme. Any molecule which acts directly on an enzyme to lower its catalytic rate is called an **inhibitor**. Some enzyme inhibitors are normal body metabolites that inhibit a particular enzyme as part of the normal metabolic control of a pathway. Other inhibitors may be foreign substances, such as drugs or toxins, where the effect of enzyme inhibition could be either therapeutic or, at the other extreme, lethal. Enzyme inhibition may be of two main types: **irreversible** or **reversible**, with reversible inhibition itself being subdivided into **competitive** and **noncompetitive** inhibition. Reversible inhibition can be overcome by removing the inhibitor from the enzyme, for example by dialysis (see Topic B6), but this is not possible for irreversible inhibition, by definition.

## Irreversible inhibition

Inhibitors which bind **irreversibly** to an enzyme often form a **covalent bond** to an amino acid residue at or near the active site, and permanently inactivate the enzyme. Susceptible amino acid residues include Ser and Cys residues which have reactive –OH and –SH groups, respectively. The compound **diisopropylphosphofluoridate** (DIPF), a component of nerve gases, reacts with a

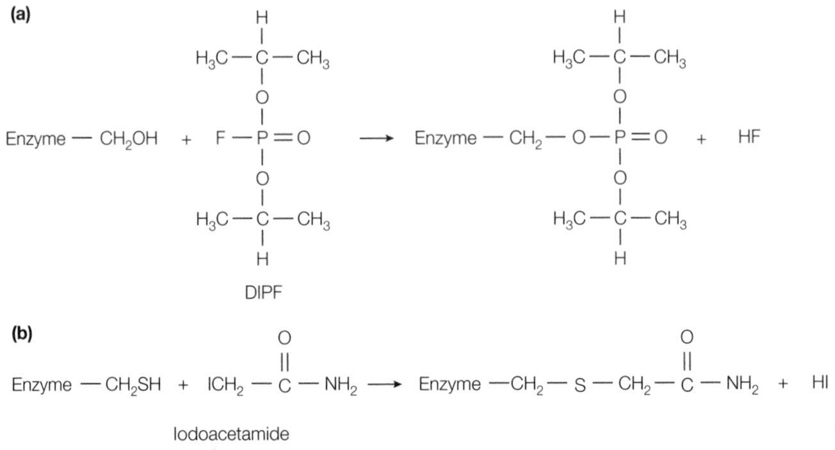

(a)

$$\text{Enzyme—CH}_2\text{OH} + \text{F—P(=O)(O—CH(CH}_3)_2)_2 \longrightarrow \text{Enzyme—CH}_2\text{—O—P(=O)(O—CH(CH}_3)_2)_2 + \text{HF}$$

DIPF

(b)

$$\text{Enzyme—CH}_2\text{SH} + \text{ICH}_2\text{—C(=O)—NH}_2 \longrightarrow \text{Enzyme—CH}_2\text{—S—CH}_2\text{—C(=O)—NH}_2 + \text{HI}$$

Iodoacetamide

*Fig. 1. Structure and mechanism of action of (a) diisopropylphosphofluoridate (DIPF) and (b) iodoacetamide.*

Ser residue in the active site of the enzyme **acetylcholinesterase**, irreversibly inhibiting the enzyme and preventing the transmission of nerve impulses (*Fig. 1a*). **Iodoacetamide** modifies Cys residues and hence may be used as a diagnostic tool in determining whether one or more Cys residues are required for enzyme activity (*Fig. 1b*). The **antibiotic penicillin** irreversibly inhibits the glycopeptide transpeptidase enzyme that forms the cross-links in the bacterial cell wall by covalently attaching to a Ser residue in the active site of the enzyme (see Topic A1).

## Reversible competitive inhibition

A **competitive inhibitor** typically has close structural similarities to the normal substrate for the enzyme. Thus it competes with substrate molecules to bind to the active site (*Fig. 2a*). The enzyme may bind either a substrate molecule or an inhibitor molecule, but not both at the same time (*Fig. 2b*). The competitive inhibitor binds **reversibly to the active site**. At **high substrate concentrations** the action of a competitive inhibitor is overcome because a sufficiently high substrate concentration will successfully compete out the inhibitor molecule in binding to the active site. Thus there is no change in the $V_{max}$ of the enzyme but the apparent affinity of the enzyme for its substrate decreases in the presence of the competitive inhibitor, and hence $K_m$ increases.

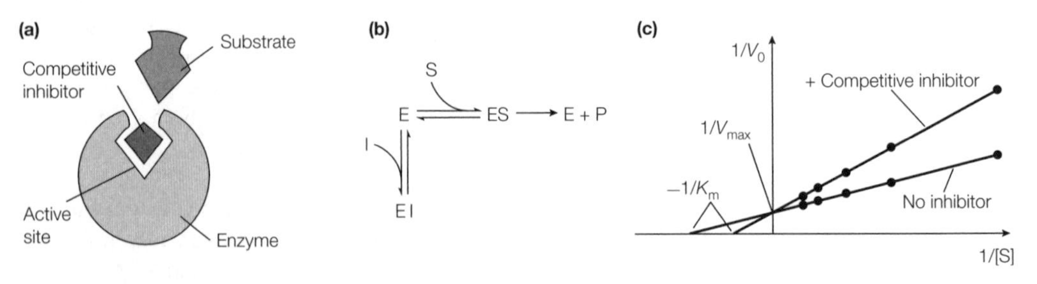


*Fig. 2. The characteristics of competitive inhibition. (a) A competitive inhibitor competes with the substrate for binding at the active site; (b) the enzyme can bind either substrate or the competitive inhibitor but not both; (c) Lineweaver–Burk plot showing the effect of a competitive inhibitor on $K_m$ and $V_{max}$.*

$$\mathrm{^{-}OOC{-}CH_2{-}CH_2{-}COO^{-}}\ (\text{Succinate}) + \mathrm{FAD} \xrightarrow[\text{dehydrogenase}]{\text{Succinate}} \mathrm{^{-}OOC{-}CH{=}CH{-}COO^{-}}\ (\text{Fumarate}) + \mathrm{FADH_2}$$

$$\mathrm{^{-}OOC{-}CH_2{-}COO^{-}}\ (\text{Malonate}) \xrightarrow[\text{dehydrogenase}]{\text{Succinate}} \text{No reaction}$$

*Fig. 3. Inhibition of succinate dehydrogenase by malonate.*

A good example of competitive inhibition is provided by **succinate dehydrogenase**. This enzyme uses **succinate** as its substrate and is competitively inhibited by **malonate** which differs from succinate in having one rather than two methylene groups (*Fig. 3*). Many drugs work by mimicking the structure of the substrate of a target enzyme, and hence act as competitive inhibitors of the enzyme. Competitive inhibition can be recognized by using a Lineweaver–Burk plot. $V_0$ is measured at different substrate concentrations in the presence of a fixed concentration of inhibitor. A competitive inhibitor increases the slope of the line on the Lineweaver–Burk plot, and alters the intercept on the *x*-axis (since **$K_m$ is increased**), but leaves the intercept on the *y*-axis unchanged (since **$V_{max}$ remains constant**; *Fig. 2c*).

## Reversible noncompetitive inhibition

A **noncompetitive inhibitor** binds **reversibly at a site other than the active site** (*Fig. 4a*) and causes a change in the overall three-dimensional shape of the enzyme that leads to a decrease in catalytic activity. Since the inhibitor binds at a different site to the substrate, the enzyme may bind the inhibitor, the substrate or both the inhibitor and substrate together (*Fig. 4b*). The effects of a noncompetitive inhibitor cannot be overcome by increasing the substrate concentration, so there is a decrease in $V_{max}$. In noncompetitive inhibition the affinity of the enzyme for the substrate is unchanged and so $K_m$ remains the same. An example of noncompetitive inhibition is the action of **pepstatin** on the enzyme **renin**.

Noncompetitive inhibition can be recognized on a Lineweaver–Burk plot, since it increases the slope of the experimental line, and alters the intercept on the *y*-axis (since **$V_{max}$ is decreased**), but leaves the intercept on the *x*-axis unchanged (since **$K_m$ remains constant**; *Fig. 4c*).

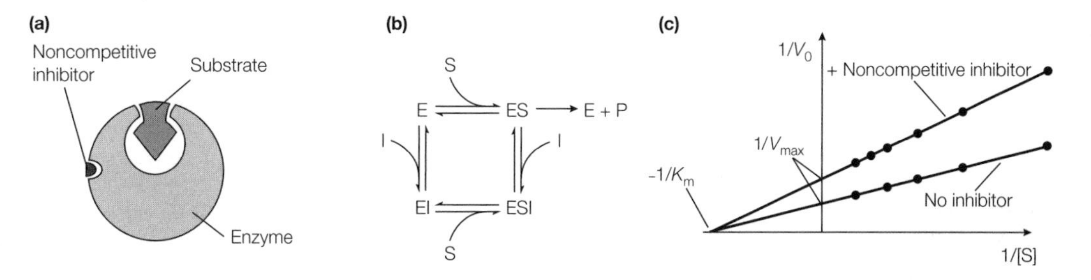


*Fig. 4. The characteristics of noncompetitive inhibition. (a) A noncompetitive inhibitor binds at a site distinct from the active site; (b) the enzyme can bind either substrate or the noncompetitive inhibitor or both; (c) Lineweaver–Burk plot showing the effect of a noncompetitive inhibitor on $K_m$ and $V_{max}$.*

# C5 REGULATION OF ENZYME ACTIVITY

## Key Notes

**Feedback regulation**

The rates of enzyme-catalyzed reactions in biological systems are altered by activators and inhibitors, collectively known as effector molecules. In metabolic pathways, the end-product often feedback-inhibits the committed step earlier on in the same pathway to prevent the build up of intermediates and the unnecessary use of metabolites and energy. For branched metabolic pathways a process of sequential feedback inhibition often operates.

**Allosteric enzymes**

A plot of $V_0$ against [S] for an allosteric enzyme gives a sigmoidal-shaped curve. Allosteric enzymes often have more than one active site which co-operatively bind substrate molecules, such that the binding of substrate at one active site induces a conformational change in the enzyme that alters the affinity of the other active sites for substrate. Allosteric enzymes are often multi-subunit proteins, with an active site on each subunit. In addition, allosteric enzymes may be controlled by effector molecules (activators or inhibitors) that bind to a site other than the active site and alter the rate of enzyme activity. Aspartate transcarbamoylase is an allosteric enzyme that catalyzes the committed step in pyrimidine biosynthesis. This enzyme consists of six catalytic subunits each with an active site and six regulatory subunits to which the allosteric effectors cytosine triphosphate (CTP) and ATP bind. Aspartate transcarbamoylase is feedback-inhibited by the end-product of the pathway, CTP, which acts as an allosteric inhibitor. In contrast, ATP an intermediate earlier in the pathway, acts as an allosteric activator.

**Reversible covalent modification**

The activity of many enzymes is altered by the reversible making and breaking of a covalent bond between the enzyme and a small nonprotein group. The most common such modification is the addition and removal of a phosphate group; phosphorylation and dephosphorylation, respectively. Phosphorylation is catalyzed by protein kinases, often using ATP as the phosphate donor, whereas dephosphorylation is catalyzed by protein phosphatases.

**Proteolytic activation**

Some enzymes are synthesized as larger inactive precursors called proenzymes or zymogens. These are activated by the irreversible hydrolysis of one or more peptide bonds. The pancreatic proteases trypsin, chymotrypsin and elastase are all derived from zymogen precursors (trypsinogen, chymotrypsinogen and proelastase, respectively) by proteolytic activation. Premature activation of these zymogens leads to the condition of acute pancreatitis. The blood clotting cascade also involves a series of zymogen activations that brings about a large amplification of the original signal.

**Regulation of enzyme synthesis and breakdown**

The amount of enzyme present is a balance between the rates of its synthesis and degradation. The level of induction or repression of the gene encoding the enzyme, and the rate of degradation of its mRNA, will alter the rate of synthesis of the enzyme protein. Once the enzyme protein has been synthesized, the rate of its breakdown (half-life) can also be altered as a means of regulating enzyme activity.

| **Related topics** | Introduction to enzymes (C1) | Enzyme inhibition (C4) |
|---|---|---|
| | Enzyme kinetics (C3) | |

## Feedback regulation

In biological systems the rates of many enzymes are altered by the presence of other molecules such as activators and inhibitors (collectively known as **effectors**). A common theme in the control of metabolic pathways is when an enzyme early on in the pathway is inhibited by an end-product of the metabolic pathway in which it is involved. This is called **feedback inhibition** and often takes place at the **committed step** in the pathway (conversion of A to B in *Fig. 1a*). The committed step is the first step to produce an intermediate which is unique to the pathway in question, and therefore normally commits the metabolite to further metabolism along that pathway. Control of the enzyme which carries out the committed step of a metabolic pathway conserves the metabolic energy supply of the organism, and prevents the build up of large quantities of unwanted metabolic intermediates further along the pathway.

As many metabolic pathways are **branched**, feedback inhibition must allow the synthesis of one product of a branched pathway to proceed even when another is present in excess. Here a process of **sequential feedback inhibition** may operate where the end-product of one branch of a pathway will inhibit the first enzyme after the branchpoint (the conversion of C to D or C to E in *Fig. 1b*). When this branchpoint intermediate builds up, it in turn inhibits the first committed step of the whole pathway (conversion of A to B in *Fig. 1b*). Since the end-product of a metabolic pathway involving multiple enzyme reactions is unlikely to resemble the starting compound structurally, the end-product will bind to the enzyme at the control point at a site other than the active site. Such enzymes are always **allosteric enzymes**.

## Allosteric enzymes

A plot of $V_0$ against [S] for an **allosteric enzyme** gives a **sigmoidal curve** rather than the hyperbolic plots predicted by the Michaelis–Menten equation (see Topic C3 and *Fig. 4*). The curve has a steep section in the middle of the substrate concentration range, reflecting the rapid increase in enzyme velocity which occurs over a narrow range of substrate concentrations. This allows allosteric enzymes to be particularly sensitive to small changes in substrate concentration within the physiological range. In allosteric enzymes, the binding of a substrate molecule to one active site affects the binding of substrate molecules to other active sites in the enzyme; the different active sites are said to

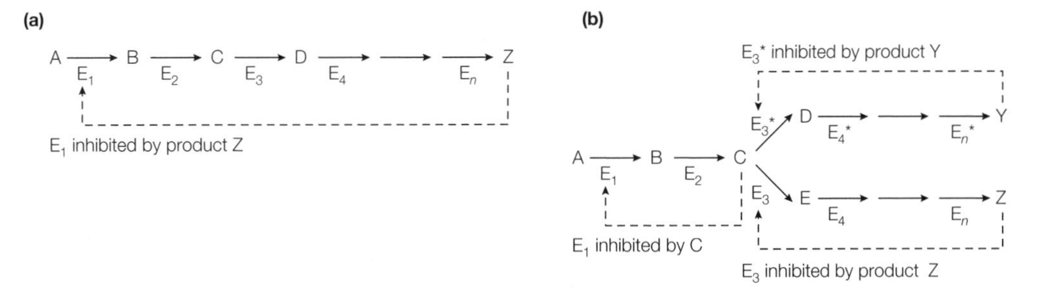


*Fig. 1. Feedback inhibition (a) and sequential feedback inhibition (b) in metabolic pathways.*

behave **cooperatively** in binding and acting on substrate molecules (cf. the binding of $O_2$ to the four subunits of hemoglobin; Topic B4). Thus allosteric enzymes are often multi-subunit proteins, with one or more active sites on each subunit. The binding of substrate at one active site **induces a conformational change** in the protein that is conveyed to the other active sites, altering their affinity for substrate molecules.

In addition, allosteric enzymes may be controlled by **effector** molecules (activators and inhibitors) that bind to the enzyme at a site other than the active site (either on the same subunit or on a different subunit), thereby causing a change in the conformation of the active site which alters the rate of enzyme activity (cf. the binding of $CO_2$, $H^+$ and 2,3-bisphosphoglycerate to hemoglobin; see Topic B4). An **allosteric activator** increases the rate of enzyme activity, while an **allosteric inhibitor** decreases the activity of the enzyme.

### *Aspartate transcarbamoylase*

**Aspartate transcarbamoylase** (aspartate carbamoyltransferase; ATCase), a key enzyme in **pyrimidine biosynthesis** (see Topic F1), provides a good example of allosteric regulation. ATCase catalyzes the formation of ***N*-carbamoylaspartate** from aspartate and carbamoyl phosphate, and is the committed step in pyrimidine biosynthesis (*Fig. 2*). The binding of the two substrates aspartate and carbamoyl phosphate is cooperative, as shown by the **sigmoidal curve** of $V_0$ against substrate concentration (*Fig. 3*).

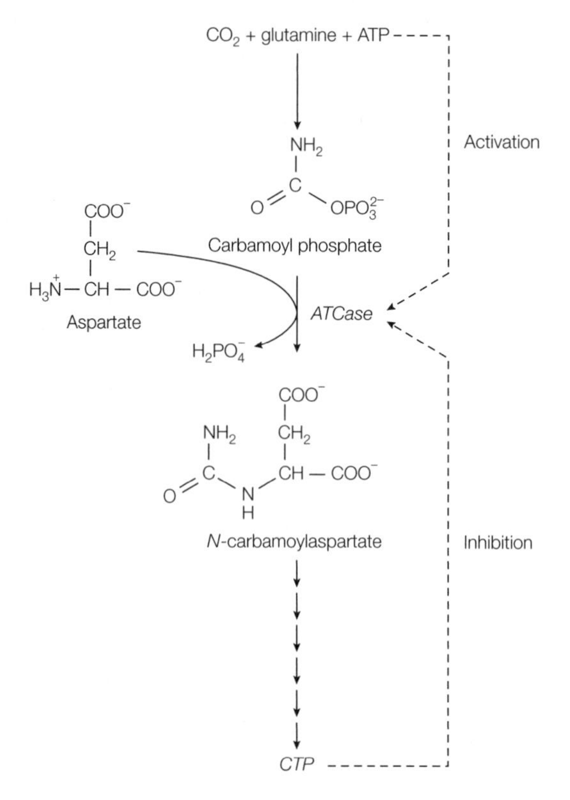


*Fig. 2. Formation of N-carbamoylaspartate by aspartate transcarbamoylase (ATCase) is the committed step in pyrimidine biosynthesis and a key control point.*

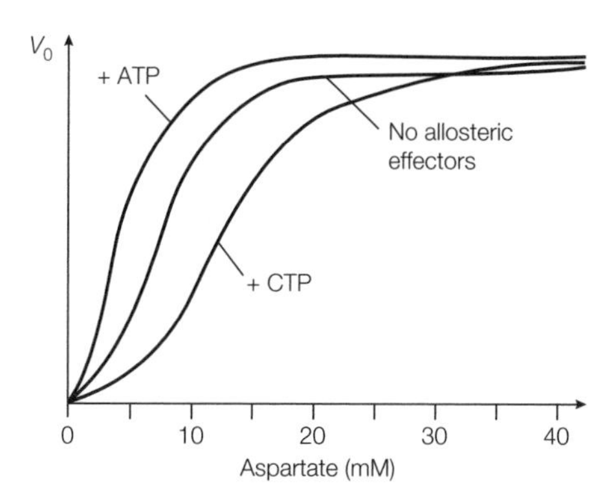


*Fig. 3. Plot of initial reaction velocity ($V_0$) against substrate concentration for the allosteric enzyme aspartate transcarbamoylase.*

ATCase consists of six **catalytic subunits** and six **regulatory subunits**. The enzyme is **feedback-inhibited** by the end-product of the pathway, **cytosine triphosphate** (**CTP**; see Topic F1) which acts as an **allosteric inhibitor** (*Fig. 2*). This molecule binds to the regulatory subunits and causes a decrease in the catalytic activity of ATCase by decreasing the affinity of the catalytic subunits for substrate molecules. In contrast, **ATP**, one of the intermediates earlier on in the pathway, acts as an **allosteric activator**, enhancing the affinity of ATCase for its substrates and leading to an increase in activity (*Fig. 2*). ATP competes with the same binding site on the regulatory subunit as CTP. High levels of ATP signal to the cell that energy is available for DNA replication, and so ATCase is activated, resulting in the synthesis of the required pyrimidine nucleotides. When pyrimidines are abundant, the high levels of CTP inhibit ATCase, preventing needless synthesis of *N*-carbamoylaspartate and subsequent intermediates in the pathway.

## Reversible covalent modification

**Reversible covalent modification** is the making and breaking of a covalent bond between a nonprotein group and an enzyme molecule. Although a range of nonprotein groups may be reversibly attached to enzymes which affect their activity, the most common modification is the addition and removal of a **phosphate group** (**phosphorylation** and **dephosphorylation**, respectively). Phosphorylation is catalyzed by **protein kinases**, often using ATP as the phosphate donor, and dephosphorylation is catalyzed by **protein phosphatases** (*Fig. 4*). The addition and removal of the phosphate group causes changes in the **tertiary structure** of the enzyme that alter its catalytic activity. One class of protein kinases transfers the phosphate specifically on to the hydroxyl group of Ser or Thr residues on the target enzyme [serine/threonine protein kinases, typified by 3′,5′-cyclic adenosine monophosphate (cAMP)-dependent protein kinase], while a second class transfers the phosphate on to the hydroxyl group of Tyr residues (tyrosine kinases). Protein phosphatases catalyze the hydrolysis of phosphate groups from proteins to regenerate the unmodified hydroxyl group of the amino acid and release $P_i$ (*Fig. 4*).

A phosphorylated enzyme may be either more or less active than its dephosphorylated form. Thus phosphorylation/dephosphorylation may be used as a rapid, reversible switch to turn a metabolic pathway on or off according to the needs of the cell. For example, **glycogen phosphorylase**, an enzyme involved

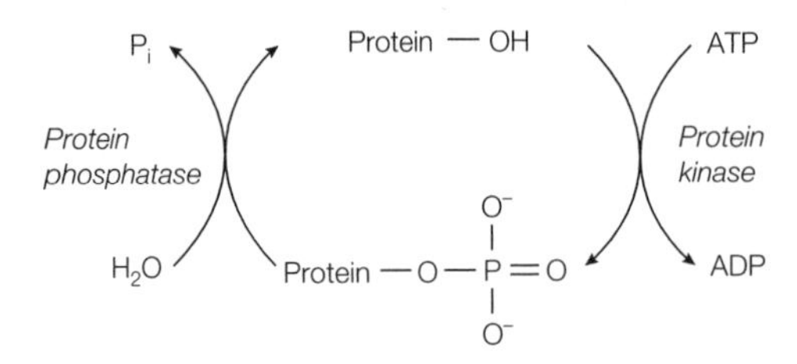


*Fig. 4. The reversible phosphorylation and dephosphorylation of an enzyme.*

in glycogen breakdown, is active in its phosphorylated form, and **glycogen synthase**, involved in glycogen synthesis, is most active in its unphosphorylated form (see Topic J7).

Other types of reversible covalent modification that are used to regulate the activity of certain enzymes include **adenylylation** (the transfer of adenylate from ATP) and **ADP-ribosylation** [the transfer of an adenosine diphosphate (ADP)-ribosyl moiety from $NAD^+$].

## Proteolytic activation

Several enzymes are synthesized as larger inactive precursor forms called **proenzymes** or **zymogens**. Activation of zymogens involves **irreversible hydrolysis** of one or more **peptide bonds**.

### *Pancreatic proteases*

The digestive enzymes **trypsin**, **chymotrypsin** and **elastase** (see Topic C1) are produced as zymogens in the **pancreas**. They are then transported to the small intestine as their zymogen forms and activated there by cleavage of specific peptide bonds. Trypsin is synthesized initially as the zymogen **trypsinogen**. It is cleaved (and hence activated) in the intestine by the enzyme **enteropeptidase** which is only produced in the intestine. Once activated, trypsin can cleave and activate further trypsinogen molecules as well as other zymogens, such as **chymotrypsinogen** and **proelastase** (*Fig. 5*).

Chymotrypsin is initially synthesized as the zymogen chymotrypsinogen, a single polypeptide chain of 245 amino acid residues (*Fig. 6*). On reaching the intestine, chymotrypsinogen is cleaved first by trypsin on the C-terminal side of Arg15 to form π-chymotrypsin which is fully active (*Fig. 6*). Two dipeptides are removed from within the polypeptide chain of π-chymotrypsin by other chymotrypsin molecules, producing a more stable form of chymotrypsin, known as δ-chymotrypsin. δ-Chymotrypsin undergoes conformational changes to produce the mature active α-chymotrypsin (*Fig. 6*). The three fragments of the

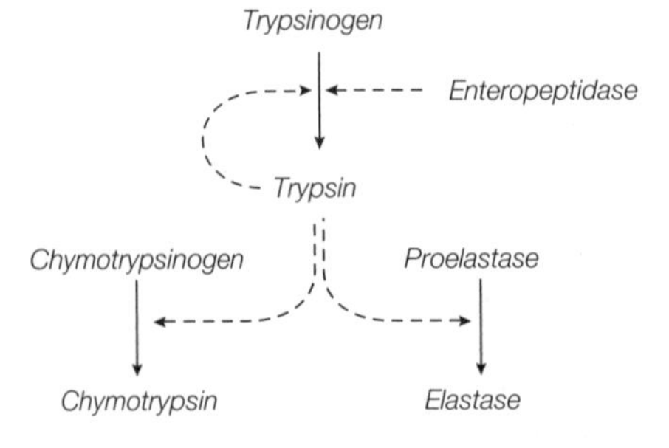


*Fig. 5. The central role of trypsin in activating the pancreatic zymogens.*

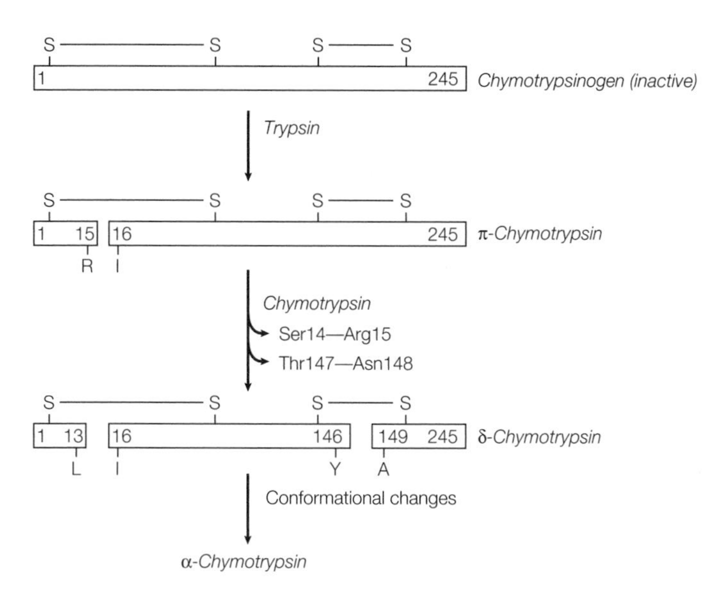


*Fig. 6. Activation of chymotrypsinogen by proteolytic cleavage.*

original single polypeptide chain are held together in α-chymotrypsin by both noncovalent interactions and covalent disulfide bonds.

As well as synthesizing and secreting the zymogens, the pancreas also synthesizes a small **trypsin-inhibitor protein**. This inhibitor protein binds very tightly to the active site of trypsin, preventing the pancreas from being destroyed by prematurely activated trypsin molecules. If this safety mechanism fails, for example because of a blocked pancreatic duct, the zymogens can become activated and literally digest the pancreas, a condition known as **acute pancreatitis**.

### *Blood clotting cascade*

Another example of the occurrence of inactive zymogens is found in the enzymes involved in the **blood clotting cascade**. Here the whole process of blood clotting is brought about by a **series of zymogen activations**. Zymogen activation may produce a **large amplification** of the initial signal, as a single activated enzyme may act on many thousands of substrate molecules to bring about further activation. Since proteolytic cleavage does not require ATP, zymogen cleavage is a particularly appropriate mechanism for activation of proteins outside cells. However, unlike the covalent modification of an enzyme (see above), zymogen activation is not reversible. Once activated, the enzyme stays active.

## Regulation of enzyme synthesis and breakdown

The amount of a particular enzyme present in a cell or tissue changes according to the rates of its **synthesis** and **degradation**.

Factors affecting the rate of synthesis include the level of **induction or repression of the gene** encoding the enzyme (see Topics G3 and G4 and also the **rate of degradation of the mRNA** produced from that gene. Many key enzymes at control points in metabolic pathways have particularly short-lived mRNAs and the rate of enzyme synthesis is thus readily controlled by factors that affect the rate of gene transcription.

The rate of degradation of an enzyme is reflected in its **half-life** – the time taken for 50% of the protein to be degraded. Most enzymes that are important in metabolic regulation have short half-lives, and are termed **labile** enzymes.

# D1 THE IMMUNE SYSTEM

## Key Notes

**Functions**

The immune system has two main functions; to recognize invading pathogens and then to trigger pathways that will destroy them. The humoral immune system relies on B lymphocytes to produce soluble antibodies that will bind the foreign antigens. The cellular immune system uses killer T lymphocytes that recognize and destroy invading cells directly.

**Primary and secondary immune responses**

The primary immune response occurs on initial contact with a foreign antigen and results in production of immunoglobulin M (IgM) and then immunoglobulin G (IgG). If the same antigen is encountered again, immunological memory leads to a secondary immune response that produces a much more rapid and larger increase in specific IgG production.

**Clonal selection theory**

A large number of antibody-producing cells exist in an animal even before it encounters a foreign antigen, each cell producing only one specific antibody and displaying this on its cell surface. An antigen binds to cells that display antibodies with appropriate binding sites and causes proliferation of those cells to form clones of cells secreting the same antibody in high concentration.

**Self-tolerance**

Cells that produce antibody that reacts with normal body components are killed early in fetal life so that the adult animal normally is unable to make antibodies against self, a condition called self-tolerance.

**Complement**

Antibodies bound to an invading microorganism activate the complement system via the classical pathway. This consists of a cascade of proteolytic reactions leading to the formation of membrane attack complexes on the plasma membrane of the microorganism that cause its lysis. Polysaccharides on the surface of infecting microorganisms can also activate complement directly in the absence of antibody via the alternative pathway.

**Related topics**

Antibody structure (D2)
Polyclonal and monoclonal antibodies (D3)
Antibody synthesis (D4)
Membrane transport: macromolecules (E4)

## Functions

There are two vital functions of the immune system; recognition of an invading pathogen (disease-producing bacteria, fungi, protozoa and viruses) as being distinct from normal body components (and hence treated as foreign) and then the triggering of pathways that lead to destruction of the invader, such as activation of complement (see below) and phagocytic cells that engulf and digest the invading organism. The immune system may also be able to recognize and destroy abnormal cells that arise spontaneously in the body which would otherwise lead to cancer, but the significance of this phenomenon in protecting against human tumors is still debatable. The key cells responsible for immunity in vertebrates are white blood cells called lymphocytes which arise from precursor (stem) cells in the bone marrow. There are two main parts of the immune system which interact to provide overall protection for the animal:

- the **humoral immune response** (*humor* is an ancient term meaning fluid) relies on the production of soluble proteins called antibodies (or immunoglobulins) by **B lymphocytes**, so called because the cells mature in the bone marrow. As a common shorthand nomenclature, B lymphocytes are often called simply **B cells**.
- The **cellular immune response** is mediated by **T lymphocytes**, so called because their maturation from stem cells occurs in the thymus. In cellular immunity it is the intact T lymphocytes themselves that are responsible for the recognition and killing of foreign invaders. These cells are the **cytotoxic T lymphocytes (CTL)**, also called **killer T cells**. Other T lymphocytes have another role; they provide essential help for B lymphocytes to produce antibodies and so are called **helper T cells**.

In both cellular and humoral immunity, recognition of the foreign invader depends upon the recognition of foreign macromolecules (proteins, carbohydrates, nucleic acids); these foreign components are called **antigens**.

## Primary and secondary immune responses

The presence of a foreign antigen stimulates the production of a specific antibody in the bloodstream which will recognize and bind tightly to it. Antibody molecules fall into five main classes, as defined by their precise structure (see Topic D2). The antibody molecules first produced after antigen injection are in the immunoglobulin M class and so are called IgM molecules. However, about 10 days after antigen injection, the amount of IgM in the bloodstream (the **titer** of antibody) declines and there is a concurrent increase in another class of antibody called immunoglobulin G (IgG); see *Fig. 1*. This is called the **primary immune response**.

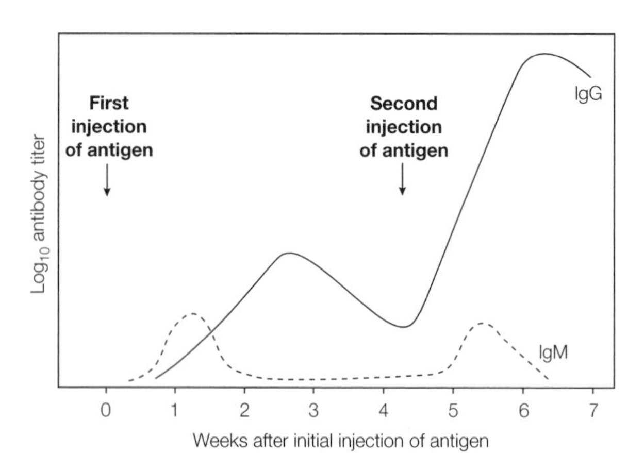


*Fig 1. The primary and secondary immune responses to injections of antigen.*

One of the most important features of the immune system is that, once an animal has encountered a particular pathogen, the system confers protection against future infection. This **immunological memory** means that if the same pathogen or antigen is encountered a second time, perhaps even decades after the previous occurrence, then the system reacts much faster and more dramatically to produce a large titer of specific IgG to counter the antigen. This is the **secondary immune response** (see *Fig. 1*) and is mediated by long-lived memory T cells and memory B cells.

### Clonal selection theory

The **clonal selection theory** explains the operation of the humoral immune system as follows.

- Even before an animal meets a foreign antigen, each immature antibody-producing cell makes one (and only one) specific kind of antibody molecule and anchors this in the plasma membrane so that it is exposed on the cell surface. A large number of such antibody-producing cells exist in the animal which collectively express a very wide range of antibody specificities.
- If a foreign antigen is encountered, this will bind to those antibody-producing cells which are displaying antibody that has an appropriate binding site for that antigen. Each cell to which the antigen binds is stimulated to divide to form a population of identical cells called a **clone**. Since all the cells of the clone are genetically identical, they all produce antibody of the same specificity. The antibody now synthesized is no longer inserted into the plasma membrane but instead is secreted. Thus, when an antigen is encountered, it 'selects' certain antibody-producing cells for cell division (**clonal selection**) and antibody production simply on the basis of which cell-surface antibodies bind that antigen. The specificity of the antibody produced is thus exquisitely tailored to the particular antigen(s) encountered.

### Self-tolerance

In fetal life, if an immature antibody-producing cell displays cell-surface antibody that binds a normal body component, then the cell dies. Thus, usually, in the adult animal, no cells exist that can make antibodies against the animal's own macromolecules. This inability to make antibodies against self is called **self-tolerance**. However, in some disease states (the so-called **autoimmune diseases**) the immune system loses its tolerance against self-antigens.

### Complement

When the recognition function of the humoral immune system has been carried out by the production of specific antibodies and their binding to foreign antigens, destruction of the invading pathogen is the next step. One main defense pathway is the **complement system** which is activated by antibodies bound to the invading microorganism and eventually causes it to lyse by punching holes in its plasma membrane.

The complement system consists of about 20 interacting soluble proteins that circulate in the blood and extracellular fluid. Immunoglobulin molecules bound to the surface of the microorganisms activate **C1**, the first component of the complement pathway. The activation occurs through the Fc portion (see Topic D2) of the bound antibody. Only bound antibody can activate complement, soluble antibody not bound to an antigen has no such effect.

The early components of the complement pathway, including C1, are proteases that activate their substrate by limited cleavage. Activated C1 now activates several molecules of the next component by proteolysis, each of which activates several molecules of the next component by proteolysis, and so on. Therefore, the early steps in complement activation consist of a **proteolytic cascade** in which more and more molecules are activated at each step. Component C3 is the key component whose cleavage leads to the assembly of

**membrane attack complexes** on the plasma membrane of the microorganisms, which create holes in the plasma membrane that lead to cell death. Various white blood cells also become activated during this process and phagocytose (see Topic E4) the pathogen.

This pathway of complement activation, which starts when antibody has bound to antigen on the microbe surface, is called the **classical pathway**. An **alternative pathway** of activation also exists which is activated directly by polysaccharides in the cell wall of microorganisms even in the absence of antibody. The alternative pathway therefore defends the body against attack in the early stages before an immune response can occur and also augments the effects of the classical pathway of complement activation when the immune response has occurred.

# D2 ANTIBODY STRUCTURE

## Key Notes

**Light and heavy chains**

Each IgG antibody molecule consists of four polypeptide chains (two identical light chains and two identical heavy chains joined by disulfide bonds) and has two antigen-binding sites (i.e. is bivalent).

**Variable and constant regions**

Each light chain and each heavy chain consists of a variable region and a constant region. Variability in the variable regions is largely confined to three hypervariable regions; the remaining parts of the variable regions are far less variable and are called the framework regions.

**Antibody domains**

Each light chain folds into two domains, one for the variable region and one for the constant region. Each IgG heavy chain folds into four domains, one for the variable region and three in the constant region.

**Fab and Fc fragments**

Papain digests IgG into two Fab fragments (each of which has an antigen-binding site, i.e. is univalent) and one Fc fragment (that carries effector sites for complement activation and phagocytosis). Pepsin digests IgG to release an $F(ab')_2$ fragment that has two antigen-binding sites.

**Five classes of immunoglobulins**

Human immunoglobulins exist as IgA, IgD, IgE, IgG and IgM classes which contain $\alpha$, $\delta$, $\epsilon$, $\gamma$ and $\mu$ heavy chains, respectively. IgM is a pentamer that binds to invading microorganisms and activates complement killing of the cells and phagocytosis. IgG is the main antibody found in the blood after antigen stimulation and also has the ability to cross the placenta. IgA mainly functions in body secretions. IgE provides immunity against some parasites but is also responsible for the clinical symptoms of allergic reactions. The role of IgD is unknown. All antibody molecules contain either kappa ($\kappa$) or lambda ($\lambda$) light chains.

**Related topics**

The immune system (D1)
Polyclonal and monoclonal antibodies (D3)
Antibody synthesis (D4)

## Light and heavy chains

Each molecule of immunoglobulin G (IgG) is Y-shaped and consists of four polypeptide chains joined together by disulfide bonds; two identical copies of **light (L) chains** about 220 amino acids long and two identical copies of **heavy (H) chains** about 440 amino acids long (*Fig. 1a*). The N-terminal ends of one heavy chain and its neighboring light chain cooperate to form an antigen binding site, so that the IgG molecule has two binding sites for antigen, that is, it is **bivalent**. Because of this, a single antibody molecule can bind two antigen molecules and so cross-link and precipitate antigens out of solution.

## Variable and constant regions

Comparison of the amino acid sequences of many immunoglobulin polypeptides has shown that each light chain has a **variable region** at its N-terminal end

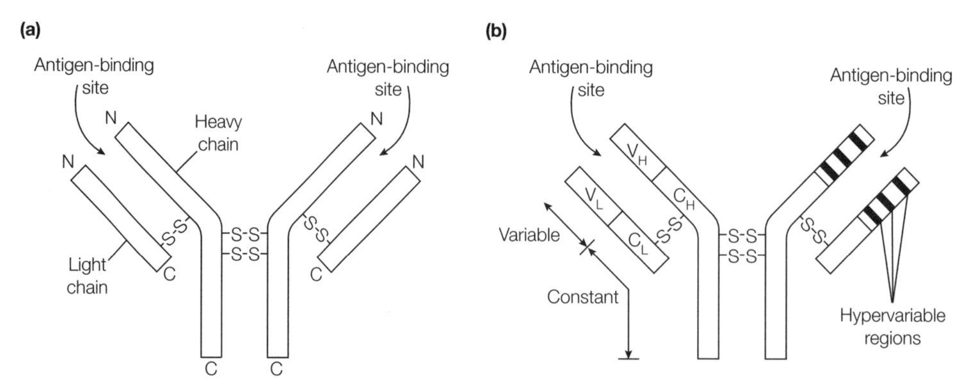


*Fig. 1. Structure of an antibody molecule. (a) Each antibody molecule consists of two identical light chains and two identical heavy chains. The molecule has two antigen-binding sites, each formed by a light chain and a heavy chain. (b) The N-terminal regions of the light and heavy chains are variable in amino acid sequence from antibody to antibody (variable regions; V regions) whilst the C-terminal regions are relatively constant in sequence (constant regions; C regions). The generic terms for these regions in the light chain are $V_L$ and $C_L$ and for the heavy chains are $V_H$ and $C_H$.*

and an invariant or **constant region** at its C terminus (*Fig. 1b*). Similarly each heavy chain has an N-terminal variable region and a C-terminal constant region. Since it is the N-terminal parts of the light and heavy chains that form the antigen-binding site, the variability in amino acid sequence of these regions explains how different sites with different specificities for antigen binding can be formed. In fact, the variability in the variable regions of both light and heavy chains is mainly localized to three **hypervariable regions** in each chain (*Fig. 1b*). In the three-dimensional structure of the immunoglobulin molecule, the hypervariable parts of the light and heavy chains are looped together to form the antigen-binding site. The remaining parts of each variable region stay reasonably constant in sequence, usually do not contact the antigen directly and are called **framework regions**.

## Antibody domains

Each light chain consists of two repeating segments of about 110 amino acids that fold into two compact three-dimensional **domains**, one representing the variable region of the light chain and the other domain representing the constant region. Each heavy chain is also made up of repeating units about 110 amino acids long. Since each IgG heavy chain is about 440 amino acids long, it forms four domains, one domain for the variable region and three domains in the constant region. The similarity of amino acid sequence between the various domains suggests that they arose in evolution by gene duplication.

## Fab and Fc fragments

Papain, a protease, cuts the IgG molecule to release the two arms of the Y-shaped molecule, each of which has one antigen-binding site and is called an **Fab fragment** (Fragment antigen binding) (*Fig. 2*). Because Fab fragments have only one antigen-binding site (i.e. are **univalent**), they cannot cross-link antigens. The released stem of the Y-shaped molecule (consisting of the identical C-terminal parts of the two H chains) is named the **Fc fragment** (so called because it readily crystallizes). The Fc fragment carries the effector sites that trigger the destruction of the antigen, for example triggering of the complement system (see Topic D1) and inducing phagocytosis of pathogens by other white blood cells. In contrast to papain, pepsin (another protease) cuts the IgG

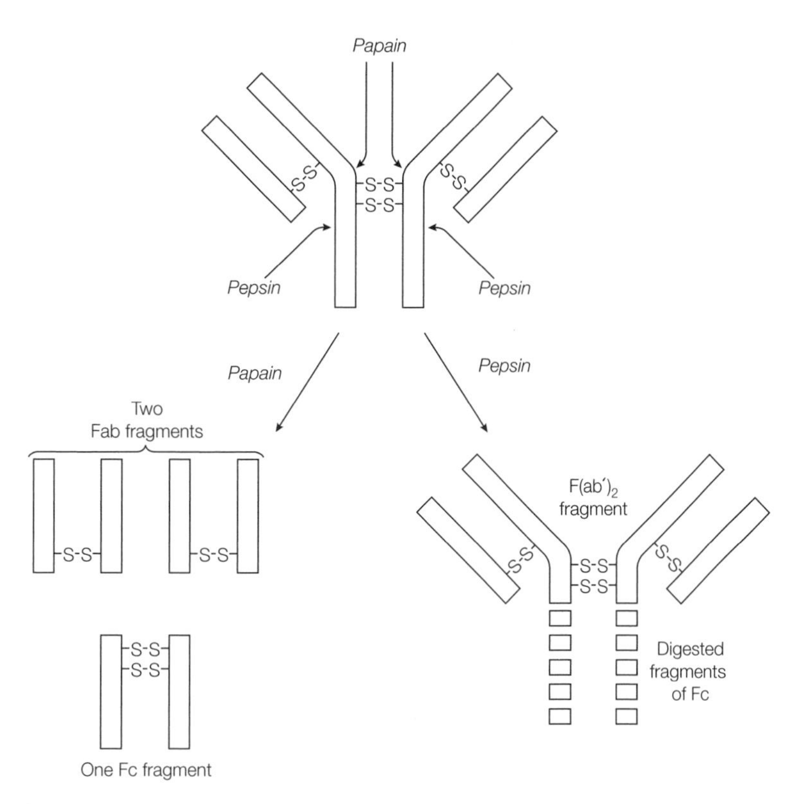


*Fig. 2. Papain digestion of an antibody molecule yields two univalent Fab fragments and an Fc fragment whereas pepsin digestion yields a bivalent F(ab')$_2$ fragment.*

molecule to release the two arms of IgG still linked together and hence this fragment has two antigen-binding sites (i.e. is bivalent) and can still cross-link antigens. This is called the **F(ab')$_2$ fragment** (*Fig. 2*).

## Five classes of immunoglobulins

Humans have five different classes of antibody molecule which differ both in structure and in function. These are called immunoglobulin A (IgA), IgD, IgE, IgG and IgM and each has its own type of heavy chain; $\alpha$, $\delta$, $\epsilon$, $\gamma$ and $\mu$, respectively. Thus IgA molecules have two identical $\alpha$ heavy chains, IgD molecules have two identical $\delta$ heavy chains, etc. The human IgG class of antibodies is further divided into four IgG subclasses; $IgG_1$, $IgG_2$, $IgG_3$ and $IgG_4$, having $\gamma_1$, $\gamma_2$, $\gamma_3$ and $\gamma_4$ heavy chains respectively.

The different heavy chains confer different properties and functions on each of the immunoglobulin classes:

- **IgM** has $\mu$ heavy chains and exists as a pentamer in combination with another polypeptide called the J chain, which is responsible for initiating the polymerization to form the pentameric structure. With its large number of antigen-binding sites, each IgM molecule binds very tightly to any pathogen that has multiple copies of the same antigen on its surface. The binding induces the Fc region to activate the complement pathway which eventually causes the death of the pathogen. IgM also activates macrophages to phagocytose pathogens. Not surprisingly given these functions, IgM is the first antibody produced when an animal responds to a new antigen.

- **IgG** is the main immunoglobulin in the bloodstream late in the primary immune response and particularly during the secondary immune response (see Topic D1). Like IgM, it can activate complement and trigger macrophages, but is the only antibody that can pass through the placenta and so provide immunological protection for the fetus. It is also secreted into the mother's milk and is taken up from the gut of the newborn animal into the bloodstream, thus providing continuing protection after birth.
- **IgA** is the main class of antibody in secretions such as tears, saliva, and in secretions of the lungs and the intestine. It is the first line of immunological defense against infection at these sites.
- **IgE** occurs in tissues where, having bound the antigen, it stimulates mast cells to release a range of factors. Some of these in turn activate white blood cells (called eosinophils) to kill various types of parasite. However, the mast cells can also release biologically active amines, including histamine, which cause dilation and increased permeability of blood vessels and lead to the symptoms seen in allergic reactions such as hay fever and asthma.
- **IgD** is found on the surface of mature B lymphocytes and in traces in various body fluids, but its exact function remains unclear.

Two different forms of light chains also exist. Antibody molecules in any of the antibody classes or sub-classes can have either two κ light chains or two λ light chains. Unlike the different heavy chains described above, no difference in biological function between κ and λ light chains is known.

# D3 POLYCLONAL AND MONOCLONAL ANTIBODIES

## Key Notes

**Polyclonal antibodies**

A preparation of antibody molecules that arises from several different clones of cells is called a polyclonal antibody. It is a mixture of antibody molecules that bind to different parts of the antigen and with different binding affinities.

**Monoclonal antibodies**

Antibody produced by a single clone of cells is a monoclonal antibody; all the antibody molecules are identical and bind to the same antigenic site with identical binding affinities. Monoclonal antibodies can be generated in large amounts by creating a cell fusion (called a hybridoma) between an antibody-producing cell and a myeloma cell.

**Related topics**

The immune system (D1)
Antibody structure (D2)
Antibody synthesis (D4)
Antibodies as tools (D5)

## Polyclonal antibodies

If an antigen is injected into an animal, a number of antibody-producing cells will bind that antigen (see Topic D1), albeit with varying degrees of affinity, and so the antibody which appears in the bloodstream will have arisen from several clones of cells, that is it will be a **polyclonal antibody**. Different antibody molecules in a preparation of polyclonal antibody will bind to different parts of the macromolecular antigen and will do so with different binding affinities. The binding region recognized by any one antibody molecule is called an **epitope**. Most antibodies recognize particular surface structures in a protein rather than specific amino acid sequences (i.e. the epitopes are defined by the conformation of the protein antigen). A preparation of polyclonal antibodies will bind to many epitopes on the protein antigen.

## Monoclonal antibodies

If a single clone of antibody-producing cells (see above and Topic D4) could be isolated, then all of the antibody produced from that clone would be identical; all antibody molecules in such a **monoclonal antibody** preparation would bind to the same antigen epitope.

The problem is that if an individual antibody-producing cell is isolated and grown in culture, its descendants have a limited lifespan that severely limits their use for the routine preparation of monoclonal antibodies. In 1975, Milstein and Köhler discovered how monoclonal antibodies of almost any desired antigen specificity can be produced indefinitely and in large quantities. Their method was to fuse a B lymphocyte producing antibody of the desired specificity with a cell derived from a cancerous lymphocyte tumor, called a myeloma cell, which is immortal. The cell fusion is called a **hybridoma**, which is both immortal and secretes the same specific antibody originally encoded by the B lymphocyte.

Monoclonal antibodies produced using this technology are now common tools in research because of their very high specificity. For example, they can be used to locate particular molecules within cells or particular amino acid sequences within proteins. If they are first bound to an insoluble matrix, they are also extremely useful for binding to and hence purifying the particular molecule from crude cell extracts or fractions (see Topic D5). They are also increasingly of use in medicine, both for diagnosis and as therapeutic tools, for example to inactivate bacterial toxins and to treat certain forms of cancer.

# D4 ANTIBODY SYNTHESIS

## Key Notes

**Somatic recombination**

No complete antibody gene exists in germ-line cells. The genes for light chains and heavy chains assemble by somatic recombination during B-lymphocyte maturation.

**Recombination of light chain genes**

In the germ-line, each light chain gene exists as multiple V and J gene segments upstream of a single C gene segment. During B-lymphocyte differentiation, one V gene segment joins with one J gene segment (VJ joining) to assemble the complete light chain gene, usually by deletion of intervening DNA.

**Recombination of heavy chain genes**

Heavy chains are encoded by multiple V, J and D gene segments which lie upstream of a single copy of C gene segments for each of the constant regions of $\mu$, $\delta$, $\gamma$, $\epsilon$ and $\alpha$ chains. During B-lymphocyte differentiation, a D gene segment joins a J segment (DJ joining) and then the recombined DJ joins a V gene segment (VDJ joining).

**Class switching**

A B lymphocyte can change the class of antibody being expressed by moving a new C gene segment into position after the recombined VDJ segment, deleting the intervening DNA. The new heavy chain has a different constant region but retains the same antigen-binding specificity of the previous heavy chain.

**Related topics**

The immune system: (D1)
Antibody structure (D2)
Polyclonal and monoclonal antibodies (D3)

## Somatic recombination

In most animals, it is possible to distinguish **germ-line** cells from **somatic** cells. The germ-line cells are those that give rise to the male and female **gametes** (sperm and ova, respectively) whilst the somatic cells form the rest of the body structures of the individual animal. The importance of the germ-line cells is that it is these that ultimately give rise to the next generation.

The human genome is thought to contain fewer than $10^5$ genes yet a human can make at least $10^{15}$ different types of antibody in terms of antigen-binding specificity. Clearly the number of genes is far too small to account for most of this antibody diversity. Thus a **germ-line hypothesis**, whereby all antibodies are encoded by genes in germ-line cells, must be incorrect. In fact, there are no complete genes in germ-line cells that encode complete light chains or complete heavy chains. Instead the genes exist in separate coding sections and are assembled during B-lymphocyte maturation by a process called **somatic recombination.** This process of assembly takes place in every B lymphocyte. By assembling different fragments of DNA, completely new immunoglobulin genes can be created and hence this gives an enormous potential reservoir of antibody diversity.

## Recombination of light chain genes

The variable (V) region of a **κ light chain** is encoded by a separate DNA sequence from that encoding the constant (C) region. These two segments lie on the same chromosome but they are sited some way apart. During maturation of the B lymphocyte, the V and C region DNAs are moved next to each other and are joined together to create a functioning light chain gene (*Fig. 1*). This somatic recombination usually occurs by deletion of the DNA between the germ-line V and C regions but can also occur in some cases by an inversion mechanism.

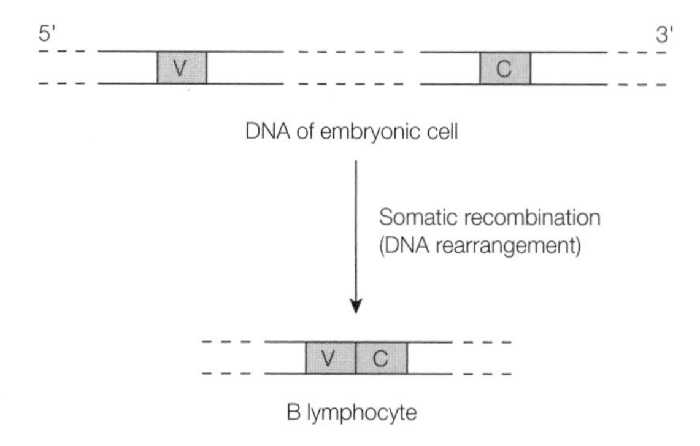


*Fig. 1. In the germ-line (embryo) DNA, sequences coding for the variable (V) region lie distant from those encoding the constant (C) region. During the differentiation of B lymphocytes, these two sequences are brought together to form an active antibody gene by deletion of the intervening DNA (somatic recombination).*

In fact, the situation is rather more complex than this simple model implies. The germ-line **V gene segment** (shown in *Fig. 1*) encodes only the first 95 amino acids of the variable region of the light chain polypeptide. The remaining few amino acids of the light chain variable region (residues 96–108) are encoded by a piece of DNA called the **J gene segment** (*Fig. 2*) This J segment (for 'joining') must not be confused with the J polypeptide in IgM pentamers (see Topic D2). In the germ-line, the J gene segment lies just upstream of the **C gene segment** and separated from it only by an intron (*Fig. 3*). Furthermore, there are multiple V gene segments (about 300) and five J gene segments (*Fig. 3*), one of them inactive.

During B-lymphocyte differentiation, one of the 300 or so V regions becomes joined precisely to one of the J gene segments to create a light chain gene. This

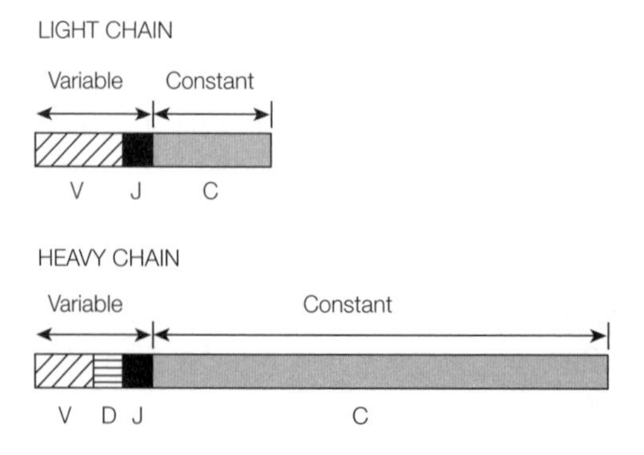


*Fig. 2. The light chain variable region is encoded by two separate gene segments, V and J. The heavy chain variable region is encoded by three gene segments, V, D and J.*

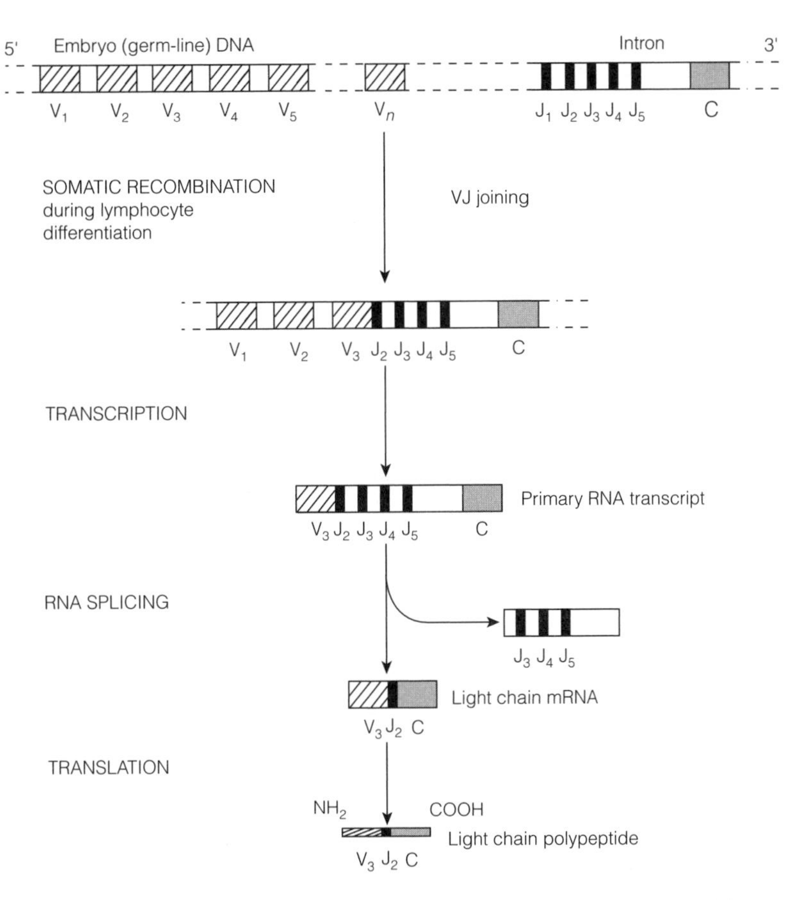


*Fig. 3. Somatic recombination to create a κ light chain gene and expression of that gene to produce κ light chain polypeptides.*

somatic recombination process is called **VJ joining**. In *Fig. 3*, $V_3$ has been chosen for recombination and has joined to $J_2$. Transcription now starts just upstream of the recombined V segment (just upstream of $V_3$ in *Fig. 3*) and continues until the end of the C segment. The other J segment sequences are also transcribed, but these sequences are lost during subsequent RNA splicing that removes the intron upstream of the C segment (*Fig. 3*). Thus the final mRNA contains only $V_3J_2C$ sequences and encodes a corresponding light chain polypeptide. A very large number of different light chains can be made depending on which one of the 300 V segments is chosen and joined to which one of the J gene segments.

The **λ light chain genes** also arise by somatic recombination during maturation of the B lymphocyte, but there are far fewer V and J gene segments than for κ chain genes. Most antibody molecules have κ light chains and not λ light chains.

**Recombination of heavy chain genes**

Heavy chains are synthesized in an analogous manner but are encoded by four gene segments, V, J, D and C (*Fig. 4*). There are about 200–1000 $V_H$ segments, about 15 active $D_H$ segments (D for diversity, H for heavy) and four $J_H$ segments. Thus whereas the variable region of a light chain polypeptide is encoded by

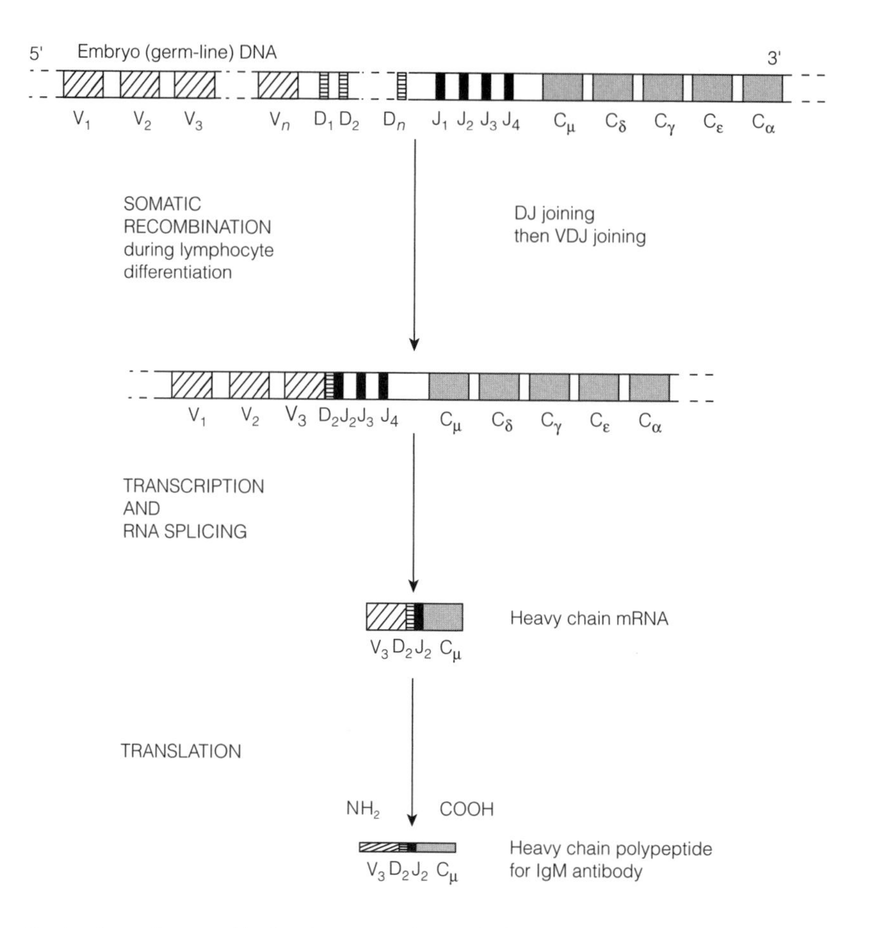


*Fig. 4. Somatic recombination to create a heavy chain gene and expression of that gene to produce heavy chain polypeptides. For simplicity, the constant regions for the various $C_\gamma$ subclasses are not shown.*

V and J segments, the variable region of a heavy chain is encoded by V, D and J segments (*Fig. 2*). In the heavy chain gene system, there are also several C segments, one for each class of heavy chain; $C_\mu$, $C_\delta$, the various $C_\gamma$ subclasses, $C_\epsilon$ and $C_\alpha$ encoding μ, δ, γ, ϵ and α heavy chain constant regions respectively (*Fig. 4*). During lymphocyte maturation, **two heavy chain gene rearrangements occur**. First a chosen $D_H$ segment joins a $J_H$ segment (**DJ joining**) and then the recombined $D_HJ_H$ joins a chosen $V_H$ segment (**VDJ joining**) (*Fig. 4*).

For the assembly of both heavy chain and light chain genes, the ends of the various DNA segments to be joined can also undergo modification during the recombination process and this modifies existing codons at these junctions or even creates new codons, thus increasing antibody diversity still further. In addition, antibody genes exhibit a higher than normal rate of mutation.

## Class switching

When the complete heavy chain gene shown in *Fig. 4* is transcribed, it generates a heavy chain for an IgM antibody since it is always the first C segment after the recombined VDJ (in this case $C_\mu$) that is transcribed. To switch to making a heavy chain for a different antibody class, say IgA, the lymphocyte DNA has to undergo yet another recombination event that moves $C_\alpha$ next to the recombined VDJ and deletes the intervening C segments. When this new gene is expressed, it will synthesize a $C_\alpha$ heavy chain for an IgA antibody instead of the earlier $C_\mu$ heavy chain for an IgM antibody. One crucial point about this **class switching** or **heavy chain switch** is that only the C region of the synthesized heavy chain changes; the variable region stays the same as before the switch. The specificity of the antibody is determined by the antigen-binding site, which is formed by the variable regions of the heavy and light chains and not by the constant regions. Thus even when class switching occurs so that the lymphocyte now makes IgA instead of IgM, the specificity of the antibody for antigens stays the same.

# D5 ANTIBODIES AS TOOLS

## Key Notes

**Immunolocalization methods**

Because of the high specificity of an antibody for its epitope, an antibody raised against a particular protein antigen can be used to determine the location of that antigen in a cell using immunofluorescence light microscopy or immuno-electron microscopy.

**ELISA**

ELISA can be used to quantify the amount of a specific protein antigen in a sample. The antibody is bound to an inert polymer support, then exposed to the sample. Unbound protein is washed away and a second antibody that reacts with the antigen at a different epitope is added. The second antibody used is one that has an enzyme attached to it that converts a colorless or nonfluorescent substrate into a colored or fluorescent product. The amount of second antibody bound, and hence the amount of protein antigen present in the original sample, is determined by quantification of the intensity of color or fluorescence produced.

**Western blotting**

Protein samples are separated by one-dimensional SDS-PAGE or two-dimensional gel electrophoresis in polyacrylamide gels. The separated proteins are then transferred (blotted) to a nitrocellulose or nylon sheet. This is incubated with specific antibody to the protein and then unbound antibody is washed away. Those proteins in the gel that bind the antibody are detected either by autoradiography (if the specific antibody was radiolabeled) or by using a second labeled antibody that binds to the primary antibody.

**Immunoaffinity chromatography**

Immunoaffinity chromatography can be used to purify protein antigens by immobilizing the relevant antibodies on an inert matrix such as polysaccharide beads. When exposed to a protein mixture, only the protein recognized by that antibody will bind to the beads and can be eluted later in pure or almost pure form. Cells bearing the antigen on their surface can also be purified using a similar procedure.

**Related topics**

Microscopy (A3)
Chromatography of proteins (B7)
Electrophoresis of proteins (B8)
The immune system (D1)
Antibody structure (D2)
Polyclonal and monoclonal antibodies (D3)

## Immunolocalization methods

The availability of an antibody (immunoglobulin) against a specific antigen offers the opportunity to use that antibody in a range of immunological methods. The site recognized by an antibody on the antigen is called the **antigenic determinant** or **epitope**. The high specificity of an antibody for its epitope allows it to be used as a reagent for determining the location of the antigen in a cell (**immunocytochemistry**), for example by coupling a fluorescent label to the antibody and then using fluorescence to localize its sites of binding by **immunofluorescence light microscopy** (see also Topic A3). Even higher resolution can be achieved using antibody to which electron-dense particles, such as ferritin or colloidal gold, have been coupled and then viewing the binding sites using electron microscopy (Topic A3). Indeed, **immuno-electron microscopy** can map the position of protein antigens within macromolecular complexes such as ribosomes.

## ELISA

Specific antibodies can also be used to quantify the amount of the corresponding antigen in a biological sample. Several types of immunological assays exist. An increasingly popular version is **enzyme-linked immunosorbent assay (ELISA)** (see *Fig. 1*) which can readily detect and quantify less than a nanogram of a specific antigenic protein. In ELISA, the specific antibody is coupled to a solid support. A convenient format for ELISA is to use a plastic tray that has molded

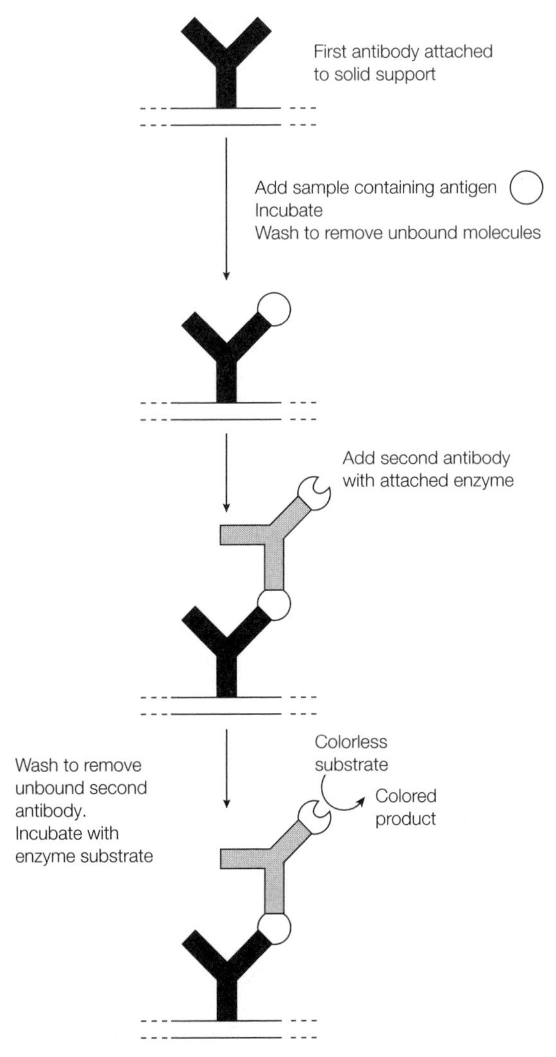


*Fig. 1. ELISA using a second antibody with an attached (conjugated) enzyme that converts a colorless substrate to a colored product.*

wells in it (a **microtiter tray**) where the antibody has been coupled to the plastic forming the wells. Samples to be assayed are added to the wells. If antigen is present that is recognized by the antibody, it becomes bound (*Fig. 1*). The wells are then washed to remove unbound protein and incubated with a second antibody that recognizes the protein but at a different epitope than the first antibody (*Fig. 1*). The second antibody is attached to an enzyme that can catalyze the conversion of a colorless or nonfluorescent substrate into a colored or fluorescent product. The intensity of the color or fluorescence produced for each sample is then measured to determine the amount of antigen present in each sample. Several machines are now commercially available that scan the wells of microtiter plates following ELISA and quantify the amount of antigen bound in each well.

## Western blotting

Western blotting can be used for detection of one or more antigens in a mixture. The sample is electrophoresed on an SDS–polyacrylamide gel (SDS-PAGE; see Topic B8) that separates the proteins on the basis of size, resulting in a series of protein bands down the gel (*Fig. 2*). Because the gel matrix does not let

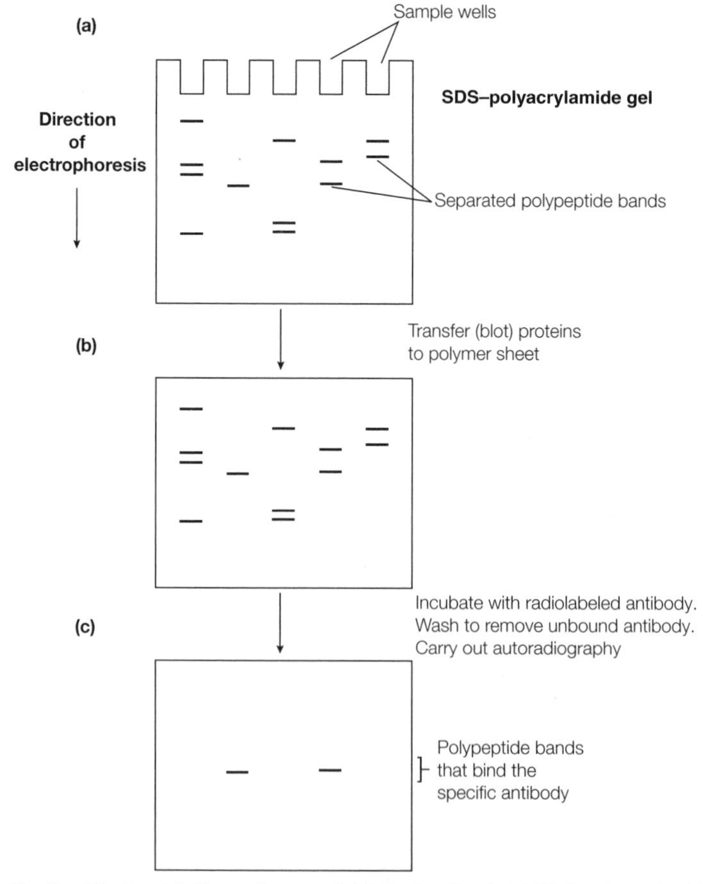


*Fig. 2. Western blotting using a radiolabeled antibody (a) SDS polyacrylamide gel electrophoresis, yielding polypeptides separated into discrete bands; (b) nitrocellulose or nylon membrane onto which the protein bands have been blotted (i.e. a Western blot); (c) autoradiograph after incubating the Western blot with radiolabeled antibody, washing away unbound antibody and placing the membrane against X-ray film.*

large proteins such as antibodies enter readily, the sample proteins must be first transferred to a more accessible medium. This process is called **blotting**. The gel is placed next to a nitrocellulose or nylon sheet and an electric field is applied so that proteins migrate from the gel to the sheet where they become bound. This particular form of blotting (i.e. blotting of proteins) is called **Western blotting** (see also Topic B8) to distinguish it from blotting of DNA (Southern blotting; see Topic I2) or RNA (Northern blotting; see Topic I2). Next, the Western blot is incubated with a protein such as casein to bind to nonspecific protein-binding sites and hence prevent spurious binding of antibody molecules in subsequent steps. This step is said to 'block' nonspecific binding sites. The Western blot is then reacted with labeled antibody, unreacted antibody is washed away and those protein bands that have bound the antibody become visible and hence are identified (*Fig. 2*). The method of visualization depends on how the antibody was labeled. If the labeling is by the incorporation of a radiolabel (e.g. $^{125}I$), then autoradiography is carried out to detect the radioactive protein bands (*Fig. 2*). Alternatively, the antibody may be detected by incubating the sheet with a second antibody that recognizes the first antibody (e.g. if the first specific antibody was raised in rabbits, the second antibody could be a goat anti-rabbit antibody). The second antibody could be radiolabeled and its binding detected by autoradiography or it could be conjugated to an enzyme that generates a colored product as in ELISA (see above). Western blotting can also be used to analyze specific antigens after two-dimensional gel electrophoresis which resolves proteins as spots, separated on the basis of both charge and size (see Topic B8).

## Immunoaffinity chromatography

**Immunoaffinity chromatography** is one example of a range of different separation procedures generally called affinity chromatography which depend on high affinity interactions between two components (see Topic B7). In immunoaffinity chromatography, a specific antibody to a protein antigen may be coupled to an inert matrix such as polysaccharide beads. These can be placed in a column and the cell sample loaded on. The protein antigen will bind to the matrix but other components will flow through the column. The protein antigen can then often be eluted in pure, or almost pure form; well over 1000-fold purification is routinely achieved in this single step. If the protein antigen is normally exposed on the plasma membrane of a desired cell type, these cells can be purified from a cell mixture by passing the mixture through the column. Only cells bearing the antigen on their surface will bind and can be eluted subsequently.

# E1 MEMBRANE LIPIDS

## Key Notes

**Membranes**

Membranes form boundaries around the cell and around distinct subcellular compartments. They act as selectively permeable barriers and are involved in signaling processes. All membranes contain varying amounts of lipid and protein and some contain small amounts of carbohydrate.

**Membrane lipids**

In membranes the three major classes of lipids are the glycerophospholipids, the sphingolipids and the sterols. The glycerophospholipids have a glycerol backbone that is attached to two fatty acid hydrocarbon chains and a phosphorylated headgroup. These include phosphatidate, phosphatidylcholine, phosphatidylethanolamine, phosphatidylglycerol, phosphatidylinositol, phosphatidylserine and diphosphatidylglycerol. The sphingolipids are based on sphingosine to which a single fatty acid chain is attached and either a phosphorylated headgroup (sphingomyelin) or one or more sugar residues (cerebrosides and gangliosides, the glycosphingolipids). The major sterol in animal plasma membranes is cholesterol, while the structurally related stigmasterol and β-sitosterol are found in plants.

**Fatty acid chains**

The fatty acid chains of glycerophospholipids and sphingolipids consist of long chains of carbon atoms which are usually unbranched and have an even number of carbon atoms (e.g. palmitate C16, stearate C18). The chains are either fully saturated with hydrogen atoms or have one or more unsaturated double bonds that are in the *cis* configuration (e.g. oleate C18:1 with one double bond).

**Lipid bilayer**

Membrane lipids are amphipathic (amphiphilic) since they contain both hydrophilic and hydrophobic regions. In the glycerophospholipids and the sphingolipids the fatty acid hydrocarbon chains are hydrophobic whereas the polar headgroups are hydrophilic. In cholesterol the entire molecule except for the hydroxyl group on carbon-3 is hydrophobic. In aqueous solution the amphipathic lipids arrange themselves into either micelles or more extensive bimolecular sheets (bilayers) in order to prevent the hydrophobic regions from coming into contact with the surrounding water molecules. The structure of the bilayer is maintained by multiple noncovalent interactions between neighboring fatty acid chains and between the polar headgroups of the lipids. In biological membranes the lipid bilayers have an asymmetrical distribution of lipids between the inner and outer leaflets.

**Membrane fluidity**

Lipids are relatively free to move within the plane of the bilayer by either rotational or lateral motion, but do not readily flip from one side of the bilayer to the other (transverse motion). Heating the bilayer changes it from a gel-like consistency to a more fluid-like consistency. Increasing the length of the fatty acid chains or decreasing the number of unsaturated double bonds in the fatty acid chains leads to a decrease in the fluidity of the membrane. In animal membranes, increasing the amount of cholesterol with its rigid fused ring system also decreases the fluidity of the membrane.

**Fluid mosaic model of membrane structure**

The fluid mosaic model is now known to be correct for the structure of biological membranes, in which the membranes are considered as two-dimensional solutions of oriented lipids and globular proteins.

**Related topics**

Prokaryotes (A1)
Eukaryotes (A2)
Membrane protein and carbohydrate (E2)
Membrane transport: small molecules (E3)
Structures and roles of fatty acids (K1)
Cholesterol (K5)

## Membranes

**Membranes** form boundaries both around the cell (the plasma membrane) and around distinct subcellular compartments (e.g. nucleus, mitochondria, lysosomes, etc.) (see Topics A1 and A2). They act as **selectively permeable barriers** allowing the inside environment of the cell or organelle to differ from that outside (see Topic E3). Membranes are involved in **signaling processes**; they contain specific receptors for external stimuli and are involved in both chemical and electrical signal generation (see Topics E5 and N3). All membranes contain two basic components: **lipids** and **proteins**. Some membranes also contain **carbohydrate**. The composition of lipid, protein and carbohydrate varies from one membrane to another. For example, the inner mitochondrial membrane has a larger amount of protein than lipid due to the presence of numerous protein complexes involved in oxidative phosphorylation and electron transfer (Topic L2), whereas the myelin sheath membrane of nerve cells, which serves to electrically insulate the cell, has a larger proportion of lipid (Topic N3).

## Membrane lipids

Lipids were originally classified as biological substances that were insoluble in water but soluble in organic solvents such as chloroform and methanol. In addition to being structural components of membranes, lipids have several other biological roles. They serve as fuel molecules (Topic K2), as concentrated energy stores (e.g. triacylglycerol) (Topic K4) and as signaling molecules (Topic E5). Within membranes there are three major types of lipid: the **glycerophospholipids**, the **sphingolipids** and the **sterols**.

*Glycerophospholipids*

The glycerophospholipids are made up of three components: a **phosphorylated headgroup**, a three-carbon **glycerol backbone** and two **hydrocarbon fatty acid chains** (*Fig. 1*). The phosphorylated headgroup is attached to carbon-3 of the glycerol backbone, while the two fatty acid chains are attached to the other two carbon atoms. The simplest glycerophospholipid is phosphatidate (diacylglycerol 3-phosphate) which has only a phosphoric acid group esterified to carbon-3 of the glycerol. Although phosphatidate itself is present in small amounts in membranes, the major glycerophospholipids are derived from it. In these other lipids the phosphate is further esterified to the hydroxyl group of one of several alcohols (choline, ethanolamine, glycerol, inositol or serine). The major glycerophospholipids found in membranes include phosphatidylcholine, phosphatidylethanolamine, phosphatidylglycerol, phosphatidylinositol and

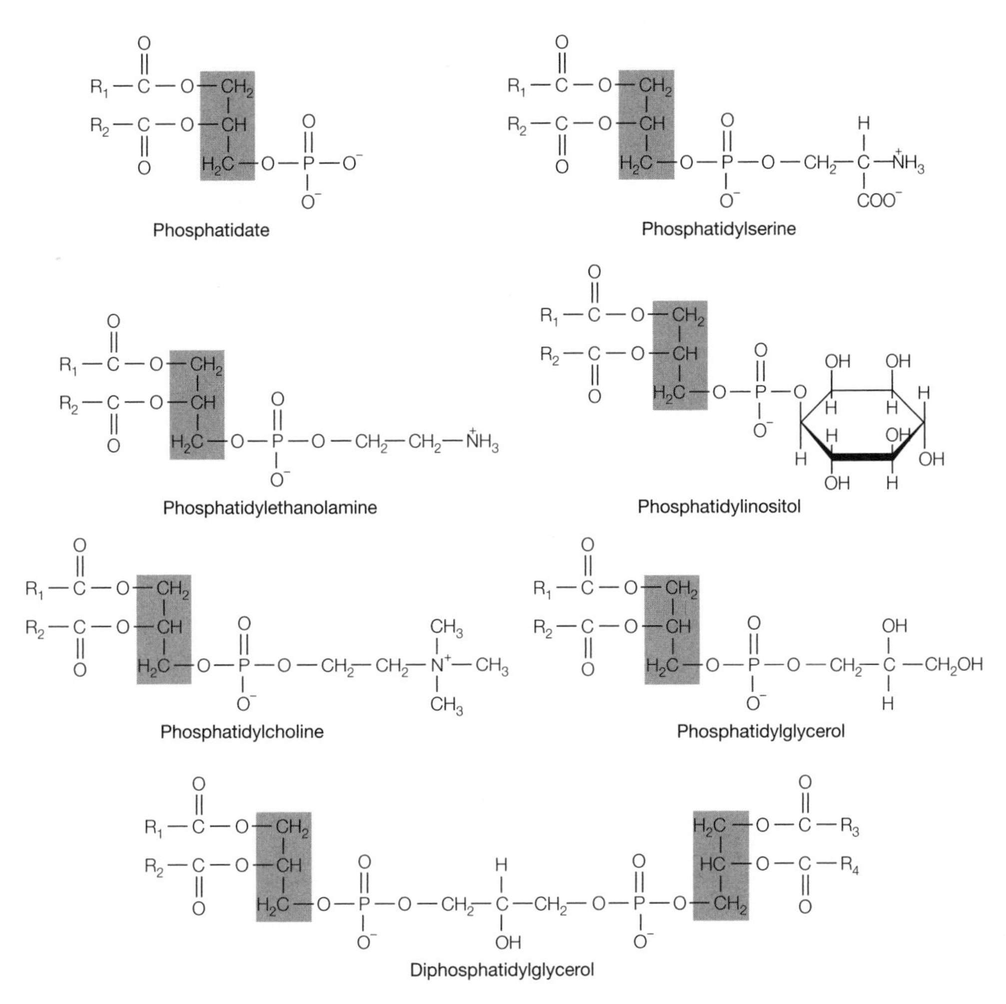


*Fig. 1. Structures of membrane glycerophospholipids. $R_1$ and $R_2$ represent hydrocarbon chains of fatty acids.*

phosphatidylserine (*Fig. 1*). Diphosphatidylglycerol (or cardiolipin) is found predominantly in the inner mitochondrial membrane.

*Sphingolipids*

**Sphingomyelins**, the commonest sphingolipids, have a **sphingosine backbone** (*Fig. 2a*) in place of the glycerol in glycerophospholipids. Like the glycerophospholipids, they also have a phosphorylated headgroup (either choline or ethanolamine) and two hydrocarbon chains (*Fig. 2a*). One of the hydrocarbon chains comes from the sphingosine molecule, the other is a fatty acid as found in the glycerophospholipids except that it is bonded via an amide bond in sphingolipids. The sphingomyelins are particularly abundant in the myelin sheath that surrounds nerve cells. The **glycosphingolipids**, such as the **cerebrosides** and **gangliosides**, are also derived from sphingosine, but in place of the phosphorylated headgroup they have one or more sugar residues. The galactocerebrosides have a single galactose residue (*Fig. 2a*) and are found predominantly in the neuronal cell membranes of the brain. The gangliosides

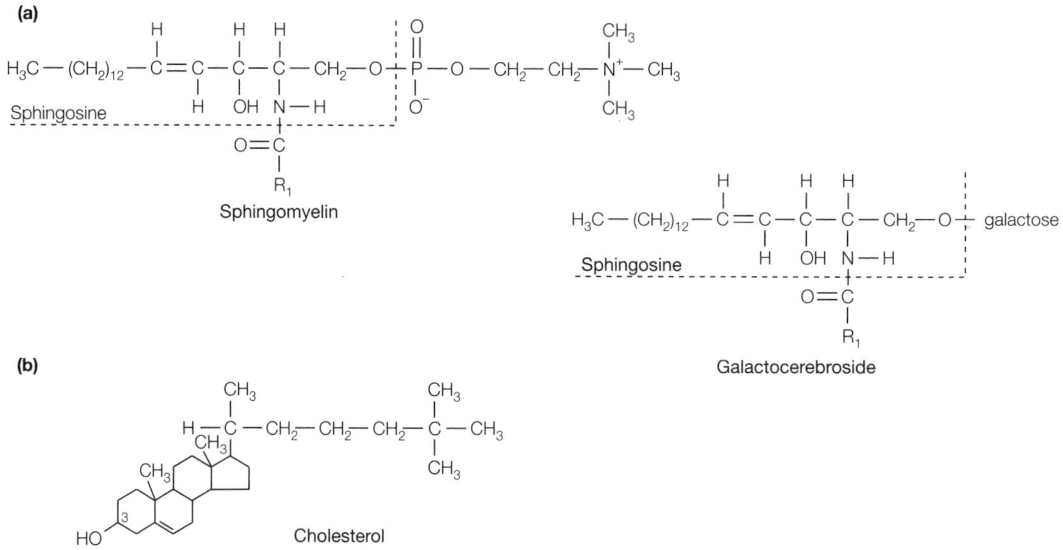

*Fig. 2. Structures of (a) the sphingolipids sphingomyelin and galactocerebroside; (b) cholesterol. $R_1$ represents the hydrocarbon chain of fatty acids.*

have several sugar residues including at least one sialic acid (*N*-acetylneuraminic acid) residue and are a major constituent of most mammalian plasma membranes, being particularly abundant in brain cells.

### *Sterols*

The sterol **cholesterol** (*Fig. 2b*) is a major constituent of animal plasma membranes but is absent from prokaryotes. The fused ring system of cholesterol means that it is more rigid than other membrane lipids. As well as being an important component of membranes, cholesterol is the metabolic precursor of the steroid hormones (see Topic K5). Plants contain little cholesterol but have instead a number of other sterols, mainly **stigmasterol** and **β-sitosterol** which differ from cholesterol only in their aliphatic side chains.

## Fatty acid chains

The two fatty acid chains of glycerophospholipids and the single fatty acid chain and the hydrocarbon chain of the sphingosine in sphingolipids consist of long chains of carbon atoms. Usually these chains have an even number of carbon atoms (e.g. palmitate, C16; stearate, C18) and are unbranched. The chains can either be fully saturated with hydrogen atoms or unsaturated and have one or more double bonds that are usually in the *cis* configuration (e.g. oleate C18:1 which has 18 carbon atoms and one double bond; arachidonic acid C20:4 which has 20 carbon atoms and four double bonds). The two fatty acid chains on a glycerophospholipid are usually not identical [e.g. 1-stearoyl-2-oleoyl-3-phosphatidylcholine (*Fig. 3*)].

## Lipid bilayer

**Amphipathic** (or amphiphilic) molecules contain both **hydrophilic** (water-loving) and **hydrophobic** (water-hating) regions. Membrane lipids are amphipathic molecules as they are made up of hydrophobic fatty acid chains and a hydrophilic polar headgroup. In the glycerophospholipids, the two hydrocarbon

```
                                  O
                                  ||
                H3C — (CH2)16 — C — O — CH2
                                        |
H3C — (CH2)7 — C = C — (CH2)7 — C — O — CH          O                    CH3
               |   |            ||      |           ||                   |
               H   H            O      H2C — O — P — O — CH2 — CH2 — N+ — CH3
                                                    |                    |
                                                    O-                   CH3
```

*Fig. 3. Structure of 1-stearoyl-2-oleoyl-3-phosphatidylcholine showing the saturated stearoyl (C18:0) and the mono-unsaturated oleoyl (C18:1) hydrocarbon chains. The hydrogen atoms surrounding the C=C bond are in the* cis *configuration.*

chains are hydrophobic whereas the glycerol backbone and the phosphorylated headgroup are hydrophilic. In the sphingolipids, the fatty acid chain and the hydrocarbon chain of the sphingosine are hydrophobic whereas the phosphorylated or sugar headgroup is hydrophilic. In the case of cholesterol, the entire molecule apart from the hydroxyl group on carbon-3 is hydrophobic in nature.

In aqueous solution, amphipathic molecules will orientate themselves in such a way as to prevent the hydrophobic region coming into contact with the water molecules. In the case of those fatty acid salts which contain only one fatty acid chain (such as sodium palmitate, a constituent of soap), the molecules form a spherical micellar structure (diameter usually $< 20$ nm) in which the hydrophobic fatty acid chains are hidden inside the **micelle** and the hydrophilic headgroups interact with the surrounding water molecules (*Fig. 4a*). Because the two fatty acid chains of phospholipids are too bulky to fit into the interior of a micelle, the favored structure for most phospholipids in aqueous solution is a two-dimensional bimolecular sheet or **lipid bilayer** (*Fig. 4b*). Such lipid bilayers, in which the phospholipid molecules are orientated with their hydrophobic chains in the interior of the structure and their hydrophilic headgroups on the surfaces, can be relatively large structures of up to about 1 $mm^2$ in area. The two layers of lipids in the bilayer are referred to as the **inner and outer leaflets**. In biological membranes the individual lipid species are asymmetrically distributed between the two leaflets. For example, in the plasma membrane of erythrocytes, sphingomyelin and phosphatidylcholine are preferentially located in the outer leaflet, whereas phosphatidylethanolamine and phosphatidylserine are mainly in the inner leaflet.

Lipid bilayers will spontaneously self-assemble in aqueous solution. The major driving force behind this is the **hydrophobic effect** – the hydrophobic fatty acid chains avoid coming into contact with the water molecules. Once

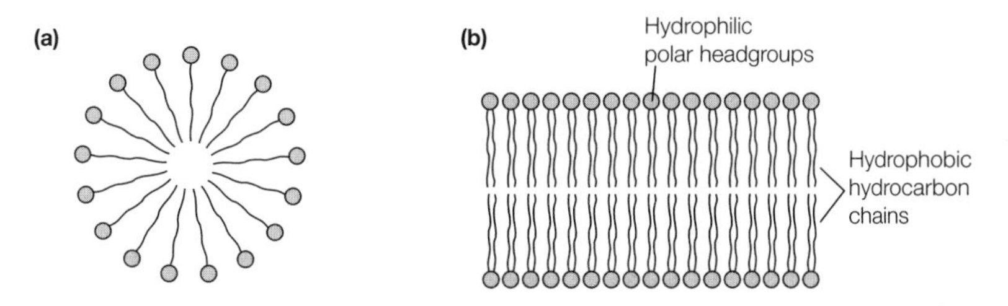


*Fig. 4. Structure of (a) a micelle and (b) a lipid bilayer.*

formed, the bilayer structure is maintained by multiple noncovalent interactions including hydrophobic interactions and van der Waals forces between the hydrocarbon chains, charge interactions and hydrogen bonding between the polar headgroups, and hydrogen bonding between the headgroups and the surrounding water molecules.

## Membrane fluidity

Because there are no covalent bonds between the lipids in the bilayer, the membrane is not a static structure but has **fluidity**. The lipids are generally free to move within the plane of the inner or outer leaflet of the bilayer by either **rotational** or **lateral movement** (*Fig. 5*). However, they cannot readily flip from one leaflet of the bilayer to the other, so-called **transverse movement**, due to the unfavorable energetics involved in moving a hydrophilic headgroup through the hydrophobic interior of the bilayer. The fluidity of the bilayer can be altered in a number of ways. Upon heating above a characteristic **transition temperature**, the lipid bilayer will change from a gel-like consistency to a more fluid-like consistency. This transition temperature depends on the length of the fatty acid chains and on their degree of unsaturation. If the length of the fatty acid chains is increased, the fluidity of the bilayer will decrease due to the larger propensity for noncovalent interactions between the hydrocarbon chains. In contrast, if the degree of unsaturation in the fatty acid chains is increased, the fluidity of the bilayer will increase. This is because the double bonds which are in the *cis* configuration kink the hydrocarbon chain and disrupt the highly ordered packing of the fatty acid chains, thus reducing the number of interactions between neighboring lipids. An important regulator of membrane fluidity in mammalian systems is **cholesterol**. At physiological temperature (37°C), increasing the amount of cholesterol in the bilayer will lead to a decrease in the fluidity of the membrane since the rigid steroid ring system interferes with the lateral movement of the fatty acid chains.

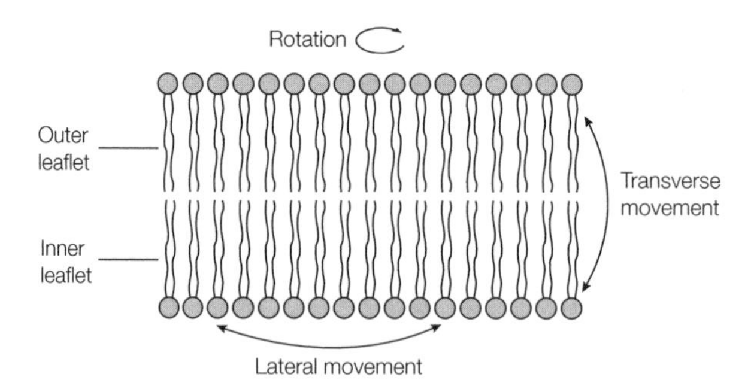


*Fig. 5. Movement of lipids in membranes.*

## Fluid mosaic model of membrane structure

In 1972 S. Jonathan Singer and Garth Nicholson proposed the **fluid mosaic model** for the overall structure of biological membranes, in which the membranes can be viewed as two-dimensional solutions of oriented lipids and globular proteins (*Fig. 6*). The integral membrane proteins can be considered as 'icebergs' floating in a two-dimensional lipid 'sea'. They proposed that the bilayer organization of the lipids would act both as a solvent for the amphipathic integral membrane proteins and as a **permeability barrier**. They also proposed that some lipids may interact with certain membrane proteins, that

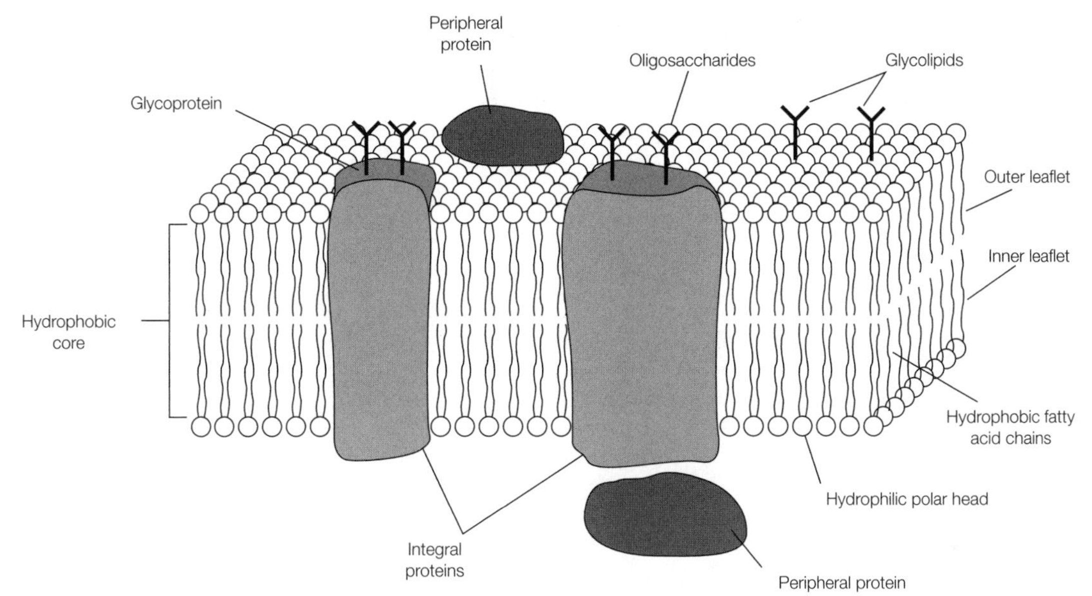


*Fig. 6. The fluid mosaic model of membrane structure.*

these interactions would be essential for the normal functioning of the protein, and that membrane proteins would be free to diffuse laterally in the plane of the bilayer unless restricted in some way, but would not be able to flip from one side of the bilayer to the other. This model is now supported by a wide variety of experimental observations (see Topic E2).

# E2 MEMBRANE PROTEIN AND CARBOHYDRATE

## Key Notes

**Integral membrane proteins**

Integral (intrinsic) membrane proteins are tightly associated with the hydrophobic core of the lipid bilayer and can be removed from it only with organic solvents or detergents that disrupt the membrane structure. Most integral proteins have one (e.g. glycophorin) or more (e.g. bacteriorhodopsin) regions of the polypeptide chain that traverse the lipid bilayer. These regions consist mainly of amino acids with hydrophobic side chains that fold into an α-helix and interact noncovalently with the surrounding lipids. Some integral proteins do not traverse the membrane but are attached covalently to a lipid that interacts with the hydrophobic interior of the bilayer. Integral proteins are asymmetrically distributed across the membrane.

**Integral membrane protein movement and distribution**

Integral proteins are usually free to move in the plane of the bilayer by lateral and rotational movement, but are not able to flip from one side of the membrane to the other (transverse movement). Immunofluorescence microscopy may be used to follow the movement of two proteins from different cells following fusion of the cells to form a hybrid heterokaryon. Immediately after fusion the two integral proteins are found segregated at either end of the heterokaryon but with time diffuse to all areas of the cell surface. The distribution of integral proteins within the membrane can be studied by electron microscopy using the freeze–fracture technique in which membranes are fractured along the interface between the inner and outer leaflets.

**Membrane protein purification and reconstitution**

The first step in the purification of integral proteins is the disruption of the membrane structure by solubilizing it with a detergent (e.g. Triton X-100 or octyl glucoside). Once solubilized, the hydrophobic region of the protein is coated with a layer of amphipathic detergent molecules, allowing the protein to remain in aqueous solution and be purified as for a soluble globular protein. Once purified, integral proteins can be reincorporated into artificial lipid vesicles (liposomes) in order to study their function.

**Peripheral membrane proteins**

Peripheral membrane proteins are only loosely bound to the membrane and can readily be removed by washing the membranes with a solution of either high ionic strength or high pH. These treatments disrupt the noncovalent ionic and hydrogen bonds holding the peripheral proteins on the surface of the membrane. No part of a peripheral protein interacts with the hydrophobic interior of the bilayer. The cytoskeleton that covers the cytosolic surface of the erythrocyte plasma membrane is made up of a number of peripheral proteins (spectrin, ankyrin, protein 4.1 and actin) and is important in maintaining and altering the shape of the cell.

**Membrane carbohydrate**

Sugar residues are found only on the extracellular side of the plasma membrane attached either to lipids to form glycolipids or to proteins to form glycoproteins. In glycoproteins the sugar residues are attached either to the hydroxyl group of Ser or Thr to form O-linked oligosaccharides, or to the amide group of Asn when the Asn residue is in the tripeptide consensus sequence Asn-Xaa-Ser or Asn-Xaa-Thr (where Xaa is any amino acid except Pro) to form N-linked oligosaccharides. The carbohydrate forms a protective coat on the outer surface of the cell and is involved in intercellular recognition.

**Related topics**

Protein structure (B3)
Protein purification (B6)
Membrane lipids (E1)
Protein glycosylation (H5)

## Integral membrane proteins

Membrane proteins are classified as either **peripheral** (extrinsic) or **integral** (intrinsic) depending on how tightly they are associated with the membrane. Integral membrane proteins are tightly bound to the membrane through interactions with the hydrophobic core of the bilayer (see Topic E1; *Fig. 5*) and can be extracted from them only by using agents that disrupt the membrane structure, such as **organic solvents** (e.g. chloroform) or **detergents**. Most integral proteins have one or more regions of the polypeptide chain that span the lipid bilayer and interact noncovalently with the hydrophobic fatty acid chains. However, some are anchored in the membrane by a covalently attached fatty acid or hydrocarbon chain. Like lipids, integral proteins are **amphipathic**, having both hydrophobic and hydrophilic regions, and are **asymmetrically distributed** across the bilayer.

### *Glycophorin*

Because **erythrocytes** (red blood cells) do not contain any subcellular organelles (they are essentially a membranous sac for carrying hemoglobin) their plasma membrane is a convenient model system for studies of membrane structure as it can readily be isolated from other membranes and intracellular components. One of the major glycoproteins in the plasma membrane of erythrocytes is **glycophorin A**; a 131 amino acid protein that was the first integral protein to be sequenced (see Topic B9). This revealed that the polypeptide chain of glycophorin consists of three domains:

1. An N-terminal region on the extracellular side of the membrane that contains all the N- and O-linked **glycosylation sites**;
2. A hydrophobic central region that is buried in the hydrophobic core of the bilayer; and
3. A C-terminal region rich in polar and charged residues that is exposed on the cytosolic side of the membrane (*Fig. 1a*).

As with the majority of transmembrane proteins, the hydrophobic membrane-spanning region consists mainly of amino acid residues with hydrophobic side-chains that are folded in an **α-helical conformation** (see Topic B3). As each amino acid residue adds 0.15 nm to the length of an α-helix, a helix of 25 residues would have a length of 3.75 nm, just enough to span the hydrophobic core of the bilayer. The hydrophobic side-chains of the residues in the helix protrude outwards from the helix axis to interact via hydrophobic bonds with

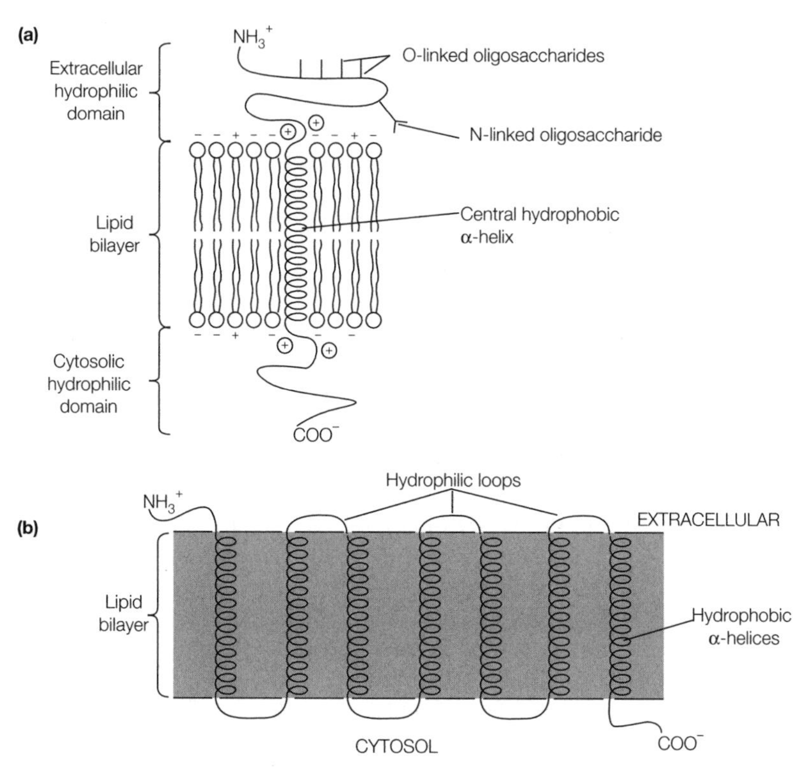


*Fig. 1. Integral membrane proteins. (a) Single membrane-spanning region (e.g. glycophorin); (b) multiple membrane-spanning regions (e.g. bacteriorhodopsin).*

the fatty acid chains. Either side of this hydrophobic α-helix are clusters of amino acids with charged side-chains which interact noncovalently with opposite charges on the polar headgroups of the membrane lipids.

*Multiple membrane-spanning proteins*

Some integral proteins have **multiple membrane-spanning α-helices**. **Bacteriorhodopsin**, a protein found in a photosynthetic bacterium, captures energy from light and uses it to pump protons across the bacterial membrane. Like numerous other integral proteins, the polypeptide chain of bacteriorhodopsin loops backwards and forwards across the lipid bilayer seven times (*Fig. 1b*). Each of the seven transmembrane α-helices is linked to the next by a short hydrophilic region of the polypeptide chain that is exposed either on the extracellular or cytosolic side of the membrane. In contrast, the **anion exchange band 3 protein** of the erythrocyte plasma membrane that transports $Cl^-$ and $HCO_3^-$ across the membrane loops backwards and forwards across the lipid bilayer 12–14 times.

*Lipid-anchored proteins*

A growing number of integral proteins in eukaryotes do not traverse the membrane but are anchored in one or other leaflet of the bilayer through covalent attachment to a hydrocarbon chain. Several proteins, including alkaline phosphatase, are stably anchored at the cell surface through covalent linkage of their C-terminal amino acid to the headgroup of a phosphatidylinositol lipid via an ethanolamine–phosphate–trimannose–glucosamine bridge, so-called

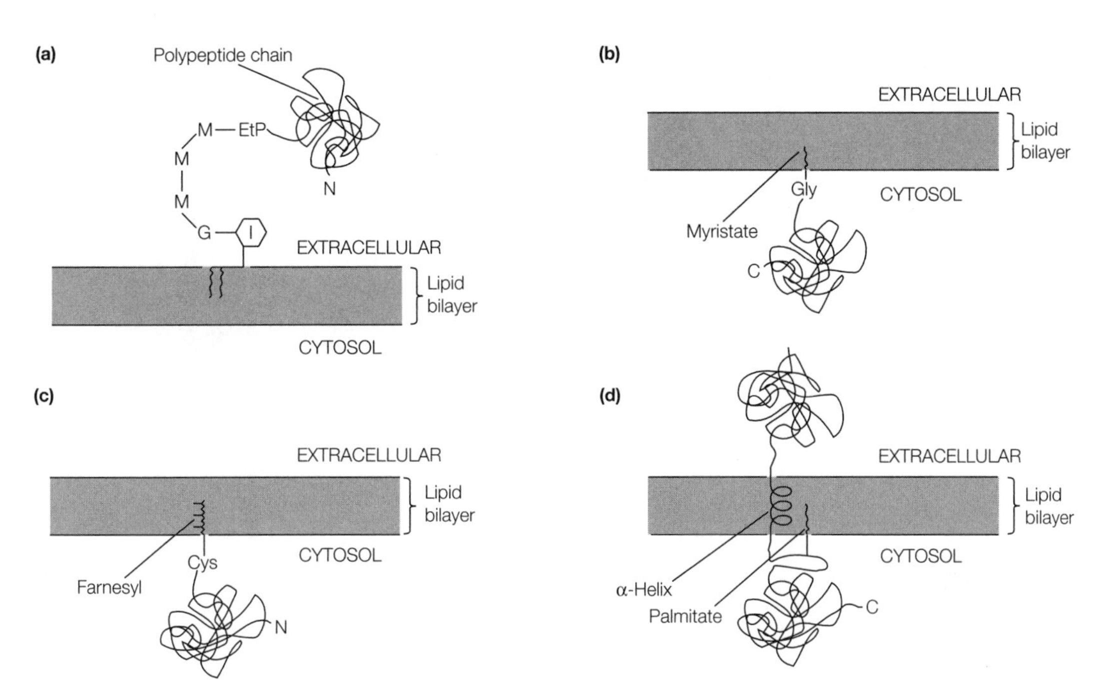


*Fig. 2. Lipid-modified proteins. (a) A glycosyl phosphatidylinositol-anchored protein (G, glucosamine; M, mannose; EtP, ethanolamine phosphate); (b) a myristoylated protein; (c) a prenylated protein; (d) a palmitoylated protein.*

**glycosyl phosphatidylinositol (GPI)-anchored** proteins (*Fig. 2a*). This complex structure is built up by sequential addition of the individual sugar residues and ethanolamine phosphate to phosphatidylinositol. A C-terminal hydrophobic signal peptide is removed from the protein in the lumen of the RER and the preformed GPI anchor added to the newly exposed C-terminal amino acid.

Other proteins are transiently attached to the cytosolic face of the membrane either by amide linkage of a myristate (C14:0) molecule to an N-terminal Gly residue (**myristoylated** proteins; *Fig. 2b*), or by thioether linkage of a 15-carbon **farnesyl** or a 20-carbon **geranylgeranyl** polyunsaturated hydrocarbon to a C-terminal Cys residue (**prenylated** proteins; *Fig. 2c*). Farnesyl and geranylgeranyl are synthesized from isopentenyl pyrophosphate, the precursor of cholesterol (see Topic K5). Some proteins are also modified on Cys residues with covalently attached palmitate (C16:0) (**palmitoylated** proteins). These include some with membrane-spanning polypeptides (*Fig. 2d*), some prenylated proteins and some myristoylated proteins.

## Integral membrane protein movement and distribution

That proteins are free to move laterally in the plane of the bilayer was shown by fusing cultured mouse cells with human cells under appropriate conditions to form a hybrid cell known as a **heterokaryon** (*Fig. 3a*). The mouse cells were labeled with mouse protein-specific antibodies to which the green-fluorescing dye fluorescein had been covalently attached, whilst the human cells were labeled with the red-fluorescing dye rhodamine. Upon cell fusion, the mouse and human proteins as seen under the fluorescence microscope (see Topic A3) were segregated on the two halves of the heterokaryon (*Fig. 3a*). After 40 min at 37°C, however, the mouse and human proteins had completely intermingled. Lowering the temperature to below 15°C inhibited this process, indicating that

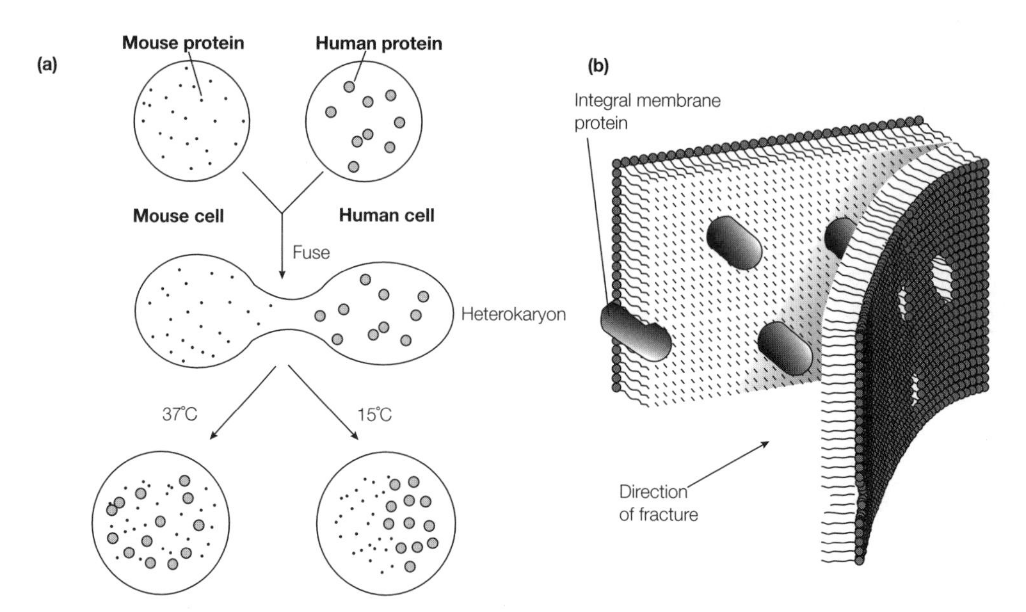


*Fig. 3. (a) Movement and (b) distribution (as shown by freeze–fracture electron microscopy) of integral membrane proteins.*

the proteins are free to diffuse laterally in the membrane and that this movement is slowed as the temperature is lowered. It should be noted, though, that some integral membrane proteins are not free to move laterally in the membrane because they interact with the cytoskeleton inside the cell.

The distribution of proteins in membranes can be revealed by electron microscopy using the **freeze–fracture technique** (*Fig. 3b*). In this technique, a membrane specimen is rapidly frozen to the temperature of liquid nitrogen (–196°C) and then fractured by a sharp blow. The bilayer often splits into monolayers, revealing the interior. The exposed surface is then coated with a film of carbon and shadowed with platinum in order for the surface to be viewed in the electron microscope (see Topic A3). The fractured surface of the membrane is revealed to have numerous randomly distributed protuberances that correspond to integral membrane proteins.

## Membrane protein purification and reconstitution

The first step in the **purification** of an integral membrane protein is to disrupt its interactions with other integral proteins and the lipids in the membrane. This is commonly achieved by adding a **detergent** which solubilizes the membrane. In order to solubilize the membrane but not denature the protein, gentle detergents such as Triton X-100 or octyl glucoside are used (*Fig. 4*), rather than the harsh detergent SDS. As the detergent molecules are themselves amphipathic they readily intercalate into the lipid bilayer and disrupt the hydrophobic interactions.

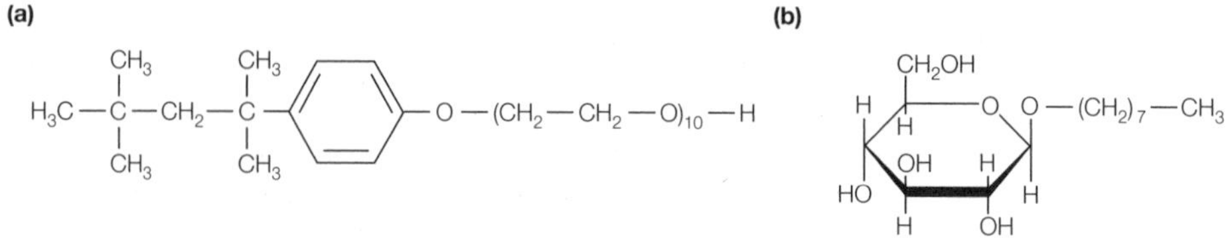

(a)

$$H_3C-C(CH_3)_2-CH_2-C(CH_3)_2-C_6H_4-O-(CH_2-CH_2-O)_{10}-H$$

(b)

$CH_2OH$, H, O, $O-(CH_2)_7-CH_3$, H, OH, H, HO, H, H, OH

*Fig. 4. Structures of (a) Triton X-100 and (b) octyl glucoside.*

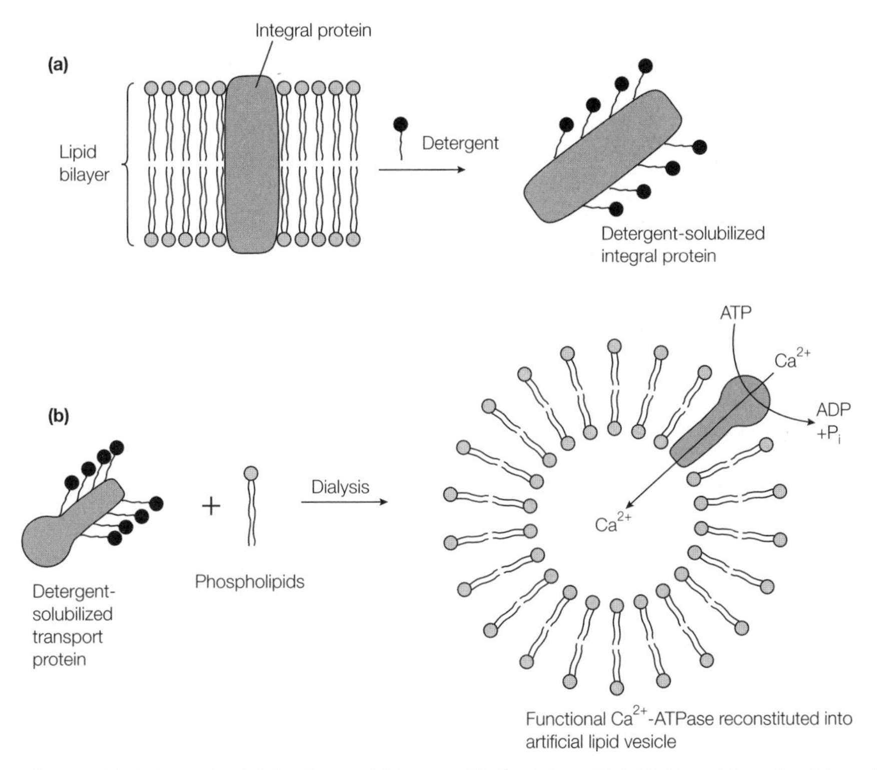


*Fig. 5. (a) Detergent solubilization and (b) reconstitution into artificial lipid vesicles of an integral membrane protein.*

Once solubilized, the hydrophobic region of the integral protein is coated with a layer of detergent molecules which enables the protein to remain in solution (*Fig. 5a*). The solubilized protein can then be purified as for a water-soluble globular protein (see Topic B6) as long as detergent is kept in the buffers to prevent aggregation and loss of the protein. Once purified, an integral protein can be reincorporated into artificial lipid vesicles (**liposomes**) in order to study its function (*Fig. 5b*). If phospholipids are added to the protein in detergent solution and the detergent dialyzed away, phospholipid vesicles containing the protein will spontaneously form. The function of the protein can then be studied. For example, if the $Ca^{2+}$-ATPase is reincorporated into lipid vesicles, its function (i.e. transport of $Ca^{2+}$ upon ATP hydrolysis) can be studied by monitoring $Ca^{2+}$ on the inside of the vesicle upon addition of $Ca^{2+}$ and ATP to the outside (*Fig. 5b*).

## Peripheral membrane proteins

Peripheral membrane proteins are less tightly bound to the lipid bilayer than integral membrane proteins and can be readily removed by washing the membranes with a solution of high ionic strength (e.g. 1 M NaCl) or high pH. These procedures leave the lipid bilayer intact but disrupt the ionic and hydrogen bond interactions that hold the peripheral proteins on the surface of the membrane. No part of a peripheral protein interacts with the hydrophobic core of the bilayer. Peripheral membrane proteins can be found either on the outer or the inner surface of the bilayer and can be associated with the membrane through noncovalent interactions with either the lipid headgroups and/or other proteins in the membrane (see Topic E1; *Fig. 5*). Once removed from the membrane, peripheral proteins behave as water-soluble globular proteins and can be purified as such (see Topic B6).

*Cytoskeleton*

The cytosolic surface of the erythrocyte plasma membrane is covered by a network of peripheral membrane proteins that make up the **cytoskeleton** (*Fig. 6*). The major component of this cytoskeleton is **spectrin** which folds into a triple-stranded **α-helical coiled coil** to form long chains. The spectrin chains are attached to the plasma membrane through interactions with two other peripheral proteins, **ankyrin** and **protein band 4.1**. Ankyrin forms a cross-link between spectrin and the cytosolic domain of the integral anion exchanger band 3 protein, while band 4.1 promotes the binding of **actin filaments** to the spectrin chains linking them to the cytosolic domain of glycophorin. The cytoskeleton gives the erythrocyte plasma membrane great strength and flexibility, and is important in maintaining and altering the shape of the cell. In other mammalian cells the cytoskeleton has fewer attachment points on the inner surface of the plasma membrane and criss-crosses throughout the cytoplasm (see Topics N1 and N2).

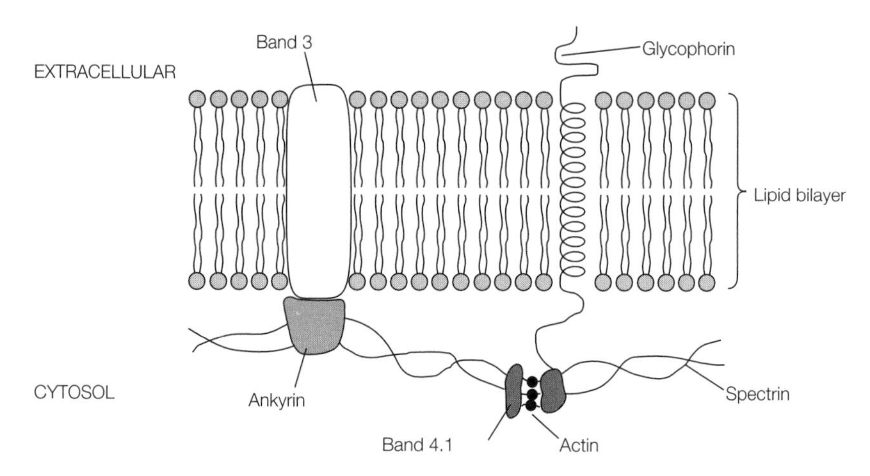


*Fig. 6. The erythrocyte cytoskeleton.*

## Membrane carbohydrate

The extracellular surface of the plasma membrane is often covered with a protective coat of **carbohydrate**. The sugar residues of this carbohydrate coat can be attached either to certain lipids such as the **glycosphingolipids** (see Topic E1), or to the polypeptide chains of peripheral or integral membrane proteins. These **glycolipids** and **glycoproteins** are abundant in the plasma membrane of eukaryotic cells but are virtually absent from most intracellular membranes, particularly the inner mitochondrial membrane and the chloroplast lamellae. In glycoproteins the sugar residues can be attached to the polypeptide chain either through the hydroxyl group in the side chain of Ser or Thr residues as **O-linked oligosaccharides**, or through the amide group in the side-chain of Asn as **N-linked oligosaccharides** (see Topic H5). In the case of the N-linked sugars, in order to be modified the Asn must lie in the tripeptide consensus sequence Asn-Xaa-Ser or Asn-Xaa-Thr, where Xaa is any amino acid except Pro. O-linked oligosaccharides usually consist of only a few (approximately four) sugar residues, whereas N-linked oligosaccharides can be quite large structures with a dozen or more different sugars attached to the Asn side-chain. The carbohydrate on the extracellular face of the membrane not only serves a protective role but is also involved in intercellular recognition and in maintaining the asymmetry of the membrane.

# E3 MEMBRANE TRANSPORT: SMALL MOLECULES

## Key Notes

**Membrane permeability**

The plasma membrane is a selectively permeable barrier. Some molecules (water, gases, urea) can pass directly through the bilayer unaided, whereas other molecules (sugars, amino acids, ions, etc.) require the presence of integral membrane transport proteins.

**Passive transport**

The movement of molecules across a membrane by passive transport does not require an input of metabolic energy. The molecule moves from a high concentration to a lower concentration. Passive transport by simple diffusion does not require the presence of integral membrane proteins. The rate of movement of a molecule (e.g. water, gases, urea) by simple diffusion is directly proportional to its concentration gradient across the membrane. Passive transport by facilitated diffusion requires the presence of specific integral membrane proteins to facilitate the movement of the molecule (e.g. glucose, other sugars, amino acids) across the membrane. The transport protein (e.g. the erythrocyte glucose transporter) is specific for a particular molecule, is saturable, displays binding kinetics, and is influenced by the temperature, pH and inhibitors.

**Active transport**

Active transport of a molecule across a membrane against its concentration gradient requires an input of metabolic energy. In the case of ATP-driven active transport, the energy required for the transport of the molecule ($Na^+$, $K^+$, $Ca^{2+}$ or $H^+$) across the membrane is derived from the coupled hydrolysis of ATP (e.g $Na^+/K^+$-ATPase). In ion-driven active transport, the movement of the molecule to be transported across the membrane is coupled to the movement of an ion (either $Na^+$ or $H^+$) down its concentration gradient. If both the molecule to be transported and the ion move in the same direction across the membrane, the process is called symport (e.g. $Na^+$/glucose transporter); if the molecule and the ion move in opposite directions it is called antiport (e.g. erythrocyte band 3 anion transporter).

**Glucose transport into intestinal epithlial cells**

The transport of glucose across the polarized epithelial lining cells of the intestine involves its symport across the apical membrane by the $Na^+$/glucose transporter, with the energy for the movement of the glucose coming from the movement of $Na^+$ down its concentration gradient. The concentration of $Na^+$ ions inside the cell is maintained at a low level by the action of a $Na^+/K^+$-ATPase on the basolateral membrane. The glucose then exits from the cell across the basolateral membrane by facilitated diffusion through the glucose transporter. The movement of $Na^+$ and glucose across the cell sets up a difference in osmotic pressure causing water to follow by simple diffusion, which forms the basis of glucose rehydration therapy.

**Related topics**

Membrane protein and carbohydrate (E2)
Nerve (N3)

## Membrane permeability

A pure phospholipid bilayer is **permeable** to water, gases ($O_2$, $CO_2$, $N_2$) and small uncharged polar molecules (e.g. urea, ethanol), but is **impermeable** to large uncharged polar molecules (e.g. glucose), ions ($Na^+$, $K^+$, $Cl^-$, $Ca^{2+}$) and charged polar molecules (e.g. amino acids, ATP, glucose 6-phosphate). The first group of molecules can cross the membrane unaided and without an input of energy, whereas the latter group require the presence of **integral membrane transport proteins** and, in some cases, an **input of energy** to travel through the otherwise impermeable membrane barrier. Thus the plasma membrane and the membranes of internal organelles are **selectively permeable barriers**, maintaining a distinct internal environment.

## Passive transport

The passive transport of molecules across a membrane does not require an input of metabolic energy. The rate of transport (diffusion) is proportional to the concentration gradient of the molecule across the membrane. There are two types of passive transport: **simple diffusion** and **facilitated diffusion**.

### *Simple diffusion*

Only relatively small uncharged or hydrophobic molecules ($H_2O$, $O_2$, $CO_2$, other gases, urea and ethanol) cross the lipid bilayer by simple diffusion. No membrane proteins are involved, so there is no specificity. The molecule in aqueous solution on one side of the membrane dissolves into the lipid bilayer, crosses it, and then dissolves into the aqueous solution on the opposite side. The rate of diffusion is directly proportional to the concentration gradient of the molecule across the membrane and the process is not saturable (*Fig. 1a*).

### *Facilitated diffusion*

Unlike simple diffusion, the facilitated (or carrier-mediated) diffusion of a molecule across a biological membrane is dependent on **specific integral membrane proteins**, often called **uniporters** (see Topic E2). The molecule binds to the protein on one side of the membrane, the protein then undergoes a **conformational change**, transports the molecule across the membrane and then releases it on the other side. Molecules transported across membranes in this way include hydrophilic molecules such as glucose, other sugars and amino acids. The transport proteins are **specific** for one particular molecule or a group of structurally similar molecules. The transport proteins are capable of being saturated, display Michaelis–Menten-type binding kinetics ($K_m$ and $V_{max}$) (*Fig. 1b*), and are influenced by temperature, pH and inhibitor molecules in a similar manner to enzymes (see Topics C1 and C3).

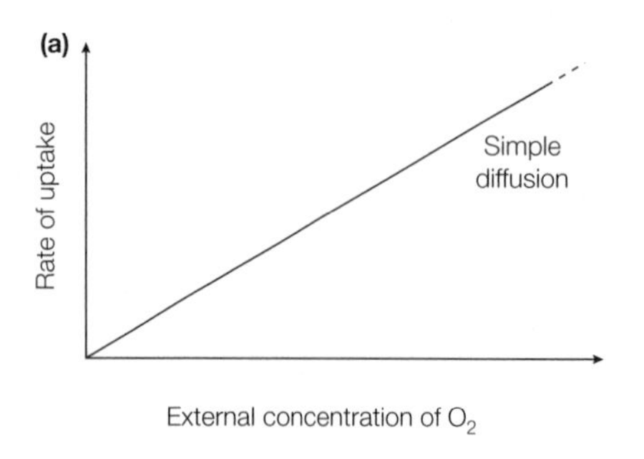


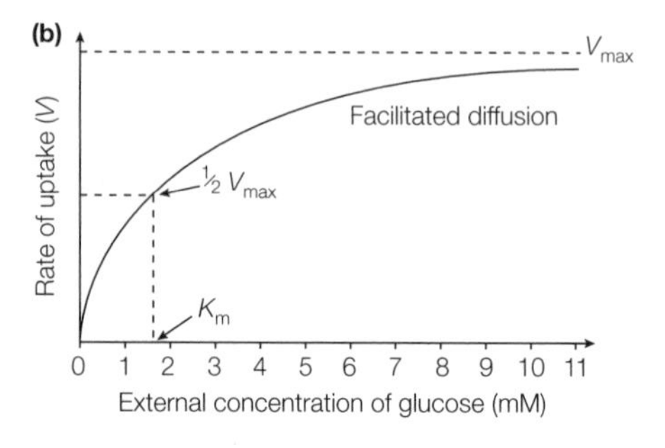


*Fig. 1. Kinetics of (a) simple and (b) facilitated diffusion.*

An example of facilitated diffusion is the uptake of glucose into erythrocytes by the glucose transporter. The **erythrocyte glucose transporter** is an integral membrane protein of mass 45 kDa that is asymmetrically orientated in the plasma membrane. This uniporter protein structure traverses the membrane with 12 α-helices (see Topic E2) which form a central pore through which the glucose molecule is passed upon **conformational changes** in the protein (*Fig. 2*). All the steps in the transport of glucose into the cell are **freely reversible**, the direction of movement of glucose being dictated by the relative concentrations of glucose on either side of the membrane. In order to maintain the concentration gradient across the membrane, the glucose is rapidly phosphorylated inside the cell by hexokinase to glucose 6-phosphate (see Topic J3) which is no longer a substrate for the glucose transporter. The erythrocyte glucose transporter is highly specific for D-glucose ($K_m$ 1.5 mM), the nonbiological L-isomer being transported at a barely measurable rate. D-Mannose and D-galactose, which differ from D-glucose in the configuration at one carbon atom (see Topic J1), are transported at intermediate rates. Thus the transporter has a higher affinity (lower $K_m$) for glucose than for other sugars.

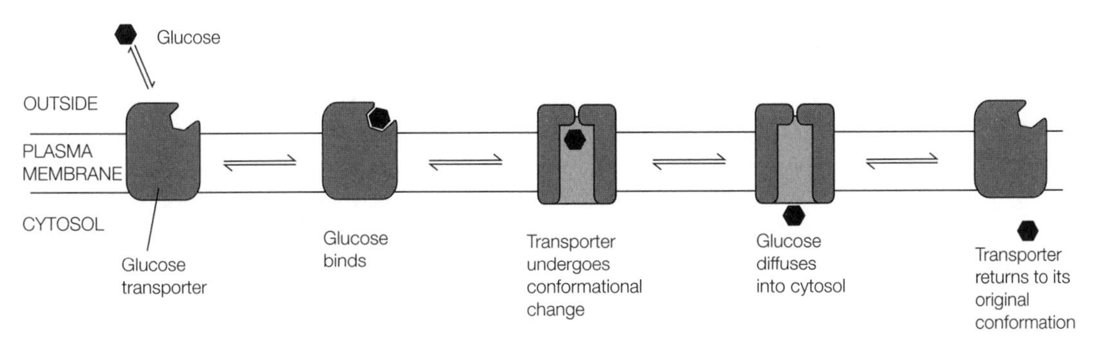


*Fig. 2. Facilitated diffusion of glucose into erythrocytes.*

## Active transport

The active transport of molecules requires an **input of metabolic energy**. This can be derived either from direct coupling to the **hydrolysis of ATP** or by coupling to the **movement of an ion** down its concentration gradient.

### *ATP-driven active transport*

In this case, the energy required for the transport of the molecule across the membrane is derived from the **coupled hydrolysis of ATP**, for example the movement of $Na^+$ and $K^+$ ions by the **$Na^+/K^+$-ATPase**. All cells maintain a high internal concentration of $K^+$ and a low internal concentration of $Na^+$. The resulting **$Na^+/K^+$ gradient** across the plasma membrane is important for the active transport of certain molecules, and the maintenance of the **membrane electrical potential** (see Topic N3). The movement across the membrane of $Na^+$, $K^+$, $Ca^{2+}$ and $H^+$, as well as a number of other molecules, is directly coupled to the hydrolysis of ATP.

### *Structure and action of the $Na^+/K^+$-ATPase*

The $Na^+/K^+$-ATPase is an integral membrane protein consisting of 110 kDa α and 55 kDa β subunits. The functional unit is either a heterotetramer ($\alpha_2\beta_2$) or, more likely, a heterodimer (αβ; *Fig. 3*). Upon hydrolysis of one molecule of ATP to ADP and $P_i$ (the $P_i$ transiently binds to an aspartyl residue in the protein), the protein undergoes a conformational change and three $Na^+$ ions are

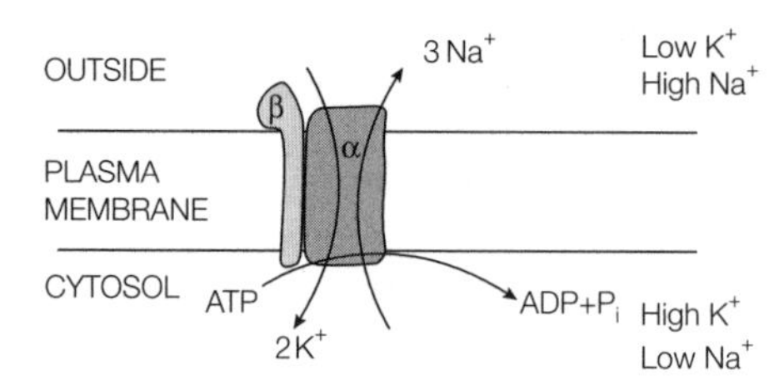


*Fig. 3. The $Na^+/K^+$-ATPase, shown as an αβ heterodimer.*

pumped out of the cell across the plasma membrane and two $K^+$ ions are pumped in the opposite direction into the cell. Both ions are being moved up their concentration gradients across the membrane; hence the requirement for an input of energy. No transport occurs unless ATP is hydrolyzed, and no ATP is hydrolyzed if there is no $Na^+$ and $K^+$ to transport (i.e. it is a **coupled system**).

***Ion-driven active transport***

The movement of the molecule to be transported across the membrane is coupled to the **movement of an ion**, usually either $Na^+$ or $H^+$. The energy for the movement of the molecule across the membrane against its concentration gradient comes from the movement of the ion down its concentration gradient. If both the molecule and the ion move in the same direction, it is termed **symport**, and the protein involved in the process is called a symporter (e.g. $Na^+$/glucose transporter; *Fig. 4a*); if the molecule and the ion move in the opposite direction, it is termed **antiport**, and the protein involved in the process is called an antiporter (e.g. erythrocyte band 3 anion transporter; *Fig. 4b*).

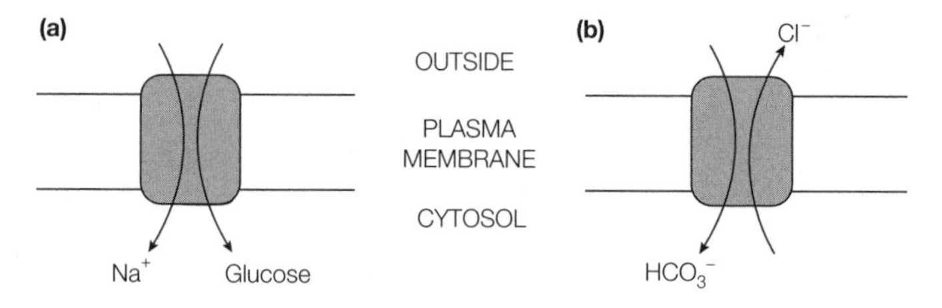


*Fig. 4. Ion-driven cotransport mechanisms. (a) Symport process involving a symporter (e.g. $Na^+$/glucose transporter); (b) antiport process involving an antiporter (e.g. erythrocyte band 3 anion transporter).*

## Glucose transport into intestinal epithelial cells

The cells lining the lumen of the intestine are **polarized**, that is they have two distinct sides or domains which have different lipid and protein compositions. The **apical** or **brush border** membrane facing the lumen is highly folded into **microvilli** to increase the surface area available for the absorption of nutrients. The rest of the plasma membrane, the **basolateral** surface, is in contact with neighboring cells and the blood capillaries (*Fig. 5*). Movement between adjacent epithelial cells is prevented by the formation of **tight junctions** around the cells near the apical domain. Thus any nutrient molecules in the lumen of the intestine have to pass through the cytosol of the epithelial cell in order to enter the blood.

Glucose (or other sugars and amino acids) are transported across the apical

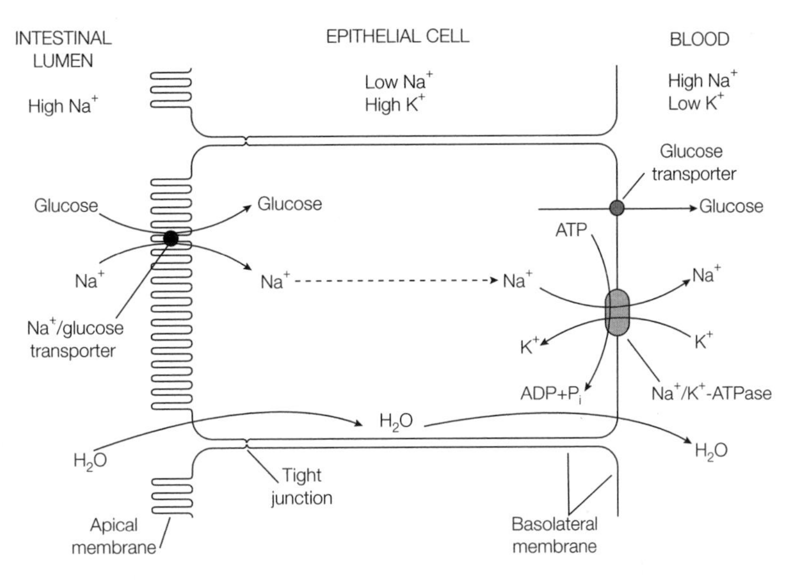


*Fig. 5. Transport of glucose and water across intestinal epithelial cells.*

membrane from a relatively low concentration in the lumen of the intestine to a relatively high concentration in the cytosol of the epithelial cell by a **glucose/Na$^+$ symporter protein** (*Fig. 5*). This is a form of **ion-driven active transport**; the energy for the movement of glucose against its concentration gradient comes from the movement of Na$^+$ down its concentration gradient. The blood flowing through the capillaries on the basolateral side of the epithelial cell maintains a concentration gradient of glucose across this membrane allowing the glucose to move out of the cell by **facilitated diffusion** through a **glucose transporter (a uniporter)** which is similar to the erythrocyte glucose transporter. The relatively low concentration of Na$^+$ inside the epithelial cell is maintained by a **Na$^+$/K$^+$-ATPase** on the basolateral membrane, an example of **ATP-driven active transport** (*Fig. 5*).

*Glucose rehydration therapy*

The movement of Na$^+$ and glucose from the lumen of the intestine across the epithelial cell to the blood sets up a difference in **osmotic pressure** across the cell. As a result, water flows through the cell, across the apical and basolateral membranes by **simple diffusion**. Hence the uptake of water requires both Na$^+$ and glucose (or amino acids) to be present in the lumen of the intestine. The presence of water alone in the lumen of the intestine is much less effective. This is the basis of **glucose rehydration therapy** as a remedy for dehydration; a solution of glucose and salt (NaCl) is administered to the patient. This is a simple, inexpensive but extremely important treatment which has saved the lives of many infants in developing countries who would have otherwise died of the effects of dehydration, usually associated with diarrhea.

# E4 Membrane transport: macromolecules

## Key Notes

**Exocytosis**

Exocytosis is the secretion of proteins out of the cell across the plasma membrane into the extracellular space. Proteins destined to be secreted are synthesized on ribosomes bound to the RER membrane and are then transported in membrane-bound vesicles to the Golgi apparatus where they are sorted and packaged up into secretory vesicles. All cells continuously secrete proteins via the constitutive pathway, whereas only specialized cells (e.g. of the pancreas, nerve cells) secrete proteins via the regulated secretory pathway in response to certain stimuli.

**Endocytosis**

Endocytosis is the uptake of macromolecules from the extracellular space into the cell across the plasma membrane via the formation of an intracellular vesicle pinching off from the plasma membrane.

**Phagocytosis**

Phagocytosis is the uptake of large particles (bacteria and cell debris). The particle binds to receptors on the surface of the phagocytic cell and the plasma membrane then engulfs the particle and ingests it via the formation of a large endocytic vesicle, a phagosome. Most protozoa utilize phagocytosis as a form of feeding, whereas in multicellular organisms only a few specialized cells (e.g. macrophages and neutrophils) can undergo phagocytosis.

**Pinocytosis**

Pinocytosis is the nonspecific uptake of extracellular fluid via small endocytic vesicles that pinch off from the plasma membrane. This is a constitutive process occurring in all eukaryotic cells.

**Receptor-mediated endocytosis**

Receptor-mediated endocytosis is the selective uptake of extracellular macromolecules (such as cholesterol) through their binding to specific cell-surface receptors. The receptor–macromolecule complex then accumulates in clathrin-coated pits and is endocytosed via a clathrin-coated vesicle.

**Clathrin-coated pits and vesicles**

Both endocytosis of material at the plasma membrane and exocytosis from the Golgi apparatus involve the formation of clathrin-coated pits and vesicles. On the cytosolic side of the membrane these structures have an electron-dense coat consisting mainly of the protein clathrin, the polypeptides of which form a three-legged structure known as a triskelion. The clathrin triskelions assemble into a basket-like convex framework that causes the membrane to invaginate at that point and eventually to pinch off and form a vesicle. In endocytosis these clathrin-coated vesicles migrate into the cell where the clathrin coats are lost before delivering their contents to the lysosomes.

**Related topics**

Eukaryotes (A2)
Membrane protein and carbohydrate (E2)
Membrane transport: small molecules (E3)
Signal transduction (E5)
Protein targeting (H4)

## Exocytosis

A cell often needs to secrete larger molecules than can be accommodated by the transport systems dealt with in Topic E3. **Exocytosis** refers to the movement of proteins out of the cell across the plasma membrane into the extracellular space. Proteins destined to be **secreted** from the cell are translated on ribosomes attached to the RER (see Topic H3). **Membrane-bound vesicles** containing these proteins then bud off from the RER, migrate through the cytosol and fuse with the membrane of the Golgi apparatus (*Fig. 1*). On transport through the endoplasmic reticulum and Golgi apparatus, various post-translational modifications to the proteins take place, such as glycosylation (see Topic H5).

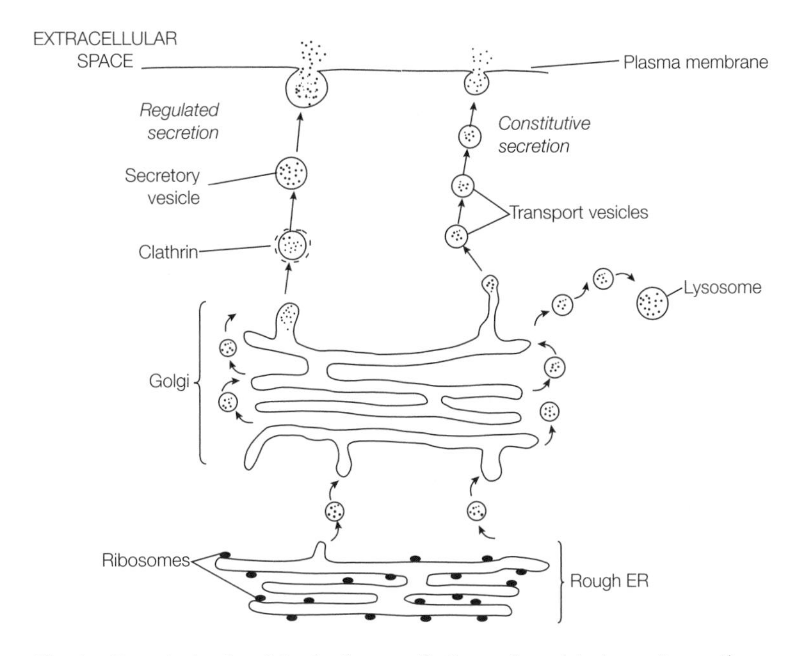


*Fig. 1. Exocytosis of proteins by the constitutive and regulated secretory pathways.*

The **Golgi apparatus** is the major sorting center of the cell where proteins and lipids are packaged into discrete vesicles and then targeted to the appropriate part of the cell (see Topic H4). For example, vesicles containing lysosomal proteins will be targeted to the lysosomes (see Topic A2). If the vesicles contain no specific targeting signal they will migrate through the cytosol to the plasma membrane, the so-called default or **constitutive secretory pathway** (*Fig. 1*). Transport vesicles destined for the plasma membrane in the constitutive secretory pathway leave the Golgi apparatus in a steady stream. The membrane proteins and lipids in these vesicles provide new material for the plasma membrane, while the soluble proteins inside the vesicles are secreted to the extracellular space. All cells have this constitutive secretory pathway.

In certain cells, however, an additional secretory pathway exists, the **regulated secretory pathway**. This pathway is found mainly in cells that are specialized for secreting products rapidly on demand in response to a particular stimulus (*Fig. 1*). For example, the hormone insulin and digestive enzymes are secreted by the pancreas, while neurotransmitters are secreted by nerve cells. In these cells such substances are initially stored in secretory vesicles which form by **clathrin-coated** budding from the Golgi apparatus (see Topic A2). The clathrin coat then dissociates and the vesicles remain in the cytosol until signaled to release their contents on fusion with the plasma membrane.

## Endocytosis

**Endocytosis** is the uptake of extracellular macromolecules across the plasma membrane into the cell. The material to be ingested is progressively enclosed by a small portion of the plasma membrane, which first invaginates and then pinches off to form an intracellular vesicle containing the ingested macromolecule. Endocytosis can be divided into three distinct types depending on the size of the ingested macromolecule and whether specific cell surface receptors are involved. These three processes are: **phagocytosis**, **pinocytosis** and **receptor-mediated endocytosis**.

## Phagocytosis

**Phagocytosis** is the ingestion of large particles such as bacteria and cell debris via large endocytic vesicles called **phagosomes**. In order to be ingested the particle must first bind to the surface of the phagocyte, usually through specialized cell-surface receptors. Once bound to the receptors, the phagocyte is stimulated to begin engulfing the particle with its plasma membrane, thereby enclosing it within a phagosome (*Fig. 2a*). The phagosome then fuses with a lysosome and the ingested particle is broken down. Utilizable material will be transported into the cytosol, while indigestible substances will remain in the lysosomes, forming residual bodies. In protozoa, phagocytosis is a form of feeding, where the ingested material is broken down in the lysosomes and utilized as food. In multicellular organisms only a few specialized cells are capable of phagocytosis. **Macrophages** and **neutrophils** (white blood cells) use phagocytosis to protect the organism against infection by ingesting invading microorganisms. Macrophages are also involved in scavenging dead and damaged cells and cell debris.

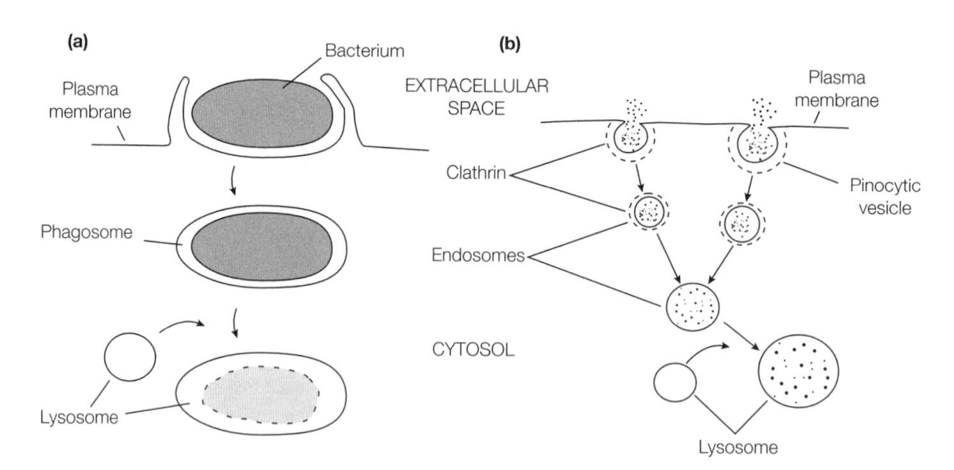


*Fig. 2. (a) Phagocytosis and (b) pinocytosis.*

## Pinocytosis

Unlike phagocytosis, which is a regulated form of endocytosis carried out by a small number of cell types, **pinocytosis** is a constitutive process that occurs continuously in all eukaryotic cells. Small areas of the plasma membrane are ingested in the form of small **pinocytic vesicles** that are later returned to the cell surface. As the pinocytic vesicle forms, a small amount of extracellular fluid is enclosed in the vesicle and also taken up into the cell (*Fig. 2b*). This nonspecific uptake of extracellular fluid is often referred to as **cell-drinking** or **fluid-phase endocytosis**. The initial endocytosis begins by the formation of a **clathrin-coated pit** (see below) at specialized regions of the plasma membrane (*Fig. 2b*). The resulting **clathrin-coated endocytic vesicle (endosome)** is then endocytosed, the clathrin coat dissociates and the endosome fuses with a lysosome (see below).

## Receptor-mediated endocytosis

**Receptor-mediated endocytosis** is the selective uptake of macromolecules from the extracellular fluid via **clathrin-coated pits and vesicles**. This process, which takes place in most animal cells, involves the macromolecule binding specifically to a **cell-surface receptor** (which is an integral membrane protein; see Topics E2 and E5). Once bound to the receptor, the receptor–macromolecule complex accumulates in a clathrin-coated pit and is then endocytosed in a clathrin-coated vesicle (*Fig. 3*, and see below). Receptor-mediated endocytosis provides a way of selectively concentrating particular macromolecules that are at low concentrations in the extracellular fluid, thereby increasing the efficiency of their uptake without having to take in large quantities of extracellular fluid. One of the best studied and understood receptor-mediated endocytic processes is the uptake of **cholesterol** by mammalian cells (see Topics K5 and K6). Many **viruses** and other **toxins** gain entry to animal cells by receptor-mediated endocytosis. Although the cells do not purposely have cell-surface receptors that recognize the viral particle, the virus has evolved to express a protein on its surface that mimics the correct ligand recognized by the receptor, thus allowing the virus to bind and be internalized.

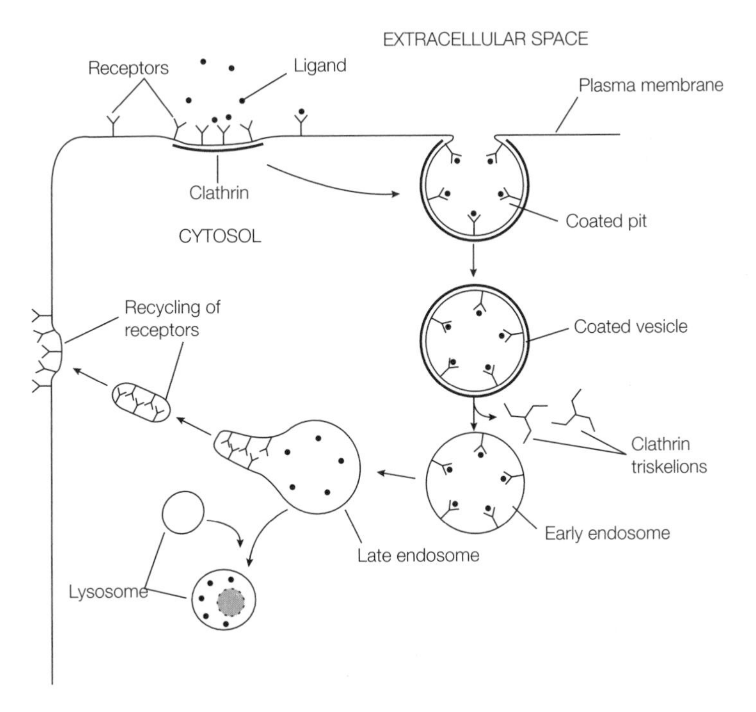


*Fig. 3. Receptor-mediated endocytosis involves clathrin-coated pits and vesicles.*

## Clathrin-coated pits and vesicles

**Clathrin-coated pits and vesicles** are involved in both the exocytosis of proteins from the Golgi apparatus and the endocytosis of material at the plasma membrane. Electron micrographs (see Topic A3) of clathrin-coated pits reveal that these pits are invaginations of the plasma membrane that are coated on their cytosolic surface with a densely packed material made up predominantly of the protein **clathrin**. This protein, which has been highly conserved throughout evolution, consists of three large and three small polypeptide chains that together form a three-legged structure called a **triskelion**. These clathrin triskelions assemble into a basket-like convex framework of hexagons and pentagons to form the coated pits. The assembly of the clathrin coat is thought to drive the membrane to invaginate at that point. As further clathrin triskelions are added to the structure a complete cage forms, pinching off a region of the membrane and forming a clathrin-coated vesicle. In endocytosis these vesicles then migrate into the cell, shed their clathrin coat and become **early endosomes** (*Fig. 3*). The early endosomes then migrate towards the Golgi apparatus and the nucleus, becoming **late endosomes** before fusing with the lysosomes.

# E5 SIGNAL TRANSDUCTION

## Key Notes

**Cell signaling**

Cells communicate with one another in multicellular organisms using extracellular signaling molecules or hormones. The hormone is secreted by the signaling cell and then binds to a receptor on the target cell, initiating a response in that cell. In endocrine signaling the hormone acts at a distant site in the body from where it was produced, in paracrine signaling the hormone acts on nearby cells, and in autocrine signaling the hormone acts on the same cell from which it was secreted.

**Hormones**

Some lipophilic hormones (e.g. the steroid hormones, thyroxine, retinoic acid and vitamin D) diffuse across the plasma membrane and interact with intracellular receptors in the cytosol or nucleus. Other lipophilic hormones (e.g. the prostaglandins) and hydrophilic hormones (e.g. the peptide hormones insulin and glucagon and the biogenic amines epinephrine and histamine) bind to receptor proteins in the plasma membrane.

**Cell-surface receptors**

Cell-surface receptors are integral membrane proteins located in the plasma membrane that bind the hormone (ligand) with high affinity and specificity. On binding the ligand, the receptor undergoes a conformational change and transmits the information into the cell (signal tranduction). Enzyme-linked receptors (e.g. the insulin receptor) have an intrinsic enzyme activity that results in the modification of an intracellular target protein. Ion channel-linked receptors change conformation to allow ions to flow across the membrane thereby altering the membrane potential. G protein-linked receptors activate G [guanosine triphosphate (GTP)-binding] proteins that in turn lead to the production of an intracellular second messenger.

**Second messengers**

Intracellular signaling molecules (second messengers) are produced in response to the activation of G proteins by cell-surface receptors. The second messengers cAMP and 3′,5′-cyclic guanosine monophosphate (cGMP) are produced by adenylate cyclase and guanylate cyclase, respectively. Activation of phospholipase C leads to the production of the second messengers inositol 1,4,5-trisphosphate ($IP_3$) and 1,2-diacylglycerol (DAG) which in turn cause the release of $Ca^{2+}$ from intracellular stores and activate protein kinase C, respectively.

**Related topics**

Membrane lipids (E1)
Membrane protein and carbohydrate (E2)
Control of glycogen metabolism (J7)
Cholesterol (K5)
Nerve (N3)

## Cell signaling

In multicellular organisms there is a need for the cells to **communicate** with one another in order to **coordinate** their growth and metabolism. The principal way by which cells communicate with each other is by means of **extracellular signaling molecules** or **hormones**. These molecules are synthesized

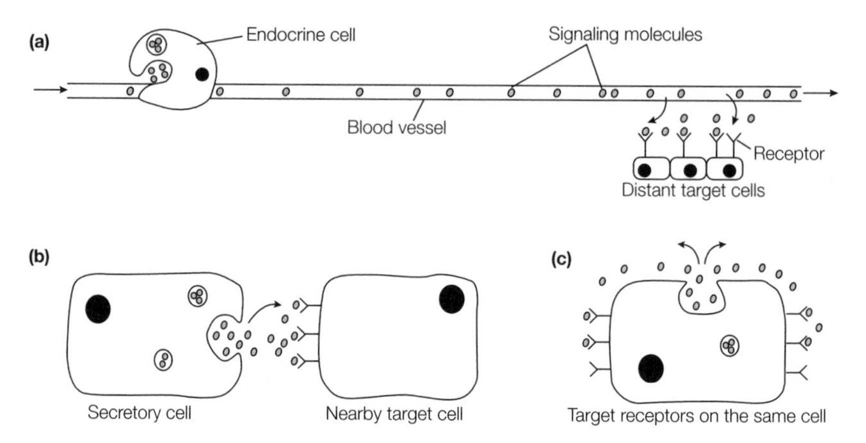


*Fig. 1. Cellular signaling. (a) Endocrine, (b) paracrine, (c) autocrine.*

and secreted by **signaling cells** and produce a specific response in **target cells** that have **specific receptors** for the signaling molecule. Different cells can respond differently to the same signaling molecule depending on the type of receptor and the intracellular reactions initiated. Cell signaling can be classified into three distinct types based on the distance over which the signaling molecule acts. In **endocrine signaling**, the signaling molecule (e.g. insulin) acts on target cells distant from its site of synthesis in cells of an endocrine organ (e.g. the pancreas; *Fig. 1a*). The endocrine cells secrete the signaling molecule into the bloodstream (if an animal) or the sap (if a plant) which carries it to the target cells elsewhere in the organism. In **paracrine signaling**, the signaling molecule affects only target cells close to the cell from which it was secreted (*Fig. 1b*). The communication from one nerve cell to another by chemical neurotransmitters is an example of paracrine signaling (see Topic N3). The third type of cell signaling is **autocrine signaling**, where a cell responds to a molecule that it has produced itself (*Fig. 1c*).

## Hormones

The signaling molecules or hormones can be classified based on their solubility and the location of their receptor.

### *Lipophilic hormones with intracellular receptors*

Small lipophilic (lipid-soluble) hormones diffuse across the plasma membrane and then interact with **intracellular receptors** in the cytosol or nucleus. The resulting hormone–receptor complex often binds to regions of the DNA and affects the transcription of certain genes (see Topic G7). Small lipophilic hormones with intracellular receptors include the **steroid hormones** which are synthesized from cholesterol (see Topic K5) (e.g. the female sex hormones estrogen and progesterone), **thyroxine** which is produced by thyroid cells and is the principal iodinated compound in animals, **retinoic acid** which is derived from vitamin A, and **vitamin D** which is synthesized in the skin in response to sunlight (see Topic K5).

### *Lipophilic hormones with cell-surface receptors*

The principal lipophilic (lipid-soluble) hormones that bind to receptors located in the plasma membrane are the **prostaglandins**, a family of structurally similar compounds that are found in both vertebrates and invertebrates. Prostaglandins are synthesized from **arachidonic acid** (a 20-carbon fatty acid with four unsaturated double bonds) (see Topic K1) and act as paracrine signaling molecules. **Aspirin** and other **anti-inflammatory agents** inhibit the synthesis of prostaglandins.

*Hydrophilic hormones with cell-surface receptors*
All hydrophilic (water-soluble) molecules (which cannot diffuse across the hydrophobic interior of the lipid bilayer) bind to receptors in the plasma membrane. There are two subclasses of hydrophilic hormones: (1) **peptide hormones** such as **insulin** and **glucagon**; and (2) small charged molecules, often **biogenic amines**, such as **epinephrine** (**adrenalin**) and **histamine** that are derived from amino acids and function as hormones and neurotransmitters (see Topic N3).

## Cell-surface receptors

Hydrophilic and some lipophilic hormones bind to cell-surface receptors. These are **integral membrane proteins** situated in the plasma membrane (see Topic E2) that bind the signaling molecule (**ligand**) with high affinity. The ligand binds to a specific site on the receptor in much the same way as a substrate binds to an enzyme (see Topic C1). Binding of the ligand to the receptor causes a **conformational change** in the receptor that initiates a sequence of reactions in the target cell (often referred to as **signal transduction**) leading to a change in cellular function. The distribution of receptors varies on different cells, and there is often more than one type of receptor for a particular ligand, allowing different target cells to respond differently to the same signaling molecule. Cell-surface receptors can be classified into three classes depending on how they transfer the information from the ligand to the interior of the cell: enzyme-linked receptors, ion channel-linked receptors and G protein-linked receptors.

*Enzyme-linked receptors*
On binding of the ligand to its extracellular face, the cell-surface receptor undergoes a conformational change and activates an **intrinsic enzyme activity** (*Fig.* 2). In the case of the **insulin receptor** which is a complex of two α- and two β-subunits held together by disulfide bonds, the polypeptide hormone insulin (the ligand) binds to the extracellular face of the α-subunits (*Fig.* 2). The receptor then undergoes a conformational change leading to the **autophos-**

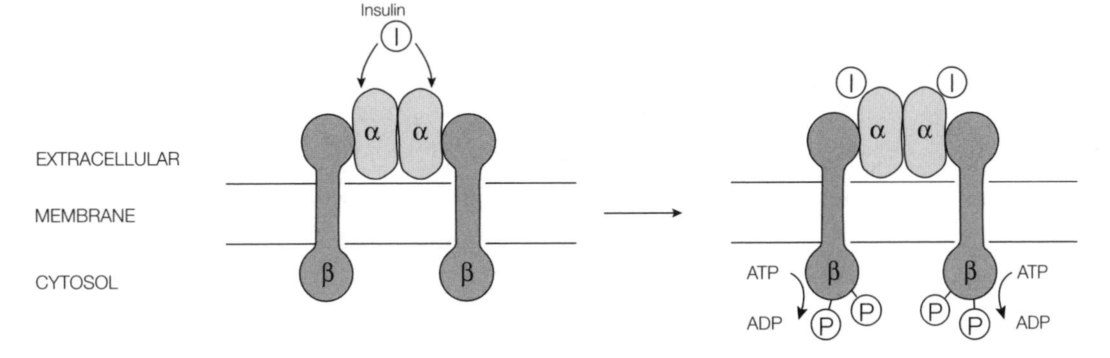


*Fig. 2. Signal transduction through an enzyme-linked receptor such as the insulin receptor.*

**phorylation** (self-phosphorylation) of the cytosolic domain of the β-subunit. Specifically the hydroxyl groups in the side chains of certain **tyrosine residues** are phosphorylated, with **ATP** being the phosphate donor. The phosphorylated receptor is then recognized by other proteins in the cytosol that in turn modulate various intracellular events, allowing the cell to respond to the hormone appropriately (see Topic J7). In addition, the β-subunit can directly phosphorylate other target proteins within the cell.

### *Ion channel-linked receptors*

Here binding of the ligand again causes a conformational change in the protein but this time such that a specific **ion channel** is opened (*Fig. 3*). This allows a certain ion to flow through that subsequently alters the **electric potential** across the membrane. For example, at the nerve–muscle junction the **neurotransmitter acetylcholine** binds to specific receptors that allow $Na^+$ ions to flow into and $K^+$ ions out of the target cell (see Topic N3).

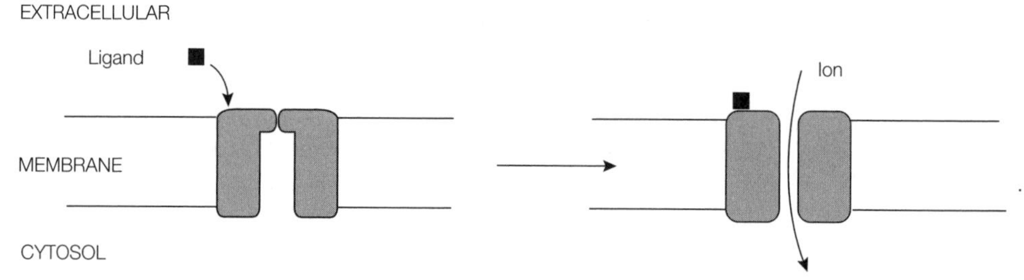


*Fig. 3. Signal transduction through an ion channel-linked receptor.*

### *G protein-linked receptors*

On binding its ligand, a G protein-linked receptor activates **G proteins [guanyl nucleotide (GTP)-binding proteins]** which in turn activate or inhibit either an enzyme that generates a specific **second messenger** or an **ion channel**, causing a change in the membrane potential (*Fig. 4a*). **Epinephrine** and **glucagon** act through interaction with G protein-linked receptors (see Topic J7). The majority of G protein-linked receptors contain **seven transmembrane α-helices**. Thus they have a similar overall shape to that of **bacteriorhodopsin** (which is not a receptor) (see Topic E2). G proteins are localized on the **cytosolic face** of the plasma membrane and act as on–off molecular switches. When it has guanosine diphosphate (GDP) bound, the G protein is in the 'off' state. The activated receptor causes it to release the GDP and exchange it for GTP, converting it to the 'on' state (*Fig. 4a* and *b*). The activated G protein with its bound GTP then dissociates from the receptor and binds to and activates an effector enzyme (e.g. adenylate cyclase) which in turn catalyzes the formation of a second messenger (e.g. cAMP). The G protein then hydrolyzes the bound GTP, causing it to revert back to the 'off' state (*Fig. 4b*). **Cholera toxin** acts by inhibiting the intrinsic GTPase activity of the G protein (*Fig. 4b*), with the result that once activated to the GTP-bound state the G protein cannot be turned off again.

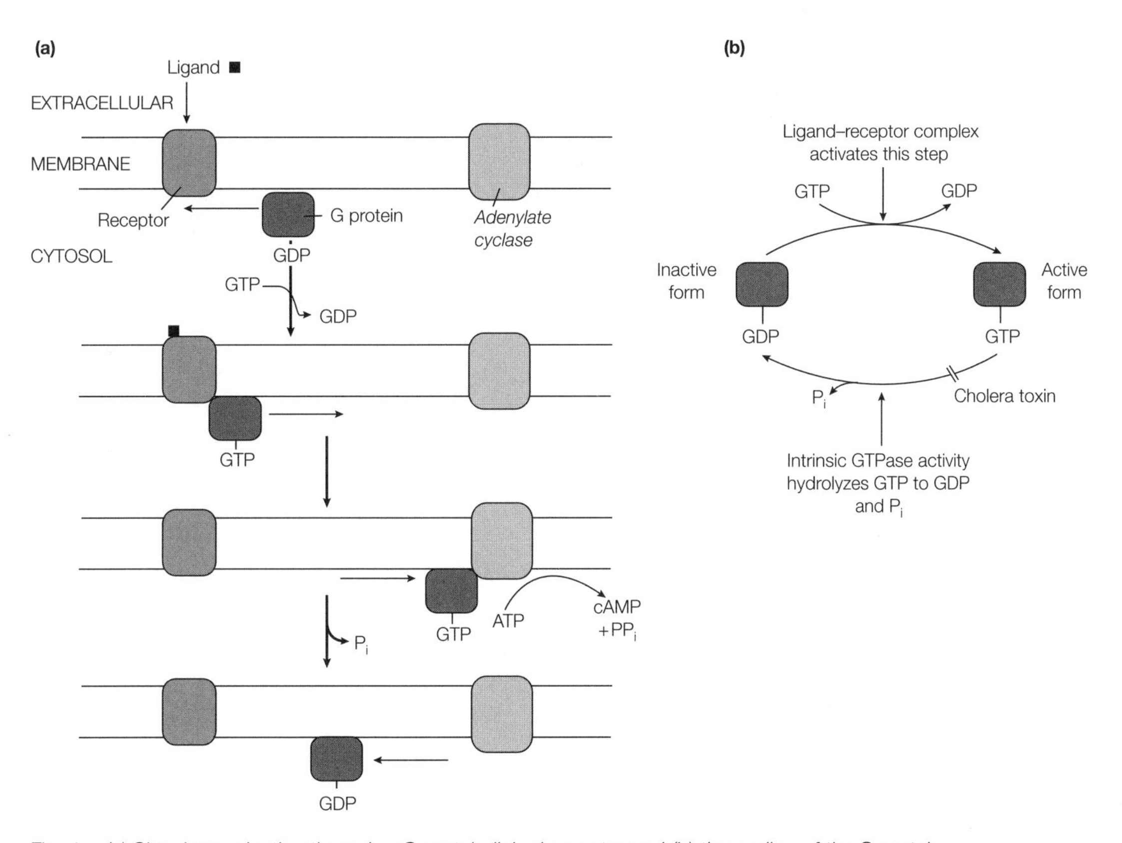


*Fig. 4. (a) Signal transduction through a G protein-linked receptor and (b) the cycling of the G protein.*

## Second messengers

The binding of ligands to many G protein-linked receptors leads to a short-lived increase in the concentration of certain **intracellular signaling molecules** called **second messengers**. (The hormone/ligand can be considered as the first messenger.) The major second messengers are 3′,5′-cyclic AMP (**cAMP**), 3′,5′-cyclic GMP (**cGMP**), **inositol 1,4,5-trisphosphate** ($IP_3$), **1,2-diacylglycerol** (DAG) and **$Ca^{2+}$**. The elevation in the level of one or other of these second messengers then leads to a rapid alteration in cellular function.

cAMP and cGMP are derived from ATP and GTP by the actions of **adenylate cyclase** (see *Fig. 4a*) and **guanylate cyclase**, respectively. For example, the action of **glucagon** on glycogen metabolism is mediated through the second messenger cAMP (see Topic J7).

$IP_3$ and DAG are derived from the membrane lipid phosphatidylinositol 4,5-bisphosphate (which is a phosphorylated derivative of phosphatidylinositol; see Topic E1) by the action of **phospholipase C** which is also located in the plasma membrane and, like adenylate cyclase, is activated by G proteins (*Fig. 5*). One of the main actions of the polar $IP_3$ is to diffuse through the cytosol and interact with $Ca^{2+}$ channels in the membrane of the ER (*Fig. 5*), causing the release of stored $Ca^{2+}$ ions which in turn mediate various cellular responses. The DAG produced by the hydrolysis of phosphatidylinositol 4,5-bisphosphate, along with $Ca^{2+}$ ions released from the ER, activates **protein kinase C**, a membrane-bound enzyme that phosphorylates various target proteins, again leading to alterations in a variety of cellular processes (*Fig. 5*).

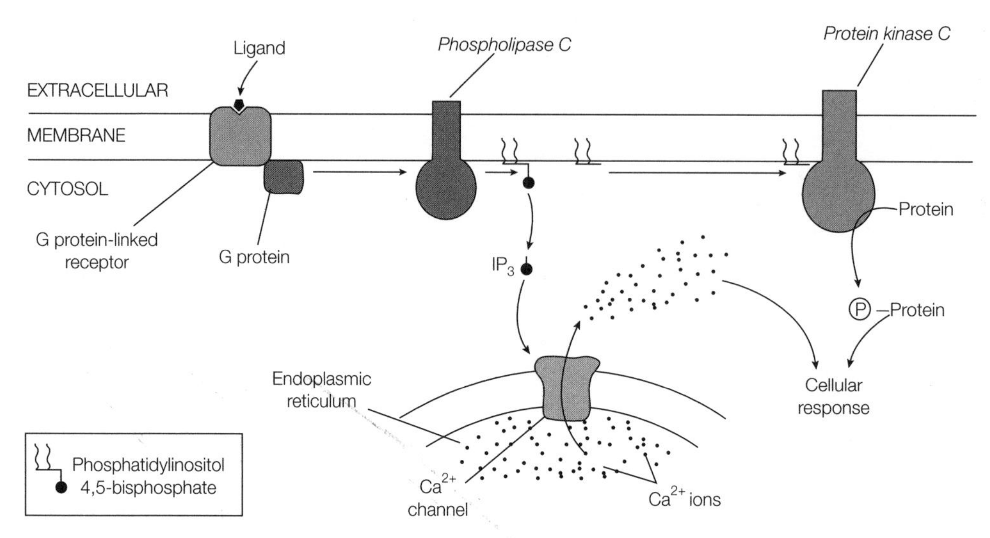


*Fig. 5. Generation of the intracellular second messengers inositol 1,4,5-trisphosphate ($IP_3$), 1,2-diacylglycerol (DAG) and $Ca^{2+}$.*

# F1 DNA STRUCTURE

## Key Notes

**Bases**

In DNA there are four bases: adenine (abbreviated A), guanine (G), thymine (T) and cytosine (C). Adenine and guanine are purines; thymine and cytosine are pyrimidines.

**Nucleosides**

A nucleoside is a pyrimidine or purine base covalently bonded to a sugar. In DNA, the sugar is deoxyribose and so this is a deoxynucleoside. There are four types of deoxynucleoside in DNA; deoxyadenosine, deoxyguanosine, deoxythymidine and deoxycytidine.

**Nucleotides**

A nucleotide is base + sugar + phosphate covalently bonded together. In DNA, where the sugar is deoxyribose, this unit is a deoxynucleotide.

**3′5′ phosphodiester bonds**

In DNA the nucleotides are covalently joined together by 3′5′ phosphodiester bonds to form a repetitive sugar–phosphate chain which is the backbone to which the bases are attached.

**DNA sequence**

The DNA sequence is the sequence of A, C, G and T along the DNA molecule which carries the genetic information.

**DNA double helix**

In a DNA double helix, the two strands of DNA are wound round each other with the bases on the inside and the sugar–phosphate backbones on the outside. The two DNA chains are held together by hydrogen bonds between pairs of bases; adenine (A) always pairs with thymine (T) and guanine (G) always pairs with cytosine (C).

**Related topics**

DNA replication in bacteria (F3)
DNA replication in eukaryotes (F4)
RNA structure (G1)
Transcription in prokaryotes (G2)
Transcription in eukaryotes: an overview (G5)

## Bases

The bases in DNA have carbon–nitrogen ring structures; because of the nitrogen atoms they are called nitrogenous bases. There are two types of ring structure. **Adenine** and **guanine** are **purines** (*Fig. 1a*), each having two joined carbon–nitrogen rings but with different side-chains. **Thymine** and **cytosine** are **pyrimidines** (*Fig. 1a*); each has only one carbon–nitrogen ring and again they differ in their side-chains.

## Nucleosides

In RNA, the nucleosides have ribose as the sugar component (see Topic G1) and so are ribonucleosides. In DNA the sugar is deoxyribose (*Fig. 1b*) (i.e. the 2′-OH group in ribose is replaced by a hydrogen atom; hence 'deoxy') and so the nucleosides are **deoxynucleosides**. For DNA these are **deoxyadenosine, deoxyguanosine, deoxythymidine** and **deoxycytidine**. In each case, the C-1 of the sugar is joined to the base via one of its nitrogen atoms. If the base is a pyrimidine, the nitrogen at the 1 position (i.e. N-1) is involved in bonding to

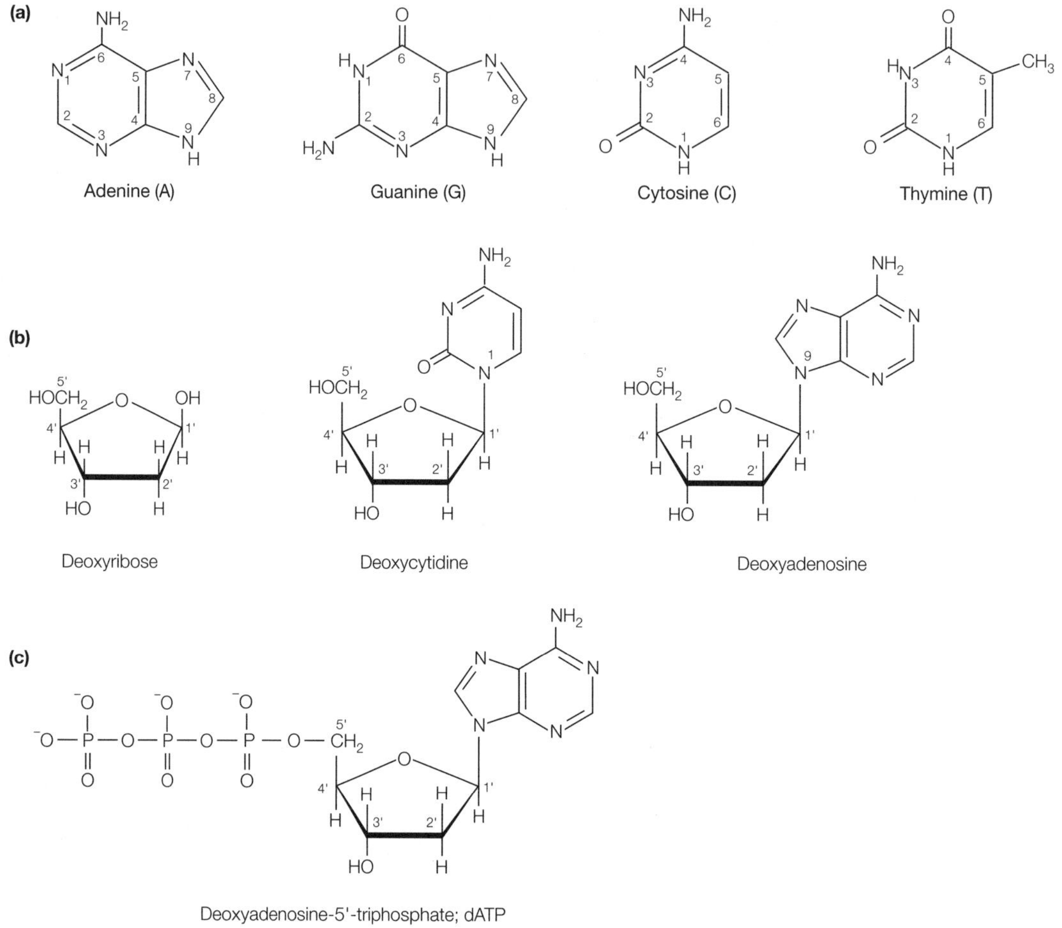

(a)

Adenine (A) Guanine (G) Cytosine (C) Thymine (T)

(b)

Deoxyribose Deoxycytidine Deoxyadenosine

(c)

Deoxyadenosine-5'-triphosphate; dATP

*Fig. 1. (a) The purines, adenine and guanine, and the pyrimidines, thymine and cytosine; (b) deoxyribose and two deoxynucleosides, deoxycytidine and deoxyadenosine; (c) a deoxynucleotide, deoxyadenosine 5′ triphosphate (dATP).*

the sugar. If the base is a purine, the bonding is to the N-9 position of the base (*Fig. 1b*).

## Nucleotides

A nucleotide is a phosphate ester of a nucleoside. It consists of a phosphate group joined to a nucleoside at the hydroxyl group attached to the C-5 of the sugar, that is it is a **nucleoside 5′-phosphate** or a **5′-nucleotide**. The primed number denotes the atom of the sugar to which the phosphate is bonded. In DNA the nucleotides have deoxyribose as the sugar and hence are called **deoxynucleotides**. Deoxynucleotides may have a single phosphate group (**deoxynucleoside 5′-monophosphates**, dNMPs), two phosphate groups (**deoxynucleoside 5′-diphosphates**, dNDPs) or three phosphate groups (**deoxynucleoside 5′-triphosphates**, dNTPs). Deoxynucleoside triphosphates are the precursors for DNA synthesis. These are deoxyadenosine 5′-triphosphate (dATP) (*Fig. 1c*), deoxyguanosine 5′-triphosphate (dGTP), deoxycytidine 5′-triphosphate (dCTP) and deoxythymidine 5′-triphosphate (dTTP). In each case the 'd' in the abbreviation (for example in dATP) indicates that the sugar in the nucleotide is deoxyribose. During DNA synthesis (see Topics F3 and F4), two of the phosphates of each deoxynucleotide are split off (as pyrophosphate) so that only a single phosphate (the α phosphate) is incorporated into DNA.

## 3′5′ phosphodiester bonds

In a DNA molecule, the different nucleotides are covalently joined to form a long polymer chain by covalent bonding between the phosphates and sugars. For any one nucleotide, the phosphate attached to the hydroxyl group at the 5′ position of the sugar is in turn bonded to the hydroxyl group on the 3′ carbon of the sugar of the next nucleotide. Since each phosphate–hydroxyl bond is an ester bond, the linkage between the two deoxynucleotides is a **3′5′ phosphodiester bond** (*Fig. 2*). Thus, in a DNA chain, all of the 3′ and 5′ hydroxyl groups are involved in phosphodiester bonds except for the first and the last nucleotide in the chain. The first nucleotide has a 5′ phosphate not bonded to any other nucleotide and the last nucleotide has a free 3′ hydroxyl. Thus each DNA chain has **polarity;** it has a **5′ end** and a **3′ end**.

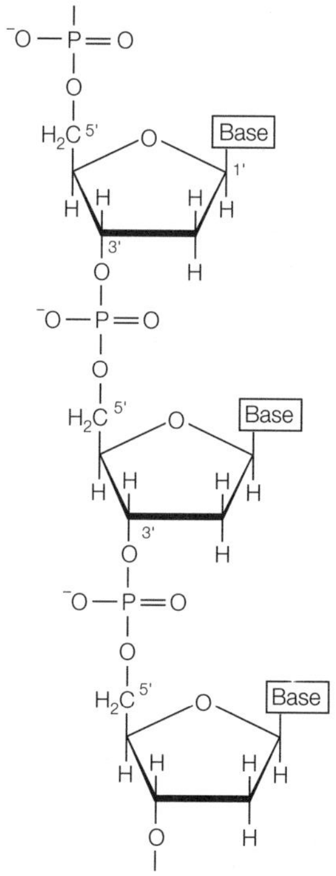

*Fig. 2. 3′5′ phosphodiester bonds formed between nucleotides in a DNA molecule.*

## DNA sequence

Each nucleotide can be thought of as a single letter in an alphabet that has only four letters, A, G, C and T. Different genes have different sequences of these four nucleotides and so code for different biological messages. Since the deoxynucleotides in DNA differ only in the bases they carry, this sequence of deoxynucleotides can be recorded simply as a **base sequence**. For example, ACTTTCAGACC is part of the base sequence of one gene and codes for part of one protein whereas TGGAACCGTCA is part of the base sequence of a different gene coding for a different protein. Traditionally the base sequence is written in the order from the 5′ end of the DNA strand to the 3′ end, that is it is **written in the 5′→3′ direction**. Given that there are four types of nucleotide, the number of different possible sequences (or messages) in a DNA strand $n$ nucleotides long is $4^n$. DNA molecules are typically many thousands of nucleotides long so that the number of possible messages is enormous.

## DNA double helix

In 1953, Watson and Crick worked out the three-dimensional structure of DNA, starting from X-ray diffraction photographs taken by Franklin and Wilkins. They deduced that DNA is composed of two strands wound round each other to form a double helix, with the bases on the inside and the sugar–phosphate backbones on the outside. In the double helix (*Fig. 3*), the two DNA strands are organized in an **antiparallel** arrangement (i.e. the two strands run in opposite directions, one strand is oriented 5′→3′ and the other is oriented 3′→5′). The bases of the two strands form hydrogen bonds to each other; A pairs with T and G pairs with C. This is called **complementary base pairing** (*Fig. 4*). Thus a large two-ringed purine is paired with a smaller single-ringed pyrimidine and the two bases fit neatly in the gap between the sugar–phosphate strands and maintain the correct spacing. There would be insufficient space for two large purines to pair and too much space for two pyrimidines to pair, which would be too far apart to bond. The G:C and A:T base pairing also maximizes the number of effective hydrogen bonds that can form between the bases; there are three hydrogen bonds between each G:C base pair and two hydrogen bonds between each A:T base pair. Thus A:T and G:C base pairs form the most stable conformation both from steric considerations and from the point of view of maximizing hydrogen bond formation.

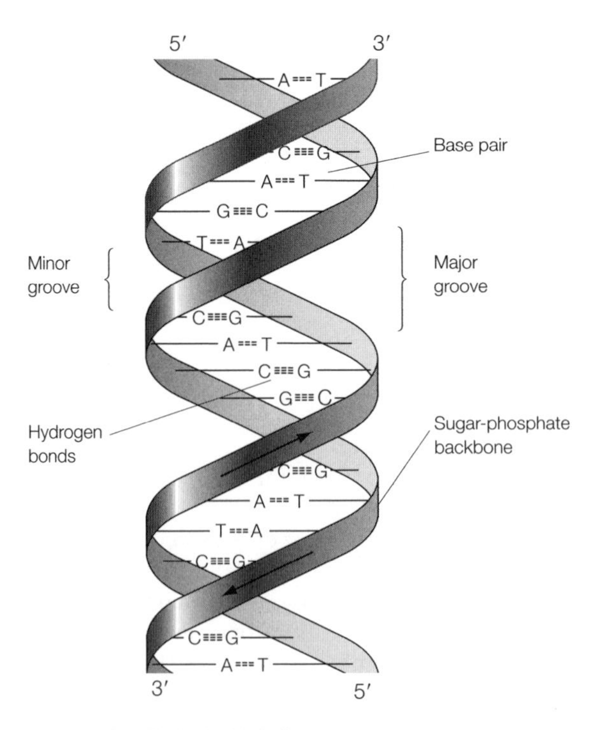


*Fig. 3. The DNA double helix.*

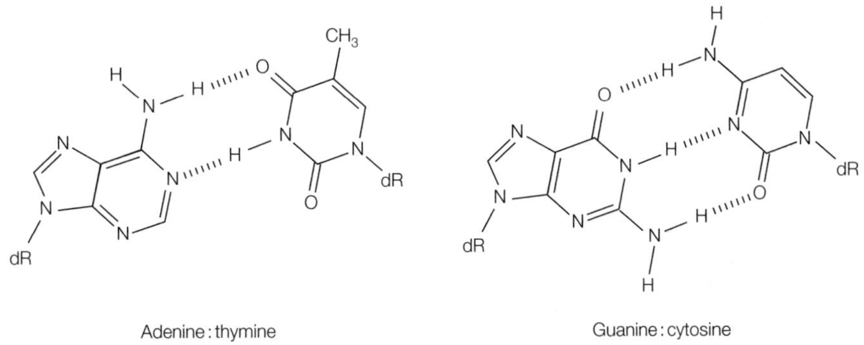

Adenine : thymine

Guanine : cytosine

*Fig. 4. The DNA base pairs. Hydrogen bonds are shown as dashed lines. dR, deoxyribose.*

# F2 CHROMOSOMES

## Key Notes

| | |
|---|---|
| **Prokaryotic chromosomes** | The DNA in a bacterium is a supercoiled double-stranded circular molecule that is packaged in the nucleoid region of the cell. The DNA is negatively supercoiled, complexed to several histone-like proteins (mainly proteins HU, HSP-1 and H-NS) and organized into about 50 domains bound to a protein scaffold. |
| **Eukaryotic chromosomes** | Eukaryotic cells contain much more DNA than prokaryotes. In the nucleus, the DNA is packaged into chromosomes that consist mainly of DNA and proteins called histones although other nonhistone proteins (NHP) are also present. Each chromosome contains a single linear double-stranded DNA molecule. |
| **Nucleosomes** | The chromosomal DNA is complexed with five types of histone (H1, H2A, H2B, H3 and H4). These are very basic proteins, rich in arginine and lysine. The amino acid sequences of histones are highly conserved in evolution. The DNA is wound round a histone octamer (two molecules each of H2A, H2B, H3 and H4) to form a nucleosome. The DNA between neighboring nucleosomes (linker DNA) binds histone H1. The packing ratio of nucleosomes is about 7. |
| **30 nm fiber** | Nucleosomes are organized into a 30 nm fiber, possibly by forming a higher order helix called a solenoid. The overall packing ratio is about 40. |
| **Radial loops** | The 30 nm fiber is attached to a central protein scaffold in each chromosome in a series of radial loops. |
| **Related topics** | DNA structure (F1) DNA replication in eukaryotes (F4) |

## Prokaryotic chromosomes

The DNA of a bacterial cell, such as *Escherichia coli*, is a circular double-stranded molecule often referred to as the **bacterial chromosome**. In *E. coli* this DNA molecule contains 4.6 million base pairs. The circular DNA is packaged into a region of the cell called the **nucleoid** (see Topic A1) where it is organized into 50 or so **loops** or **domains** that are bound to a central **protein scaffold**, attached to the cell membrane. *Fig. 1a* illustrates this organization, although only six loops are shown for clarity. Within this structure, the DNA is actually not a circular double-stranded DNA molecule such as that shown in *Fig. 1b* but is negatively **supercoiled**, that is, it is twisted upon itself (*Fig. 1c*) and is also complexed with several **DNA-binding proteins**, the most common of which are proteins **HU**, **HLP-1** and **H-NS**. These are **histone-like proteins** (see below for a description of histones).

## Eukaryotic chromosomes

The genomic DNA of a eukaryotic cell is contained within a specialized organelle, the nucleus. A typical human cell contains 1000 times more DNA

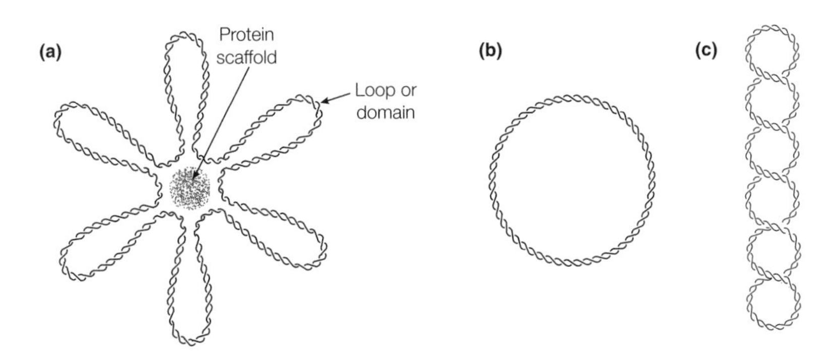


*Fig. 1. (a) The association of circular bacterial DNA with a protein scaffold; (b) a circular double-stranded DNA molecule; (c) supercoiled DNA.*

than the bacterium *E. coli.* This very large amount of eukaryotic nuclear DNA is tightly packaged in chromosomes. With the exception of the sex chromosomes, diploid eukaryotic organisms such as humans have two copies of each chromosome, one inherited from the father and one from the mother. Chromosomes contain both DNA and protein. Most of the protein on a weight basis is **histones**, but there are also many thousands of other proteins found in far less abundance and these are collectively called **nonhistone proteins** (NHP). This nuclear DNA–protein complex is called **chromatin**. The mitochondria and chloroplasts of eukaryotic cells also contain DNA but, unlike the nuclear DNA, this consists of double-stranded circular molecules resembling bacterial chromosomes.

In the nucleus, each chromosome contains a single linear double-stranded DNA molecule. The length of the packaged DNA molecule varies. In humans, the shortest DNA molecule in a chromosome is about 1.6 cm and the longest is about 8.4 cm. During the **metaphase** stage of mitosis, when the chromosomes align on the mitotic spindle ready for segregation, they are at their most condensed and range in size from only 1.3 μm to 10 μm long. Thus the **packing ratio**, that is the ratio of the length of the linear DNA molecule to the length of the metaphase chromosome, is about $10^4$. In the time period between the end of one mitosis and the start of the next (i.e. **interphase**), the chromatin is more disperse. Here the packing ratio is in the range $10^2$–$10^3$. Overall, the extensive packaging of DNA in chromosomes results from three levels of folding involving nucleosomes, 30 nm filaments and radial loops.

## Nucleosomes

The first level of packaging involves the binding of the chromosomal DNA to histones. Overall, in chromosomes, the ratio of DNA to histones on a weight basis is approximately 1:1. There are five main types of histones called H1, H2A, H2B, H3 and H4. Histones are very basic proteins; about 25% of their amino acids are lysine or arginine so histones have a large number of positively charged amino acid side chains. These positively charged groups therefore bind to the negatively charged phosphate groups of DNA. Not surprisingly given their importance in packaging DNA, the amino acid sequences of histones have been highly conserved in evolution. The most conserved are histones H3 and H4; for example, H3 and H4 from peas and cows differ in only four and two amino acids respectively! Histone H1 is the least conserved histone, which reflects its somewhat

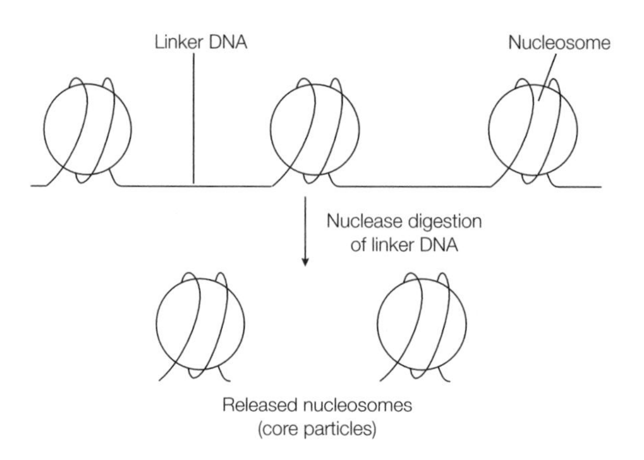


*Fig. 2. 'Beads-on-a-string' structure of chromatin.*

different role in packaging DNA compared with the other histones (see below). In sperm heads, DNA is particularly highly condensed and here the histones are replaced with small basic proteins called **protamines**.

When chromosomes are gently 'decondensed', they have the appearance under the electron microscope of 'beads on a string' (*Fig. 2*). The 'beads' are called **nucleosomes** and consist of DNA complexed with histones. The 'string' is linear double-stranded DNA between adjacent nucleosomes and is called **linker DNA** (*Fig. 2*). The average distance between nucleosomes, that is the length of the linker DNA, is typically about 55 base pairs (bp) but varies from organism to organism in the range 8–114 bp. Even in a single nucleus, the distance between adjacent nucleosomes varies depending on, for example, the presence of other sequence-specific DNA-binding proteins. If a chromatin preparation is incubated with micrococcal nuclease, an enzyme that degrades DNA, the linker DNA is destroyed leaving **nucleosome core particles** in which the histones protect the associated DNA from digestion. Each nucleosome core particle contains a double-stranded DNA fragment 146 bp long bound to a complex of eight histones, the **histone octamer**, consisting of two molecules each of histones H2A, H2B, H3 and H4 (*Fig. 3*). The DNA is wound round the outside of the histone octamer in about 1.8 turns of a left-handed supercoil. DNA–histone contacts are made along the inside face of this superhelix. Overall the packing ratio is about 7, that is the DNA length is shortened about sevenfold by winding around the nucleosome.

## 30 nm fiber

If nuclei are lysed very gently, the chromatin is seen to exist as a 30 nm diameter fiber. This diameter is larger than a single nucleosome and suggests that the nucleosomes are organized into a higher order structure. The fiber is formed by a histone H1 molecule binding to the linker DNA of each nucleosome at the point where it enters and leaves the nucleosome (*Fig. 3*). The histone H1 molecules interact with each other, pulling the nucleosomes together. One possibility is that the nucleosomes wind up into a higher order helix with six nucleosomes per turn to form a **solenoid** (*Fig. 4*). This would give a fiber three nucleosomes wide, which is indeed the diameter observed. In such a solenoid the linear length of the DNA has been reduced by a further factor of 6 (equiv-

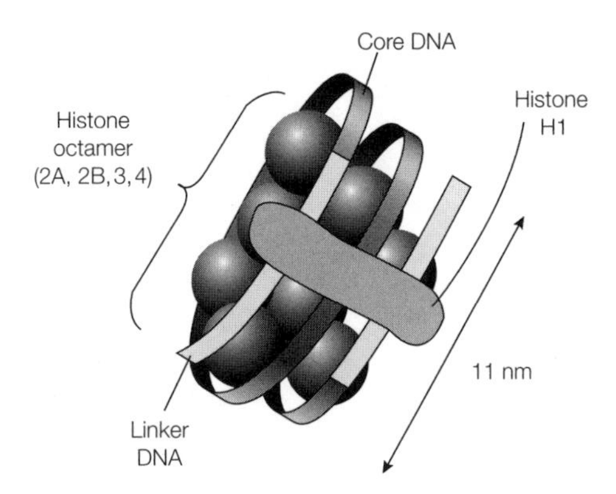


*Fig. 3. Schematic diagram of a nucleosome consisting of the DNA double helix wound 1.8 times round a histone octamer (two molecules each of histones H2A, H2B, H3 and H4).*

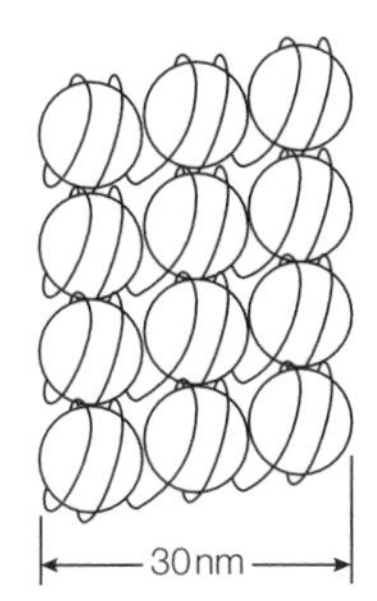


*Fig. 4. Schematic diagram of proposed solenoid structure of chromatin to yield a 30 nm fiber. The structure consists of six nucleosomes per turn of the helix and hence would be three nucleosomes wide. In the diagram, only three nucleosomes of each turn are visible; the other three nucleosomes per turn are hidden from view.*

alent to six nucleosomes per turn of the solenoid). Coupled with the packing ratio of 7 for the nucleosome itself (see above), this gives a packing ratio for the solenoid of approximately 6 × 7 (i.e. about 40).

**Radial loops**

When chromosomes are depleted of histones, they are seen to have a central fibrous '**protein scaffold**' (or **nuclear matrix**) to which the DNA is attached in loops (*Fig. 5*). Therefore, *in vivo* it seems likely that the next order of packaging involves the attachment of the 30 nm fiber to multiple locations on this central protein scaffold in a series of radial loops. Little is known as to how this structure is organized.

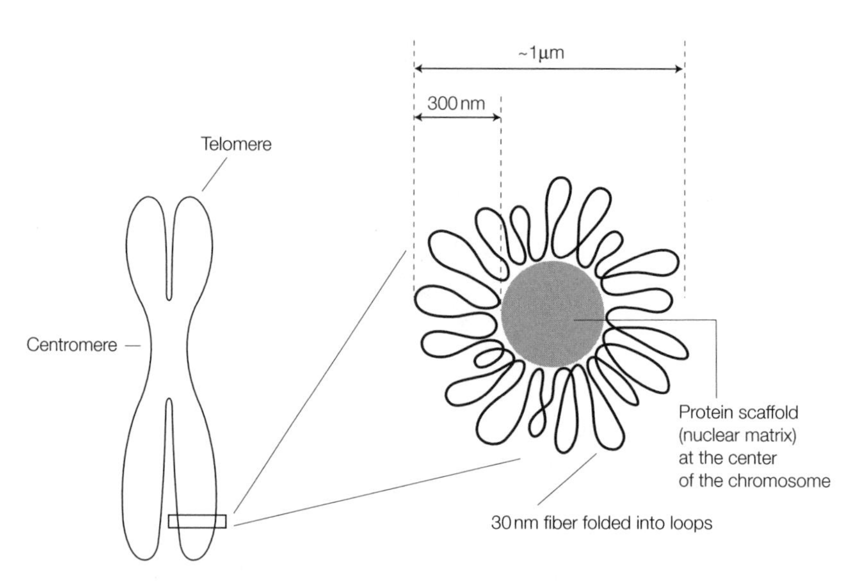


*Fig. 5. Attachment of the 30 nm fiber to a central protein scaffold with the loops arranged radially around the scaffold. The diagram on the right shows a representational cross-section through a chromosome.*

# F3 DNA REPLICATION IN BACTERIA

## Key Notes

**DNA polymerases**

*E. coli* DNA polymerase I requires all four deoxynucleoside 5′ triphosphates (dNTPs) as precursors, $Mg^{2+}$, a DNA template and a primer with a 3′-OH end. DNA synthesis occurs in a 5′ → 3′ direction. DNA polymerase I also has a 3′→5′ exonuclease (proof-reading) activity and a 5′→3′ exonuclease activity. *E. coli* DNA polymerases II and III lack the 5′→3′ exonuclease activity.

**Replication forks**

Replication starts at a single origin, is bi-directional and semi-conservative. Each replication bubble (or eye) consists of two replication forks.

**Okazaki fragments**

DNA synthesis proceeds in a 5′→3′ direction on each strand of the parental DNA. On the strand with 3′→5′ orientation (the leading strand) the new DNA is synthesized continuously. On the strand that has 5′→3′ orientation (the lagging strand) the DNA is synthesized discontinuously as a series of short Okazaki fragments that are then joined together.

**RNA primer**

DNA replication requires an RNA primer that is synthesized by an RNA polymerase called primase. This is extended by DNA polymerase III, which makes the DNA for both the leading and lagging strands. DNA polymerase degrades the primer and replaces it with DNA. DNA ligase then joins DNA ends.

**Accessory proteins**

A helicase unwinds the DNA double helix and single-stranded DNA-binding (SSB) protein stabilizes the single-stranded regions during replication. DNA topoisomerase I is needed to allow the helix to unwind without causing extensive rotation of the chromosome. DNA topoisomerase II separates the two daughter DNA circles following replication.

**Related topics**

DNA structure (F1) | DNA replication in eukaryotes (F4)

## DNA polymerases

**DNA polymerase I** from *E. coli* catalyzes the stepwise addition of deoxyribonucleotides to the 3′-OH end of a DNA chain:

$$(\text{DNA})_{n \text{ residues}} + \text{dNTP} \rightarrow (\text{DNA})_{n+1} + \text{PP}_i$$

The enzyme has the following requirements:

- all four dNTPs (dATP, dGTP, dTTP and dCTP) must be present to be used as precursors; $Mg^{2+}$ is also required;
- a DNA **template** is essential, to be copied by the DNA polymerase;
- a **primer** with a free 3′-OH that the enzyme can extend.

DNA polymerase I is a template-directed enzyme, that is it recognizes the next nucleotide on the DNA template and then adds a complementary

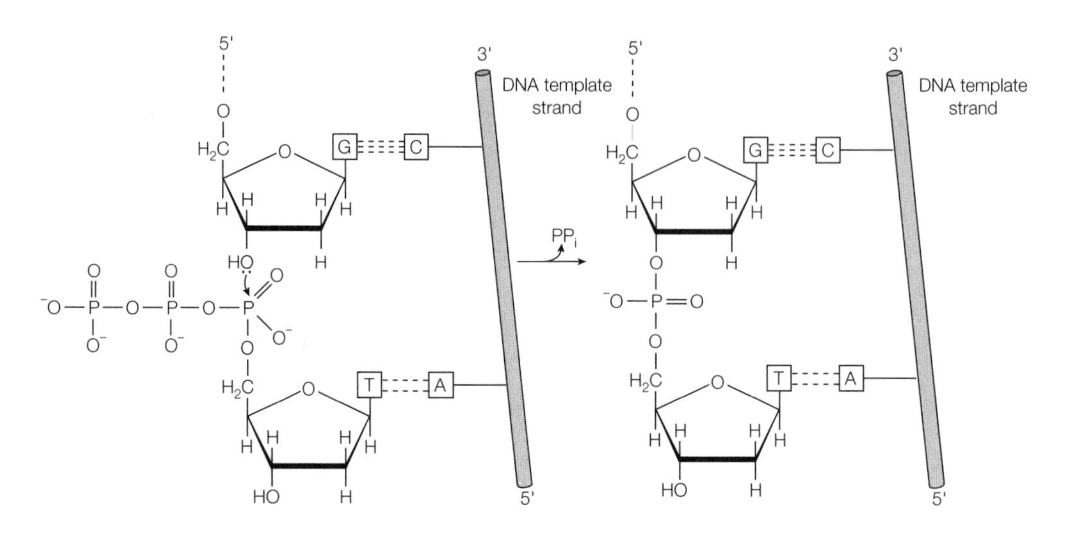


*Fig. 1. DNA synthesis. In this schematic diagram, the incoming dTTP hydrogen bonds with the adenine on the template DNA strand and a 3′5′ phosphodiester bond is formed, releasing pyrophosphate.*

nucleotide to the 3′-OH of the primer, creating a 3′5′ phosphodiester bond, and releasing pyrophosphate. The reaction is shown in *Fig. 1*. It involves nucleophilic attack of the 3′-OH of the primer on the α-phosphate group of the incoming nucleotide. The primer is extended in a 5′→3′ direction.

DNA polymerase I also corrects mistakes in DNA by removing mismatched nucleotides (i.e. it has **proof-reading activity**). Thus, during polymerization, if the nucleotide that has just been incorporated is incorrect (mismatched), it is removed using a 3′→5′ exonuclease activity. This gives very high fidelity; an error rate of less than $10^{-8}$ per base pair. DNA polymerase also has a 5′→3′ exonuclease activity; it can hydrolyze nucleic acid starting from the 5′ end of a chain. This activity plays a key role in removing the RNA primer used during replication (see below). Thus, overall, DNA polymerase I has three different active sites on its single polypeptide chain; 5′→3′ polymerase, 3′→5′ exonuclease and 5′→3′ exonuclease.

*E. coli* also contains two other DNA polymerases, **DNA polymerase II** and **DNA polymerase III**. As with DNA polymerase I, these enzymes also catalyze the template-directed synthesis of DNA from deoxynucleotidyl 5′-triphosphates, need a primer with a free 3′-OH group, synthesize DNA in the 5′→3′ direction, and have 3′→5′ exonuclease activity. Neither enzyme has 5′→3′ exonuclease activity.

## Replication forks

When the bacterial circular chromosome is replicated, replication starts at a **single origin**. The double helix opens up and both strands serve as template for the synthesis of new DNA. DNA synthesis then proceeds outward in both directions from the single origin (i.e. it is **bi-directional**; *Fig. 2*). The products of the reaction are two daughter double-stranded DNA molecules each of which has one original template strand and one strand of newly synthesized DNA. Thus, replication is **semi-conservative**. The region of replicating DNA associated with the single origin is called a **replication bubble** or **replication eye** and consists of two **replication forks** moving in opposite directions around the DNA circle (*Fig. 2*).

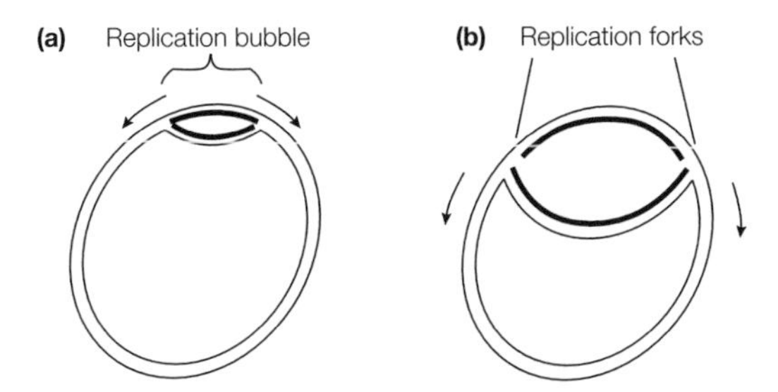


*Fig. 2. Replication of the bacterial circular chromosome. Replication starts from a single origin and proceeds bi-directionally (a) moving around the chromosome with time (b). The two replication forks eventually meet and fuse. The two circular daughter DNA molecules produced each have one original template DNA strand (thin line) and one new strand (thick line).*

## Okazaki fragments

Double-stranded DNA is **antiparallel** (see Topic F1); one strand runs 5′→3′ and the complementary strand runs 3′→5′. As the original double-stranded DNA opens up at a replication fork, new DNA is made against each template strand. Superficially, therefore, one might expect new DNA to be made 5′→3′ for one daughter strand and 3′→5′ for the other daughter strand. However, all DNA polymerases make DNA only in the 5′→3′ direction and never in the 3′→5′ direction. What actually happens is that on the template strand with 3′→5′ orientation, new DNA is made in a continuous piece in the correct 5′→3′ direction. This new DNA is called the **leading strand** (*Fig. 3*). On the other template strand (that has a 5′→3′ orientation), DNA polymerase synthesizes short pieces of new DNA (about 1000 nucleotides long) in the 5′→3′ direction (*Fig. 3*) and then joins these pieces together. The small fragments are called **Okazaki fragments** after their discoverer. The new DNA strand which is made by this discontinuous method is called the **lagging strand**.

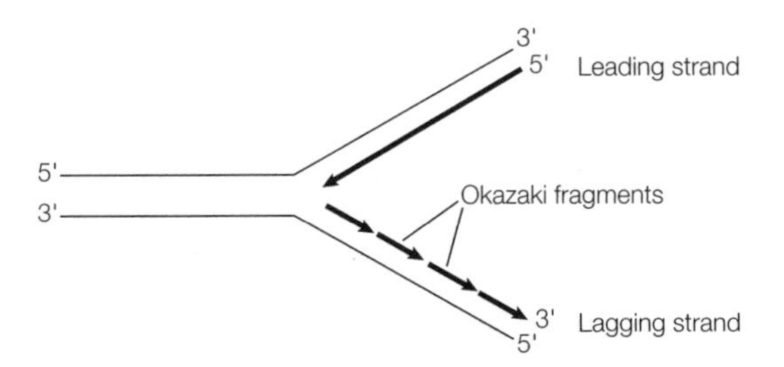


*Fig. 3. Synthesis of DNA at a replication fork. As the parental DNA (thin line) opens up, each of the two parental strands acts as a template for new DNA synthesis (thick lines). The leading strand is synthesized continuously but the lagging strand is synthesized as short (Okazaki) DNA fragments that are then joined together.*

## RNA primer

DNA polymerase cannot start DNA synthesis without a primer. Even on the lagging stand, each Okazaki fragment requires an RNA primer before DNA synthesis can start. The primer used in each case is a short (approximately five nucleotides long) piece of RNA and is synthesized by an RNA polymerase called **primase** (*Fig. 4a*). Primase can make RNA directly on the single-stranded DNA template because, like all RNA polymerases, it does not require a primer to begin synthesis. The RNA primer made by primase (*Fig. 4b*) is then extended by DNA polymerase III (*Fig. 4c*). DNA polymerase III synthesizes DNA for

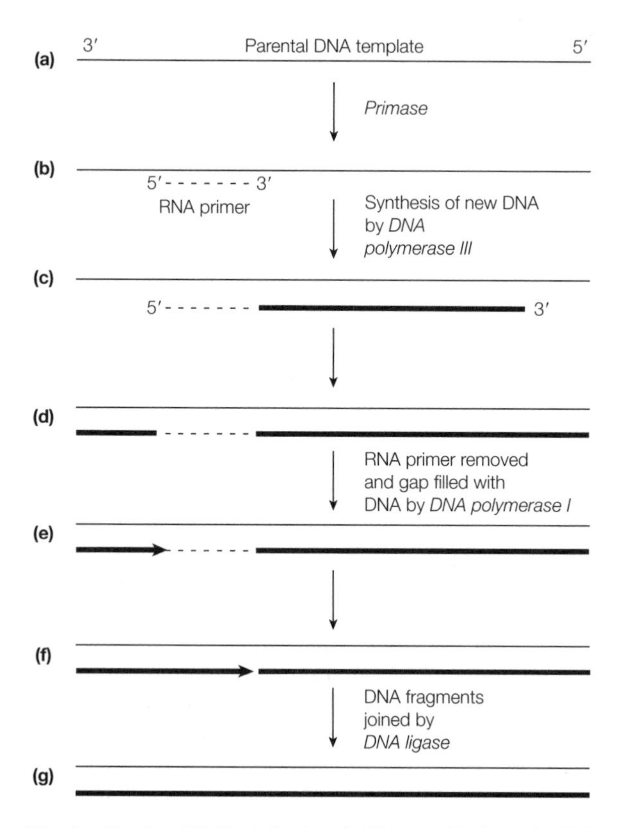


*Fig. 4. Details of DNA replication. (a) Primase binds to the DNA template strand (thin line) and (b) synthesizes a short RNA primer (dotted line); (c) DNA polymerase III now extends the RNA primer by synthesizing new DNA (thick line); (d) during synthesis of the lagging stand, adjacent Okazaki fragments are separated by the RNA primers; (e) the RNA primers are now removed and the gaps filled with DNA by DNA polymerase I (f) generating adjacent DNA fragments that are then (g) joined by DNA ligase.*

both the leading and lagging strand. After DNA synthesis by DNA polymerase III, DNA polymerase I uses its 5′→3′ exonuclease activity to remove the RNA primer and then fills the gap with new DNA (*Fig. 4e* and *f*). DNA polymerase III cannot carry out this task because it lacks the 5′→3′ activity of DNA polymerase I. Finally, DNA ligase joins the ends of the DNA fragments together (*Fig. 4g*).

## Accessory proteins

DNA polymerases I and III, primase and DNA ligase are not the only proteins needed for replication of the bacterial chromosome. The DNA template is a double helix with each strand wound tightly around the other and hence the two strands must be unwound during replication. How is this **unwinding problem** solved? A **DNA helicase** is used to unwind the double helix (using ATP as energy source) and **SSB (single-stranded DNA-binding) protein** prevent the single-stranded regions from base-pairing again so that each of the two DNA strands is accessible for replication. In principle, for a replication fork to move along a piece of DNA, the DNA helix would need to unwind ahead of it, causing the DNA to rotate rapidly. However, the bacterial chromosome is circular and so there are no ends to rotate. The solution to the problem is that an enzyme called **topoisomerase I** breaks a phosphodiester bond in one DNA strand (a single-strand break) a small distance ahead of the fork, allowing the DNA to rotate freely (swivel) around the other (intact) strand. The phosphodiester bond is then re-formed by the topoisomerase.

After the bacterial circular DNA has been replicated, the result is two double-stranded circular DNA molecules that are interlocked. **Topoisomerase II** separates them as follows. This enzyme works in a similar manner to topoisomerase I but causes a transient break in each strand (a double-strand break) of a double-stranded DNA molecule. Thus topoisomerase II binds to one double-stranded DNA circle and causes a transient double-strand break that acts as a 'gate' through which the other DNA circle can pass (*Fig. 5*). Topoisomerase II then re-seals the strand breaks.

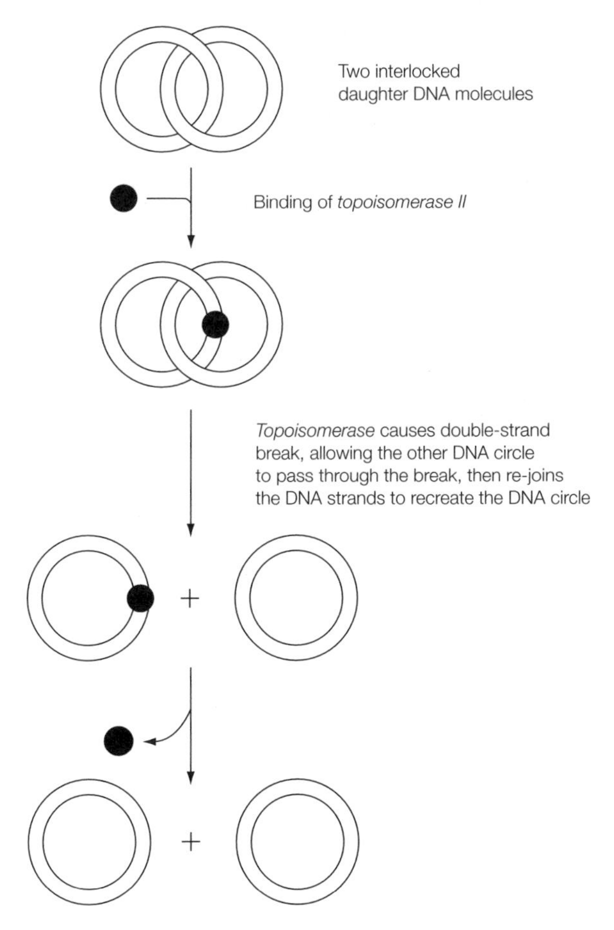


*Fig. 5. Separation of daughter DNA circles by topoisomerase II.*

# F4 DNA REPLICATION IN EUKARYOTES

## Key Notes

**Cell cycle**

In eukaryotes, the cell cycle consists of $G_1$, S, $G_2$ and M phases. Most differences in the cycle times of different cells are due to differences in the length of the $G_1$ phase. Quiescent cells are said to be in the $G_0$ phase.

**Multiple replicons**

DNA replication occurs only in the S phase. It occurs at many chromosomal origins, is bi-directional and semi-conservative. Sets of 20–80 replicons act as replication units that are activated in sequence.

**Five DNA polymerases**

DNA polymerases $\alpha$ and $\delta$ replicate chromosomal DNA, DNA polymerases $\beta$ and $\epsilon$ repair DNA, and DNA polymerase $\gamma$ replicates mitochondrial DNA.

**Leading and lagging strands**

DNA polymerase $\alpha$ synthesizes the lagging strand, via Okazaki fragments, and DNA polymerase $\delta$ synthesizes the leading strand. The RNA primers are synthesized by DNA polymerase $\alpha$ which carries a primase subunit.

**Telomere replication**

Telomerase, a DNA polymerase that contains an integral RNA that acts as its own primer, is used to replicate DNA at the ends of chromosomes (telomeres).

**Replication of chromatin**

Nucleosomes do not dissociate from the DNA during DNA replication; rather they must open up to allow the replication apparatus to pass.

**Related topics**

DNA structure (F1)    DNA replication in bacteria (F3)

## Cell cycle

The life of a eukaryotic cell can be defined as a **cell cycle** (*Fig. 1*). Mitosis and cell division occur in the **M phase** which lasts for only about 1 h. This is followed by the **$G_1$ phase** (G for gap), then the **S phase** (S for synthesis), during which time the chromosomal DNA is replicated, and finally the **$G_2$ phase** in which the cells prepare for mitosis. Eukaryotic cells in culture typically have cell cycle times of 16–24 h but the cell cycle time can be much longer (> 100 days) for some cells in a multicellular organism. Most of the variation in cell cycle times occurs by differences in the length of the $G_1$ phase. Some cells *in vivo*, such as neurons, stop dividing completely and are said to be quiescent, locked in a **$G_0$ phase**.

## Multiple replicons

In eukaryotes, replication of chromosomal DNA occurs only in the S phase of the cell cycle. As for bacterial DNA (see Topic F3), eukaryotic DNA is replicated **semi-conservatively**. Replication of each linear DNA molecule in a chromosome starts at **many origins**, one every 3–300 kb of DNA depending on the species and tissue, and proceeds **bi-directionally** from each origin. The use of multiple

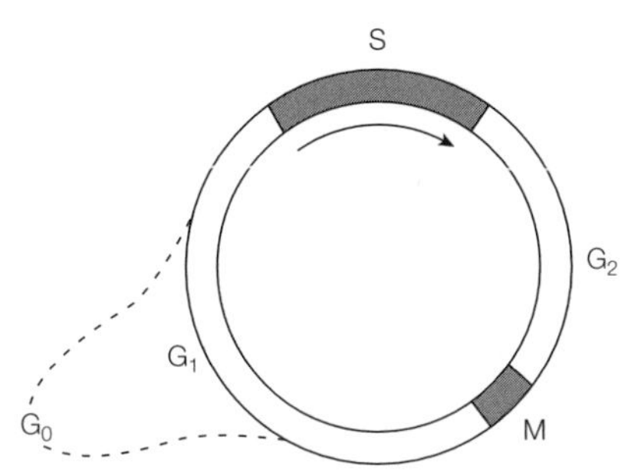


*Fig. 1. The eukaryotic cell cycle. The S phase is typically 6–8 h long, $G_2$ is a phase in which the cell prepares for mitosis and lasts for 2–6 h, mitosis itself (M) is short and takes only about 1 h. The length of $G_1$ is very variable and depends on the cell type. Cells can enter $G_0$, a quiescent phase, instead of continuing with the cell cycle.*

origins is essential in order to ensure that the chromosomal DNA is replicated within the necessary time period. At each origin, a **replication bubble** forms consisting of two **replication forks** moving in opposite directions. The DNA replicated under the control of a single origin is called a **replicon**. DNA synthesis proceeds until replication bubbles merge together (*Fig. 2*).

All of the regions of a chromosome are not replicated simultaneously. Rather, many replication eyes will be found on one part of the chromosome and none on another section. Thus replication origins are activated in clusters, called **replication units**, consisting of 20–80 origins. During S phase, the different

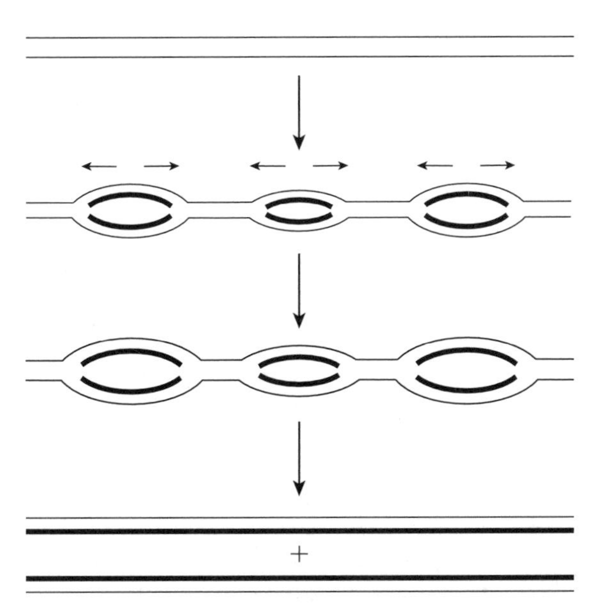


*Fig. 2. Replication of eukaryotic chromosomal DNA. Replication begins at many origins and proceeds bi-directionally at each location. Eventually the replication eyes merge together to produce two daughter DNA molecules, each of which consists of one parental DNA strand (thin line) and one newly synthesized DNA strand (thick line).*

replication units are activated in a set order until eventually the whole chromosome has been replicated. Transcriptionally active genes appear to be replicated early in S phase, whilst chromatin that is condensed and not transcriptionally active is replicated later.

**Five DNA polymerases**

Eukaryotic cells contain five different DNA polymerases; α, β, γ, δ and ε. The DNA polymerases involved in replication of chromosomal DNA are α and δ. DNA polymerases β and ε are involved in DNA repair. All of these polymerases except DNA polymerase γ are located in the nucleus; DNA polymerase γ is found in mitochondria and replicates mitochondrial DNA.

**Leading and lagging strands**

The basic scheme of replication of double-stranded chromosomal DNA in eukaryotes follows that for bacterial DNA replication (see Topic F3); a leading strand and a lagging strand are synthesized, the latter involving discontinuous synthesis via Okazaki fragments. However, in eukaryotes, replication forks move much slower than in prokaryotes (about one-tenth of the rate) and the two new DNA strands are made by different polymerases; DNA polymerase α catalyzes synthesis of the lagging strand, via Okazaki fragments, and DNA polymerase δ synthesizes the leading strand. The RNA primers required are made by DNA polymerase α which carries a **primase subunit**. Whereas the δ enzyme has 3′→5′ exonuclease activity and so can proof-read the DNA made, DNA polymerase α has no such activity and so the new lagging strand DNA made by DNA polymerase α is probably proof-read by a separate accessory protein.

**Telomere replication**

The replication of a linear DNA molecule in a eukaryotic chromosome creates a problem that does not exist for the replication of bacterial circular DNA molecules. The normal mechanism of DNA synthesis (see above) means that the 3′ end of the lagging strand is not replicated. This creates a gap at the end of the chromosome and therefore a shortening of the double-stranded replicated portion. The effect is that the chromosomal DNA would become shorter and shorter after each replication. Various mechanisms have evolved to solve this problem. In many organisms the solution is to use an enzyme called **telomerase** to replicate the chromosome ends (**telomeres**).

Each telomere contains many copies of a repeated hexanucleotide sequence that is G-rich; in *Tetrahymena* it is GGGTTG. Telomerase carries, as an integral part of its structure, a short RNA molecule that is complementary to part of this G-rich sequence. The exact mechanism of action of telomerase is not clear; *Fig. 3* shows one possible model. The RNA molecule of telomerase is envisaged to hydrogen-bond to the telomere end. Then, using the RNA as a template, telomerase copies the RNA template (hence this enzyme is a reverse transcriptase; see Topic I4) and adds six deoxynucleotides to the telomere DNA end. Telomerase then dissociates from the DNA, re-binds at the new telomere end and repeats the extension process. It can do this hundreds of times before finally dissociating. The newly extended DNA strand can then act as a template for normal DNA replication to form double-stranded chromosomal DNA. The two processes, of the DNA ends shortening through normal replication and of lengthening using telomerase, are very roughly in balance so that each chromosome stays approximately the same length.

**Replication of chromatin**

Once DNA is bound to histones to form nucleosomes, histones rarely leave the DNA. Thus when a chromosome is replicated, the histones stay in place but

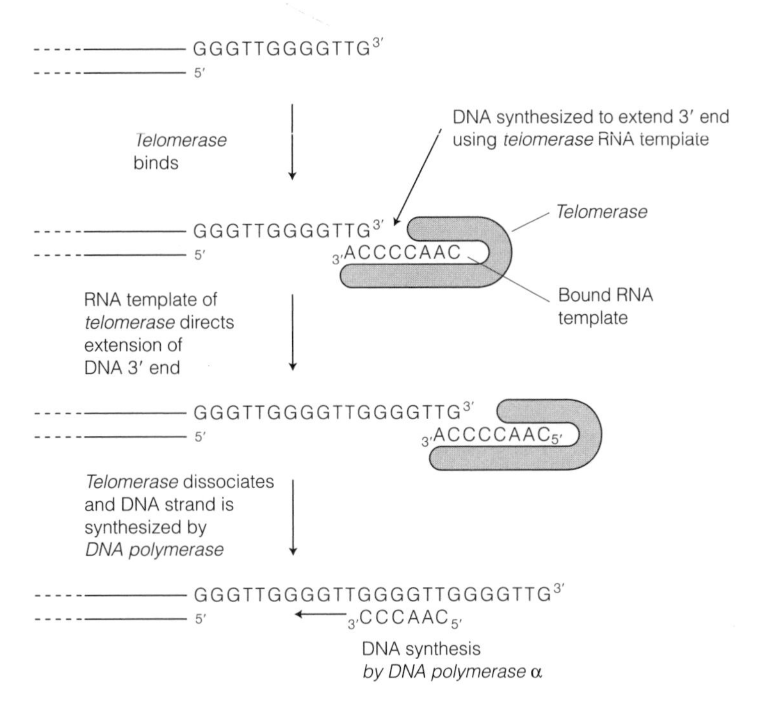


*Fig. 3. Replication of telomeric DNA. Telomerase has a bound RNA molecule that is used as template to direct DNA synthesis and hence extension of the ends of chromosomal DNA.*

somehow must allow the replication machinery to pass through and make new DNA. One suggestion is that the nucleosome histone octamer transiently unfolds into two half-nucleosomes to allow the replication machinery access to the DNA. The new DNA must also be packaged into nucleosomes and so histones are also synthesized during the S phase of the cell cycle. Experiments indicate that the old nucleosomes stay with the daughter DNA molecule containing the leading strand whilst new nucleosomes assemble on the daughter molecule containing the lagging strand.

# G1 RNA STRUCTURE

## Key Notes

**Covalent structure**

RNA is a polymer chain of ribonucleotides joined by 3′5′ phosphodiester bonds. The covalent structure is very similar to that for DNA except that uracil replaces thymine and ribose replaces deoxyribose.

**RNA secondary structure**

RNA molecules are largely single-stranded but there are regions of self-complementarity where the RNA chain forms internal double-stranded regions.

**Related topics**

DNA structure (F1)
Transcription in prokaryotes (G2)
The *lac* operon (G3)
The *trp* operon (G4)
Transcription in eukaryotes: an overview (G5)
Transcription of protein-coding genes in eukaryotes (G6)
Regulation of transcription by RNA Pol II (G7)

## Covalent structure

Like DNA (see Topic F1), RNA is a long polymer consisting of nucleotides joined by 3′5′ phosphodiester bonds. However, there are some differences:

- the bases in RNA are adenine (abbreviated A), guanine (G), uracil (U) and cytosine (C). Thus thymine in DNA is replaced by **uracil** in RNA, a different pyrimidine (*Fig. 1a*). However, like thymine (see Topic F1), uracil can form base pairs with adenine.
- The sugar in RNA is **ribose** rather than deoxyribose as in DNA (*Fig. 1b*).

The corresponding **ribonucleosides** are **adenosine**, **guanosine**, **cytidine** and **uridine**. The corresponding **ribonucleotides** are **adenosine 5′-triphosphate** (ATP), **guanosine 5′-triphosphate** (GTP), **cytidine 5′-triphosphate** (CTP) and **uridine 5′-triphosphate** (UTP).

As with DNA, the nucleotide sequence of RNA is also written as a base sequence in the 5′→3′ direction. Thus GUCAAGCCGGAC is the sequence of one short RNA molecule.

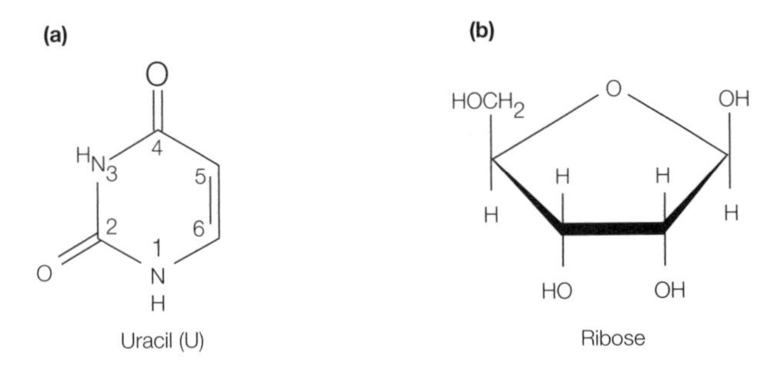


*Fig. 1. (a) Uracil, (b) ribose.*

## RNA secondary structure

Most RNA molecules are single-stranded but an RNA molecule may contain regions which can form complementary base-pairing where the RNA strand loops back on itself (*Fig. 2*). If so, the RNA will have some double-stranded regions. Ribosomal RNAs (rRNAs) and transfer RNAs (tRNAs) (see Topics G9 and G10, respectively) exhibit substantial secondary structure, as do some messenger RNAs (mRNAs).

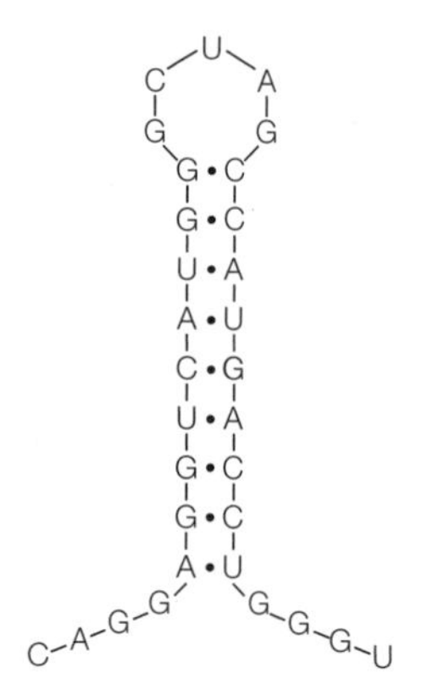


*Fig. 2. An example of self-complementarity in RNA forming an internal double-stranded region; hydrogen bonding between bases is shown by the symbol •.*

# G2 TRANSCRIPTION IN PROKARYOTES

## Key Notes

**Three phases of transcription**

Transcription by *E. coli* RNA polymerase occurs in three phases; initiation, elongation and termination. Initiation involves binding of the enzyme to a promoter upstream of the gene. During elongation, the antisense DNA strand is used as the template so that the RNA made has the same base sequence as the sense (coding) strand, except that U replaces T. A termination signal is eventually encountered that halts synthesis and causes release of the completed RNA.

**Promoters and initiation**

RNA polymerase holoenzyme (containing $\alpha_2\beta\beta'\omega\sigma$ subunits) initiates transcription by binding to a 40–60 bp region that contains two conserved promoter elements, the –10 sequence (Pribnow box) with the consensus TATAAT and the –35 sequence with the consensus TTGACA. The $\sigma$ factor is essential for initiation. No primer is required. Promoters vary up to 1000-fold in their efficiency of initiation which depends on the exact sequence of the key promoter elements as well as flanking sequences.

**Elongation**

Following initiation, the $\sigma$ subunit dissociates from RNA polymerase to leave the core enzyme ($\alpha_2\beta\beta'\omega$) that continues RNA synthesis in a $5' \rightarrow 3'$ direction using the four ribonucleoside 5′ triphosphates as precursors. The DNA double helix is unwound for transcription, forming a transcription bubble, and is then rewound after the transcription complex has passed.

**Termination**

A common termination signal is a hairpin structure formed by a palindromic GC-rich region, followed by an AT-rich sequence. Other signals are also used which require the assistance of rho ($\rho$) protein for effective termination.

**RNA processing**

Messenger RNA transcripts of protein-coding genes in prokaryotes require little or no modification before translation. Ribosomal RNAs and transfer RNAs are synthesized as precursor molecules that require processing by specific ribonucleases to release the mature RNA molecules.

**Related topics**

DNA structure (F1)
RNA structure (G1)
The *lac* operon (G3)
The *trp* operon (G4)
Transcription in eukaryotes: an overview (G5)
Transcription of protein-coding genes in eukaryotes (G6)
Regulation of transcription by RNA Pol II (G7)
Processing of eukaryotic pre-mRNA (G8)
Ribosomal RNA (G9)
Transfer RNA (G10)

## Three phases of transcription

Gene transcription by *E. coli* RNA polymerase takes place in three phases: **initiation, elongation** and **termination**. During initiation, RNA polymerase recognizes a specific site on the DNA, upstream from the gene that will be transcribed, called a **promoter site** and then unwinds the DNA locally. During elongation the RNA polymerase uses the **antisense (−) strand** of DNA as template and synthesizes a complementary RNA molecule using ribonucleoside 5′ triphosphates as precursors. The RNA produced has the same sequence as the non-template strand, called the **sense (+) strand** (or **coding strand**) except that the RNA contains U instead of T. At different locations on the bacterial chromosome, sometimes one strand is used as template, sometimes the other, depending on which strand is the coding strand for the gene in question. The correct strand to be used as template is identified for the RNA polymerase by the presence of the promoter site. Finally, the RNA polymerase encounters a termination signal and ceases transcription, releasing the RNA transcript and dissociating from the DNA.

## Promoters and initiation

In *E. coli*, all genes are transcribed by a single large RNA polymerase with the subunit structure $\alpha_2\beta\beta'\omega\sigma$. This complete enzyme, called the **holoenzyme**, is needed to initiate transcription since the σ factor is essential for recognition of the promoter; it decreases the affinity of the core enzyme for nonspecific DNA binding sites and increases its affinity for the promoter. It is common for prokaryotes to have several σ factors that recognize different types of promoter (in *E. coli*, the most common σ factor is $\sigma^{70}$).

The holoenzyme binds to a promoter region about 40–60 bp in size and then initiates transcription a short distance downstream (i.e. 3′ to the promoter). Within the promoter lie two 6-bp sequences that are particularly important for promoter function and which are therefore highly conserved between species. Using the convention of calling the first nucleotide of a transcribed sequence as +1, these two **promoter elements** lie at positions –10 and –35, that is about 10 and 35 bp, respectively, upstream of where transcription will begin (*Fig. 1*).

- The **–10 sequence** has the consensus TATAAT. Because this element was discovered by Pribnow, it is also known as the **Pribnow box**. It is an important recognition site that interacts with the σ factor of RNA polymerase.
- The **–35 sequence** has the consensus TTGACA and is important in DNA unwinding during transcriptional initiation.

The actual sequence between the –10 sequence and the –35 sequence is not conserved (i.e. it varies from promoter to promoter) but the distance between these two sites is extremely important for correct functioning of the promoter.

Promoters differ by up to 1000-fold in their efficiency of initiation of transcription so that genes with strong promoters are transcribed very frequently

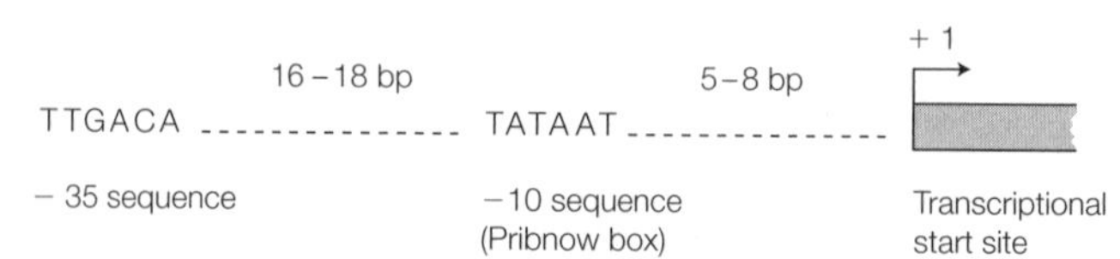


*Fig. 1. Prokaryotic promoter showing the –10 sequence and –35 sequence. By convention, the first nucleotide of the template DNA that is transcribed into RNA is denoted +1, the transcriptional start site.*

whereas genes with weak promoters are transcribed far less often. The –10 and –35 sequences of strong promoters correspond well with the consensus sequences shown in *Fig. 1* whereas weaker promoters may have sequences that differ from these at one or more nucleotides. The nature of the sequences around the transcriptional start site can also influence the efficiency of initiation. RNA polymerase does not need a primer to begin transcription (cf. DNA polymerases, Topics F3 and F4); having bound to the promoter site, the RNA polymerase begins transcription directly.

## Elongation

After transcription initiation, the σ factor is released from the transcriptional complex to leave the **core enzyme** ($\alpha_2\beta\beta'\omega$) which continues elongation of the RNA transcript. Thus the core enzyme contains the catalytic site for polymerization, probably within the β subunit. The first nucleotide in the RNA transcript is usually pppG or pppA. The RNA polymerase then synthesizes RNA in the 5′ → 3′ direction, using the four ribonucleoside 5′ triphosphates (ATP, CTP, GTP, UTP) as precursors. The 3′-OH at the end of the growing RNA chain attacks the α phosphate group of the incoming ribonucleoside 5′ triphosphate to form a 3′5′ phosphodiester bond (*Fig. 2*). The complex of RNA polymerase, DNA template and new RNA transcript is called a **ternary complex** (i.e. three components) and the region of unwound DNA that is undergoing transcription is called the **transcription bubble** (*Fig. 3*). The RNA transcript forms a transient RNA–DNA hybrid helix with its template strand but then peels away from the DNA as transcription proceeds. The DNA is unwound ahead of the transcription bubble and after the transcription complex has passed, the DNA rewinds.

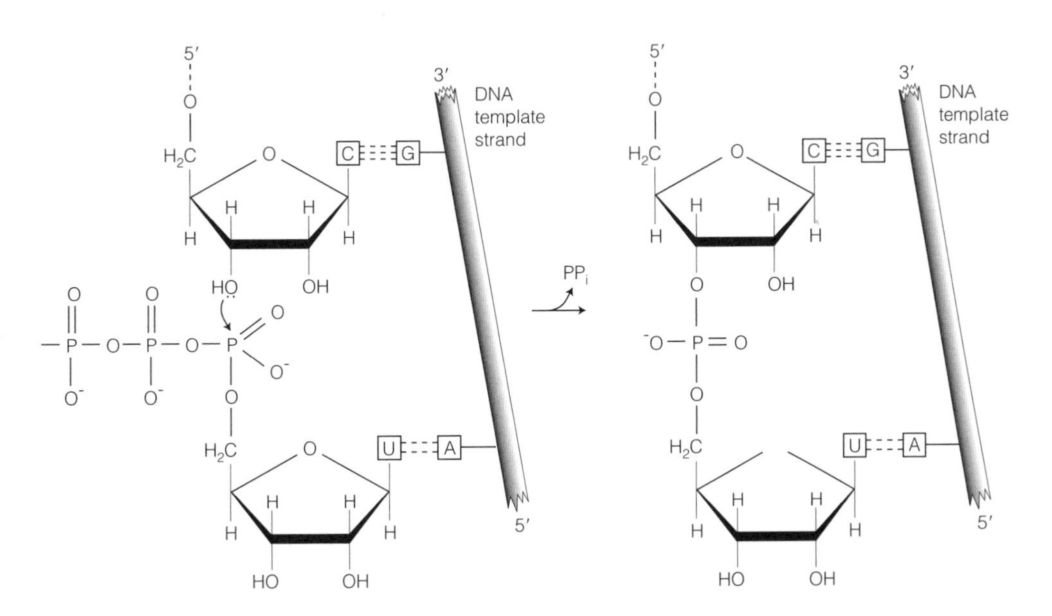


*Fig. 2. Transcription by RNA polymerase. In each step the incoming ribonucleotide selected is that which can base-pair with the next base of the DNA template strand. In the diagram, the incoming nucleotide is rUTP to base-pair with the A residue of the template DNA. A 3′5′ phosphodiester bond is formed, extending the RNA chain by one nucleotide, and pyrophosphate is released. Overall the RNA molecule grows in a 5′ to 3′ direction.*

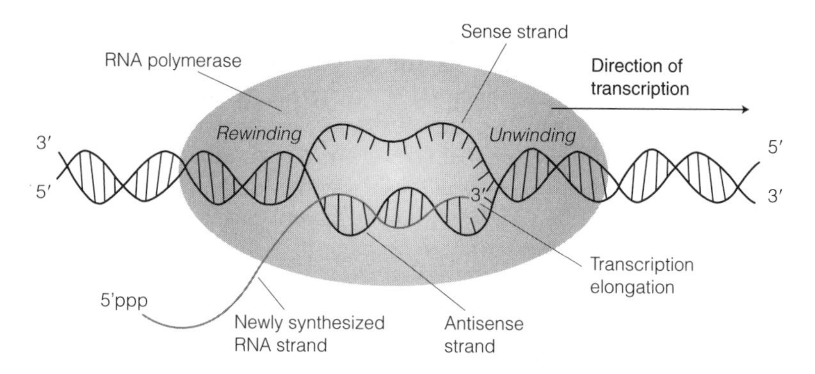


*Fig. 3. A transcription bubble. The DNA double helix is unwound and RNA polymerase then synthesizes an RNA copy of the DNA template strand. The nascent RNA transiently forms an RNA–DNA hybrid helix but then peels away from the DNA which is subsequently rewound into a helix once more.*

## Termination

Transcription continues until a termination sequence is reached. The most common termination signal is a GC-rich region that is a **palindrome**, followed by an AT-rich sequence. The RNA made from the DNA palindrome is self-complementary and so base-pairs internally to form a **hairpin structure** rich in GC base pairs followed by four or more U residues (*Fig. 4*). However, not all termination sites have this hairpin structure. Those that lack such a structure require an additional protein, called **rho** (ρ), to help recognize the termination site and stop transcription.

## RNA processing

In prokaryotes, RNA transcribed from protein-coding genes (**messenger RNA, mRNA**), requires little or no modification prior to translation. In fact, many mRNA molecules begin to be translated even before RNA synthesis has finished. However, **ribosomal RNA (rRNA)** and **transfer RNA (tRNA)** are synthesized as precursor molecules that do require post-transcriptional processing (see Topics G9 and G10, respectively).

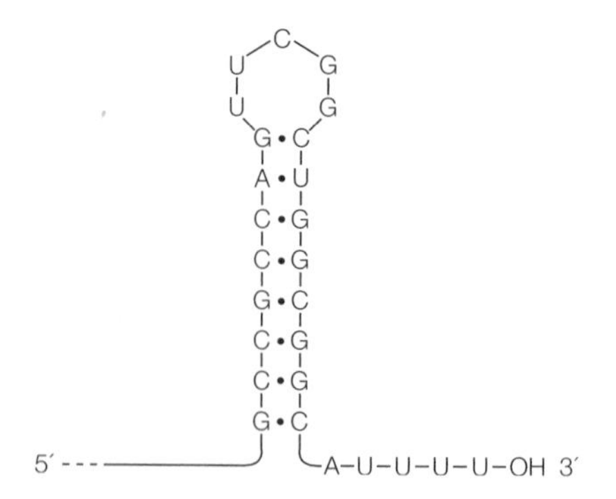


*Fig. 4. A typical hairpin structure formed by the 3′ end of an RNA molecule during termination of transcription.*

# G3 THE *LAC* OPERON

## Key Notes

**Induction of the *lac* operon**

The *lac* operon contains *lacZ*, *lacY* and *lacA* genes encoding β-galactosidase, galactose permease, and thiogalactoside transacetylase, respectively, preceded by an operator site ($O_{lac}$) and a promoter ($P_{lac}$). The operon is transcribed by RNA polymerase to produce a single polycistronic mRNA that is then translated to produce all three enzymes. These enzymes are involved in lactose metabolism. When lactose is absent, *E. coli* makes only small amounts of these enzymes but the presence of lactose induces synthesis of large amounts of all three enzymes. The mechanism of induction is that the background level of β-galactosidase converts some lactose to allolactose which then acts as an inducer and turns on transcription of the *lac* operon. IPTG can also act as an inducer. Transcription of the operon is controlled by the *lac* repressor protein encoded by the *lacI* gene.

**The lac repressor**

The *lac*I gene has its own promoter ($P_{lacI}$) to which RNA polymerase binds and initiates transcription. In the absence of an inducer, the *lacI* gene is transcribed, producing lac repressor mRNA and hence lac repressor protein monomers. These monomers assemble to form active tetramers which bind to the *lac* operator site, $O_{lac}$, and prevent transcription of the *lac* operon. In the presence of an inducer (such as allolactose or IPTG), the inducer binds to the repressor and changes its conformation, reducing its affinity for the *lac* operator. Thus the repressor now dissociates and allows RNA polymerase to transcribe the *lac* operon.

**CRP/CAP**

Catabolite activator protein, CAP (also called cAMP receptor protein, CRP) is an activator required for high level transcription of the *lac* operon. The active molecule is a CRP dimer that binds 3′5′ cyclic AMP to form a CRP–cAMP complex. CRP–cAMP binds to the *lac* promoter and increases the binding of RNA polymerase, stimulating transcription of the *lac* operon. CRP dimer without cAMP cannot bind to this DNA. The action of CRP depends upon the carbon source available to the bacterium. When glucose is present, the intracellular level of cAMP falls, CRP cannot bind to the *lac* promoter and the *lac* operon is only weakly transcribed. When glucose is absent, the level of intracellular cAMP rises, the CRP–cAMP complex stimulates transcription of the *lac* operon and allows lactose to be used as an alternative carbon source.

**Positive and negative regulation**

In negative regulation of prokaryotic gene expression, bound repressor prevents transcription of the structural genes. In positive regulation of gene expression, an activator binds to DNA and increases the rate of transcription. Through the *lac* repressor and CRP/CAP protein, the *lac* operon is subject to both negative and positive control.

**Related topics**

DNA structure (F1)
RNA structure (G1)
Transcription in prokaryotes (G2)
The *trp* operon (G4)
Transcription in eukaryotes: an overview (G5)
Transcription of protein-coding genes in eukaryotes (G6)
Regulation of transcription by RNA Pol II (G7)
Processing of eukaryotic pre-RNA (G8)
Ribosomal RNA (G9)
Transfer RNA (G10)

## Induction of the *lac* operon

Many protein-coding genes in bacteria are clustered together in **operons** which serve as transcriptional units that are co-ordinately regulated. One of the most studied of these is the ***lac* operon** in *E. coli*. This codes for key enzymes involved in lactose metabolism: **galactoside permease** (also known as **lactose permease**; it transports lactose into the cell across the cell membrane) and **β-galactosidase** (which hydrolyzes lactose to glucose and galactose). It also codes for a third enzyme, **thiogalactoside transacetylase** but the role of this enzyme is not clear. Normally *E. coli* cells make very little of any of these three proteins but when lactose is available it causes a large and coordinated increase in the amount of each enzyme. Thus each enzyme is an **inducible enzyme** and the process is called **induction**. The mechanism is that the few molecules of β-galactosidase in the cell before induction convert the lactose to allolactose which then turns on transcription of these three genes in the *lac* operon. Thus allolactose is an **inducer**. Another inducer of the *lac* operon is **isopropylthiogalactoside** (IPTG). Unlike allolactose, this inducer is not metabolized by *E. coli* and so is useful for experimental studies of induction.

It was Jacob and Monod in 1961 who proposed the operon model for the regulation of transcription. The *lac* operon is a good example of how operons work (*Fig. 1*). The operon model proposes three elements:

- a set of **structural genes** (i.e. genes encoding the proteins to be regulated);
- an **operator site**, which is a DNA sequence that regulates transcription of the structural genes;
- a **regulator gene** which encodes a protein that recognizes the operator sequence.

In the *lac* operon, the structural genes are the *lacZ*, *lacY* and *lacA* genes encoding β-galactosidase, the permease and the transacetylase, respectively. They are transcribed to yield a single **polycistronic mRNA** that is then translated to produce all three enzymes (*Fig. 1*). The existence of a polycistronic mRNA ensures that the amounts of all three gene products are regulated coordinately. Transcription occurs from a single promoter ($P_{lac}$) that lies upstream of these structural genes (*Fig. 1*) and binds RNA polymerase (see Topic G2). However, also present are an operator site ($O_{lac}$) between the promoter and the structural genes, and a *lac*I gene that codes for the **lac repressor** protein.

## The lac repressor

The *lac*I gene has its own promoter ($P_{lacI}$) that binds RNA polymerase and leads to transcription of lac repressor mRNA and hence production of lac repressor protein monomers. Four identical repressor monomers come together to form

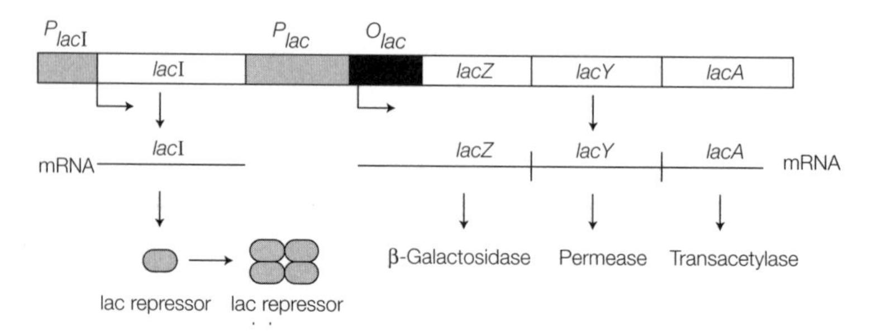


*Fig. 1. Structure of the* lac *operon.*

the active tetramer which can bind tightly to the *lac* operator site, $O_{lac}$. The $O_{lac}$ sequence is **palindromic**, that is it has the same DNA sequence when one strand is read 5′ to 3′ and the complementary strand is read 5′ to 3′. This symmetry of the operator site is matched by the symmetry of the repressor tetramer.

In the absence of an inducer such as allolactose or IPTG, the *lac*I gene is transcribed and the resulting repressor protein binds to the operator site of the *lac* operon, $O_{lac}$, and prevents transcription of the *lacZ*, *lacY* and *lacA* genes (*Fig. 2*). During induction, the inducer binds to the repressor. This causes a change in conformation of the repressor that greatly reduces its affinity for the *lac* operator site. The lac repressor now dissociates from the operator site and allows the RNA polymerase (already in place on the adjacent promoter site) to begin transcribing the *lacZ*, *lacY* and *lacA* genes (*Fig. 3*). This yields many copies of the polycistronic mRNA and, after translation, large amounts of all three enzymes.

If inducer is removed, the lac repressor rapidly binds to the *lac* operator site and transcription is inhibited almost immediately. The *lacZYA* RNA transcript is very unstable and so degrades quickly such that further synthesis of the β galactosidase, permease and transacetylase ceases.

## CRP/CAP

High level transcription of the *lac* operon requires the presence of a specific activator protein called **catabolite activator protein (CAP)**, also called **cAMP**

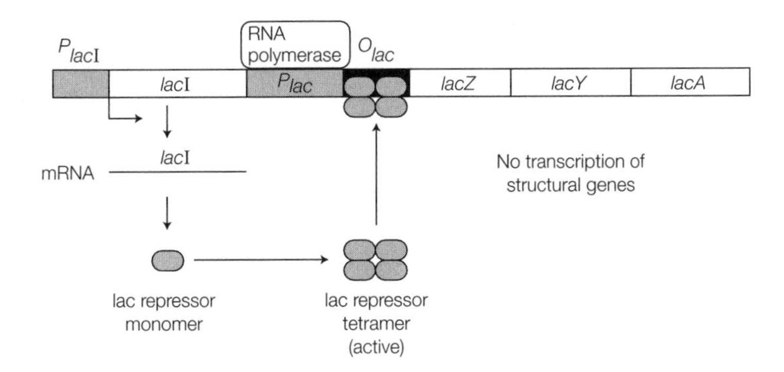


*Fig. 2. Repression of transcription by the* lac *repressor in the absence of inducer.*

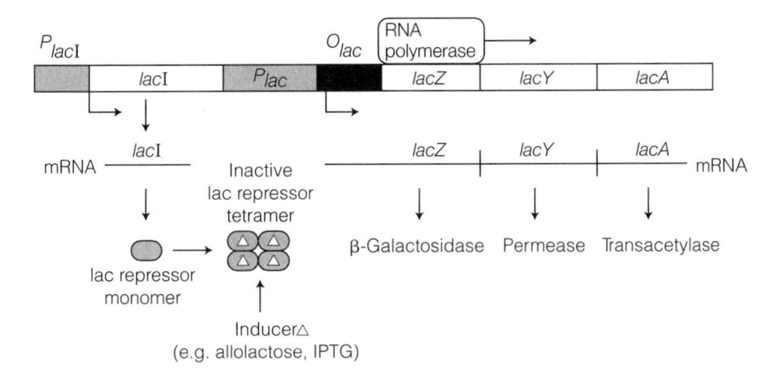


*Fig. 3. Inducer inactivates the lac repressor and so allows transcription of the structural genes.*

**receptor protein** (**CRP**). This protein, which is a dimer, cannot bind to DNA unless it is complexed with 3′5′ cyclic AMP (cAMP). The CRP–cAMP complex binds to the *lac* promoter just upstream from the binding site for RNA polymerase. It increases the binding of RNA polymerase and so stimulates transcription of the *lac* operon.

Whether or not the CRP protein is able to bind to the *lac* promoter depends on the carbon source available to the bacterium (*Fig. 2*). When glucose is present, *E. coli* does not need to use lactose as a carbon source and so the *lac* operon does not need to be active. Thus the system has evolved to be responsive to glucose. Glucose inhibits **adenylate cyclase**, the enzyme that synthesizes cAMP from ATP. Thus, in the presence of glucose the intracellular level of cAMP falls, so CRP cannot bind to the *lac* promoter, and the *lac* operon is only weakly active (even in the presence of lactose). When glucose is absent, adenylate cyclase is not inhibited, the level of intracellular cAMP rises and binds to CRP. Therefore, when glucose is absent but lactose is present, the CRP–cAMP complex stimulates transcription of the *lac* operon and allows the lactose to be used as an alternative carbon source. In the absence of lactose, the lac repressor of course ensures that the *lac* operon remains inactive. These combined controls ensure that the *lacZ*, *lacY* and *lacA* genes are transcribed strongly only if glucose is absent and lactose is present.

## Positive and negative regulation

The *lac* operon is a good example of **negative control** (**negative regulation**) of gene expression in that bound repressor prevents transcription of the structural genes. **Positive control** (**positive regulation**) of gene expression is when the regulatory protein binds to DNA and increases the rate of transcription. In this case the regulatory protein is called an activator. The CAP/CRP involved in regulating the *lac* operon is a good example of an activator. Thus the *lac* operon is subject to both negative and positive control.

# G4 THE *TRP* OPERON

## Key Notes

**Organization of the *trp* operon**

The *trp* operon contains five structural genes encoding enzymes for tryptophan biosynthesis, a *trp* promoter ($P_{trp}$) and a *trp* operator sequence ($O_{trp}$). The operon is transcribed only when tryptophan is scarce.

**Repression**

When tryptophan is lacking, a trp repressor protein (encoded by the *trpR* operon) is synthesized. The trp repressor dimer is inactive, cannot bind to the *trp* operator and so the *trp* operon is transcribed to produce the enzymes that then synthesize tryptophan for the cell. When tryptophan is present, tryptophan synthesis is not needed. In this situation, acting as a co-repressor, tryptophan binds to the repressor and activates it so that the repressor now binds to the *trp* operator and stops transcription of the *trp* operon.

**Attenuation**

The *trp* operon is also controlled by attenuation. A leader sequence in the polycistronic mRNA upstream of the coding region of the *trpE* structural gene encodes a 14 amino acid leader peptide including two tryptophan residues. The RNA leader sequence can form several possible stem-loop secondary structures, one of which can act as a transcription terminator whilst a different stem-loop can act as an anti-terminator. In the presence of tryptophan, ribosomes bind to the *trp* polycistronic mRNA that is being transcribed, following closely behind the RNA polymerase, and begin to translate the leader sequence. When tryptophan is present, translation of the leader sequence occurs to completion. The positioning of the ribosome prevents formation of the anti-terminator stem-loop but allows the terminator loop to form which then inhibits further transcription of the *trp* operon. If tryptophan is scarce, the ribosome pauses when attempting to translate the two trp codons in the leader sequence, which leaves the leader sequence available to form the anti-terminator stem-loop. Transcription of the *trp* operon is then allowed to continue.

**Attenuation vs. repression**

The *trp* operon is regulated by both repression (which determines whether transcription will occur or not) and attenuation (which fine tunes transcription). Other operons for amino acid biosynthetic pathways may also be regulated by both repression and attenuation or only by attenuation.

**Related topics**

DNA structure (F1)
RNA structure (G1)
Transcription in prokaryotes (G2)
The *lac* operon (G3)
Transcription in eukaryotes: an overview (G5)
Transcription of protein-coding genes in eukaryotes (G6)
Regulation of transcription by RNA Pol II (G7)
Processing of eukaryotic pre-mRNA (G8)
Ribosomal RNA (G9)
Transfer RNA (G10)

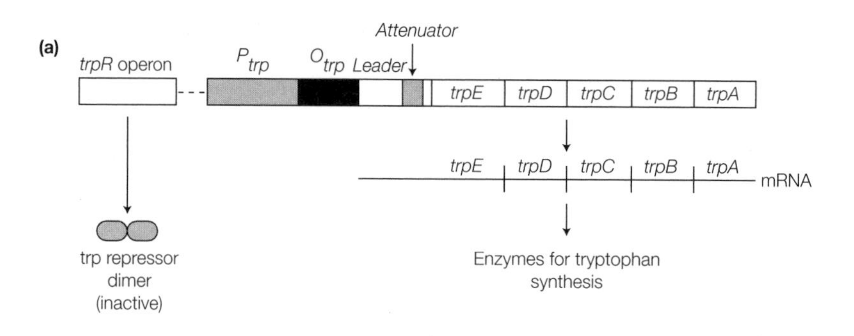


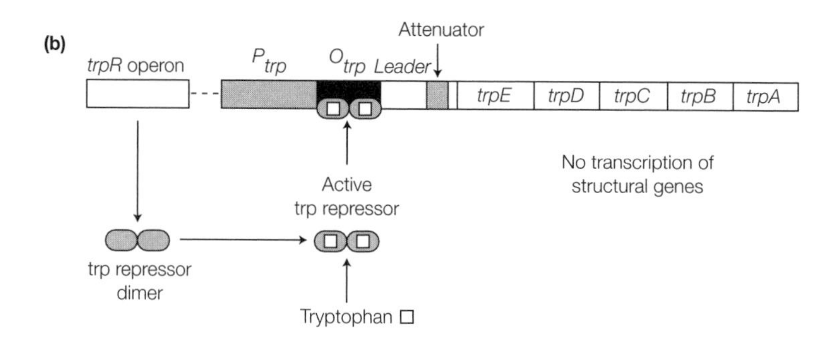


*Fig. 1. Regulation of the* trp *operon (a) transcription in the absence of tryptophan (b) no transcription in the presence of tryptophan.*

## Organization of the *trp* operon

The tryptophan (*trp*) operon (*Fig. 1*) contains five structural genes encoding enzymes for tryptophan biosynthesis with an upstream *trp* promoter ($P_{trp}$) and *trp* operator sequence ($O_{trp}$). The *trp* operator region partly overlaps the *trp* promoter. The operon is regulated such that transcription occurs when tryptophan in the cell is in short supply.

## Repression

In the absence of tryptophan (*Fig. 1a*), a trp repressor protein encoded by a separate operon, *trpR*, is synthesized and forms a dimer. However, this is inactive and so is unable to bind to the *trp* operator and the structural genes of the *trp* operon are transcribed. When tryptophan is present (*Fig. 1b*), the enzymes for tryptophan biosynthesis are not needed and so expression of these genes is turned off. This is achieved by tryptophan binding to the repressor to activate it so that it now binds to the operator and stops transcription of the structural genes. In this role, tryptophan is said to be a **co-repressor**. This is negative control, because the bound repressor prevents transcription, but note that the *lac* operon (see Topic G3) and *trp* operon show two ways in which negative control can be achieved; either (as in the *lac* operon) by having an active bound repressor that is inactivated by a bound ligand (the inducer) or (as in the *trp* operon) by having a repressor that is inactive normally but activated by binding the ligand. As in the case of the *lac* operator (Topic G3), the core binding site for the *trp* repressor in the *trp* operator is palindromic.

## Attenuation

A second mechanism, called **attenuation**, is also used to control expression of the *trp* operon. The 5′ end of the polycistronic mRNA transcribed from the *trp* operon has a **leader sequence** upstream of the coding region of the *trpE* struc-

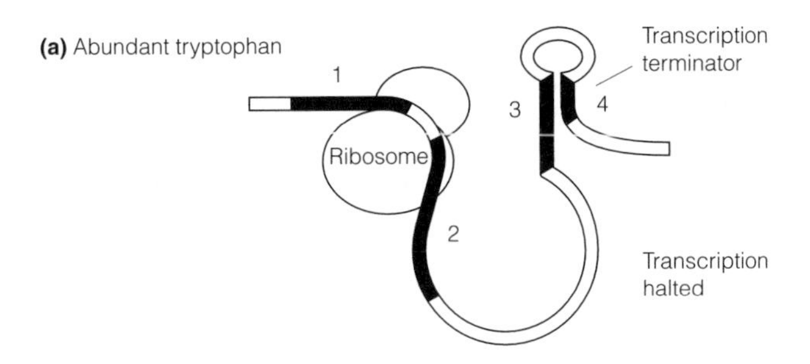


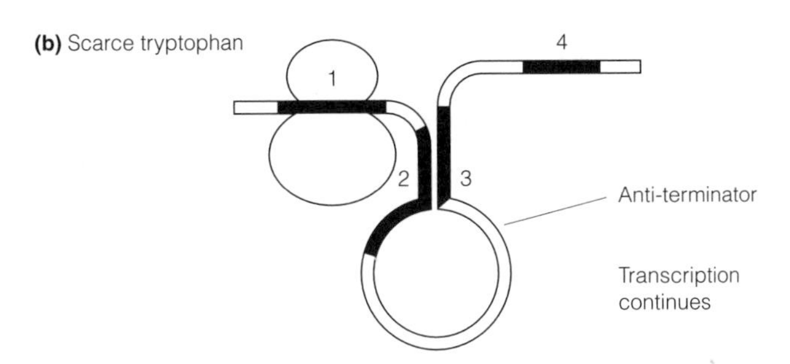


*Fig. 2. Attenuation of the* trp *operon. (a) When tryptophan is plentiful, sequences 3 and 4 base-pair to form a 3:4 structure that stops transcription (b) when tryptophan is in short supply, the ribosome stalls at the trp codons in sequence 1, leaving sequence 2 available to interact with sequence 3. Thus a 3:4 transcription terminator structure cannot form and transcription continues.*

tural gene (*Fig. 1*). This leader sequence encodes a 14 amino acid **leader peptide** containing two tryptophan residues.

The function of the leader sequence is to fine tune expression of the *trp* operon based on the availability of tryptophan inside the cell. It does this as follows. The leader sequence contains four regions (*Fig. 2*, numbered 1–4) that can form a variety of base-paired **stem-loop** (**'hairpin'**) secondary structures. Now consider the two extreme situations: the presence or absence of tryptophan. Attenuation depends on the fact that, in bacteria, ribosomes attach to mRNA as it is being synthesized and so translation starts even before transcription of the whole mRNA is complete. When tryptophan is abundant (*Fig. 2a*), ribosomes bind to the *trp* polycistronic mRNA that is being transcribed and begin to translate the leader sequence. Now, the two trp codons for the leader peptide lie within sequence 1, and the translational Stop codon (see Topic H1) lies between sequence 1 and 2. During translation, the ribosomes follow very closely behind the RNA polymerase and synthesize the leader peptide, with translation stopping eventually between sequences 1 and 2. At this point, the position of the ribosome prevents sequence 2 from interacting with sequence 3. Instead sequence 3 base-pairs with sequence 4 to form a 3:4 stem loop which acts as a **transcription terminator.** Therefore, when tryptophan is present, further transcription of the *trp* operon is prevented. If, however, tryptophan is in short supply (*Fig. 2b*), the ribosome will pause at the two *trp* codons contained within sequence 1. This leaves sequence 2 free to base pair with sequence 3 to form a 2:3 structure (also called the **anti-terminator**), so the 3:4 structure cannot form and transcription continues to the end of the *trp* operon. Hence the availability of tryptophan controls whether transcription of this operon will stop early (attenuation) or continue to synthesize a complete polycistronic mRNA.

Historically, attenuation was discovered when it was noticed that deletion of a short sequence of DNA between the operator and the first structural gene, *trpE*, increased the level of transcription. This region was named the **attenuator** (see *Fig. 1*) and is the DNA that encodes that part of the leader sequence that forms the transcription terminator stem-loop.

## Attenuation vs. repression

Overall, for the *trp* operon, repression via the *trp* repressor determines whether transcription will occur or not and attenuation then fine tunes transcription. Attenuation occurs in at least six other operons that encode enzymes for amino acid biosynthetic pathways. In some cases, such as the *trp* operon, both repression and attenuation operate to regulate expression. In contrast, for some other operons such as the *his*, *thr* and *leu* operons, transcription is regulated only by attenuation.

# G5 Transcription in eukaryotes: an overview

## Key Notes

**Three RNA polymerases** — In eukaryotes, RNA is synthesized by three RNA polymerases: RNA Pol I is a nucleolar enzyme that transcribes rRNAs, RNA Pol II is located in the nucleoplasm and transcribes mRNAs and most snRNAs, RNA Pol III is also nucleoplasmic and transcribes tRNA and 5S rRNA, as well as U6 snRNA and the 7S RNA of the signal recognition particle (SRP).

**RNA synthesis** — Each RNA polymerase transcribes only one strand, the antisense (−) strand, of a double-stranded DNA template, directed by a promoter. Synthesis occurs 5′ → 3′ and does not require a primer.

**RNA polymerase subunits** — Each of the three RNA polymerases contains 12 or more subunits, some of which are similar to those of *E. coli* RNA polymerase. However, four to seven subunits in each enzyme are unique to that enzyme.

**Related topics**

DNA structure (F1)
RNA structure (G1)
Transcription in prokaryotes (G2)
The *lac* operon (G3)
The *trp* operon (G4)
Transcription of protein-coding genes in eukaryotes (G6)
Regulation of transcription by RNA Pol II (G7)
Processing of eukaryotic pre-mRNA (G8)
Ribosomal RNA (G9)
Transfer RNA (G10)

## Three RNA polymerases

Unlike prokaryotes where all RNA is synthesized by a single RNA polymerase, the nucleus of a eukaryotic cell has three RNA polymerases responsible for transcribing different types of RNA.

- **RNA polymerase I (RNA Pol I)** is located in the nucleolus and transcribes the **28S**, **18S** and **5.8S rRNA** genes.
- **RNA polymerase II (RNA Pol II)** is located in the nucleoplasm and transcribes **protein-coding genes**, to yield pre-mRNA, and also the genes encoding **small nuclear RNAs (snRNAs)** involved in mRNA processing (see Topic G8), except for U6 snRNA.
- **RNA polymerase III (RNA Pol III)** is also located in the nucleoplasm. It transcribes the genes for **tRNA**, **5S rRNA, U6 snRNA**, and the **7S RNA** associated with the signal recognition particle (SRP) involved in the translocation of proteins across the endoplasmic reticulum membrane (see Topic H4).

## RNA synthesis

The basic mechanism of RNA synthesis by these eukaryotic RNA polymerases is the same as for the prokaryotic enzyme (see Topic G2), that is:

- the initiation of RNA synthesis by RNA polymerase is directed by the presence of a promoter site on the 5′ side of the transcriptional start site;
- the RNA polymerase transcribes one strand, the **antisense** (−) **strand**, of the DNA template;
- RNA synthesis does not require a primer;
- RNA synthesis occurs in the 5′ → 3′ direction with the RNA polymerase catalyzing a nucleophilic attack by the 3′-OH of the growing RNA chain on the α phosphorus atom on an incoming ribonucleoside 5′ triphosphate.

## RNA polymerase subunits

Each of the three eukaryotic RNA polymerases contains 12 or more subunits and so these are large complex enzymes. The genes encoding some of the subunits of each eukaryotic enzyme show DNA sequence similarities to genes encoding subunits of the core enzyme ($\alpha_2\beta\beta'\omega$) of *E. coli* RNA polymerase (see Topic G2). However, four to seven other subunits of each eukaryotic RNA polymerase are unique in that they show no similarity either with bacterial RNA polymerase subunits or with the subunits of other eukaryotic RNA polymerases.

# G6 TRANSCRIPTION OF PROTEIN-CODING GENES IN EUKARYOTES

## Key Notes

**Gene organization**

Most protein-coding genes in eukaryotes consist of coding sequences called exons interrupted by noncoding sequences called introns. The number of introns and their size varies from gene to gene. The primary transcript (pre-mRNA) undergoes processing reactions to yield mature mRNA.

**Initiation of transcription**

Most promoter sites for RNA polymerase II have a TATA box located about 25 bp upstream of the transcriptional start site. RNA polymerase binding to the promoter requires the formation of a transcription initiation complex involving several general (basal) transcription factors that assemble in a strict order. Some protein-coding genes lack a TATA box and have an initiator element instead, centered around the transcriptional start site. The initiation of transcription of these genes requires an additional protein to recognize the initiator element and facilitate formation of the transcription initiation complex; many of the same transcription factors for initiation of TATA box promoters are also involved here. Yet other promoters lack either a TATA box or an initiator element and transcription starts within a broad region of DNA rather than at a defined location.

**Elongation and termination**

Elongation continues until transcription comes to a halt at varying distances downstream of the gene, releasing the primary RNA transcript, pre-mRNA. This molecule then undergoes processing reactions to yield mRNA.

**Related topics**

DNA structure (F1)
RNA structure (G1)
Transcription in prokaryotes (G2)
The *lac* operon (G3)
The *trp* operon (G4)
Transcription in eukaryotes: an overview (G5)
Regulation of transcription by RNA Pol II (G7)
Processing of eukaryotic pre-mRNA (G8)
Ribosomal RNA (G9)
Transfer RNA (G10)

## Gene organization

In marked contrast to prokaryotic genes where proteins are encoded by a continuous sequence of triplet codons, the vast majority of protein-coding genes in eukaryotes are **discontinuous**. The coding sections of the gene (called **exons**) are interrupted by noncoding sections of DNA (called **introns**; *Fig. 1*). Nevertheless, the triplet codons within the exons and the order of exons themselves in the gene is still **colinear** with the amino acid sequence of the encoded polypeptide. The number of introns in a protein-coding gene varies (histone mRNAs lack introns) and they range in size from about 80 bp to over 10 000 bp. The primary transcript is a **pre-mRNA** molecule which must be processed to yield mature mRNA ready for translation. During RNA

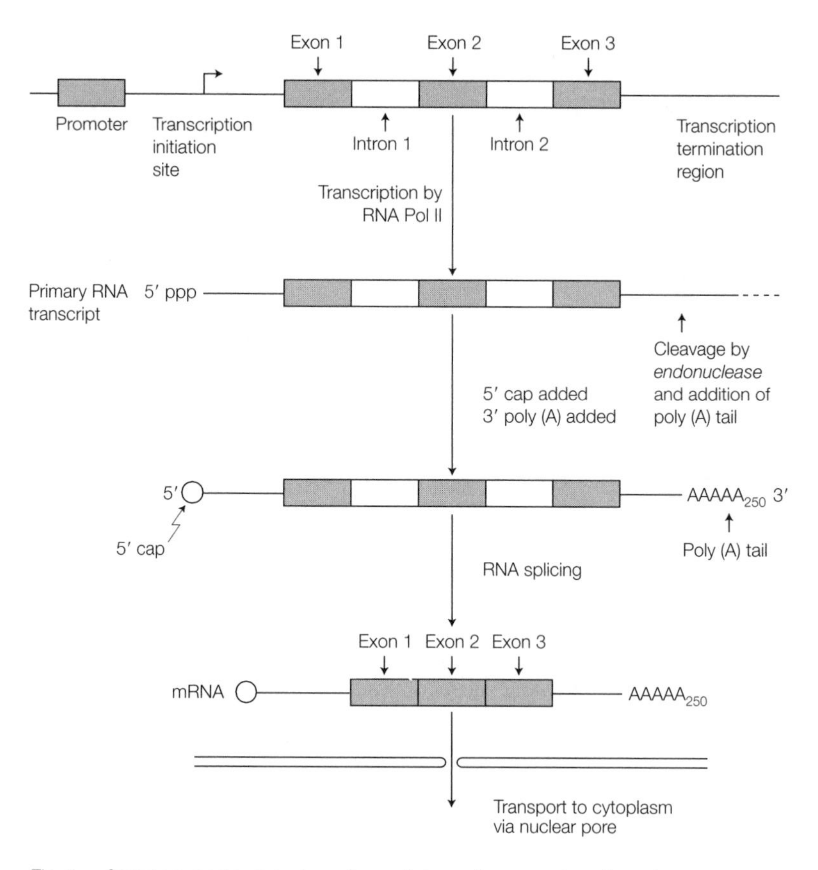


*Fig. 1. Structure and expression of a protein coding gene in eukaryotes.*

processing, the pre-mRNA receives a 5′ cap and (usually but not always) a poly(A) tail of 200–250 A residues, and the intron sequences are removed by RNA splicing. These RNA processing reactions are covered in detail in Topic G6.

## Initiation of transcription

Most promoter sites for RNA polymerase II include a highly conserved sequence located about 25–35 bp upstream (i.e. to the 5′ side) of the start site which has the consensus TATA(A/T)A(A/T) and is called the **TATA box** (*Fig.* 2). Since the start site is denoted as position +1, the TATA box position is said to be located at about position −25. The TATA box sequence resembles the −10 sequence (see Topic G2) in prokaryotes (TATAAT) except that it is located further upstream. Both elements have essentially the same function, namely recognition by the RNA polymerase in order to position the enzyme

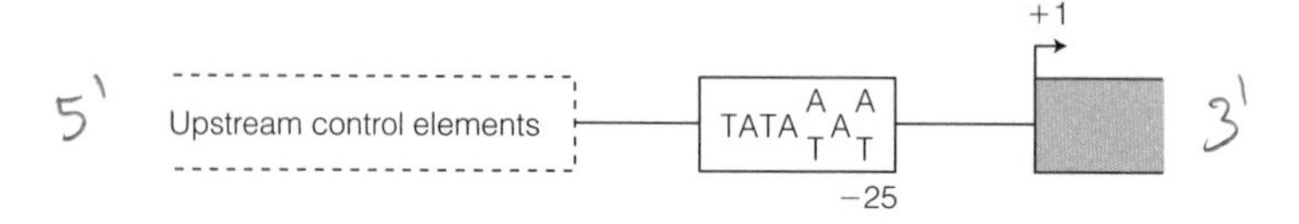


*Fig. 2. A typical promoter for RNA Pol II. The TATA box is located approximately 25 bp upstream of the transcriptional start site (denoted as +1).*

at the correct location to initiate transcription. The sequence around the TATA box is also important in that it influences the efficiency of initiation. Transcription is also regulated by **upstream control elements** that lie 5′ to the TATA box (*Fig. 2* and Topic G7).

Some eukaryotic protein-coding genes lack a TATA box and have an **initiator element** instead, centered around the transcriptional initiation site. This does not have a strong consensus between genes but often includes a C at position −1 and an A at position +1. Yet other promoters have neither a TATA box nor an initiator element; these genes tend to be transcribed at low rates and initiate transcription somewhere within a broad region of DNA (about 200 bp or so) rather than at a defined transcriptional start site.

In order to initiate transcription, RNA polymerase II requires the assistance of several other proteins or protein complexes, called **general** (or **basal) transcription factors**, which must assemble into a complex on the promoter in order for RNA polymerase to bind and start transcription (*Fig. 3*). These all have the generic name of **TFII** (for Transcription Factor for RNA polymerase II). The first event in initiation is the binding of the **transcription factor IID (TFIID)** protein complex to the TATA box. The key subunit of TFIID is **TBP (TATA box binding protein)**. Other subunits in the TFIID complex are called **TBP-associated factors ($TAF_{II}$s)**. The order of events is that TBP binds to the TATA box and then at least eight $TAF_{II}$s bind to form TFIID. As soon as the TFIID complex has bound, **TFIIA** binds and stabilizes the TFIID-TATA box interaction. Next, **TFIIB** binds to TFIID. However, TFIIB can also bind to RNA polymerase II and so acts as a bridging protein. Thus, RNA polymerase II, which has already complexed with **TFIIF**, now binds. This is followed by the binding of **TFIIE, H** and **J**. This final protein complex contains at least 40 polypeptides and is called the **transcription initiation complex**. It can now begin to transcribe the gene, although at only a relatively low rate, and is the basal transcription apparatus. For a high rate of transcription, other transcription factors are required which bind to additional sequence elements and interact with this initiation complex (see Topic G7).

Those protein-coding genes that have an initiator element instead of a TATA box (see above) appear to need another protein(s) that binds to the initiator element and facilitates the binding of TBP. The other transcription factors then bind to form the transcription initiation complex in a similar manner to that described above for genes possessing a TATA box promoter.

## Elongation and termination

Elongation of the RNA chain continues until termination occurs. Unlike RNA polymerase in prokaryotes, RNA polymerase II does not terminate transcription at specific sites but rather transcription stops at varying distances downstream of the gene. The RNA molecule made from a protein-coding gene by RNA polymerase II is called a **primary transcript**. Unlike the situation in prokaryotes, the primary transcript from a eukaryotic protein-coding gene is a precursor molecule, **pre-mRNA**, that needs extensive **RNA processing** in order to yield mature mRNA ready for translation. Several RNA processing reactions are involved: capping, 3′ cleavage and polyadenylation and RNA splicing (see *Fig. 1* and Topic G8).

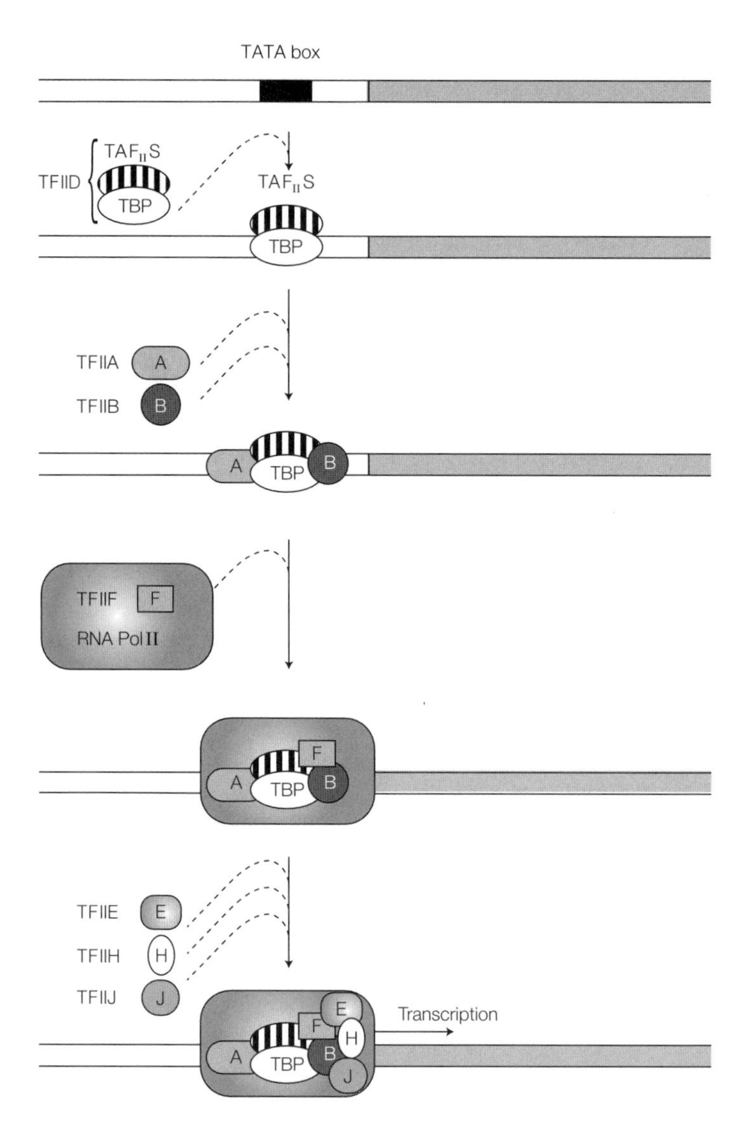


*Fig. 3. Initiation of transcription by RNA polymerase II. TFIID binds to the TATA box followed in order by the binding of TFIIA, TFIIB and a pre-formed complex of TFIIF.RNA polymerase II. Subsequently TFIIE, TFIIH and TFIIJ bind in order and transcription then starts about 25 bp downstream from the TATA box. Note that the placement of the various factors in this diagram is arbitrary; their exact positions in the complex are not yet known.*

# G7 Regulation of transcription by RNA Pol II

## Key Notes

**Mechanism of regulation**

Many genes are active in all cells but some are transcribed only in specific cell types, at specific times and/or only in response to specific external stimuli. Transcriptional regulation occurs via transcription factors that bind to short control elements associated with the target genes and then interact with each other and with the transcription initiation complex to increase or decrease the rate of transcription of the target gene.

**Upstream regulatory elements**

Many transcription factors bind to control elements located upstream within a few hundred base-pairs of the protein-coding gene. The SP1 box and CAAT box are examples of such regulatory elements found upstream of most protein-coding genes, but some upstream regulatory elements are associated with only a few genes and are responsible for gene-specific transcriptional regulation (e.g. hormone response elements).

**Enhancers**

Enhancers are positive transcriptional control elements typically 100–200 bp long that can be located either upstream or downstream of the target gene, are active in either orientation, and can activate transcription from the target gene even when located a long distance away (sometimes 10–50 kb). The transcription factors bound to these long-distance elements interact with the transcription initiation complex by looping out of the DNA.

**Transcription factors have multiple domains**

Transcription factors that increase the rate of transcription usually have at least two domains of protein structure, a DNA-binding domain that recognizes the specific DNA control element to bind to, and an activation domain that interacts with other transcription factors or the RNA polymerase. Many transcription factors operate as dimers (homodimers or heterodimers) held together via dimerization domains. Some transcription factors interact with small ligands via a ligand-binding domain.

**DNA binding domains**

DNA binding domains contain characteristic protein motifs. The helix-turn-helix motif contains two α-helices separated by a short β-turn. When the transcription factor binds to DNA, the recognition helix lies in the major groove of the DNA double helix. The second type of motif, the zinc finger, consists of a peptide loop with either two cysteines and two histidines (the $C_2H_2$ finger) or four cysteines (the $C_4$ finger) at the base of the loop that tetrahedrally co-ordinate a zinc ion. The zinc finger secondary structure is two β-strands and one α-helix. Transcription factors often contain several zinc fingers; in each case the α-helix binds in the major groove of the DNA double helix. Some transcription factors (e.g. bZIP proteins, basic HLH proteins) contain basic domains that interact with the target DNA.

**Dimerization domains**

A leucine zipper has a leucine every seventh amino acid and forms an α-helix with the leucines presented on the same side of the helix every second turn, giving a hydrophobic surface. Two transcription factor monomers can interact via the hydrophobic faces of their leucine zipper motifs to form a dimer. The helix-loop-helix (HLH) motif contains two α-helices separated by a nonhelical loop. The C-terminal α-helix has a hydrophobic face; two transcription factor monomers, each with an HLH motif, can dimerize by interaction between the hydrophobic faces of the two C-terminal α-helices.

**Activation domains**

No common structural motifs are known for the activation domains of transcription factors. Activation domains that are rich in acidic amino acids, glutamines or prolines have been reported.

**Repressors**

Repressor proteins that inhibit the transcription of specific genes in eukaryotes may bind either to control elements near to the target gene or to silencers that may be located a long distance away. The repressor may inhibit transcription of the target gene directly or may do so by interfering with the function of an activator protein required for efficient gene transcription.

**Related topics**

DNA structure (F1)
RNA structure (G1)
Transcription in prokaryotes (G2)
The *lac* operon (G3)
The *trp* operon (G4)
Transcription in eukaryotes: an overview (G5)
Transcription of protein-coding genes in eukaryotes (G6)
Processing of eukaryotic pre-mRNA (G8)
Ribosomal RNA (G9)
Transfer RNA (G10)

## Mechanism of regulation

A number of protein-coding genes are active in all cells and are required for so-called 'house-keeping' functions, such as the enzymes of glycolysis (Topic J3), the citric acid cycle (Topic L1) and the proteins of the electron transport chain (Topic L2). However, some genes are active only in specific cell types and are responsible for defining the specific characteristics and function of those cells; for example immunoglobulin genes in lymphocytes, myosin in muscle cells. In addition, the proteins expressed by any given cell may change over time (for example during early development) or in response to external stimuli, such as hormones. Eukaryotic cells can regulate the expression of protein-coding genes at a number of levels but a prime site of regulation is transcription.

Transcriptional regulation in a eukaryotic cell (i.e. which genes are transcribed and at what rate) is mediated by **transcription factors**, other than the general transcription factors, which recognize and bind to short regulatory DNA sequences (**control elements**) associated with the gene. These sequences are also called ***cis*-acting elements** (or simply ***cis*-elements**) since they are on the same DNA molecule as the gene being controlled (*cis* is Latin for 'on this side'). The protein transcription factors that bind to these elements are also known as ***trans*-acting factors** (or simply ***trans*-factors**) in that the genes encoding them can be on different DNA molecules (i.e. on different chromosomes). The transcription factors which regulate specific gene transcription do so by interacting with the proteins of the transcription initiation complex and may either increase (activate) or decrease (repress) the rate of transcription of the target gene.

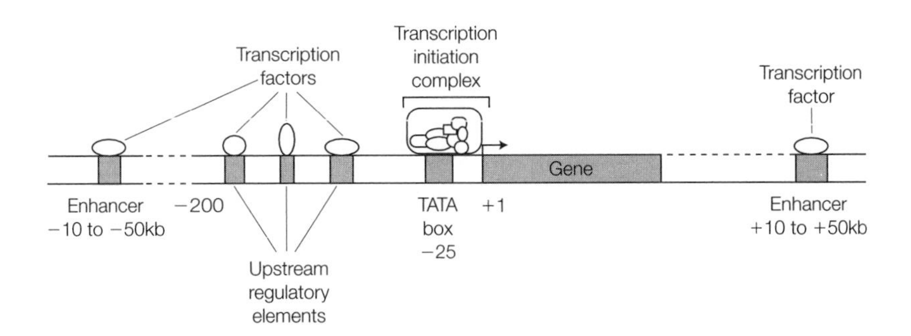


*Fig. 1. Control regions that regulate transcription of a typical eukaryotic protein-coding gene. Although shown as distinct entities here for clarity,* in vivo *the different regulatory proteins bound to the control elements and distant enhancers interact with each other and with the general transcription factors of the transcription initiation complex to modulate the rate of transcriptional initiation.*

Typically each protein-coding gene in a eukaryotic cell has several control elements in its promoter (*Fig. 1*) and hence is under the control of several transcription factors which interact with each other and with the transcription initiation complex by protein:protein interaction to determine the rate of transcription of that gene.

## Upstream regulatory elements

Many transcription factors bind to control elements within a few hundred base-pairs of the protein-coding gene being regulated. Positive control elements that lie upstream of the gene, usually within 200 bp of the transcriptional start site (*Fig. 1*), are often called **upstream regulatory elements** (**UREs**) and function to increase the transcriptional activity of the gene well above that of the basal promoter. Some of these elements, for example the **SP1 box** and the **CAAT box**, are found in the promoters of many eukaryotic protein-coding genes; indeed genes often have several copies of one or both elements. The SP1 box has the core sequence GGGCGG, and binds **transcription factor SP1** which then interacts with one of the $TAF_{II}$ proteins that bind to TBP to form TFIID (see Topic G6). In contrast, some upstream regulatory elements are associated only with a few specific genes and are responsible for limiting the transcription of those genes to certain tissues or in response to certain stimuli such as steroid hormones. For example, steroid hormones control metabolism by entering the target cell and binding to specific **steroid hormone receptors** in the cytoplasm. The binding of the hormone releases the receptor from an inhibitor protein that normally keeps the receptor in the cytoplasm. The hormone–receptor complex, now free of inhibitor, dimerizes and travels to the nucleus where it binds to a transcriptional control element, called a **hormone response element,** in the promoters of target genes. Then, like other transcription factors, the bound hormone–receptor complex interacts with the transcription initiation complex to increase the rate of transcription of the gene. The result is a hormone-specific transcription of a subset of genes in target cells that contain the appropriate steroid hormone receptor. Here, the hormone receptor is itself a transcription factor that is activated by binding the hormone ligand. Unlike steroid hormones, **polypeptide hormones**, such as **insulin** and **cytokines**, do not enter the target cell but instead bind to protein receptors located at the cell surface. The binding reaction triggers a cascade of protein activations, often involving protein phosphorylation, which relay the signal inside the cell (**signal transduction**). Again the response may be that specific transcription factors are activated and stimulate the transcription of selected genes, but here the activation is mediated via the signal transduction pathway and does not involve direct binding of the hormone or cytokine to the transcription factor. Many additional examples of transcriptional activation of specific genes by transcription factors exist in eukaryotes.

## Enhancers

Although many positive control elements lie close to the gene they regulate, others can be located long distances away (sometimes 10–50 kb) either upstream or downstream of the gene (*Fig. 1*). A long-distance positive control sequence of this kind is called an **enhancer** if the transcription factor(s) that binds to it increases the rate of transcription. An enhancer is typically 100–200 bp long and contains several sequence elements that act together to give the overall enhancer activity. When they were first discovered, enhancers were viewed as a distinct class of control element in that they:

- can activate transcription over long distances
- can be located upstream or downstream of the gene being controlled
- are active in either orientation with respect to the gene.

However, it is now clear that some upstream promoter elements and enhancers show strong similarities physically and functionally so that the distinction is not as clear as was once thought. For enhancers located a long distance away from the gene being controlled, interaction between transcription factors bound to the enhancer and to promoter elements near the gene occurs by looping out of the DNA between the two sets of elements (*Fig. 2*).

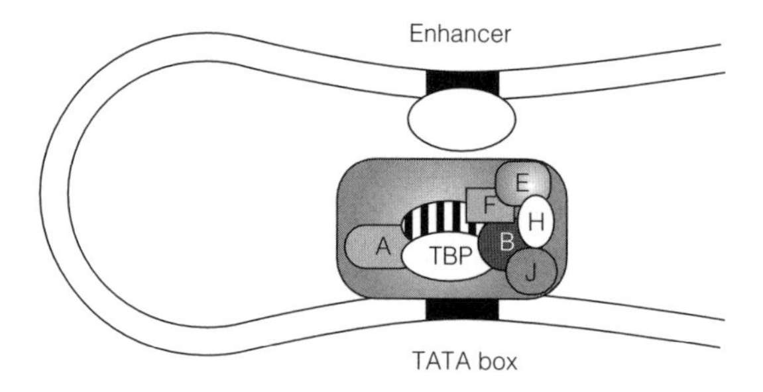


*Fig. 2. Looping out of DNA allowing the interaction of enhancer-bound factor(s) with the transcription initiation complex.*

## Transcription factors have multiple domains

In most cases, the transcription factors in eukaryotes that bind to enhancer or promoter sequences are activator proteins that induce transcription. These proteins usually have at least two distinct domains of protein structure, a **DNA-binding domain** that recognizes the specific DNA sequence to bind to, and an **activation domain** responsible for bringing about the transcriptional activation by interaction with other transcription factors and/or the RNA polymerase molecule. Many transcription factors operate as dimers, either **homodimers** (identical subunits) or **heterodimers** (dissimilar subunits) with the subunits held together via **dimerization domains**. DNA binding domains and dimerization domains contain characteristic protein structures (**motifs**) that are described below. Finally, some transcription factors (e.g. steroid hormone receptors) are responsive to specific small molecules (**ligands**) which regulate the activity of the transcription factor. In these cases, the ligand binds at a **ligand-binding domain**.

## DNA binding domains

*Helix-turn-helix*

This motif consists of two α-helices separated by a short (four-amino acid) peptide sequence that forms a **β-turn** (*Fig. 3a*). When the transcription factor binds to DNA, one of the helices, called the **recognition helix**, lies in the major groove of the DNA double helix (*Fig. 3b*). The helix-turn-helix motif was originally discovered in certain transcription factors that play major roles in *Drosophila* early development. These proteins each contain a 60-amino acid DNA-binding region called a **homeodomain** (encoded by a DNA sequence called a **homeobox**). The homeodomain has four α-helices in which helices II and III are the classic helix-turn-helix motif. Since the original discovery, the helix-turn-helix motif has been found in a wide range of transcription factors, including many that have no role in development.

*Zinc finger*

Two types of zinc finger have been reported, called the **$C_2H_2$ finger** and the **$C_4$ finger**. The $C_2H_2$ zinc finger is a loop of 12 amino acids with two cysteines and two histidines at the base of the loop that tetrahedrally coordinate a zinc ion (*Fig. 4a*). This forms a compact structure of two β-strands and one α-helix (*Fig. 4b*). The α-helix contains a number of conserved basic amino acids and interacts directly with the DNA, binding in the major groove of the double helix. Transcription factors that contain zinc fingers often contain several such motifs; usually at least three zinc fingers are needed for tight DNA binding of the protein. Indeed RNA polymerase III transcription factor A (**TFIIIA**; see Topic G9) contains nine zinc fingers! The SP1 transcription factor, which binds to the SP1 box, has three zinc fingers.

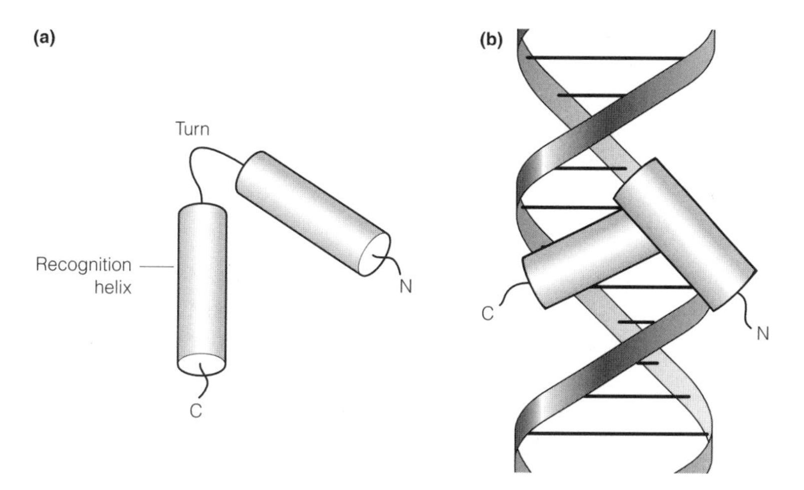


*Fig. 3. (a) Helix-turn-helix motif of a DNA-binding protein; (b) binding of the helix-turn-helix to target DNA showing the recognition helix lying in the major groove of the DNA.*

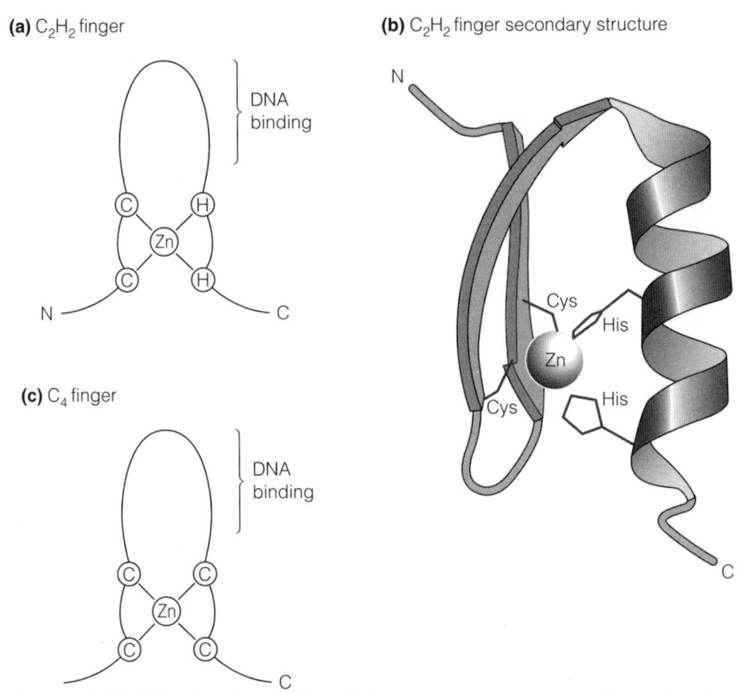


*Fig. 4. (a) A $C_2H_2$ zinc finger; (b) $C_2H_2$ zinc finger secondary structure. From A. Travers,* DNA–Protein Interactions, *Chapman & Hall, 1993. Reprinted with permission of A. Travers. (c) A $C_4$ zinc finger.*

The $C_4$ zinc finger is also found in a number of transcription factors, including steroid hormone receptor proteins. This motif forms a similar structure to that of $C_2H_2$ zinc finger but has four cysteines co-ordinated to the zinc ion instead of two cysteines and two histidines (see *Fig. 4c*).

### *Basic domains*

DNA binding domains called **basic domains** (rich in basic amino acids), occur in transcription factors in combination with leucine zipper or helix-loop-helix (HLH) dimerization domains (see below). The combination of basic domain and dimerization domain gives these proteins their names of **basic leucine zipper proteins (bZIP)** or **basic HLH proteins**, respectively. In each case the dimerization means that two basic domains (one from each monomer) interact with the target DNA.

## Dimerization domains

### *Leucine zippers*

The leucine zipper motif contains a leucine every seventh amino acid in the primary sequence and forms an α-helix with the leucines presented on the same side of the helix every second turn, giving a hydrophobic surface. The transcription factor dimer is formed by the two monomers interacting via the hydrophobic faces of their leucine zipper motifs (*Fig. 5a*). In the case of bZIP proteins, each monomer also has a basic DNA binding domain located N-terminal to the leucine zipper. Thus the bZIP protein dimer has two basic domains. These actually face in opposite directions which allows them to bind

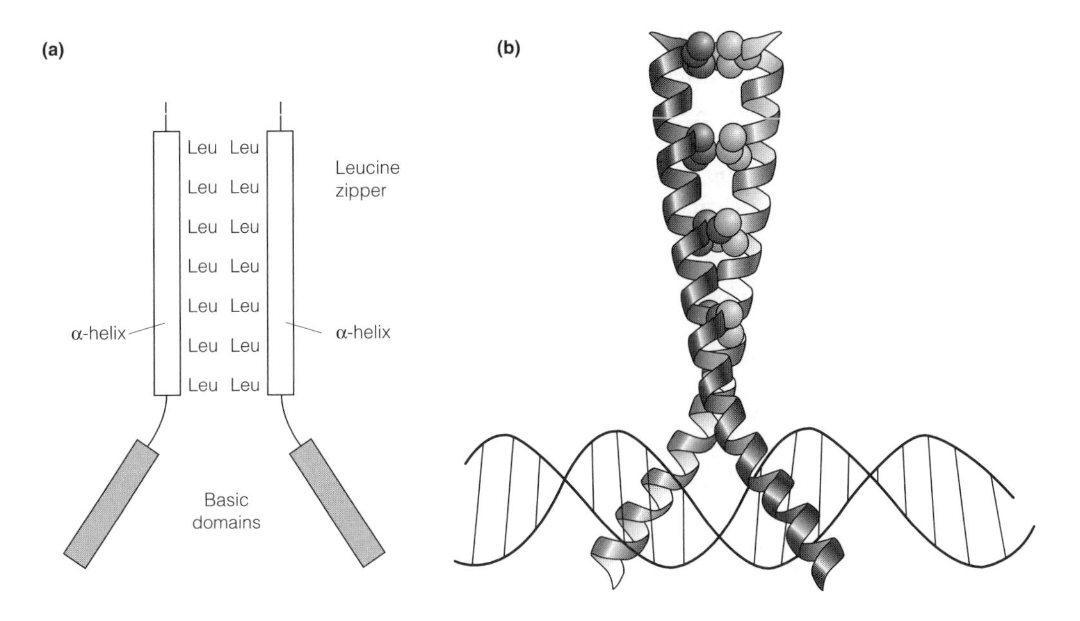


*Fig. 5. (a) A bZIP protein dimer showing the leucine zipper dimerization domain and the two basic domains; (b) folded structure of a bZIP protein showing the basic domains binding in the major groove of the target DNA. Reprinted from A. Travers,* DNA–Protein Interactions, *Chapman & Hall, 1993. With permission from A. Travers.*

to DNA sequences that have inverted symmetry. They bind in the major groove of the target DNA (*Fig. 5b*). The leucine zipper domain also acts as the dimerization domain in transcription factors that use DNA binding domains other than the basic domain. For example, some homeodomain proteins, containing the helix-turn-helix motif for DNA binding, have leucine zipper dimerization domains. In all cases, the dimers that form may be homodimers or heterodimers.

***Helix-loop-helix motif***

The helix-loop-helix (HLH) dimerization domain is quite distinct from the helix-turn-helix motif described above (which is involved in DNA binding *not* dimerization) and must not be confused with it. The HLH domain consists of two α-helices separated by a nonhelical loop. The C-terminal α-helix has hydrophobic amino acids on one face. Thus two transcription factor monomers, each with an HLH motif, can dimerize by interaction between the hydrophobic faces of the two C-terminal α-helices. Like the leucine zipper (see above), the HLH motif is often found in transcription factors that contain basic DNA binding domains. Again, like the leucine zipper, the HLH motif can dimerize transcription factor monomers to form either homodimers or heterodimers. This ability to form heterodimers markedly increases the variety of active transcription factors that are possible and so increases the potential for gene regulation.

## Activation domains

Unlike DNA binding domains and dimerization domains, no common structural motifs have yet been identified in the activation domains of diverse transcription factors. However, most activation domains so far reported appear to fall into one of three classes:

- **acidic activation domains** are rich in acidic amino acids (aspartic and glutamic acids). For example, mammalian glucocorticoid receptor proteins contain this type of activation domain;
- **glutamine-rich domains** (e.g. as in SP1 transcription factor);
- **proline-rich domains** (e.g. *c-jun* transcription factor).

## Repressors

Gene repressor proteins that inhibit the transcription of specific genes in eukaryotes also exist. They may act by binding either to control elements within the promoter region near the gene or at sites located a long distance away from the gene, called **silencers**. The repressor protein may inhibit transcription directly. One example is the **mammalian thyroid hormone receptor** which, in the absence of thyroid hormone, represses transcription of the target genes. However, other repressors inhibit transcription by blocking activation. This can be achieved in one of several ways: by blocking the DNA binding site for an activator protein, by binding to and masking the activation domain of the activator factor, or by forming a non-DNA binding complex with the activator protein. Several examples of each mode of action are known.

# G8 PROCESSING OF EUKARYOTIC PRE-mRNA

## Key Notes

**Overview**

The primary RNA transcript from a protein-coding gene in a eukaryotic cell must be modified by several RNA processing reactions in order to become a functional mRNA molecule. The 5′ end is modified to form a 5′ cap structure. Most pre-mRNAs are then cleaved near the 3′ end and a poly(A) tail is added. Intron sequences are removed by RNA splicing.

**5′ processing: capping**

Immediately after transcription, the 5′ phosphate is removed, guanosyl transferase adds a G residue linked via a 5′–5′ covalent bond, and this is methylated to form a 7-methylguanosine ($m^7G$) cap (methylated in N–7 position of the base). The ribose residues of either the adjacent one or two nucleotides may also be methylated by methyl group addition to the 2′ OH of the sugar. The cap protects the 5′ end of the mRNA against ribonuclease degradation and also functions in the initiation of protein synthesis.

**3′ processing: cleavage and polyadenylation**

Most pre-mRNA transcripts are cleaved post-transcriptionally near the 3′ end between a polyadenylation signal (5′-AAUAAA-3′) and 5′-YA-3′ (where Y = a pyrimidine). A GU-rich sequence may also be located further downstream. Specific proteins bind to these sequence elements to form a complex. One of the bound proteins, poly(A) polymerase, then adds a poly(A) tail of up to 250 A residues to the new 3′ end of the RNA molecule and poly(A) binding protein molecules bind to this. The poly(A) tail protects the 3′ end of the final mRNA against nuclease degradation and also increases translational efficiency of the mRNA. Some pre-mRNAs (e.g. histone pre-mRNAs) are cleaved near the 3′ end but no poly(A) tail is added.

**RNA splicing**

Intron sequences are removed by RNA splicing that cleaves the RNA at exon–intron boundaries and ligates the ends of the exon sequences together. The cleavage sites are marked by consensus sequences that are evolutionarily conserved. In most cases the intron starts with GU and ends with AG, a polypyrimidine tract lies upstream of the AG, and a conserved branchpoint sequence is located about 20–50 nt upstream of the 3′ splice site. The splicing reaction involves two transesterification steps which ligate the exons together and release the intron as a branched lariat structure containing a 2′5′ bond with a conserved A residue in the branchpoint sequence. The RNA splicing reactions require snRNPs and accessory proteins that assemble into a spliceosome at the intron to be removed. The RNA components of the snRNPs are complementary to the 5′ and 3′ splice site sequences and to other conserved sequences in the intron and so can base-pair with them. Some introns start with AU and end with AC, instead of GU and AG respectively. The splicing of these 'AT-AC introns' requires a different set of snRNPs than those used for splicing of the major form of intron, except both classes of intron use U5 snRNP.

**Alternative processing**

Some pre-mRNAs contain more than one set of sites for 3′ end cleavage and polyadenylation, such that the use of alternative sites can lead to mRNA products that contain different 3′ noncoding regions (which may influence the lifetime of the mRNA) or have different coding capacities. Alternative splice pathways also exist whereby the exons that are retained in the final mRNA depends upon the pathway chosen, allowing several different proteins to be synthesized from a single gene.

**RNA editing**

The sequence of an mRNA molecule may be changed after synthesis and processing by RNA editing. Individual nucleotides may be substituted, added or deleted. In human liver, apolipoprotein B pre-mRNA does not undergo editing and subsequent translation yields apolipoprotein B100. In cells of the small intestine, RNA editing converts a single C residue in apolipoprotein B pre-mRNA to U, changing a codon for glutamine (CAA) to a termination codon (UAA). Translation of the edited mRNA yields the much shorter protein, apolipoprotein B48, with a restricted function in that it lacks a protein domain for receptor binding. Many other examples of editing occur, including trypanosome mitochondrial mRNAs, where RNA editing results in over half of the uridines in the final mRNA being acquired through the editing process.

**Related topics**

DNA structure (F1)
RNA structure (G1)
Transcription in prokaryotes (G2)
The *lac* operon (G3)
The *trp* operon (G4)
Transcription in eukaryotes: an overview (G5)
Transcription of protein-coding genes in eukaryotes (G6)
Regulation of transcription by RNA Pol II (G7)
Ribosomal RNA (G9)
Transfer RNA (G10)

## Overview

In eukaryotes, the product of transcription of a protein-coding gene is pre-mRNA (see Topic G6) which requires processing to generate functional mRNA. Several processing reactions occur. The 5′ end of the **primary RNA transcript, pre-mRNA,** is modified by the addition of a **5′ cap** (a process known as **capping**) and the 3′ end of most (but not all) pre-mRNAs is also modified by cleavage and then the addition of 200–250 A residues to form a **poly(A) tail** (a process called **polyadenylation**). The pre-mRNA sequence includes both coding (**exon**) and noncoding (**intron**) regions (see Topic G6) The latter need to be removed and the exon sequences joined together by **RNA splicing** to generate a continuous coding sequence for translation. All of these mRNA processing reactions occur in the nucleus so that, at any one time, there is a population of pre-mRNAs of different sizes reflecting both the sizes of the protein-coding genes from which they were transcribed and the extent of processing that has occurred. This population of RNA molecules is called **heterogeneous nuclear RNA (hnRNA)**. HnRNA is not naked but has specific proteins bound to it forming **heterogeneous nuclear ribonucleoprotein (hnRNP) complexes**. The proteins are probably involved both in the various processing reactions and subsequent transport of mRNA from the nucleus.

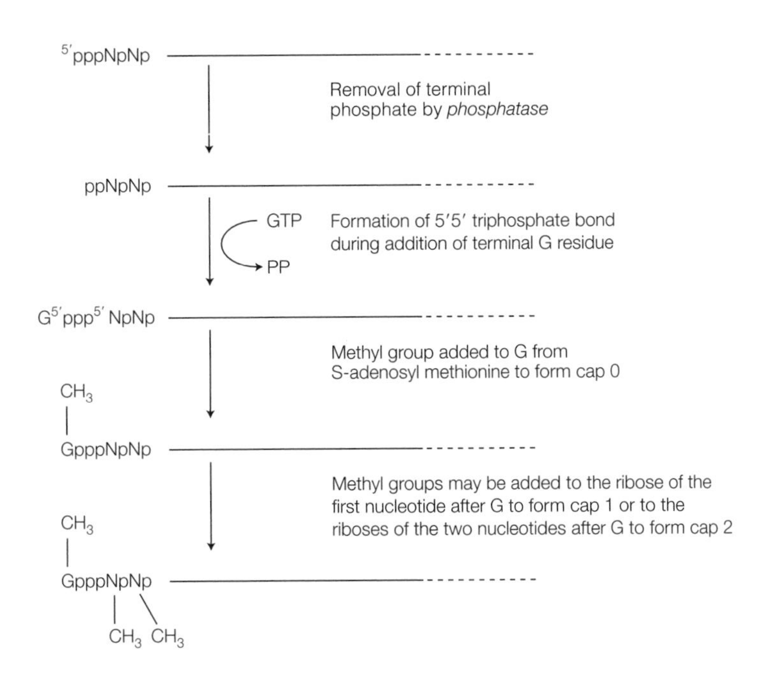


*Fig. 1. Steps involved in the formation of the 5′ cap.*

## 5′ processing: capping

Capping of pre-mRNA occurs immediately after synthesis and involves the addition of **7-methylguanosine ($m^7G$)** to the 5′ end (*Fig. 1*). To achieve this, the terminal 5′ phosphate is first removed by a phosphatase. **Guanosyl transferase** then catalyzes a reaction whereby the resulting diphosphate 5′ end attacks the α phosphorus atom of a GTP molecule to add a G residue in an unusual 5′5′ triphosphate link (*Fig. 1*). The G residue is then methylated by adding a methyl group to the N–7 position of the guanine ring, using **S-adenosyl methionine** as methyl donor. This structure, with just the $m^7G$ in position, is called a **cap 0 structure**. The ribose of the adjacent nucleotide (nucleotide 2 in the RNA chain) or the riboses of both nucleotides 2 and 3 may also be methylated to give **cap 1** or **cap 2** structures respectively. In these cases, the methyl groups are added to the 2′ OH groups of the ribose sugars (*Fig. 1*).

The cap protects the 5′ end of the primary transcript against attack by ribonucleases that have specificity for 3′5′ phosphodiester bonds and so cannot hydrolyze the 5′5′ bond in the cap structure. In addition, the cap plays a role in the initiation step of protein synthesis in eukaryotes. Only RNA transcripts from eukaryotic protein-coding genes become capped; prokaryotic mRNA and eukaryotic rRNA and tRNAs are uncapped.

## 3′ processing: cleavage and polyadenylation

Most eukaryotic pre-mRNAs undergo polyadenylation which involves cleavage of the RNA at its 3′ end and the addition of up to 250 A residues to form a poly(A) tail. The cleavage and polyadenylation reactions require the existence of a **polyadenylation signal sequence (5′-AAUAAA-3′)** located near the 3′ end of the pre-mRNA followed by a sequence **5′-YA-3′** (where Y = a pyrimidine) in the next 11–20 nt (*Fig. 2*). A **GU-rich sequence** is also often present further downstream. Several specific proteins (**cleavage factors**) recognize and bind to these sequence elements to form a protein complex that cleaves the RNA

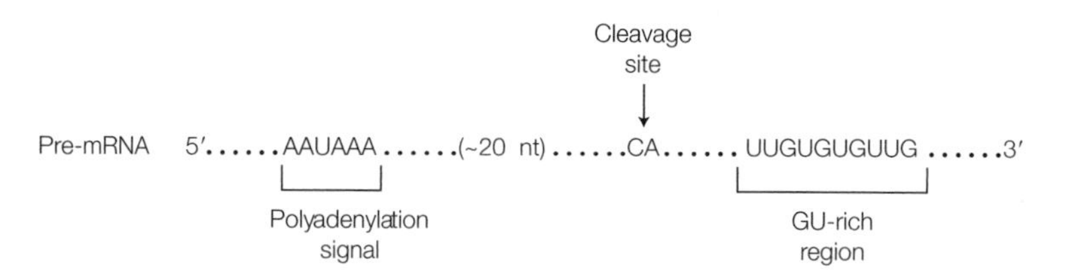


*Fig. 2. Conserved sequences for polyadenylation.*

between the AAUAAA sequence and the GU-rich sequence, at a site about 20 nt downstream of the AAUAAA sequence (*Fig. 2*). Another of the bound proteins, **poly(A) polymerase (PAP)**, then adds up to 250 A residues to the new 3′ end of the RNA molecule using ATP as precursor. The **poly(A) tail** immediately binds several copies of a poly(A) binding protein. The poly(A) tail protects the 3′ end of the final mRNA against ribonuclease digestion and hence stabilizes the mRNA. In addition, it increases the efficiency of translation of the mRNA. However, some mRNAs, notably histone pre-mRNAs, lack a poly(A) tail. Nevertheless, histone pre-mRNA is still subject to 3′ processing. It is cleaved near the 3′ end by a protein complex that recognizes specific signals, one of which is a stem-loop structure, to generate the 3′ end of the mature mRNA molecule.

## RNA splicing

The next step in RNA processing is the precise removal of intron sequences and joining the ends of neighboring exons to produce a functional mRNA molecule, a process called **RNA splicing**. The exon–intron boundaries are marked by specific sequences (*Fig. 3*). In most cases, at the 5′ boundary between the exon and the intron (**the 5′ splice site**), the intron starts with the sequence GU and at the 3′ exon–intron boundary (**the 3′ splice site)** the intron ends with the sequence AG. Each of these two sequences lies within a longer consensus sequence. A **polypyrimidine tract** (a conserved stretch of about 11 pyrimidines) lies upstream of the AG at the 3′ splice site (*Fig. 3*). A key signal sequence is the **branchpoint sequence** which is located about 20–50 nt upstream of the 3′ splice site. In vertebrates this sequence is 5′-CURAY-3′ where R = purine and Y = pyrimidine (in yeast this sequence is 5′-UACUAAC-3′).

RNA splicing occurs in two steps (*Fig. 4*). In the first step, the 2′ OH of the A residue at the branch site (indicated as $\dot{A}$ in *Fig. 3*) attacks the 3′5′ phosphodiester bond at the 5′ splice site causing that bond to break and the 5′ end of the intron to loop round and form an unusual 2′5′ bond with the A residue in the

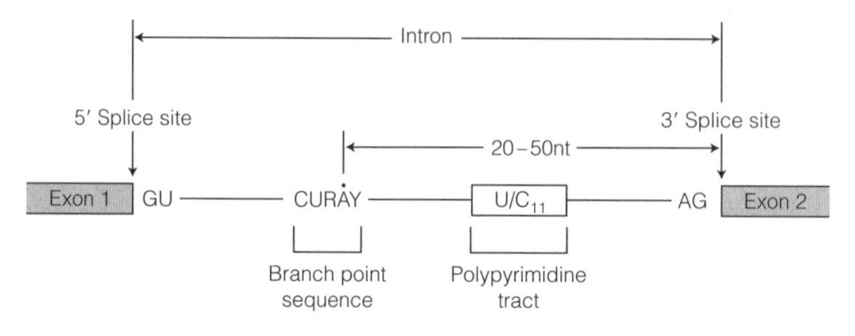


*Fig. 3. Conserved sequences for RNA splicing. The residue marked as $\dot{A}$ in the branchpoint sequence is the site of formation of the 2′5′ branch.*

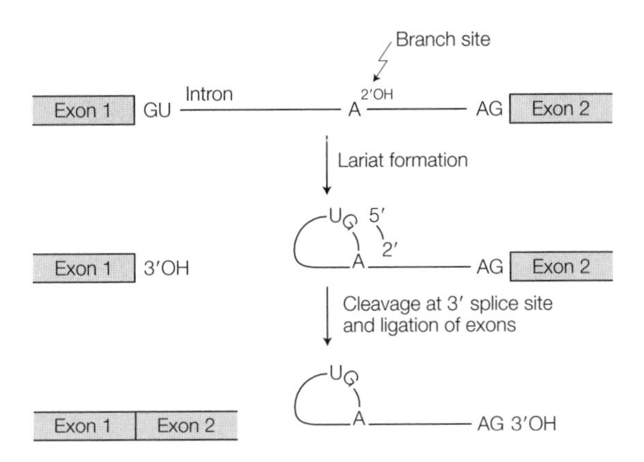


*Fig. 4. The two steps of RNA splicing.*

branchsite sequence. Because this A residue already has 3′5′ bonds with its neighbors in the RNA chain, the intron becomes branched at this point to form what is known as a **lariat** intermediate (named as such since it resembles a cowboy's lasso). The new 3′-OH end of exon 1 now attacks the phosphodiester bond at the 3′ splice site causing the two exons to join and release the intron, still as a lariat. In each of the two splicing reactions, one phosphate–ester bond is exchanged for another (i.e. these are two **transesterification reactions**). Since the number of phosphate–ester bonds is unchanged, no energy (ATP) is consumed.

RNA splicing requires the involvement of several **small nuclear RNAs (snRNAs)** each of which is associated with several proteins to form a small nuclear ribonucleoprotein particle or **snRNP** (pronounced '**snurp**'). Because snRNAs are rich in U residues, they are named U1, U2, etc. The RNA components of the snRNPs have regions that are complementary to the 5′ and 3′ splice site sequences and to other conserved sequences in the intron and so can base-pair with them. The U1 snRNP binds to the 5′ splice site and U2 snRNP binds to the branchpoint sequence (*Fig. 5*). A **tri-snRNP complex** of U4, U5 and U6 snRNPs then binds, as do other accessory proteins, so that a multicomponent complex (called a **spliceosome**) is formed at the intron to be removed and causes the intron to be looped out (*Fig. 5*). Thus, through interactions between the snRNAs and the pre-mRNA, the spliceosome brings the upstream and downstream exons together ready for splicing. The spliceosome next catalyzes the two-step splicing reaction to remove the intron and ligate together the two exons. The spliceosome then dissociates and the released snRNPs can take part in further splicing reactions at other sites on the pre-mRNA.

Although the vast majority of pre-mRNA introns start with GU at the 5′ splice site and end with AG at the 3′ splice site, some introns (possibly as many as 1%) have different splice site consensus sequences. In these cases, the intron starts with AU and ends with AC instead of GU and AG, respectively (*Fig. 6*). Since RNA splicing involves recognition of the splice site consensus sequences by key snRNPs (see above), and since these sequences are different in the minor intron class, U1, U2, U4, U6 snRNPs do not take part in splicing these so-called '**AT-AC introns**' (the AT-AC refers of course to the corresponding DNA sequence). Instead, U11, U12, $U4_{atac}$ and $U6_{atac}$ snRNPs are involved, replacing the roles of U1, U2, U4 and U6 respectively, and assemble to form the '**AT-AC spliceosome**'. U5 snRNP is required for splicing both classes of intron.

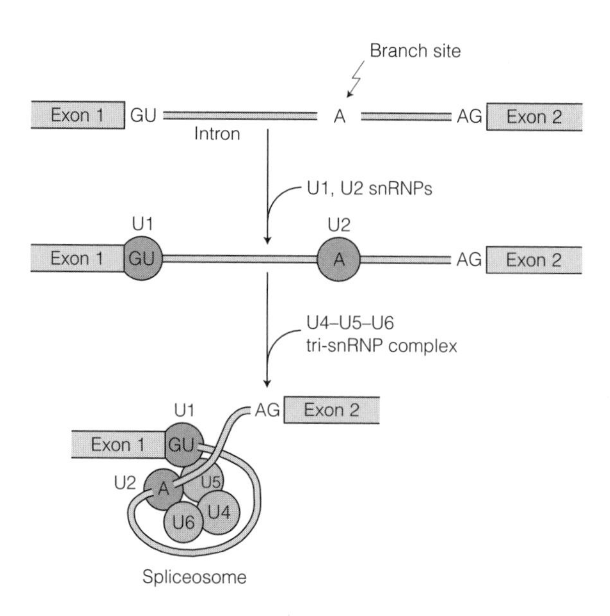


*Fig. 5. Formation of the spliceosome.*

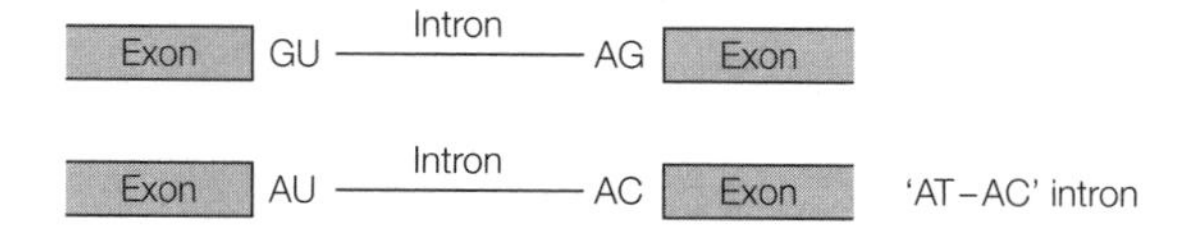


*Fig. 6. Comparison of the conserved splice site sequences of the majority of introns (top diagram) with those for AT-AC introns (bottom diagram).*

In some cases, RNA precursor molecules are known to undergo splicing in the absence of protein; the intron excises itself (**self-splicing**, see Topic G9).

## Alternative processing

### *Alternative polyadenylation sites*

Certain pre-mRNAs contain more than one set of signal sequences for 3′ end cleavage and polyadenylation. In some cases, the location of the alternative polyadenylation sites is such that, depending on the site chosen, particular exons may be lost or retained in the subsequent splicing reactions (*Fig. 7*). Here the effect is to change the coding capacity of the final mRNA so that different proteins are produced depending on the polyadenylation site used. In other

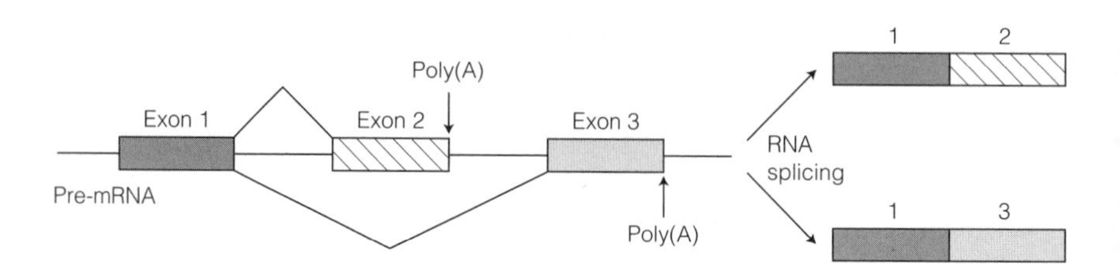


*Fig. 7. Use of alternative polyadenylation sites.*

cases, the alternative sites both lie within the 3′ noncoding region of the pre-mRNA so that the same coding sequences are included in the final mRNA irrespective of which site is used but the 3′ noncoding region can vary. Since the 3′ noncoding sequence may contain signals to control mRNA stability, the choice of polyadenylation site in this situation can affect the lifetime of the resulting mRNA.

*Alternative splicing*

Many cases are now known where different tissues splice the primary RNA transcript of a single gene by alternative pathways, where the exons that are lost and those that are retained in the final mRNA depend upon the pathway chosen (*Fig. 8*). Presumably some tissues contain regulatory proteins that promote or

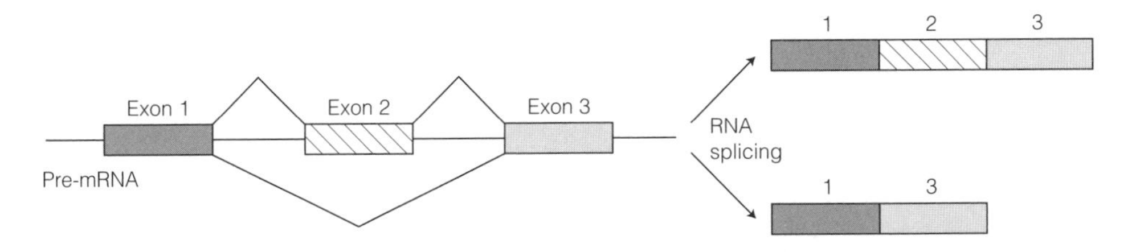


*Fig. 8. Alternative RNA splicing pathways. In the simple example shown, the transcript can be spliced by alternative pathways leading to two mRNAs with different coding capacities, i.e. exons 1, 2 and 3 or just exons 1 and 3. For genes containing many exons, a substantial number of alternative splice pathways may exist which are capable of generating many possible mRNAs from the single gene.*

suppress the use of certain splice sites to direct the splicing pathway selected. These **alternative splicing pathways** are very important since they allow cells to synthesize a range of functionally distinct proteins from the primary transcript of a single gene.

**RNA editing**

RNA editing is the name given to several reactions whereby the nucleotide sequence on an mRNA molecule may be changed by mechanisms other than RNA splicing. Individual nucleotides within the mRNA may be changed to other nucleotides, deleted entirely or additional nucleotides inserted. The effect of RNA editing is to change the coding capacity of the mRNA so that it encodes a different polypeptide than that originally encoded by the gene. An example of RNA editing in humans is **apolipoprotein B mRNA**. In liver, the mRNA does not undergo editing and the protein produced after translation is called **apolipoprotein B100** (*Fig. 9a*). In cells of the small intestine, RNA editing (*Fig. 9b*) causes the conversion of a single C residue in the mRNA to U and, in so doing, changes a codon for glutamine (CAA) to a termination codon (UAA). Subsequent translation of the edited mRNA yields the much shorter **apolipoprotein B48** (48% of the size of apolipoprotein B100). This is not a trivial change; apolipoprotein B48 lacks a protein domain needed for receptor binding which apolipoprotein B100 possesses and hence the functional activities of the two proteins are different. Many other cases of RNA editing are also known. Trypanosome mitochondrial mRNAs, for example, undergo extensive RNA editing which results in over half of the uridines in the final mRNA being acquired through the editing process.

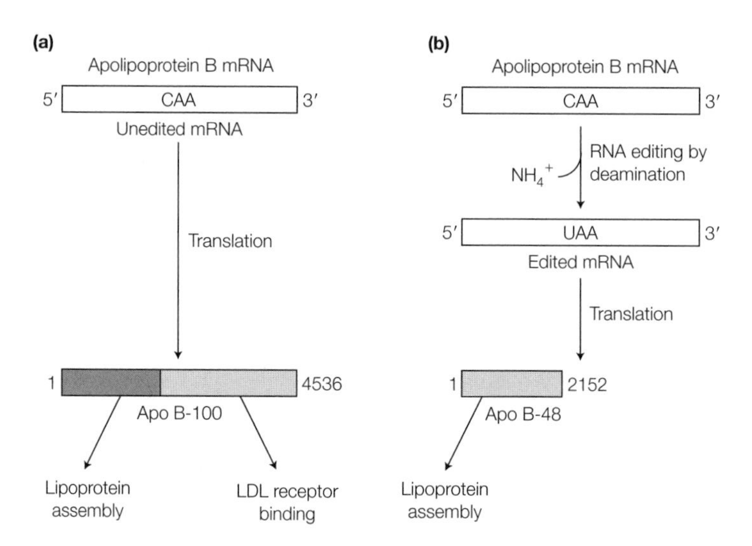


*Fig. 9. RNA editing. (a) Unedited apolipoprotein B mRNA is translated to yield ApoB-100, a 4536-amino acid long polypeptide with structural domains for lipoprotein assembly and receptor binding functions; (b) translation of the edited mRNA yields the shorter ApoB-48 which lacks the receptor binding domain.*

# G9 Ribosomal RNA

## Key Notes

**Ribosomes**

A prokaryotic 70S ribosome comprises two subunits (50S and 30S). The 50S subunit has 23S and 5S rRNAs complexed with 34 polypeptides whereas the 30S subunit contains 16S rRNA and 21 polypeptides. A eukaryotic 80S ribosome comprises two subunits (60S and 40S). The 50S subunit has 28S, 5.8S and 5S rRNAs complexed with approx. 49 polypeptides whereas the 40S subunit contains 18S rRNA and about 33 polypeptides.

**Transcription and processing of prokaryotic rRNA**

*E. coli* has seven rRNA transcription units, each containing one copy each of the 23S, 16S and 5S rRNA genes as well as one to four tRNA genes. Transcription produces a 30S pre-rRNA transcript. This folds up to form stem-loop structures, ribosomal proteins bind, and a number of nucleotides become methylated. The modified pre-rRNA transcript is then cleaved at specific sites by RNase III and the ends are trimmed by ribonucleases M5, M6 and M23 to release the mature rRNAs.

**Synthesis of eukaryotic 28S, 18S and 5.8S rRNA**

The 28S, 18S and 5.8S rRNA genes are present as multiple copies clustered together as tandem repeats. These rRNA transcription units are transcribed, in the nucleolus, by RNA Pol I. The promoter contains a core element that straddles the transcriptional start site and an upstream control element (UCE) about 50–80 bp in size, located at about position −100. Transcription factors, one of which is TATA binding protein (TBP) bind to these control elements and, together with RNA Pol I, form a transcription initiation complex. Transcription produces a 45S pre-rRNA which has external transcribed spacers (ETSs) at the 5′ and 3′ ends and internal transcribed spacers (ITSs) internally separating rRNA sequences. The pre-rRNA folds up to form a defined secondary structure with stem-loops, ribosomal proteins bind to selected sequences, and methylation of over 100 nucleotides occurs, guided by interaction of the pre-rRNA with snoRNAs (as snoRNPs). The 45S pre-rRNA molecule is then cleaved by ribonucleases, releasing 32S and 20S precursor rRNAs that are processed further to generate mature 28S, 18S and 5.8S rRNAs.

**Ribozymes**

In *Tetrahymena*, the pre-rRNA molecule contains an intron that is removed by self-splicing (in the presence of guanosine, GMP, GDP or GTP) without the need for involvement of any protein. This was the first ribozyme discovered but many have since been reported.

**Synthesis of eukaryotic 5S rRNA**

Eukaryotic cells contain multiple copies of the 5S rRNA gene. Unlike other eukaryotic rRNA genes, the 5S rRNA genes are transcribed by RNA Pol III. Two control elements, an A box and a C box, lie downstream of the transcriptional start site. The C box binds TFIIIA which then recruits TFIIIC. TFIIIB now binds and interacts with RNA Pol III to form the transcription initiation complex. Transcription produces a mature 5S rRNA that requires no processing.

**Related topics**

DNA structure (F1)
RNA structure (G1)
Transcription in prokaryotes (G2)
The *lac* operon (G3)
The *trp* operon (G4)
Transcription in eukaryotes: an overview (G5)
Transcription of protein-coding genes in eukaryotes (G6)
Regulation of transcription by RNA Pol II (G7)
Processing of eukaryotic pre-mRNA (G8)
Transfer RNA (G10)

## Ribosomes

Each ribosome consists of two subunits, a small subunit and a large subunit, each of which is a multicomponent complex of **ribosomal RNAs (rRNAs)** and **ribosomal proteins** (*Fig. 1*). One way of distinguishing between particles such as ribosomes and ribosomal subunits is to place the sample in a tube within a centrifuge rotor and spin this at very high speed. This causes the particles to sediment to the tube bottom. Particles that differ in mass, shape and/or density

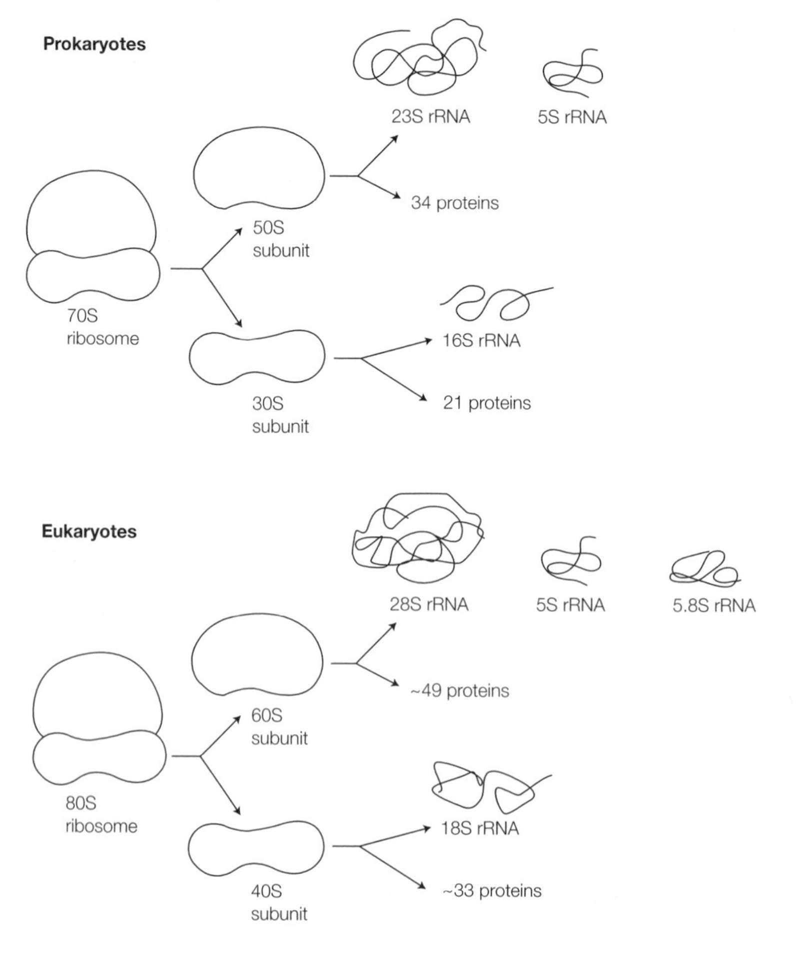


*Fig. 1. Composition of ribosomes in prokaryotic and eukaryotic cells.*

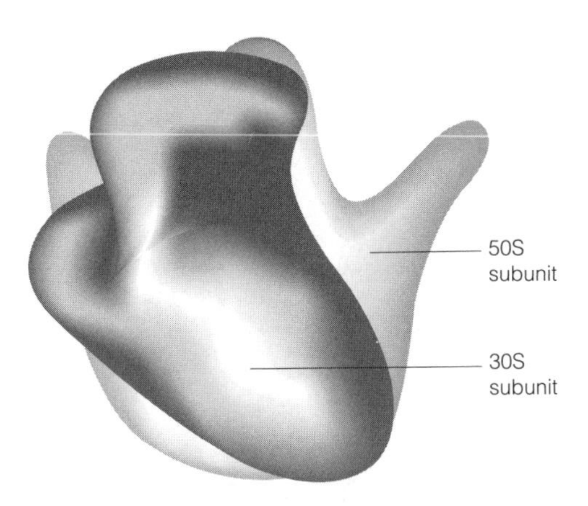


*Fig. 2. The prokaryotic 70S ribosome.*

sediment at different velocities (sedimentation velocities). Thus a particle with twice the mass of another will always sediment faster provided both particles have the same shape and density. The sedimentation velocity of any given particle is also directly proportional to the gravitational forces (the centrifugal field) experienced during the centrifugation, which can be increased simply by spinning the rotor at a higher speed. However, it is possible to define a **sedimentation coefficient** that depends solely on the size, shape and density of the particle and is independent of the centrifugal field. Sedimentation coefficients are usually measured in **Svedberg units (S)**. A prokaryotic ribosome has a sedimentation coefficient of 70S whereas the large and small subunits have sedimentation coefficients of 50S and 30S, respectively (note that S values are not additive). The 50S subunit contains two rRNAs (23S and 5S) complexed with 34 polypeptides whereas the 30S subunit contains 16S rRNA and 21 polypeptides (*Fig. 1*). In eukaryotes the ribosomes are larger and more complex; the ribosome monomer is 80S and consists of 60S and 40S subunits. The 60S subunit contains three rRNAs (28S, 5.8S and 5S) and about 49 polypeptides and the 40S subunit has 18S rRNA and about 33 polypeptides (*Fig. 1*).

A wide range of studies have built up a detailed picture of the fine structure of ribosomes, mapping the location of the various RNA and protein components and their interactions. The overall shape of a 70S ribosome, gained through electron microscopy studies, is shown in *Fig. 2*.

## Transcription and processing of prokaryotic rRNA

In *E. coli* there are seven rRNA transcription units scattered throughout the genome, each of which contains one copy of each of the 23S, 16S and 5S rRNA genes and one to four copies of various tRNA genes (*Fig. 3*). This gene assembly is transcribed by the single prokaryotic RNA polymerase to yield a single **30S pre-rRNA transcript** (about 6000 nt in size). This arrangement ensures that stoichiometric amounts of the various rRNAs are synthesized for ribosome assembly. Following transcription, the 30S pre-rRNA molecule forms internal base-paired regions to give a series of stem-loop structures and ribosomal proteins bind to form a **ribonucleoprotein (RNP) complex**. A number of the nucleotides in the folded pre-rRNA molecule are now methylated, on

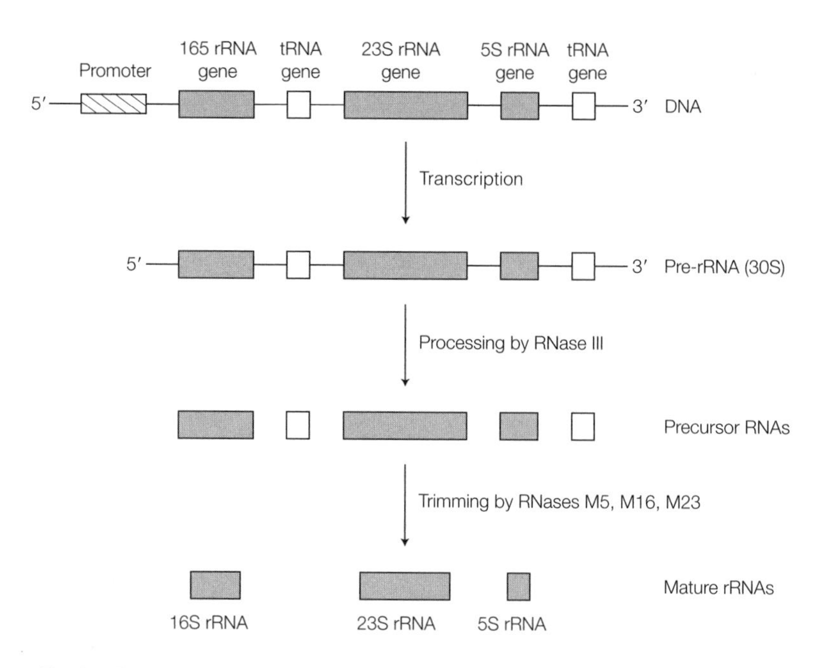


*Fig. 3. Transcription and processing of prokaryotic rRNA.*

the ribose moieties, using S-adenosylmethionine as the methyl donor (see Topic M3). Next the pre-rRNA molecule is cleaved at specific sites by **RNase III** to release precursors of the 23S, 16S and 5S rRNAs. The precursors are then trimmed at their 5′ and 3′ ends by ribonucleases **M5**, **M16** and **M23** (which act on the 5S, 16S and 23S precursor RNAs respectively) to generate the mature rRNAs.

## Synthesis of eukaryotic 28S, 18S and 5.8S rRNA

In eukaryotes, the genes for 28S, 18S and 5.8S rRNA are typically **clustered** together and **tandemly repeated** in that one copy each of 18S, 5.8S and then 28S genes occur, followed by untranscribed spacer DNA, then another set of 18S, 5.8S and 28S genes occur and so on (*Fig. 4a*). In humans, there are about 200 copies of this rRNA transcription unit arranged as five clusters of about 40 copies on separate chromosomes. These rRNA transcription units are transcribed by **RNA polymerase I (RNA Pol I)** in a region of the nucleus known as the **nucleolus** (see Topic A2). The nucleolus contains loops of DNA extending from each of the rRNA gene clusters on the various chromosomes and hence each cluster is called a **nucleolar organizer**.

The rRNA promoter consists of a **core element** which straddles the transcriptional start site (designated as position +1) from residues −31 to +6 plus an **upstream control element** (**UCE**) about 50–80 bp in size and located about 100 bp upstream from the start site (i.e. at position –100; *Fig. 4b*). A transcription factor called **upstream binding factor** (**UBF**) binds both to the UCE as well as to a region next to and overlapping with the core element. Interestingly, **TATA box binding protein** (**TBP**; see Topic G6), also binds to the rRNA promoter (in fact, TBP is required for initiation by all three eukaryotic RNA polymerases). The UBF and TBP transcription factors interact with each other and with RNA Pol I to form a **transcription initiation complex**. The RNA Pol I then transcribes

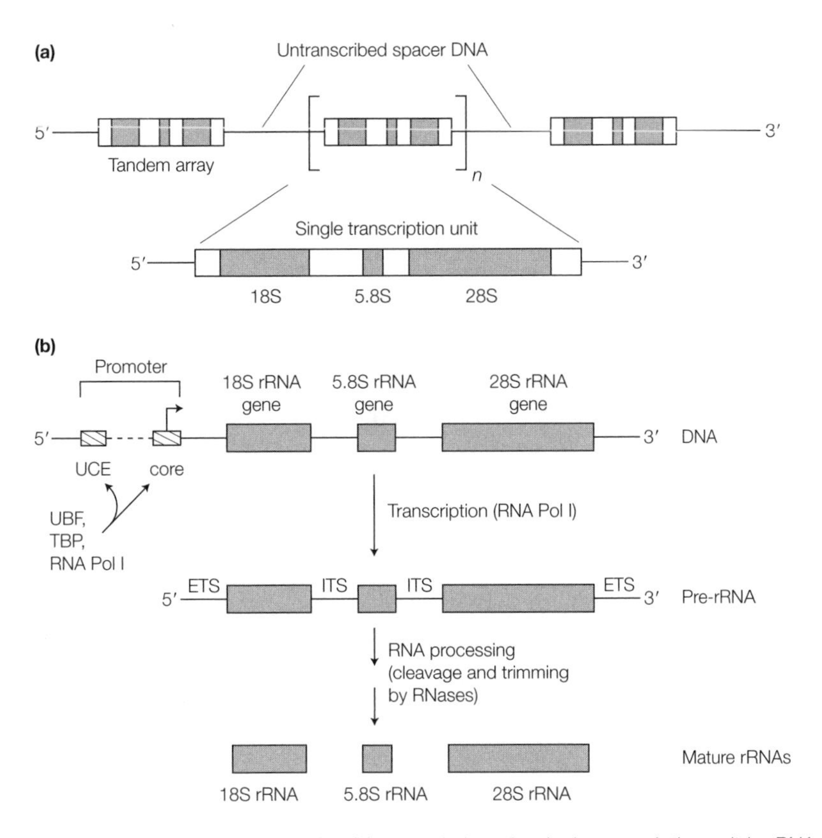


*Fig. 4. (a) rRNA transcription units; (b) transcription of a single transcription unit by RNA Pol I and processing of pre-rRNA.*

the whole transcription unit of 28S, 18S and 5.8S genes to synthesize a single large pre-rRNA molecule (*Fig. 4b*).

In humans, the product of transcription is a **45S pre-rRNA** which has non-rRNA **external transcribed spacers (ETSs)** at the 5′ and 3′ ends and non-rRNA **internal transcribed spacers (ITSs)** internally separating the rRNA sequences (*Fig. 4b*). This 45S molecule is processed in a similar pattern to that observed in prokaryotes for pre-rRNA, i.e. the pre-rRNA folds up to form a defined secondary structure with stem-loops, ribosomal proteins bind to selected sequences, and methylation of ribose moieties occurs (at over 100 nucleotides). The 45S pre-rRNA molecule is then cleaved by ribonucleases, first in the ETSs and then in the ITSs, to release precursor rRNAs which are cleaved further and trimmed by other ribonucleases to release the mature 28S, 18S and 5.8S rRNAs (*Fig. 4b*). In eukaryotes, selection of the sites in pre-rRNA that will be methylated depends upon small RNAs found in the nucleolus called **small nucleolar RNAs (snoRNAs)** that exist in ribonucleoprotein complexes called **snoRNPs**. The snoRNAs contain long regions (10–21 nt) that are complementary to specific regions of the pre-rRNA molecule, form base-pairs with the pre-rRNA at these sites and then guide where methylation of specific ribosome residues (2′O methylation) will occur. A number of pseudouridine (ψ) residues are also produced during processing of eukaryotic pre-rRNA and again snoRNAs are involved in guiding this event.

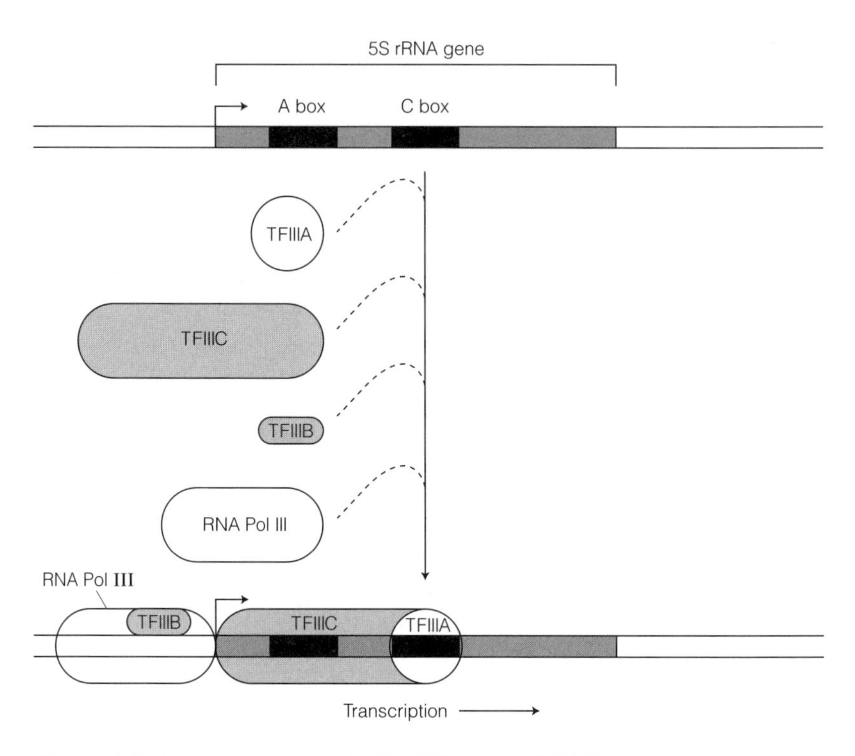


*Fig. 5. Initiation of transcription of a 5S rRNA gene by RNA Pol III.*

## Ribozymes

In at least one eukaryote, *Tetrahymena*, the pre-rRNA molecule contains an intron. Removal of the intron during processing of the pre-rRNA does not require the assistance of any protein! Instead, in the presence of guanosine, GMP, GDP or GTP, the intron excises itself, a phenomenon known as **self-splicing**. This was the first demonstration of **ribozymes**, that is, **catalytic RNA** molecules that catalyze specific reactions. The list of ribozymes is growing. For example, self-splicing introns have been discovered in some eukaryotic mRNAs and even peptidyl transferase, a key enzyme activity in protein synthesis, is now known to be a ribozyme (see Topic H2).

## Synthesis of eukaryotic 5S rRNA

In eukaryotes, the 5S rRNA gene is also present in multiple copies (2000 in human cells, all clustered together at one chromosomal site). Unlike other eukaryotic rRNA genes, the 5S rRNA genes are transcribed by **RNA polymerase III (RNA Pol III)**. The promoters of tRNA genes, which are also transcribed by RNA Pol III, contain control elements called the A box and B box located downstream of the transcriptional start site (see Topic G10). A similar situation exists for 5S rRNA genes in that the promoter has two control elements located downstream of the transcriptional start site, an **A box** and a **C box** (*Fig. 5*). The C box binds **transcription factor IIIA (TFIIIA)** which then in turn interacts with **TFIIIC** to cause it to bind, a process which probably also involves recognition of the A box. Once TFIIIC has bound, **TFIIIB** binds and interacts with RNA Pol III, causing that to bind also to form the **transcription initiation complex**. One of the three subunits of TFIIB is **TATA box binding protein (TBP**; see Topic G6), the transcription factor required for transcription by all three eukaryotic RNA polymerases. Following transcription, the 5S rRNA transcript requires no processing. It migrates to the nucleolus and is recruited into ribosome assembly.

# G10 Transfer RNA

## Key Notes

**tRNA structure**

Each tRNA has a cloverleaf secondary structure containing an anticodon arm, a D (or DHU) arm, a T or TΨC arm, and an amino acid acceptor stem to which the relevant amino acid becomes covalently bound, at the 3′ OH group. Some tRNAs also have a variable (or optional) arm. The three-dimensional structure is more complex because of additional interactions between the nucleotides.

**Transcription and processing of tRNA in prokaryotes**

*E. coli* contains clusters of up to seven tRNA genes separated by spacer regions, as well as tRNA genes within ribosomal RNA transcription units. Following transcription, the primary RNA transcript folds up into specific stem-loop structures and is then processed by ribonucleases D, E, F and P in an ordered series of reactions to release the individual tRNA molecules.

**Transcription and processing of tRNA in eukaryotes**

In eukaryotes, tRNA genes are present as multiple copies and are transcribed by RNA Pol III. Several tRNA genes may be transcribed to yield a single pre-tRNA that is then processed to release individual tRNAs. The tRNA promoter includes two control elements, called the A box and the B box, located within the tRNA gene itself and hence downstream of the transcriptional start site. Transcription initiation requires transcription factor IIIC (TFIIIC), which binds to the A and B boxes and TFIIIB that binds upstream of the A box. The primary RNA transcript folds up into stem-loop structures and non-tRNA sequence is removed by ribonuclease action. Unlike prokaryotes, in eukaryotes the CCA sequence at the 3′ end of the tRNA is added after the trimming reactions (by tRNA nucleotidyl transferase). Unlike prokaryotes, pre-tRNA molecules in eukaryotes may also contain a short intron in the loop of the anticodon arm. The intron is removed by tRNA splicing reactions involving endonuclease cleavage at both ends of the intron and then ligation of the cut ends of the tRNA.

**Modification of tRNA**

Following synthesis, nucleotides in the tRNA molecule may undergo modification to create unusual nucleotides such as 1-methylguanosine ($m^1G$), pseudouridine (Ψ), dihydrouridine (D), inosine (I) and 4-thiouridine ($S^4U$).

**Related topics**

DNA structure (F1)
RNA structure (G1)
Transcription in prokaryotes (G2)
The *lac* operon (G3)
The *trp* operon (G4)
Transcription in eukaryotes: an overview (G5)
Transcription of protein-coding genes in eukaryotes (G6)
Regulation of transcription by RNA Pol II (G7)
Processing of eukoaryotic pre-mRNA (G8)
Ribosomal RNA (G9)

## tRNA structure

Transfer RNA (tRNA) molecules play an important role in protein synthesis (Topics H2 and H3). Each tRNA becomes covalently bonded to a specific amino acid to form **aminoacyl-tRNA** which recognizes the corresponding codon in

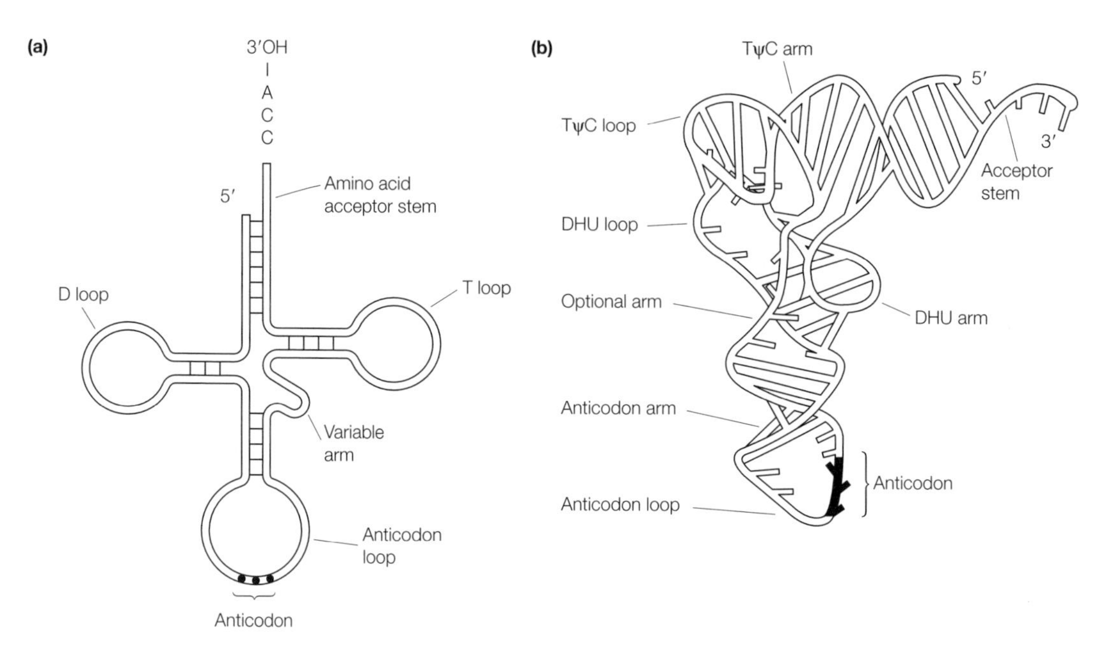


*Fig. 1. (a) Cloverleaf secondary structure of tRNA; (b) tertiary structure of tRNA (from* Genetics: a Molecular Approach, *second edition, T.A. Brown, Kluwer Academic Publishers, with permission).*

mRNA and ensures that the correct amino acid is added to the growing polypeptide chain. The tRNAs are small molecules, only 74–95 nt long, which form distinctive **cloverleaf** secondary structures (*Fig. 1a*) by internal base-pairing. The stem-loops of the cloverleaf are known as **arms**:

- the **anticodon arm** contains in its **loop** the three nucleotides of the **anticodon** which will form base-pairs with the complementary codon in mRNA during translation;
- the **D** or **DHU arm** (with its **D loop**) contains **dihydrouracil**, an unusual pyrimidine;
- the **T** or **TΨC arm** (with its **T loop**) contains another unusual base, **pseudouracil** (denoted Ψ) in the sequence TΨC;
- Some tRNAs also have **a variable arm** (**optional arm**) which is 3–21 nt in size.

The other notable feature is the **amino acid acceptor stem**. This is where the amino acid becomes attached, at the 3′ OH group of the 3′-CCA sequence.

The three dimensional structure of tRNA (*Fig. 1b*) is even more complex because of additional interactions between the various units of secondary structure.

## Transcription and processing of tRNA in prokaryotes

The rRNA transcription units in *E. coli* contain some tRNA genes that are transcribed and processed at the time of rRNA transcription (Topic G9). Other tRNA genes occur in clusters of up to seven tRNA sequences separated by spacer regions. Following transcription by the single prokaryotic RNA polymerase, the primary RNA transcript folds up into the characteristic stem-loop structures (*Fig. 2*) and is then processed in an ordered series of cleavages

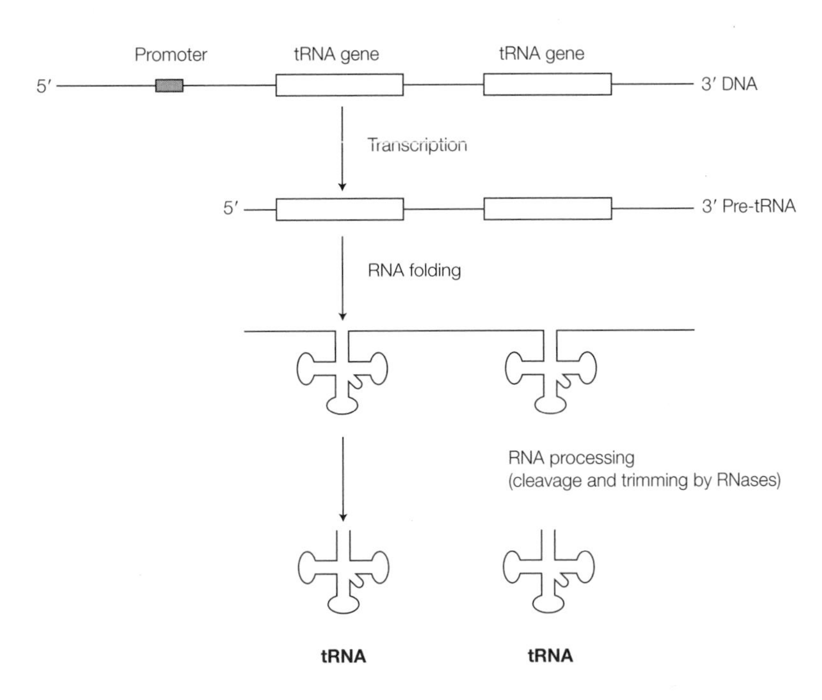


*Fig. 2. Transcription and processing of tRNA in prokaryotes.*

by ribonucleases (RNases) which release and trim the tRNAs to their final lengths. The cleavage and trimming reactions at the 5′ and 3′ ends of the precursor tRNAs involves **RNases D**, **E**, **F** and **P** working in the sequence shown in *Fig. 2*. RNases E, F and P are **endonucleases**, cutting the RNA internally, whilst RNase D is an **exonuclease**, trimming the ends of the tRNA molecules.

## Transcription and processing of tRNA in eukaryotes

In eukaryotes, the tRNA genes exist as multiple copies and are transcribed by **RNA polymerase III (RNA Pol III)**. As in prokaryotes, several tRNAs may be transcribed together to yield a single **pre-tRNA molecule** that is then processed to release the mature tRNAs. The promoters of eukaryotic tRNA genes are unusual in that the transcriptional control elements are located downstream (i.e. on the 3′ side) of the transcriptional start site (at position +1). In fact they lie within the gene itself. Two such elements have been identified, called the **A box** and **B box** (*Fig. 3*). Transcription of the tRNA genes by RNA Pol III requires **transcription factor IIIC (TFIIIC)** as well as **TFIIIB**. TFIIIC binds to the A and B boxes whilst TFIIIB binds upstream of the A box. TFIIIB contains three subunits, one of which is **TBP (TATA binding protein)**, the polypeptide required by all three eukaryotic RNA polymerases.

After synthesis, the pre-tRNA molecule folds up into the characteristic stem-loops structures (*Fig. 1*) and non-tRNA sequence is cleaved from the 5′ and 3′ ends by ribonucleases. In prokaryotes, the CCA sequence at the 3′ end of the tRNA (which is the site of bonding to the amino acid) is enclosed by the tRNA gene but this is not the case in eukaryotes. Instead, the CCA is added to the 3′ end after the trimming reactions by **tRNA nucleotidyl transferase**. Another difference between prokaryotes and eukaryotes is that eukaryotic pre-tRNA molecules often contain a short **intron** in the loop of the anticodon arm (*Fig. 4*).

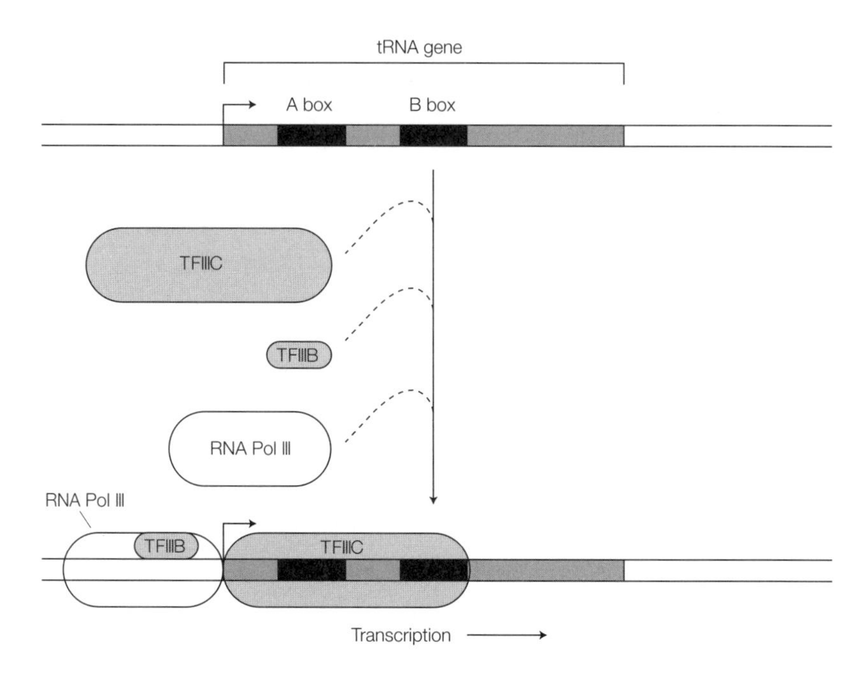


*Fig. 3. Initiation of transcription of a tRNA gene by RNA Pol III.*

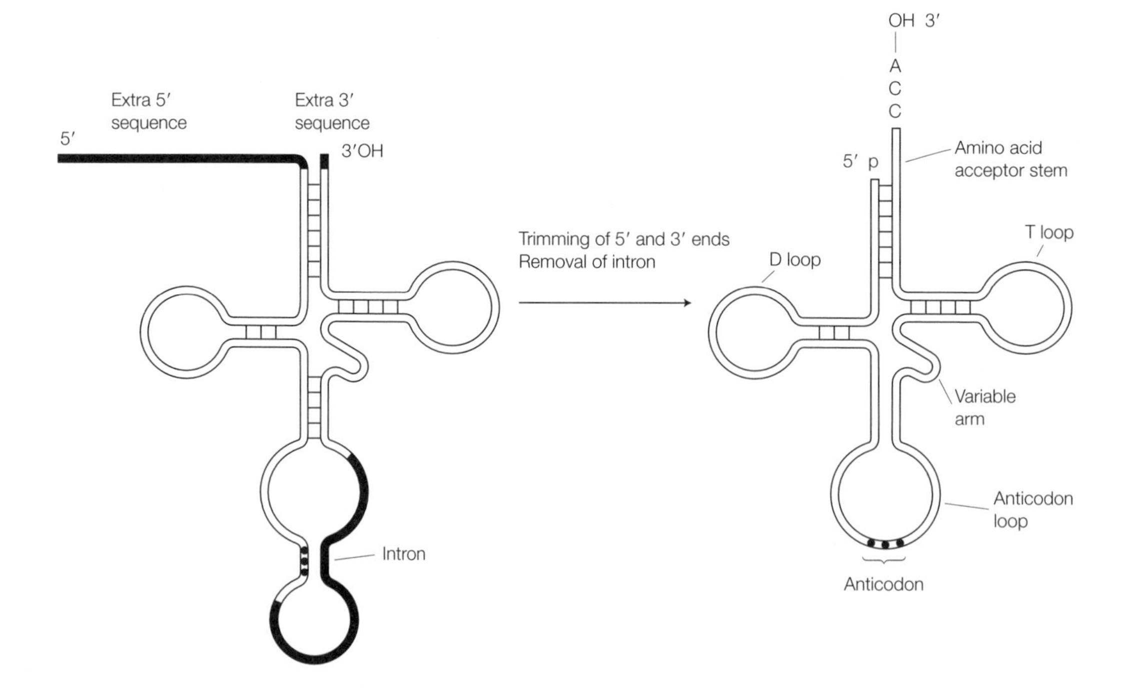


*Fig. 4. Processing of a typical eukaryotic pre-tRNA molecule.*

This intron must be removed in order to create a functional tRNA molecule. Its removal occurs by cleavage by an endonuclease at each end of the intron and then ligation together of the tRNA ends. This RNA splicing pathway for intron removal is totally different from that used to remove introns from pre-mRNA molecules in eukaryotes (Topic G8) and must have evolved independently.

## Modification of tRNA

Transfer RNA molecules are notable for containing unusual nucleotides (*Fig. 5*) such as **1-methylguanosine ($m^1G$)**, **pseudouridine (Ψ)**, **dihydrouridine (D)**, **inosine (I)** and **4-thiouridine ($S^4U$)**. These are created by modification of guanosine and uridine after tRNA synthesis. For example, inosine is generated by deamination of guanosine.

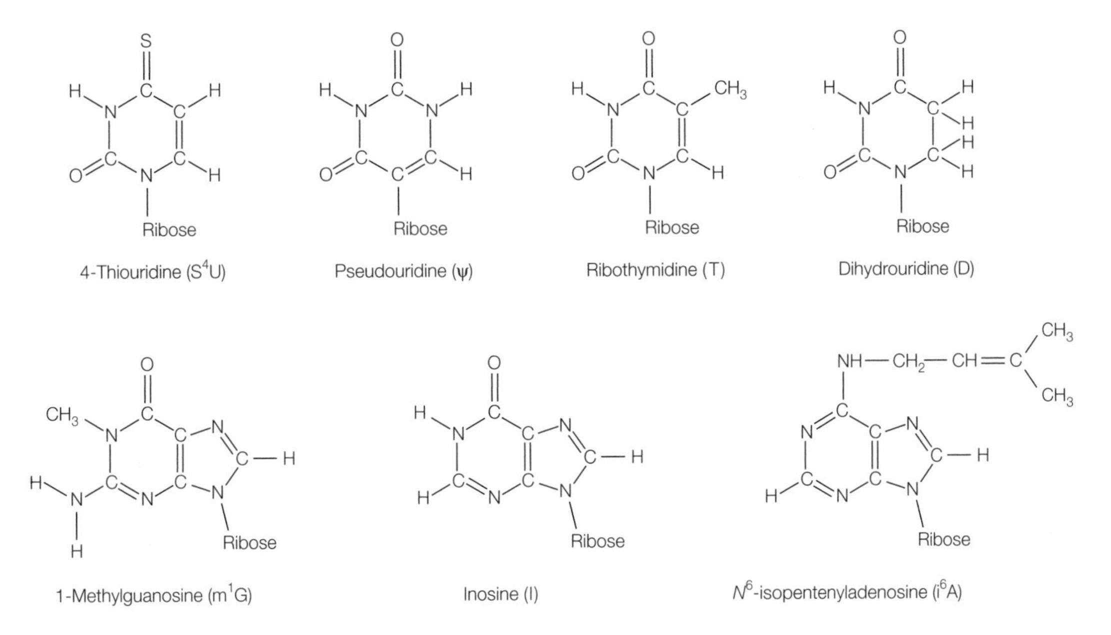


*Fig. 5. Some modified nucleosides found in tRNA molecules.*

# H1 The genetic code

## Key Notes

**The genetic code is a triplet code**

The genetic code is the rules that specify how the nucleotide sequence of an mRNA is translated into the amino acid sequence of a polypeptide. The nucleotide sequence is read as triplets called codons. The codons UAG, UGA and UAA do not specify amino acids and are called termination codons or Stop codons. AUG codes for methionine and also acts as an initiation (Start) codon.

**The genetic code is degenerate**

Most amino acids in proteins are specified by more than one codon (i.e. the genetic code is degenerate). Codons that specify the same amino acid (synonyms) often differ only in the third base, the wobble position, where base-pairing with the anticodon may be less stringent than for the first two positions of the codon.

**Universality of the genetic code**

The genetic code is not universal but is the same in most organisms. Exceptions are found in mitochondrial genomes where some codons specify different amino acids to that normally encoded by nuclear genes. In mitochondria, the UGA codon does not specify termination of translation but instead encodes for tryptophan. Similarly, in certain protozoa UAA and UAG encode glutamic acid instead of acting as termination codons.

**Reading frames**

The mRNA sequence can be read by the ribosome in three possible reading frames. Usually only one reading frame codes for a functional protein since the other two reading frames contain multiple termination codons. In some bacteriophage, overlapping genes occur which use different reading frames.

**Open reading frames**

An open reading frame (ORF) is a run of codons that starts with ATG and ends with a termination codon, TGA, TAA or TAG. Coding regions of genes contain relatively long ORFs unlike noncoding DNA where ORFs are comparatively short. The presence of a long open reading frame in a DNA sequence therefore may indicate the presence of a coding region. Computer analysis of the ORF can be used to deduce the sequence of the encoded protein.

**Related topics**

RNA structure (G1)
Ribosomal RNA (G9)
Transfer RNA (G10)
Translation in prokaryotes (H2)
Translation in eukaryotes (H3)

## The genetic code is a triplet code

During translation, the sequence of an mRNA molecule is read from its 5′ end by ribosomes which then synthesize an appropriate polypeptide. Both in prokaryotes and eukaryotes, the DNA sequence of a single gene is **colinear** with the amino acid sequence of the polypeptide it encodes. In other words, the nucleotide sequence of the coding DNA strand, 5′ to 3′, specifies in exactly the same order the amino acid sequence of the encoded polypeptide, N-terminal to C-terminal. The relationship between the nucleotide sequence of the mRNA and the amino acid sequence of the polypeptide is called **the genetic code**. The sequence of the mRNA is read in groups of three nucleotides called **codons**, with each codon specifying a particular amino acid (*Fig. 1*). However, three codons, UAG, UGA and UAA, do not encode an amino acid. Whenever one of these codons is encountered by a ribosome, it leads to termination of protein synthesis. Therefore these three codons are called **termination codons** or **stop codons**. The codon AUG codes for methionine. Although methionine is found at internal positions in polypeptide chains, all eukaryotic polypeptides also start with methionine (see Topic H3) and all prokaryotic polypeptides start with a modified methionine (*N*-formyl methionine; see Topic H2). Therefore the first AUG codon that is read by the ribosome in an mRNA is called the **initiation codon** or **start codon**.

## The genetic code is degenerate

Since RNA is composed of four types of nucleotides, there are $4^3 = 64$ possible codons, that is 64 possible triplets of nucleotides with different sequences. However, only 20 amino acids are commonly found in proteins (see Topic B1) so that, in most cases, a single amino acid is coded for by several different codons (see *Fig. 1*). The genetic code is therefore said to be **degenerate**. In fact, only methionine and tryptophan are represented by a single codon. As a result of the genetic code's degeneracy, a mutation that changes only a single nucleotide in DNA (**point mutation**), and hence changes only a single nucleotide in the corresponding mRNA, often has no effect on the amino acid sequence of the encoded polypeptide.

**Codon sequence**

| 1st base (5′end) ↓ | 2nd base: U | C | A | G | 3rd base (3′end) ↓ |
|---|---|---|---|---|---|
| U | Phe | Ser | Tyr | Cys | U |
| | Phe | Ser | Tyr | Cys | C |
| | Leu | Ser | **Stop** | **Stop** | A |
| | Leu | Ser | **Stop** | Trp | G |
| C | Leu | Pro | His | Arg | U |
| | Leu | Pro | His | Arg | C |
| | Leu | Pro | Gln | Arg | A |
| | Leu | Pro | Gln | Arg | G |
| A | Ile | Thr | Asn | Ser | U |
| | Ile | Thr | Asn | Ser | C |
| | Ile | Thr | Lys | Arg | A |
| | Met | Thr | Lys | Arg | G |
| G | Val | Ala | Asp | Gly | U |
| | Val | Ala | Asp | Gly | C |
| | Val | Ala | Glu | Gly | A |
| | Val | Ala | Glu | Gly | G |

*Fig. 1. The genetic code.*

Codons that specify the same amino acid are called **synonyms**. Most synonyms differ only in the third base of the codon; for example GUU, GUC, GUA and GUG all code for valine. During protein synthesis, each codon is recognized by a triplet of bases, called an **anticodon**, in a specific tRNA molecule (see Topics G10 and H2). Each base in the codon base pairs with its complementary base in the anticodon. However, the pairing of the third base of a codon is less stringent than for the first two bases (i.e. there is some '**wobble base-pairing**') so that in some cases a single tRNA may base-pair with more than one codon. For example, phenylalanine tRNA, which has the anticodon GAA, recognizes both of the codons UUU and UUC. The third position of the codon is therefore also called the wobble position.

## Universality of the genetic code

For many years it was thought that the genetic code is 'universal', namely that all living organisms used the same code. Now we know that the genetic code is almost the same in all organisms but there are a few differences. Mitochondria contain DNA, as double-stranded DNA circles, and the mitochondrial genome codes for about 10–20 proteins. Surprisingly, in **mitochondrial mRNAs**, some codons have different meanings from their counterparts in mRNA in the cytosol. A few examples are given below (N denotes any of the four nucleotides A, G, C or U):

| | |
|---|---|
| mitochondria | AUA = Met not Ile |
| mitochondria | UGA = Trp not **Stop** |
| some animal mitochondria | AGA and AGG = **Stop** not Arg |
| plant mitochondria | CGG = Trp not Arg |
| yeast mitochondria | CUN = Thr not Leu |

Some unicellular organisms are also now known to use a variant genetic code. For example:

| | |
|---|---|
| some ciliated protozoa | UAA and UAG = Glu not **Stop**. |

## Reading frames

Since the sequence of an mRNA molecule is read in groups of three nucleotides (codons) from the 5′ end, it can be read in three possible **reading frames**, depending on which nucleotide is used as the first base of the first codon (*Fig. 2*). Usually, only one reading frame (reading frame 3 in *Fig. 2*) will produce a functional protein since the other two reading frames will include several termination (**Stop**) codons. The correct reading frame is set *in vivo* by recognition by the ribosome of the initiation codon, AUG, at the start of the coding

5′ 3′

Reading frame 1
U U A U G A G C G C U A A A U
Leu **Stop** Ala Leu Asn

Reading frame 2
U U A U G A G C G C U A A A U
Tyr Glu Arg **Stop**

Reading frame 3
U U A U G A G C G C U A A A U
Met Ser Ala Lys

*Fig. 2. Three potential reading frames for any given mRNA sequence depending on which nucleotide is 'read' first.*

sequence. Usually one sequence of bases encodes only a single protein. However, in some bacteriophage DNAs, several genes overlap, with each gene being in a different reading frame. This organization of **overlapping genes** generally occurs when the genome size is smaller than can accommodate the genes necessary for phage structure and assembly using only one reading frame.

## Open reading frames

In many cases these days, the protein encoded by a particular gene is deduced by cloning (see Section I) and then sequencing the corresponding DNA. The DNA sequence is then scanned using a computer program to identify runs of codons that start with ATG and end with TGA, TAA or TAG. These runs of codons are called **open reading frames (ORFs)** and identify potential coding regions. Because genes carry out important cellular functions, the sequence of coding DNA (and of important regulatory sequences) is more strongly conserved in evolution that that of noncoding DNA. In particular, mutations that lead to the creation of termination codons within the coding region, and hence premature termination during translation, are selected against. This means that the coding regions of genes often contain comparatively long ORFs whereas in noncoding DNA, triplets corresponding to termination codons are not selected against and ORFs are comparatively short. Thus, when analyzing the ORFs displayed for a particular cloned DNA, it is usually true that a long ORF is likely to be coding DNA whereas short ORFs may not be. Nevertheless, one must be aware that some exons can be short and so some short ORFs may also be coding DNA. Computer analysis may be able to detect these by screening for the conserved sequences at exon/intron boundaries and the splice branchpoint sequence (see Topic G8). Finally, by referring to the genetic code, computer analysis can predict the protein sequence encoded by each ORF. This is the **deduced protein sequence**.

# H2 TRANSLATION IN PROKARYOTES

## Key Notes

**Overview**

During translation the mRNA is read in a 5′ to 3′ direction and protein is made in an N-terminal to C-terminal direction. Translation relies upon aminoacyl-tRNAs that carry specific amino acids and recognize the corresponding codons in mRNA by anticodon–codon base-pairing. Translation takes place in three phases; initiation, elongation and termination.

**Synthesis of aminoacyl-tRNA**

Each tRNA molecule has a cloverleaf secondary structure consisting of three stem loops, one of which bears the anticodon at its end. The amino acid is covalently bound to the 3′ OH group at the 3′ end by aminoacyl synthetase to form aminoacyl-tRNA. The reaction, called amino acid activation, occurs in two steps and requires ATP to form an intermediate, aminoacyl-adenylate.

**Initiation of protein synthesis**

Each ribosome has three binding sites for tRNAs; an A site where the incoming aminoacyl-tRNA binds, a P site where the tRNA linked to the growing polypeptide chain is bound, and an E site which binds tRNA prior to its release from the ribosome. Translation in prokaryotes begins by the formation of a 30S initiation complex between the 30S ribosomal subunit, mRNA, initiation factors and fMet $tRNA_f^{Met}$. The 30S subunit binds to the Shine–Dalgarno sequence which lies 5′ to the AUG Start codon and is complementary to the 16S rRNA of the small ribosomal subunit. The ribosome then moves in a 3′ direction along the mRNA until it encounters the AUG codon. The 50S ribosomal subunit now binds to the 30S initiation complex to form the 70S initiation complex. In this complex, the anticodon of the fMet $tRNA_f^{Met}$ is base-paired to the AUG initiation codon (start codon) in the P site.

**Elongation**

The elongation cycle consists of three steps: aminoacyl-tRNA binding, peptide bond formation, and translocation. In the first step, the aminoacyl-tRNA corresponding to the second codon binds to the A site on the ribosome as an aminoacyl-tRNA/EF-Tu/GTP complex. After binding, the GTP is hydrolyzed and EF-Tu/GDP is released. The EF-Tu is regenerated via the EF-Tu–EF-Ts exchange cycle. Peptide bond formation is catalyzed by peptidyl transferase between the C-terminus of the amino acyl moiety in the P site and the amino group of the aminoacyl-tRNA in the A site. In the final (translocation) step, EF-G/GTP binds to the ribosome, the deacylated tRNA moves from the P site to the E-site, the dipeptidyl-tRNA in the A site moves to the P site, and the ribosome moves along the mRNA to place the next codon in the A site. The GTP is hydrolyzed to GDP and inorganic phosphate. When the next aminoacyl-tRNA binds to the A site in the next round of elongation, the deacylated tRNA is released from the E-site.

**Termination**

The appearance of a UAA or UAG termination (stop) codon in the A site causes release factor RF1 to bind whereas RF2 recognizes UGA. RF3 assists RF1 and RF2. The release factors trigger peptidyl transferase to transfer the polypeptide to a water molecule instead of to aminoacyl-tRNA. The polypeptide, mRNA, and free tRNA leave the ribosome and the ribosome dissociates into its subunits ready to begin a new round of translation.

**Related topics**

RNA structure (G1)
Transcription in prokaryotes (G2)
The *lac* operon (G3)
The *trp* operon (G4)
Transcription in eukaryotes: an overview (G5)
Transcription of protein-coding genes in eukaryotes (G6)
Regulation of transcription by RNA Pol II (G7)
Processing of eukaryotic pre-mRNA (G8)
Ribosomal RNA (G9)
Transfer RNA (G10)
The genetic code (H1)
Translation in eukaryotes (H3)

## Overview

A ribosome binds to an mRNA molecule and reads the nucleotide sequence from the 5′ to 3′ direction, synthesizing the corresponding protein from amino acids in an N-terminal (amino-terminal) to C-terminal (carboxyl terminal) direction. The amino acids used are covalently bound to tRNA (transfer RNA) molecules to form aminoacyl-tRNAs. Each aminoacyl-tRNA bears a triplet of bases, called an **anticodon**. The ribosome reads each triplet codon of the mRNA in turn and an aminoacyl-tRNA molecule with an anticodon that is complementary to the codon binds to it via hydrogen bonding. A peptide bond is then formed between the incoming amino acid and the growing end of the polypeptide chain.

Overall, protein synthesis (or **translation**) takes place in three stages; **initiation, elongation** and **termination**. During initiation, the mRNA–ribosome complex is formed and the first codon (always AUG) binds the first aminoacyl-tRNA (called **initiator tRNA**). During the elongation phase, the other codons are read sequentially and the polypeptide grows by addition of amino acids to its C-terminal end. This process continues until a termination codon (Stop codon), which does not have a corresponding aminoacyl-tRNA with which to base-pair, is reached. At this point, protein synthesis ceases (termination phase) and the finished polypeptide is released from the ribosome. Usually at any one time, many ribosomes are translating an mRNA simultaneously, forming a structure called a **polyribosome** or **polysome**.

## Synthesis of aminoacyl-tRNA

A tRNA molecule is about 74–95 nucleotides long, making these some of the smallest RNA molecules in the cell. Each tRNA molecule has a **cloverleaf secondary structure** with the anticodon accessible at the end of the anticodon stem loop (see *Fig. 1* and Topic G10). During synthesis of the aminoacyl-tRNA, the amino acid is covalently bound to the A residue of the CCA sequence at the 3′ end (*Fig. 1*). Each tRNA molecule carries only a single amino acid. However, because of the redundancy of the genetic code (see Topic H1), several codons may encode the same amino acid and so there will also exist several types of tRNA with corresponding anticodons all bearing the same amino acid. The correct nomenclature is, for example, $tRNA^{Gly}$ for the tRNA that will accept glycine whereas the corresponding aminoacyl-tRNA is $Gly\text{-}tRNA^{Gly}$, and is the aminoacyl-tRNA shown in *Fig. 1*.

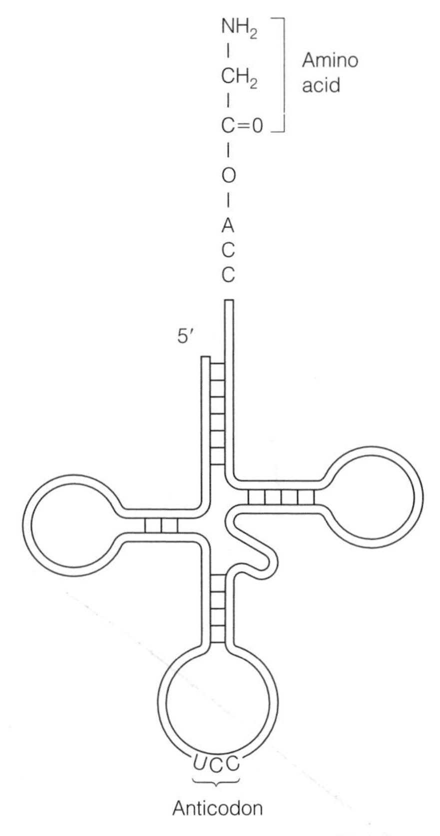


*Fig. 1. Structure of an aminoacyl-tRNA.*

Synthesis of aminoacyl-tRNAs is crucially important for two reasons. First each amino acid must be covalently linked to a tRNA molecule in order to take part in protein synthesis, which depends upon the 'adaptor' function of tRNA to ensure that the correct amino acids are incorporated. Second, the covalent bond that is formed between the amino acid and the tRNA is a high energy bond that enables the amino acid to react with the end of the growing polypeptide chain to form a new peptide bond. For this reason, the synthesis of aminoacyl-tRNA is also referred to as **amino acid activation**. Amino acids that are not linked to tRNAs cannot be added to the growing polypeptide.

The attachment of an amino acid to a tRNA is catalyzed by an enzyme called **aminoacyl-tRNA synthetase**. A separate aminoacyl-tRNA synthetase exists for every amino acid, making 20 synthetases in total. The synthesis reaction occurs in two steps. The first step is the reaction of an amino acid and ATP to form an **aminoacyl-adenylate** (also known as **aminoacyl-AMP**).

$$^{+}H_3N-\underset{H}{\overset{R}{C}}-C\begin{matrix}\nearrow O \\ \searrow O^-\end{matrix} + ATP \rightleftharpoons {}^{+}H_3N-\underset{H}{\overset{R}{C}}-\overset{O}{\overset{\|}{C}}-O-\underset{O^-}{\overset{O}{\overset{\|}{P}}}-O-\text{ribose}-\text{adenine} + PP_i$$

Aminoacyl adenylate
(Aminoacyl-AMP)

In the second step, without leaving the enzyme, the aminoacyl group of aminoacyl-AMP is transferred to the 3′ end of the tRNA molecule to form aminoacyl-tRNA:

$$\text{aminoacyl-AMP} + \text{tRNA} \rightarrow \text{aminoacyl-tRNA} + \text{AMP}$$

The overall reaction is:

$$\text{amino acid} + \text{ATP} + \text{tRNA} \rightarrow \text{aminoacyl-tRNA} + \text{AMP} + \text{PP}_i$$

and is driven by the subsequent hydrolysis of the pyrophosphate to inorganic phosphate.

## Initiation of protein synthesis

Each prokaryotic ribosome, shown schematically in *Fig. 2* (see Topic G9 for details of ribosome structure), has three binding sites for tRNAs. The **aminoacyl-tRNA binding site** (or **A site**) is where, during elongation, the incoming aminoacyl-tRNA binds. The **peptidyl-tRNA binding site** (or **P site**) is where the tRNA linked to the growing polypeptide chain is bound. The **exit site** (or **E site**) is a binding site for tRNA following its role in translation and prior to its release from the ribosome.

The first codon translated in all mRNAs is AUG which codes for methionine. This AUG is called the **start codon** or **initiation codon**. Naturally, other AUG codons also occur internally in an mRNA where they encode methionine residues internal to the protein. Two different tRNAs are used for these two types of AUG codon; $\text{tRNA}_f^{\text{Met}}$ is used for the initiation codon and is called the **initiator tRNA** whereas $\text{tRNA}_m^{\text{Met}}$ is used for internal AUG codons. In prokaryotes the first amino acid of a new protein is ***N*-formylmethionine** (abbreviated **fMet**). Hence the aminoacyl-tRNA used in initiation is **fMet-$\text{tRNA}_f^{\text{Met}}$**. It is essential that the correct AUG is used as the initiation codon since this sets the correct reading frame for translation (see Topic H1). A short sequence rich in purines (5′-AGGAGGU-3′), called the **Shine–Dalgarno sequence**, lies 5′ to the AUG initiation codon (*Fig. 3*) and is complementary to part of the 16S rRNA in the small ribosomal subunit. Therefore this is the binding site for the 30S ribosomal subunit

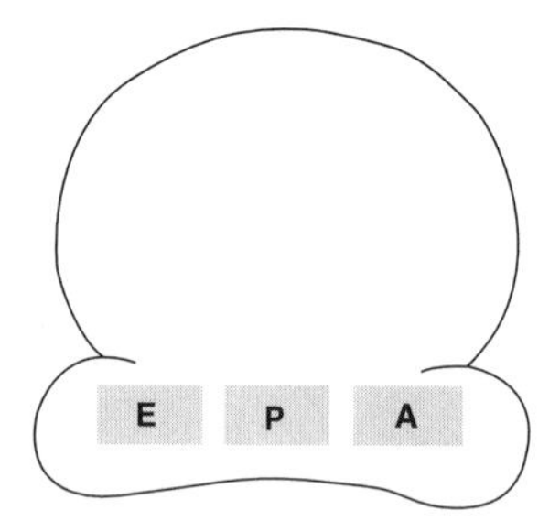


*Fig. 2. Schematic of a prokaryotic 70S ribosome showing the peptidyl-tRNA site (P site), aminoacyl-tRNA site (A site) and exit site (E site).*

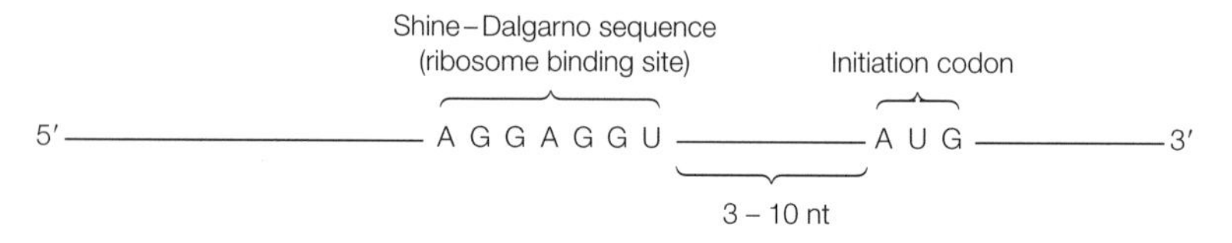


*Fig. 3. The Shine–Dalgarno sequence in prokaryotic mRNA.*

which then migrates in a 3′ direction along the mRNA until it encounters the AUG initiation codon. Thus the Shine–Dalgarno sequence delivers the ribosomal subunit to the correct AUG for initiation for translation.

Initiation of protein synthesis is catalyzed by proteins called **initiation factors (IFs)**. In prokaryotes, three initiation factors (IF1, IF2 and IF3) are essential. Because of the complexity of the process, the exact order of binding of IF1, IF2, IF3, fMet-$tRNA_f^{Met}$ and mRNA is still unclear. One current model is shown in *Fig. 4* and is described below.

- Initiation begins with the binding of IF1 and IF3 to the small (30S) ribosomal subunit.
- The small subunit then binds to the mRNA via the Shine–Dalgarno sequence and moves 3′ along the mRNA until it locates the AUG initiation codon.
- the initiator tRNA charged with *N*-formylmethionine and in a complex with IF2 and GTP (fMet-$tRNA_f^{Met}$/IF2/GTP) now binds.
- IF3 is released.
  The complex of mRNA, fMet-$tRNA_f^{Met}$, IF1, IF2 and the 30S ribosomal subunit is called the **30S initiation complex**.
- The large (50S) ribosomal subunit now binds, with the release of IF1 and IF2 and hydrolysis of GTP, to form a **70S initiation complex**.

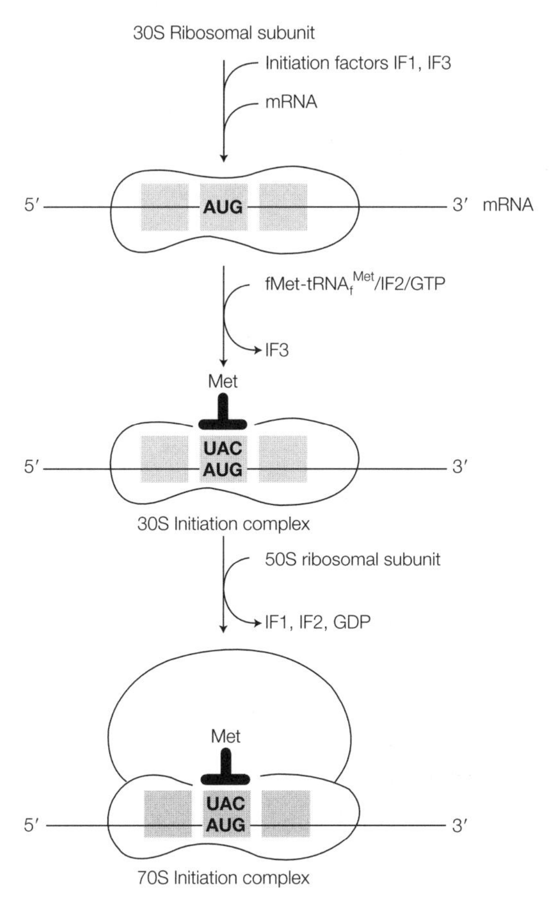


*Fig. 4. Initiation of protein synthesis in prokaryotic cells.*

One important point to note is that, unlike all other aminoacyl-tRNA molecules (which bind to the A site; see below), the binding of fMet-$tRNA_f^{Met}$ occurs directly into the P site.

## Elongation

At the start of the first round of elongation (*Fig. 5*), the initiation codon (AUG) is positioned in the P site with fMet-$tRNA_f^{Met}$ bound to it via codon–anticodon base-pairing. The next codon in the mRNA is positioned in the A site. Elongation of the polypeptide chain occurs in three steps called the **elongation cycle,** namely **aminoacyl-tRNA binding, peptide bond formation** and **translocation**:

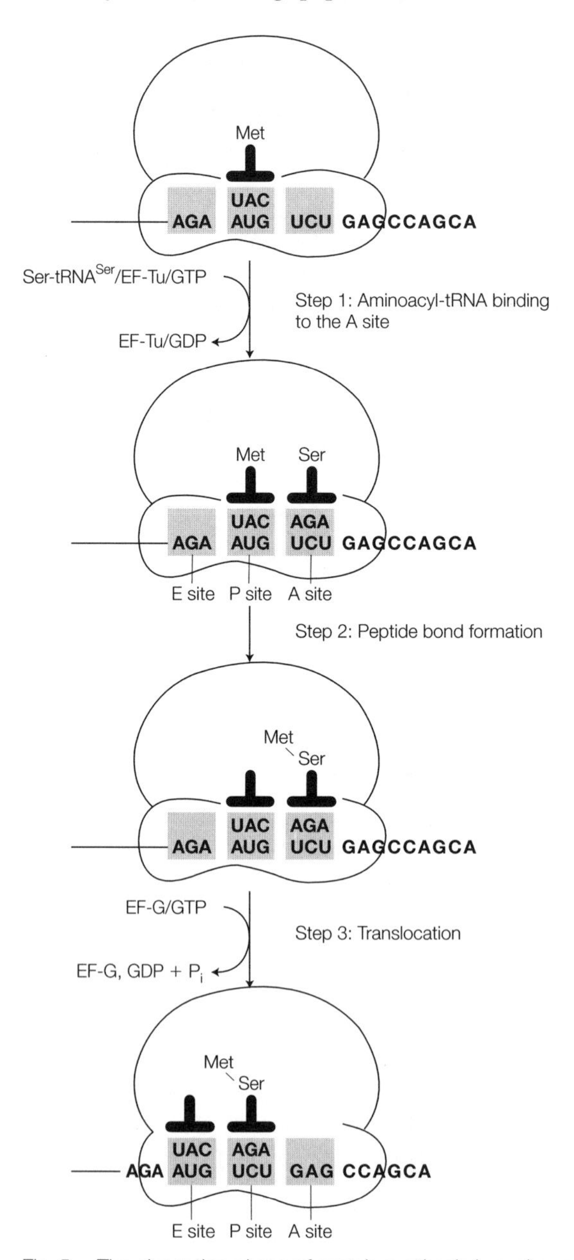


*Fig. 5. The elongation phase of protein synthesis in prokaryotes.*

- *Aminoacyl-tRNA binding*: in this first step, the corresponding aminoacyl-tRNA for the second codon binds to the A site via codon–anticodon interaction (*Fig. 5*). Binding of the aminoacyl-tRNA requires **elongation factor EF-Tu** and GTP which bind as an aminoacyl-tRNA/EF-Tu/GTP complex. Following binding, the GTP is hydrolyzed and the EF-Tu is released, now bound to GDP (*Fig. 5*). Before the EF-Tu molecule can catalyze the binding of another charged tRNA to the ribosome, it must be regenerated by a process involving another elongation factor, **EF-Ts**. This regeneration is called the **EF-Tu–EF-Ts exchange cycle** (*Fig. 6*). First, EF-Ts binds to EF-Tu and displaces the GDP. Then GTP binds to the EF-Tu and displaces EF-Ts. The EF-Tu-GTP is now ready to take part in another round of elongation.
- *Peptide bond formation*: the second step, peptide bond formation, is catalyzed by **peptidyl transferase**, part of the large ribosomal subunit. In this reaction the carboxyl end of the amino acid bound to the tRNA in the P site is uncoupled from the tRNA and becomes joined by a peptide bond to the amino group of the amino acid linked to the tRNA in the A site (*Fig. 5*). A protein with peptidyl transferase activity has never been isolated. The reason is now clear; in *E. coli* at least, the peptidyl transferase activity is associated with part of the 23S rRNA in the large ribosomal subunit. In other words, peptidyl transferase is a **ribozyme**, a catalytic activity that resides in an RNA molecule (see also Topic G9).
- *Translocation*: in the third step, a complex of **elongation factor EF-G** (also called **translocase**) and GTP (i.e. EF-G/GTP) binds to the ribosome. Three concerted movements (*Fig. 3c*) now occur, collectively called translocation; the deacylated tRNA moves from the P site to the E site, the dipeptidyl-tRNA in the A site moves to the P site, and the ribosome moves along the mRNA (5′ to 3′) by three nucleotides to place the next codon in the A site. During the translocation events, GTP is hydrolyzed to GDP and inorganic phosphate, and EF-G is released ready to bind more GTP for another round of elongation.

After translocation, the A site is empty and ready to receive the next aminoacyl-tRNA. The A site and the E site cannot be occupied simultaneously. Thus the deacylated tRNA is released from the E site before the next aminoacyl-tRNA binds to the A site to start a new round of elongation. Elongation continues, adding one amino acid to the C-terminal end of the growing polypeptide for each codon that is read, with the peptidyl-tRNA moving back and forth from the P site to the A site as it grows.

## Termination

Eventually, one of three **termination codons** (also called **Stop codons**) becomes positioned in the A site (*Fig. 7*). These are UAG, UAA and UGA. Unlike other codons, prokaryotic cells do not contain aminoacyl-tRNAs complementary to Stop codons. Instead, one of two **release factors** (**RF1** and **RF2**) binds instead. RF1 recognizes UAA and UAG whereas RF2 recognizes UGA. A third release factor, **RF3**, is also needed to assist RF1 or RF2. Thus either RF1 + RF3 or RF2 + RF3 bind depending on the exact termination codon in the A site. RF1 (or RF2) binds at or near the A site whereas RF3/GTP binds elsewhere on the ribosome. The release factors cause the peptidyl transferase to transfer the polypeptide to a water molecule instead of to aminoacyl-tRNA, effectively cleaving the bond between the polypeptide and tRNA in the P site. The polypeptide, now leaves the ribosome, followed by the mRNA and free tRNA, and the ribosome dissociates into 30S and 50S subunits ready to start translation afresh.

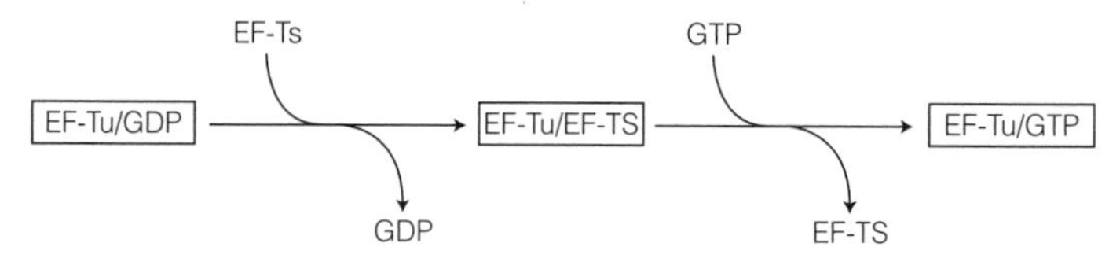


*Fig. 6. The EF-Tu–EF-Ts exchange cycle.*

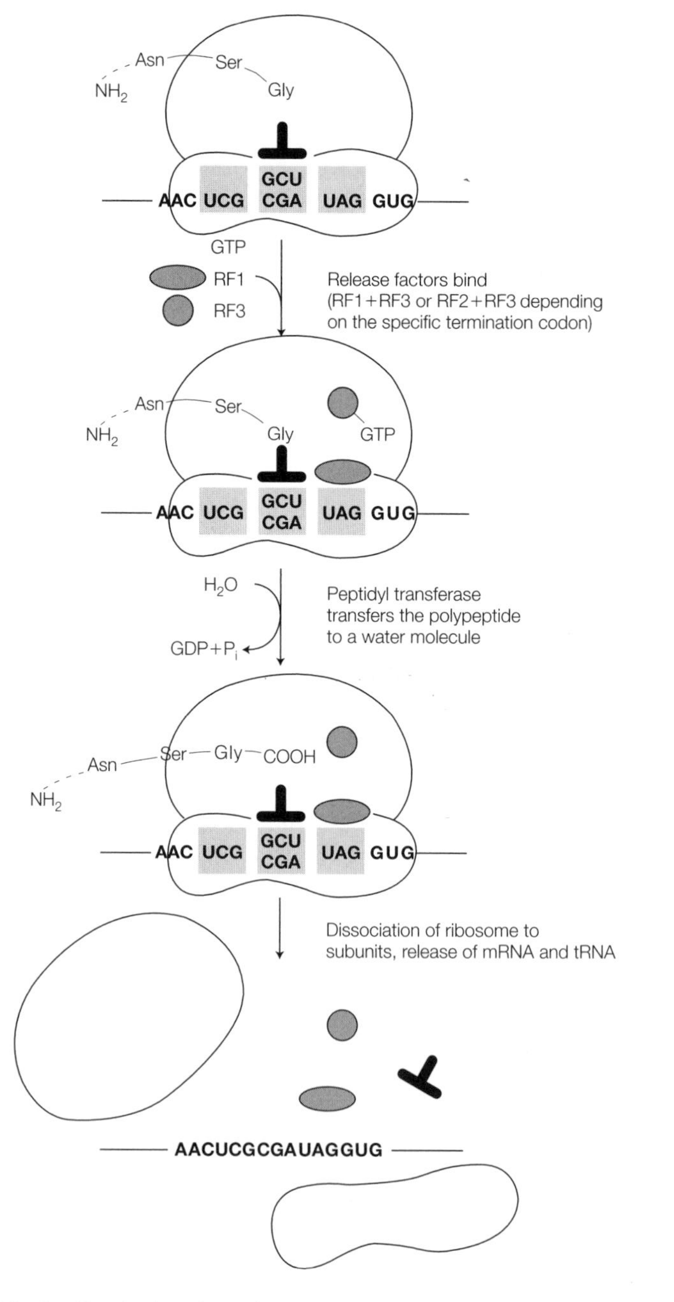


*Fig. 7. Termination of protein synthesis in prokaryotic cells.*

# H3 Translation in Eukaryotes

## Key Notes

**Initiation**

Eukaryotic ribosomes are larger (80S) and more complex than prokaryotic ribosomes (70S). Initiation is basically similar in prokaryotes and eukaryotes except that in eukaryotes at least nine initiation factors are involved (cf. three factors in prokaryotes), the initiating amino acid is methionine (cf. *N*-formylmethionine in prokaryotes), eukaryotic mRNAs do not contain Shine–Dalgarno sequences (so the AUG initiation codon is detected by the ribosome scanning instead), and eukaryotic mRNA is monocistronic (cf. some polycistronic mRNAs in prokaryotes). Initiation in eukaryotes involves the formation of a 48S preinitiation complex between the 40S ribosomal subunit, mRNA, initiation factors and Met-$tRNA_i^{met}$. The ribosome then scans the mRNA to locate the AUG initiation codon. The 60S ribosomal subunit now binds to form the 80S initation complex.

**Elongation**

Elongation in eukaryotes requires three eukaryotic initiation factors that have similar functions to the corresponding prokaryotic proteins.

**Termination**

A single eukaryotic release factor recognizes all three termination codons and requires ATP for activity.

**Related topics**

RNA structure (G1)
Transcription in prokaryotes (G2)
The *lac* operon (G3)
The *trp* operon (G4)
Transcription in eukaryotes: an overview (G5)
Transcription of protein-coding genes in eukaryotes (G6)
Regulation of transcription by RNA Pol II (G7)
Processing of eukaryotic pre-mRNA (G8)
Ribosomal RNA (G9)
Transfer RNA (G10)
The genetic code (H1)
Translation in prokaryotes (H2)

## Initiation

The overall mechanism of protein synthesis in eukaryotes is basically the same as in prokaryotes, with three phases defined as initation, elongation and termination. However, there are some significant differences, particularly during initiation.

- Whereas a prokaryotic ribosome has a sedimentation coefficient (see Topic G9) of 70S and subunits of 30S and 50S, a eukaryotic ribosome has a sedimentation coefficient of 80S with subunits of 40S and 60S (see Topic G9). The composition of eukaryotic ribosomal subunits is also more complex than prokaryotic subunits (see Topic G9) but the function of each subunit is essentially the same as in prokaryotes.
- In eukaryotes, each mRNA is **monocistronic**, that is, discounting any subsequent post-translational cleavage reactions that may occur, the mRNA encodes a single protein. In prokaryotes, many mRNAs are **polycistronic**, that is they encode several proteins. Each coding sequence in a prokaryotic mRNA has its own initiation and termination codons.

- Initiation of protein synthesis in eukaryotes requires at least nine distinct eukaryotic **initiation factors** (**eIFs**) (see *Table 1*) compared to the three initiation factors (IFs) in prokaryotes (see Topic H2).
- In eukaryotes, the initiating amino acid is **methionine**, not *N*-formylmethionine as in prokaryotes.
- As in prokaryotes, a special initiator tRNA is required for initiation and is distinct from the tRNA that recognizes and binds to codons for methionine at internal positions in the mRNA. When charged with methionine ready to begin initiation, this is known as **Met-tRNA$_i^{met}$**.
- The main difference between initiation of translation in prokaryotes and eukaryotes is that in bacteria, a Shine–Dalgarno sequence (see Topic H2) lies 5′ to the AUG initiation codon and is the binding site for the 30S ribosomal subunit, marking this AUG as the one to use for initiation rather than any other AUG internal in the mRNA. The initiation complex is assembled directly over this initiation codon. In contrast, most eukaryotic mRNAs do not contain Shine–Dalgarno sequences. Instead, a 40S ribosomal subunit attaches at the 5′ end of the mRNA and moves downstream (i.e. in a 5′ to 3′ direction) until it finds the AUG initiation codon. This process is called **scanning**. However, some eukaryotic viruses with RNA genomes (such as human poliovirus and rhinoviruses that cause the common cold) do not have a 5′ cap. Instead, when these RNAs are translated by the host cell, the eukaryotic ribosomes bind at **internal ribosome entry sites** (**IRES**) which function in a similar manner to the Shine–Dalgarno sequences in prokaryotic mRNAs.

The full details of initiation in eukaryotes are still not fully known but the process occurs broadly as follows:

- the first step is the formation of a **pre-initiation complex** consisting of the 40S small ribosomal subunit, Met-tRNA$_i^{met}$, eIF2 and GTP;
- the pre-initiation complex now binds to the 5′ end of the eukaryotic mRNA, a step that requires **eIF4F** (also called **cap binding complex**) and **eIF3**. This interaction involves the complex binding to the 5′ cap of the mRNA. Interestingly, the efficiency of initiation is influenced by the presence of a poly(A) tail at the 3′ end of the mRNA, probably via **poly(A) binding protein** that is bound to the tail, implying that the mRNA may bend back on itself to allow this interaction to occur.
- The complex now moves along the mRNA in a 5′ to 3′ direction until it locates the AUG initiation codon. The 5′ untranslated regions of eukaryotic mRNAs

*Table 1. Comparison of protein synthesis factors in prokaryotes and eukaryotes*

| Prokaryotic | Eukaryotic | Function |
|---|---|---|
| Initiation factors | | |
| IF1, IF3 | eIF3, eIF4C, eIF6 | Binding to ribosome subunits |
| | eIF4B, eIF4F | Binding to mRNA |
| IF2 | eIF2, eIF2B | Initiator tRNA delivery |
| | eIF5 | Displacement of other factors |
| Elongation factors | | |
| EF-Tu | eEF1α | Aminoacyl tRNA delivery to ribosome |
| EF-Ts | eEF1βγ | Recycling of EF-Tu or eEF1α |
| EF-G | eEF2 | Translocation |
| Termination factors | | |
| RF1 | | Polypeptide chain release |
| RF2 | eRF | |
| RF3 | | |

vary in length but can be several hundred nucleotides long and may contain secondary structures such as hairpin loops. These secondary structures are probably removed by initiation factors of the scanning complex. The initiation codon is usually recognizable because it is often (but not always) contained in a short sequence called the **Kozak consensus** (5′-ACC**AUG**G-3′).

- Once the complex is positioned over the initiation codon, the 60S large ribosomal subunit binds to form an **80S initiation complex**, a step that requires the hydrolysis of GTP and leads to the release of several initiation factors.

## Elongation

- The elongation stage of translation in eukaryotes requires three elongation factors, **eEF1α**, **eEFIβγ** and **eEF2**, which have similar functions to their prokaryotic counterparts EF-Tu, EF-Ts and EF-G (see *Table 1*).
- Although most codons encode the same amino acids in both prokaryotes and eukaryotes, the mRNAs synthesized within the organelles of some eukaryotes use a variant of the genetic code (see Topic H1).
- During elongation in bacteria, the deacylated tRNA in the P site moves to the E site prior to leaving the ribosome (see Topic H2). In contrast, although the situation is still not completely clear, in eukaryotes the deacylated tRNA appears to be ejected directly from the ribosome.

## Termination

Termination in eukaryotes is carried out by a single eukaryotic release factor (**eRF**) that recognizes all three termination codons and requires ATP for activity (*Table 1*).

# H4 Protein targeting

## Key Notes

**Overview**

Both in prokaryotes and eukaryotes, newly synthesized proteins must be delivered to a specific subcellular location or exported from the cell for correct activity. This phenomenon is called protein targeting.

**Secretory proteins**

Secretory proteins have an N-terminal signal peptide which targets the protein to be synthesized on the rough endoplasmic reticulum (RER). During synthesis it is translocated through the RER membrane into the lumen. Vesicles then bud off from the RER and carry the protein to the Golgi complex, where it becomes glycosylated. Other vesicles then carry it to the plasma membrane. Fusion of these transport vesicles with the plasma membrane then releases the protein to the cell exterior.

**Plasma membrane proteins**

Plasma membrane proteins are also synthesized on the RER but become inserted into the RER membrane (and hence ultimately the plasma membrane) rather than being released into the RER lumen. The plasma membrane protein may pass once through the plasma membrane (Type I and Type II integral membrane proteins) or may loop back and forth, passing through many times (Type III integral membrane protein). The orientation of the protein in the membrane is determined by topogenic sequences within the polypeptide chain. Type I proteins have a cleaved N-terminal signal sequence and a hydrophobic stop-transfer sequence, Type II have an uncleaved N-terminal signal sequence that doubles as the membrane-anchoring sequence, and Type III have multiple signal sequences and stop-transfer sequences.

**Proteins of the endoplasmic reticulum**

Proteins destined for the ER have an N-terminal signal peptide, are synthesized on the RER, are translocated into the RER lumen and transported by vesicles to the Golgi. Once there, a C-terminal amino acid sequence (KDEL) is recognized by a Golgi receptor protein that causes other vesicles to return the protein to the ER.

**Lysosomal proteins**

Lysosomal proteins are targeted to the lysosomes via the addition of a mannose 6-phosphate signal that is added in the *cis*-compartment of the Golgi and is recognized by a receptor protein in the *trans*-compartment of the Golgi. The protein is then transported by specialized vesicles to a late endosome that later matures into a lysosome. The mannose 6-phosphate receptor recycles back to the Golgi for re-use.

**Mitochondrial and chloroplast proteins**

Most mitochondria and chloroplast proteins are made on free cytosolic ribosomes, released into the cytosol and then taken up into the organelle. Uptake into the mitochondrial matrix requires a matrix-targeting sequence and occurs at sites where the outer and inner mitochondrial membranes come into contact. The process is mediated by hsp70 and hsp60 proteins and requires both ATP hydrolysis and an electrochemical gradient across the inner mitochondrial membrane. Targeting of proteins to other compartments of mitochondria or chloroplasts requires two signals.

| Nuclear proteins | Proteins destined for import into the nucleus typically require a nuclear localization signal, four to eight amino acids long, located internally in the protein. Uptake occurs via nuclear pores and requires ATP hydrolysis. |
|---|---|
| **Related topics** | Translation in prokaryotes (H2) Protein glycosylation (H5) Translation in eukaryotes (H3) |

## Overview

Cells must ensure that each newly synthesized protein is sorted to its correct location where it can carry out the appropriate function. This process is called **protein targeting**. In a eukaryotic cell, the protein may be destined to stay in the cytosol, for example an enzyme involved in glycolysis (see Topic J3). Alternatively it may need to be targeted to an organelle (such as a mitochondrion, lysosome, peroxisome, chloroplast or the nucleus) or be inserted into the plasma membrane or exported out of the cell. In bacteria such as *E. coli*, the protein may stay in the cytosol, be inserted into the plasma membrane or the outer membrane, be sent to the space between these two membranes (the periplasmic space) or be exported from the cell. In both prokaryotes and eukaryotes, if a protein is destined for the cytosol, it is made on free ribosomes in the cytosol and released directly into the cytosol. If it is destined for other final locations, specific protein-targeting mechanisms are involved.

## Secretory proteins

Proteins destined to be secreted from the eukaryotic cell are synthesized by ribosomes bound to the **rough endoplasmic reticulum (RER)**. As the protein is synthesized, it is translocated across the RER membrane into the lumen of the RER where it folds into its final conformation. The ER then buds off vesicles that carry the protein to the **Golgi apparatus** (see Topic A2) also called the **Golgi complex** (*Fig. 1*). The Golgi has a ***cis* face** (where vesicles enter) and a ***trans* face** (where vesicles leave). Thus the RER vesicles fuse with the ***cis* compartment** of the Golgi, releasing the protein into the Golgi lumen. The protein then moves through the Golgi complex to the ***trans* compartment**, being modified *en route* by the addition of carbohydrate residues (glycosylation, see Topic H5). Finally, vesicles bud from the *trans* compartment and carry the glycosylated secretory proteins to the plasma membrane where the vesicles fuse,

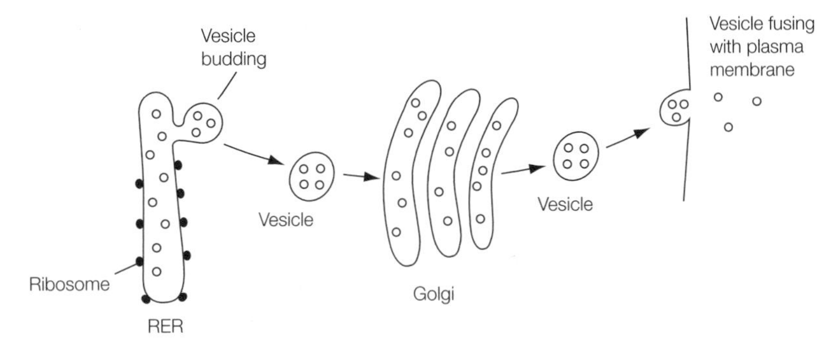


*Fig. 1. Synthesis and exocytosis of secretory proteins; see text for details. The ribosomes attached to the RER are shown as filled-in circles whereas the open circles in the lumen of the ER, vesicles and Golgi complex represent secretory protein molecules.*

releasing their contents to the cell exterior. This fusion and extracellular release of protein is also called **exocytosis** (see Topic E4).

*Signal hypothesis*

A typical secretory protein differs from a cytosolic protein by having a sequence about 13–35 amino acids long at its N-terminal end called a **signal sequence** or **signal peptide**. The signal peptides of different secretory proteins differ in amino acid sequence but there are some common features, for example the center of the sequence usually consists of 10–15 hydrophobic amino acids. The **signal hypothesis** was proposed from early work in this area and predicted that the signal peptide directs the secretory protein to the ER membrane and so targets the protein to cross into the ER lumen and be exported. The signal hypothesis has been shown to apply to protein secretion in animal, plant and bacterial cells. A simplified version of the mechanism is shown in *Fig. 2*.

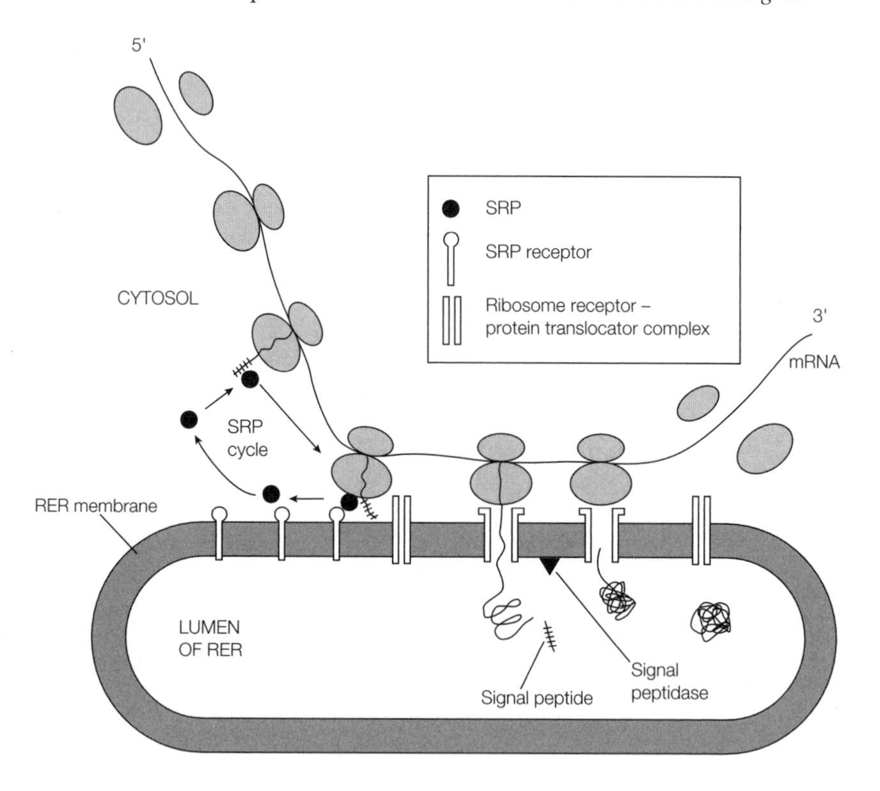


*Fig. 2. A simplified version of the signal hypothesis (see the text for details).*

The mRNA for the secretory protein binds to a free cytoplasmic ribosome and protein synthesis begins. The first part of the protein made is the N-terminal signal peptide. A **signal recognition particle (SRP)**, which is a complex of a 7S RNA and six proteins, binds to the signal peptide and stops further protein synthesis. This stops the secretory protein from being released prematurely into the cytosol. The ribosome–mRNA–SRP complex now binds to an **SRP receptor**, a protein on the surface of the ER. The ER membrane also contains a **ribosome receptor** protein associated with a **protein translocator**. In a concerted series of reactions, the ribosome is held tightly by the ribosome receptor protein, the SRP binds to the SRP receptor and is released from the signal peptide, and translation now continues once more, the nascent polypeptide passing through a pore in the membrane created by the protein translocator. As it passes through the pore, the signal peptide is cleaved off by a **signal peptidase** on the lumenal face of the ER (*Fig. 2*) and degraded, releasing the rest of the protein into the lumen. The protein is then transported through the Golgi to the cell exterior as described above. Since transport across the RER membrane occurs during protein synthesis, the process is said to be **co-translational**. The released SRP is cycled via its receptor ready for binding to another signal peptide (the **SRP cycle**).

## Plasma membrane proteins

Integral plasma membrane proteins are also synthesized by ribosomes on the RER, but become inserted in the RER membrane rather than transported into the lumen. During transport to the Golgi and then to the cell surface, these proteins stay anchored in the membrane, the final vesicles which fuse with the plasma membrane then becoming new plasma membrane (*Fig. 3*). Note that, after insertion in the RER membrane, one part of the protein faces in towards the RER lumen but eventually this faces outward on the cell surface. It is this part of the protein that receives the carbohydrate during glycosylation in the RER and Golgi complex so that the carbohydrate is exposed on the cell surface.

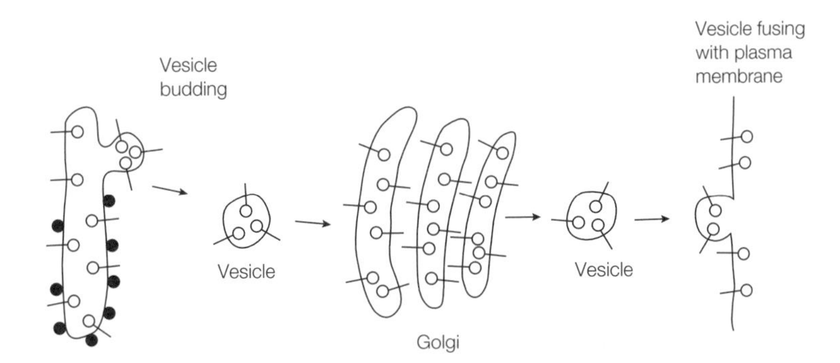


*Fig. 3. Synthesis of plasma membrane proteins; see text for details. The ribosomes attached to the RER are shown as filled-in circles whereas the newly synthesized plasma membrane proteins are shown as O—.*

Transfer of the plasma membrane protein across the ER membrane occurs during synthesis by a mechanism similar to that for secretory proteins. However, by definition, the protein is destined to remain anchored in the membrane and not enter the RER lumen entirely. There are several ways in which this is achieved, depending on the type of membrane protein. Some integral membrane proteins are **single membrane-spanning proteins**, that is the polypeptide chain crosses the membrane only once, whereas in other cases the protein is a **multiple membrane-spanning protein** (see Topic E2). The orientation of the protein in the membrane and the number of times it spans the lipid bilayer depend on specific **topogenic sequences** within the polypeptide chain. These topogenic sequences are regions of predominantly hydrophobic amino acids, and fall into three types: **N-terminal signal sequences, internal signal sequences** and **stop-transfer sequences**.

In the single membrane-spanning **Type I integral membrane proteins** (*Fig. 4a*), in addition to the N-terminal signal sequence which is cleaved from

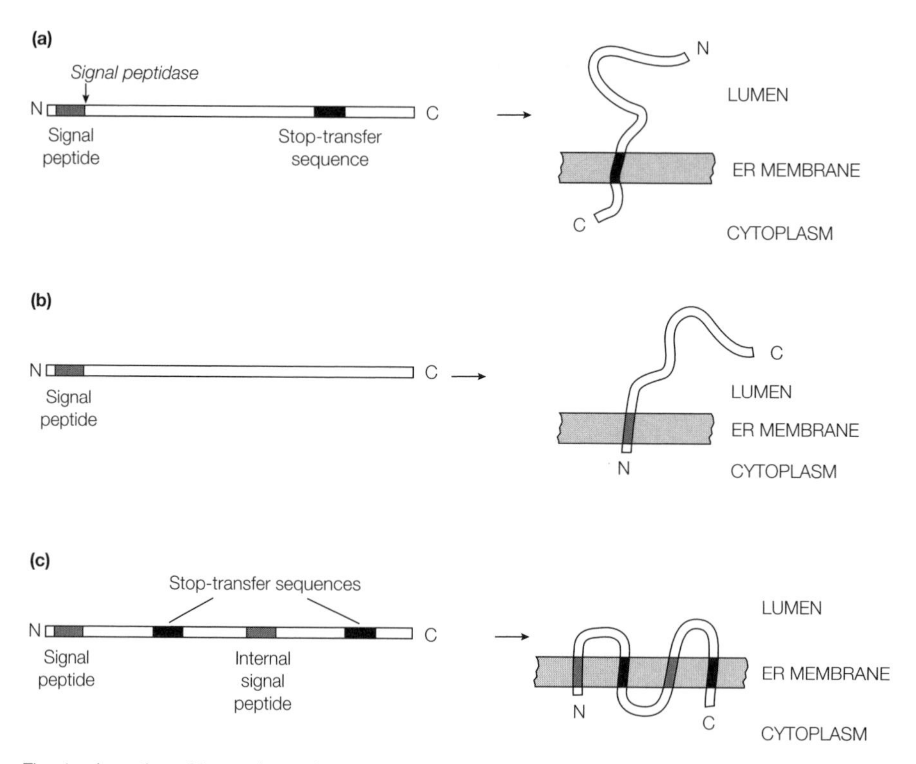


*Fig. 4. Insertion of integral membrane proteins into the ER membrane during synthesis. (a) Type I integral membrane protein with a cleavable N-terminal signal sequence and a stop-transfer sequence; (b) Type II integral membrane protein with an uncleaved N-terminal signal sequence; (c) Type III integral membrane protein with multiple signal and stop-transfer sequences.*

the protein by signal peptidase as in secretory proteins, there is a second hydrophobic sequence located internally in the protein. Thus the protein starts to cross the RER membrane during synthesis just like a secretory protein, but then transfer is stopped before the entire protein is translocated and the protein stays inserted in the membrane through the interaction of the hydrophobic stop-transfer sequence with the hydrophobic interior of the bilayer. In the single membrane-spanning **Type II membrane proteins** (*Fig. 4b*), there is just an N-terminal signal sequence as found in secretory proteins. However, in this case the signal sequence is not cleaved from the membrane protein by signal peptidase and doubles as the membrane anchor. Multiple membrane-spanning **Type III integral membrane proteins** (*Fig. 4c*), which cross the membrane several times, have multiple internal signal peptides and stop-transfer sequences to organize this arrangement during synthesis. The final orientation of the N terminus and the C terminus depends on whether the N-terminal signal sequence is cleaved and whether the final topogenic sequence is an internal signal sequence or a stop-transfer sequence, respectively. Some proteins lack an N-terminal signal sequence and have just an internal signal sequence.

## Proteins of the endoplasmic reticulum

The endoplasmic reticulum (ER) contains many proteins that have the role of assisting nascent proteins to fold correctly into their native conformation. Some of these are called **chaperones**. ER-resident proteins are made on the RER, pass into the lumen (as do secretory proteins), and are then transported to the Golgi by vesicles. However, these proteins contain a **retention signal** of Lys-Asp-Glu-Leu (or **KDEL** using the one-letter amino acid code) at the C terminus. On reaching the Golgi complex, receptors bind to the KDEL sequence and return the protein to the ER via vesicles.

## Lysosomal proteins

Lysosomal enzymes and lysosomal membrane proteins are synthesized on the RER and transported to the *cis* compartment of the Golgi complex. Here they become glycosylated and **mannose 6-phosphate** is added to the protein. The mannose 6-phosphate is the signal that targets the lysosomal protein to its correct destination. It is recognized by mannose 6-phosphate receptor proteins in the *trans* compartment of the Golgi which bind to the lysosomal protein and package it in transport vesicles that bud from the Golgi apparatus (*Fig. 5*). The transport vesicles then fuse with sorting vesicles, the contents of which are acidic. The low pH causes dissociation of the lysosomal protein from its receptor and a phosphatase removes the phosphate from the mannose 6-phosphate, preventing it from re-binding to the receptor. Vesicles bud from the sorting vesicle to return the receptor to the Golgi for re-use (**receptor recycling**) and the lysosomal protein is now delivered to the lysosome by vesicle fusion with it (*Fig. 5*).

Not all lysosomal proteins take the normal route of protein targeting; some end up being exported by the cell and must be retrieved. This **scavenger pathway** works as follows. The lysosomal glycoprotein binds to mannose 6-phosphate receptors in the plasma membrane and is internalized again by endocytosis (*Fig. 5*). This process, called **receptor-mediated endocytosis**, creates an endocytic vesicle (or **endosome**) that then delivers the lysosomal protein to the lysosome by fusion (see Topic E4).

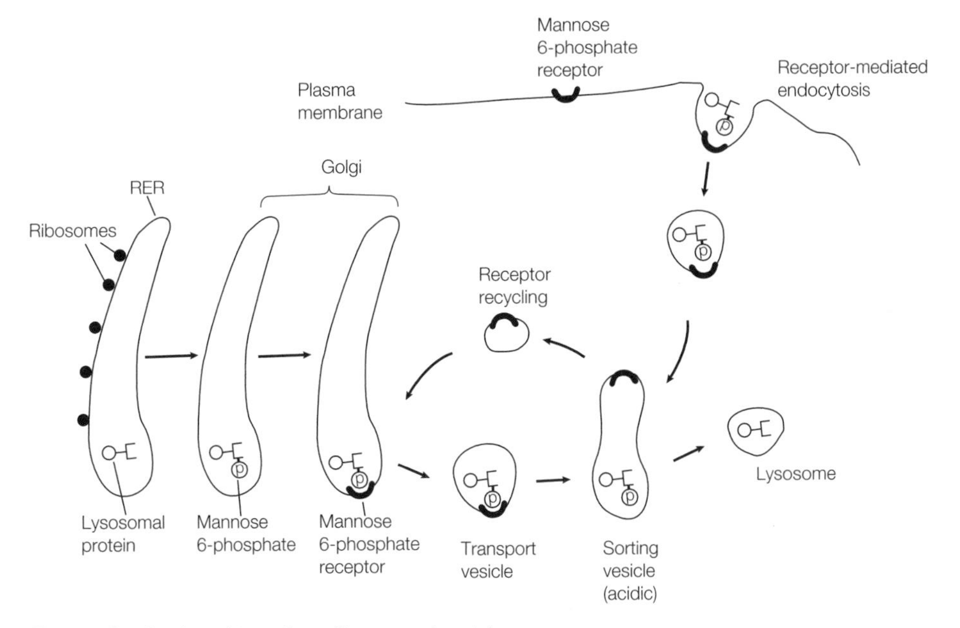


*Fig. 5. Synthesis and targeting of lysosomal proteins.*

## Mitochondrial and chloroplast proteins

Mitochondria and chloroplasts contain their own DNA, ribosomes, mRNA, etc., and carry out protein synthesis, but very few mitochondrial or chloroplast proteins are made in this way. Rather, the large majority of mitochondrial and chloroplast proteins are encoded by the nuclear genome, are synthesized in the cytosol on free ribosomes, released after synthesis and then imported into the organelle. Thus, the process is **post-translational**. The protein may need to be targeted to any one of several locations; for mitochondria this could be the **outer mitochondrial membrane**, the **inner membrane**, the **intermembrane space** or the mitochondrial **matrix**. Chloroplasts have the same subcompartments plus two other potential destinations, the **thylakoid membrane** and the **thylakoid space** (see Topics A2 and L3). Most is known about mitochondrial protein uptake.

Proteins are targeted to the mitochondrial matrix by an N-terminal sequence. This **matrix-targeting sequence** is typically 15–35 amino acids long and rich in serine, threonine and positively charged amino acids. After synthesis by cytosolic ribosomes, the protein is released into the cytosol but is kept in an unfolded state by chaperone proteins called the **hsp70** family of proteins which bind to it during synthesis. This is necessary since folded proteins cannot be imported into mitochondria. The hsp70 then transfers the unfolded protein to an **import receptor** in the outer mitochondrial membrane that is believed to slide along the membrane until it reaches a site where the inner membrane and outer membrane are in contact (a **contact site**). At this point it passes into the matrix via a protein translocator formed from the components of both membranes (*Fig. 6*). As it passes through the pore, the cytoplasmic hsp70 is released, the signal peptide is cleaved off by a signal peptidase, and the protein is bound in the matrix by **mitochondrial hsp70**. The hsp70 is then replaced by

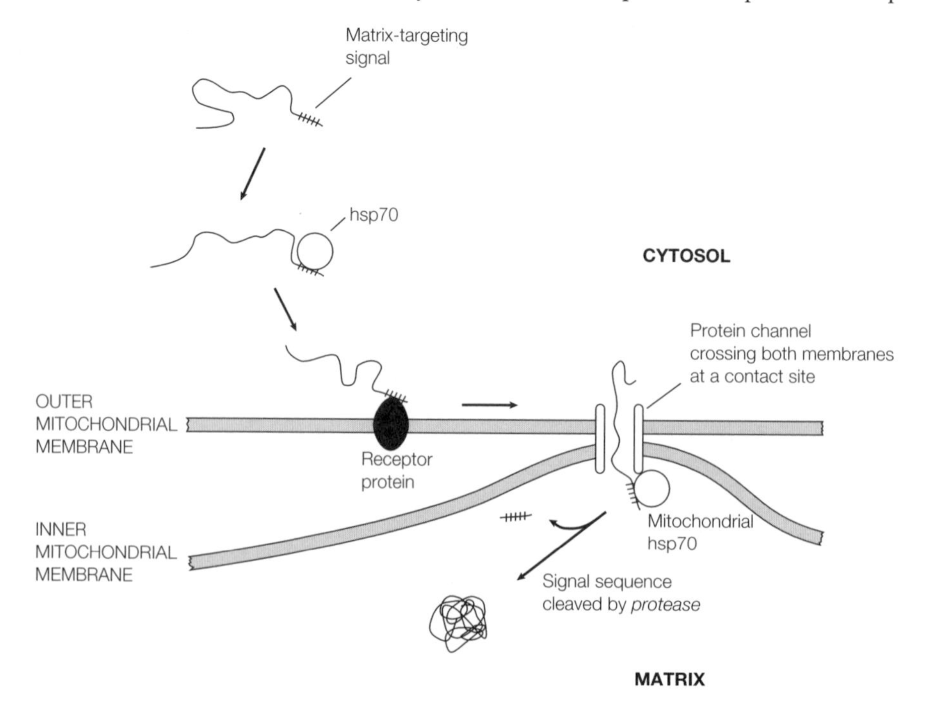


*Fig. 6. Uptake of proteins into the mitochondrial matrix; see text for details.*

**mitochondrial hsp60** which assists the protein to fold correctly into its final active state. Import of proteins into the mitochondrion requires energy from the electrochemical gradient across the inner membrane (see Topic L2) as well as ATP hydrolysis. Protein import into the mitochondrial inner membrane and intermembrane space needs two signals; the protein is first imported into the matrix as described above and then a second signal sequence directs the protein back into the inner membrane or across it into the intermembrane space.

Protein import into chloroplasts follows similar mechanisms to those in mitochondria but the signals used must be different since mitochondria and chloroplasts are present together in some plant cells and yet proteins become targeted to the correct destination.

## Nuclear proteins

The nucleus has an inner and an outer membrane (see Topic A2) and is perforated by 3000–4000 **nuclear pores**. Each pore consists of a **nuclear pore complex** of more than 100 different proteins organized in a hexagonal array. Although small molecules can pass through the pore by free diffusion, large proteins entering the nucleus require **a nuclear localization signal**. This is four to eight amino acids long and is rich in the positively charged amino acids lysine and arginine as well as usually containing proline. The protein is taken through the pore in an ATP-requiring step and enters the nucleus without cleavage of the localization signal.

# H5 Protein glycosylation

## Key Notes

**Three types of protein glycosylation**

Many proteins synthesized by the ribosomes of the RER contain short chains of carbohydrates (oligosaccharides) and are called glycoproteins. The oligosaccharides are of two main types; O-linked (to the OH side chain of Ser or Thr) and N-linked (to the $NH_2$ side chain of Asn). Some proteins are attached to the plasma membrane by a third type of carbohydrate structure called a glycosyl phosphatidylinositol (GPI) anchor.

**Synthesis of O-linked oligosaccharides**

O-linked oligosaccharides are synthesized by the sequential addition of monosaccharides to the protein as it passes through the Golgi complex.

**Synthesis of N-linked oligosaccharides**

All N-linked oligosaccharides have a common pentasaccharide core structure of three mannose residues and two *N*-acetylglucosamine (GlcNAc) residues. The oligosaccharide is initially synthesized on a dolichol phosphate carrier that is anchored to the RER membrane. This is then transferred to the protein and subsequently trimmed during passage of the protein through the RER and Golgi complex. Additional monosaccharides are added in the Golgi to produce either a high mannose type oligosaccharide or a complex type oligosaccharide.

**Related topics**

Membrane protein and carbohydrate (E2)
Translation in prokaryotes (H2)
Translation in eukaryotes (H3)
Protein targeting (H4)
Monosaccharides and disaccharides (J1)
Structures and roles of fatty acids (K1)

## Three types of protein glycosylation

Most proteins made by ribosomes on the rough endoplasmic reticulum (RER) are glycoproteins, that is they contain short chains of carbohydrates (oligosa charides) covalently linked to them during passage through the RER and Golgi complex. Two main types of oligosaccharide linkage exist:

- **O-linked oligosaccharides** are commonly attached to the protein via O-glycosidic bonds to OH groups of serine or threonine side chains (*Fig. 1a*).
- **N-linked oligosaccharides** are linked to the protein via N-glycosidic bonds, to the $NH_2$ groups of asparagine side chains (*Fig. 1b*) where the asparagine occurs in the sequence Asn-X-Ser (or Thr) where X is any amino acid except Pro and possibly Asp.

In addition, several proteins are now known to be attached to the plasma membrane via a specific structure that involves carbohydrate, namely a **glycosyl phosphatidylinositol (GPI) anchor**. This is covered in Topic E2.

## Synthesis of O-linked oligosaccharides

The synthesis of O-linked oligosaccharides occurs by the sequential addition of monosaccharide units to the newly synthesized protein as it passes through the Golgi complex. First, *N*-acetylgalactosamine (GalNAc) is transferred to the

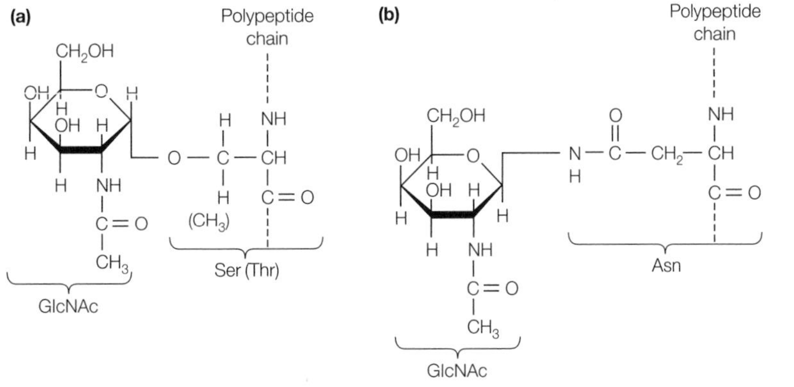

Fig. 1. Structures of oligosaccharide linkages. (a) O-linked glycosidic bond between GlcNAc and Ser (Thr) residue. (b) N-linked glycosidic bond between GlcNAc and an Asn residue.

relevant Ser or Thr residue of the protein by GalNAc transferase, an enzyme that uses UDP-GalNAc as the precursor (*Fig. 2*). Other monosaccharides [galactose, *N*-acetylglucosamine (GlcNAc), sialic acid, fucose] are then added using the corresponding sugar nucleotides as precursors. The exact type and number (up to about 10) of monosaccharides added depends on the protein substrate.

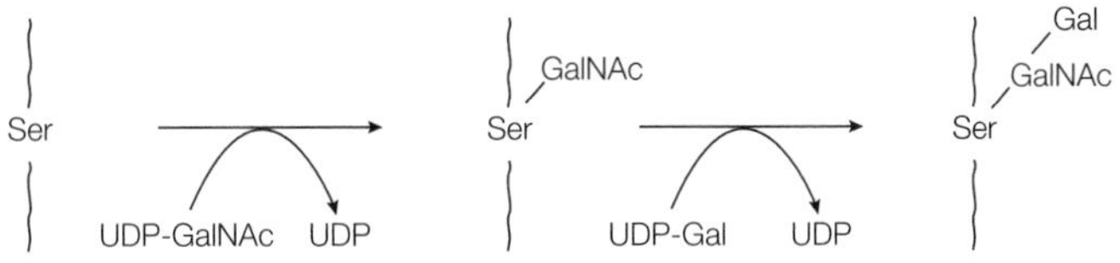

Fig. 2. Synthesis of O-linked oligosaccharide. The example shown is an O-linked oligosaccharide in human immunoglobulin A (IgA).

## Synthesis of N-linked oligosaccharides

All N-linked oligosaccharides are based on a common **pentasaccharide core** structure consisting of three mannose residues and two GlcNAc residues (*Fig. 3*).

There are two types of N-linked oligosaccharides. In the **high mannose type oligosaccharides**, the R group in *Fig. 3* is a variable number of mannose

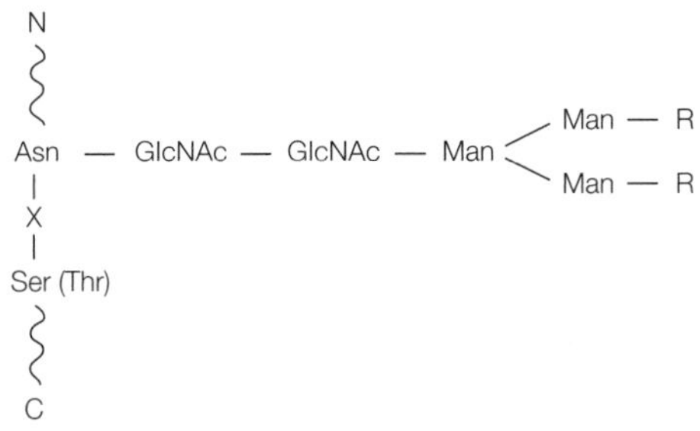

Fig. 3. Structure of the common pentasaccharide core of N-linked oligosaccharides.

residues. In the **complex type oligosaccharides**, the R group consists of a variety of other sugars such as GlcNAc, galactose, sialic acid and fucose. N-linked oligosaccharides are *not* synthesized by adding monosaccharides directly to the protein but instead the oligosaccharide is made on a lipid carrier called **dolichol phosphate**. This consists of 22 isoprene (C5) units (see Topic K5) with a terminal phosphate group and it is anchored to the RER membrane.

Synthesis of the oligosaccharide starts by the dolichol phosphate accepting monosaccharides from the cytosolic face of the RER membrane (*Fig. 4*) but when the $(Man)_5(GlcNac)_2$-dolichol phosphate intermediate has formed, this flips orientation and now accepts further monosaccharides from the lumenal side of the RER membrane (*Fig. 4*). All of these subsequent transfers are from dolichol phosphate-linked monosaccharides that are made on the cytoplasmic side of the RER membrane then likewise flipped across to act as donors. The final oligosaccharide, with the composition $(Glc)_3(Man)_9(GlcNAc)_2$, called **G-oligosaccharide**, is linked to the dolichol by a high-energy pyrophosphate bond. This provides the energy for transfer of the oligosaccharide to the protein, a reaction catalyzed by a membrane-bound **oligosaccharide transferase** enzyme and which occurs in the RER (*Fig. 5*). Collectively these reactions are called **core glycosylation**. Whilst the protein is still in the RER, the three glucose residues and a mannose are quickly removed (*Fig. 5*). Interestingly, glucose residues are added back to the protein if it is unfolded or wrongly folded. Thus, only when the protein is correctly folded are all glucose residues finally removed and the

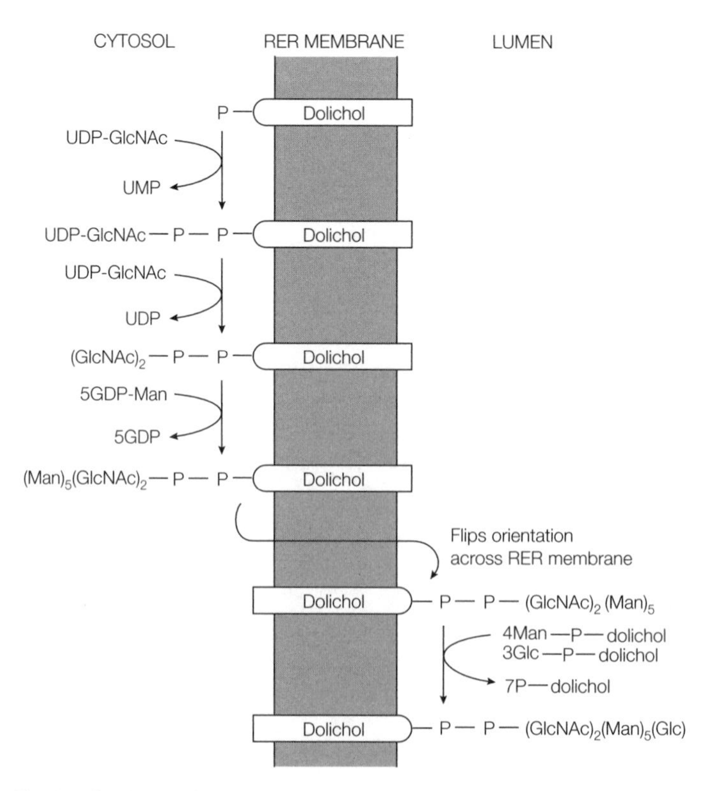


*Fig. 4. Synthesis of N-linked oligosaccharides on a dolichol phosphate carrier in the RER membrane.*

protein may continue along the modification pathway. The folded glycoprotein is now transported to the Golgi complex via vesicles. As it moves through the Golgi complex, the '**trimming**' or '**processing**' of the oligosaccharide continues with another five mannose residues being removed (*Fig. 5*). Mannose residues and other monosaccharides are also added to the oligosaccharide in the Golgi to generate either the high mannose or complex type of oligosaccharide, a process known as **terminal glycosylation**.

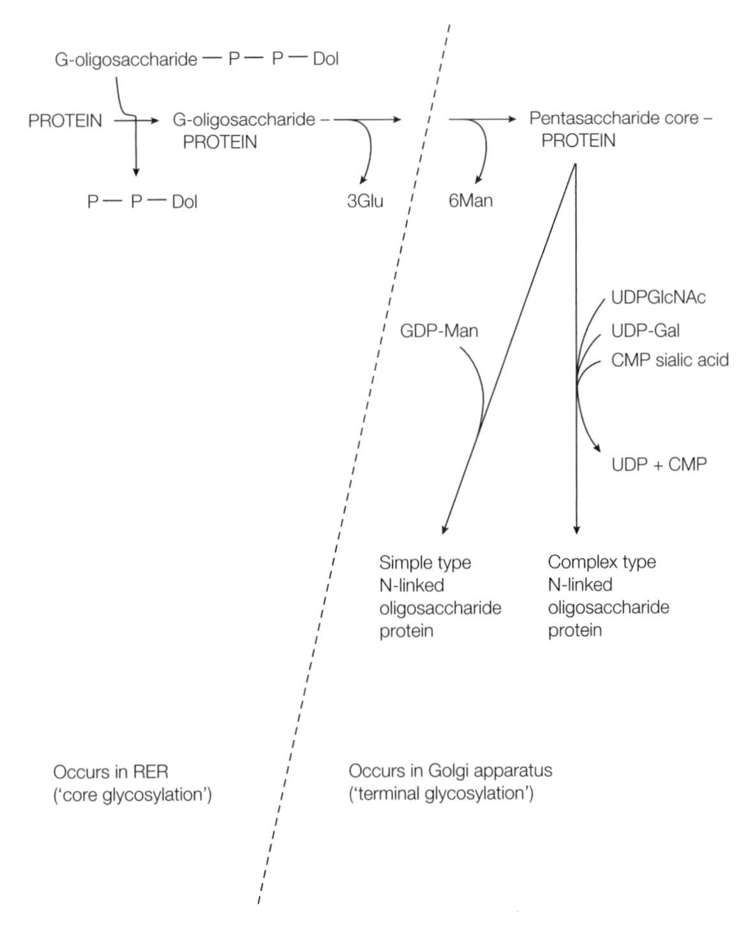


*Fig. 5. Transfer of the oligosaccharide to protein and further processing in the RER and Golgi.*

# I1 RESTRICTION ENZYMES

## Key Notes

**Overview**

Restriction enzymes allow DNA to be cut at specific sites; nucleic acid hybridization allows the detection of specific nucleic acid sequences; DNA sequencing can be used to easily determine the nucleotide sequence of a DNA molecule.

**Restriction enzyme digestion**

Restriction enzymes recognize specific recognition sequences and cut the DNA to leave cohesive ends or blunt ends. The ends of restricted DNA molecules can be joined together by ligation to create new recombinant DNA molecules.

**Nomenclature**

Restriction enzymes have a three-letter name based on the genus and species name of the bacterium from which they were isolated, together with a roman numeral designed to indicate the identity of the enzyme in cases when the bacterium contains several different restriction enzymes.

**Gel electrophoresis**

DNA fragments in a restriction digest can be separated by size by electrophoresis in polyacrylamide or agarose gel. Polyacrylamide gel is used to separate smaller DNA molecules whilst agarose gel has larger pore sizes and so can separate larger DNA fragments.

**Restriction maps**

A map showing the position of cut sites for a variety of restriction enzymes is called the restriction map for that DNA molecule. Restriction maps allow comparison between DNA molecules without the need to determine the nucleotide sequence and are also much used in recombinant DNA experiments.

**Restriction fragment length polymorphisms**

A restriction fragment length polymorphism (RFLP) is a common difference between the DNA of individuals in a population (i.e. a polymorphism) that affects the sizes of fragments produced by a specific restriction enzyme. If the RFLP lies near a gene, changes in which can cause a human genetic disease, it can be used as a marker for that gene. In the past, RFLPs have proved valuable both for screening patients for the gene defect and also in studies directed at cloning the gene. However, RFLPs are becoming less commonly used in such work as the genes themselves are identified. The polymerase chain reaction (PCR) is becoming the method of choice for screening.

**Related topics**

DNA structure (F1)
Nucleic acid hybridization (I2)
DNA cloning (I3)

## Overview

The ability to isolate, analyze and change genes at will has now become almost commonplace through recombinant DNA technology. The enormous advances brought about both in understanding gene structure and function and in practical applications of that knowledge depended originally on the development of several major new techniques. Some of the most important were the ability to cut DNA at specific sites using restriction endonucleases (**restriction enzymes**), procedures that allow the detection of specific DNA (and RNA) sequences with great accuracy (**nucleic acid hybridization**; see Topic I2), methods for preparing specific DNA sequences in large amounts in pure form (**DNA cloning**; see Topic I3) and rapid **DNA sequencing** (see Topic I5) whereby the nucleotide sequences of isolated genes and control regions can readily be determined. More recently, the development of the **polymerase chain reaction (PCR)** (see Topic I6) has revolutionized the field of molecular biology. This section covers these and some other key topics, but a more extensive description of recombinant DNA technology is provided in the companion book *Instant Notes in Molecular Biology*.

## Restriction enzyme digestion

Restriction enzymes recognize specific nucleotide sequences (**recognition sequences**) in double-stranded DNA, that are usually four, five or six nucleotides long, and then cut both strands of the DNA at specific locations. There are basically three ways in which the DNA can be cut; a staggered cut to leave a **5′ overhang** (i.e. a short single-stranded region of DNA is left that has a 5′ end and overhangs the end of the double-stranded DNA), a staggered cut to leave a **3′ overhang**, or a cut in the same place on both strands to leave a **blunt end** (*Fig. 1*). For enzymes that cut in the staggered manner, the single-stranded tails are called **cohesive ends** because they allow any two DNA fragments produced by the same restriction enzyme to form complementary base pairs (*Fig. 1*). The cut ends can then be joined together (**ligated**) by an enzyme called **DNA ligase**. The new DNA molecule that has been made by joining the DNA fragments is called a **recombinant DNA molecule** (*Fig. 2*). Blunt-ended DNA molecules can also be joined together by DNA ligase but the reaction is far less favorable.

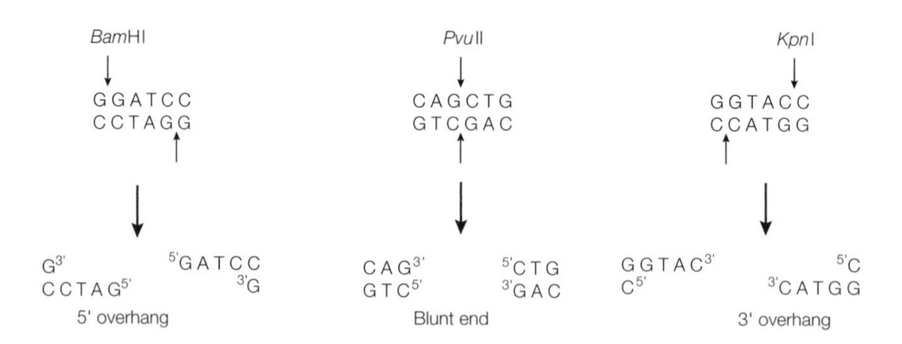


*Fig. 1. The three types of cleavage by commonly used restriction enzymes.*

## Nomenclature

Restriction enzymes are isolated from bacteria, where they play a role in protecting the host cell against virus infection. Over 100 restriction enzymes have now been isolated and have been named according to the bacterial species from which they were isolated. The first three letters of the enzyme name are the first letter of the genus name and the first two letters of the species name. Since each bacterium may contain several different restriction enzymes, a roman numeral is also used to identify each enzyme. *Eco*RI, for example, was the first enzyme isolated from *Escherichia coli*.

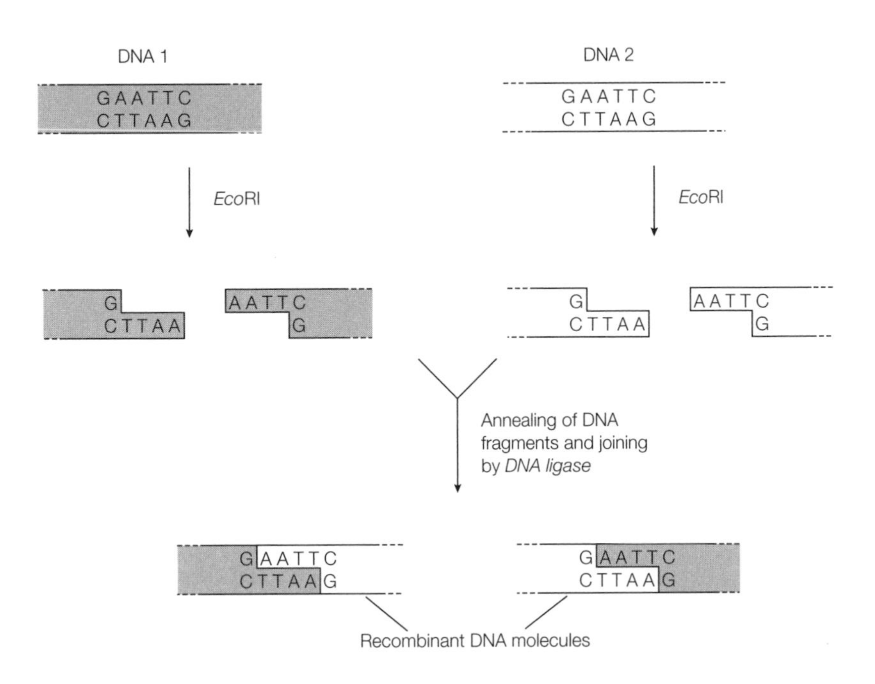


*Fig. 2. Using a restriction enzyme to create recombinant DNA.*

## Gel electrophoresis

When a DNA molecule is cut by a restriction enzyme, the DNA fragments (called **restriction fragments**) from that **restriction digest** can be separated by **gel electrophoresis** (*Fig. 3*). Electrophoresis on a polyacrylamide gel will separate small DNA fragments of less than about 500 bp in size, but agarose gels (which have larger pores) are needed to separate larger DNA fragments. The DNA digest separates into a series of bands representing the restriction fragments. Since small fragments travel further in the gel than larger fragments, the size of each fragment can be determined by measuring its migration distance relative to standard DNA fragments of known size. The DNA can be located after gel electrophoresis by staining with ethidium bromide that binds to the DNA and fluoresces a bright orange. Alternatively, if the DNA is labeled with a radioisotope such as $^{32}P$, the bands can be detected after electrophoresis by laying the gel against an X-ray film (**autoradiography**) whereby the radioactivity causes silver grains to be formed in the film emulsion, giving black images corresponding to the radioactive bands.

## Restriction maps

Any double-stranded DNA will be cut by a variety of restriction enzymes that have different recognition sequences. By separating the restriction fragments and measuring their sizes by gel electrophoresis, it is possible to deduce where on the DNA molecule each restriction enzyme cuts. A **restriction map** of the DNA molecule can be drawn showing the location of these cut sites (**restriction sites**) (*Fig. 4*). It is then easy to compare two DNA molecules (for example, to examine the evolutionary relationship between two species) by looking at their restriction maps without the need to determine the nucleotide sequence of each DNA. Restriction maps are also important experimentally during recombinant DNA work, both to plan where individual DNA molecules should best be cut and to monitor the progress of the experiment.

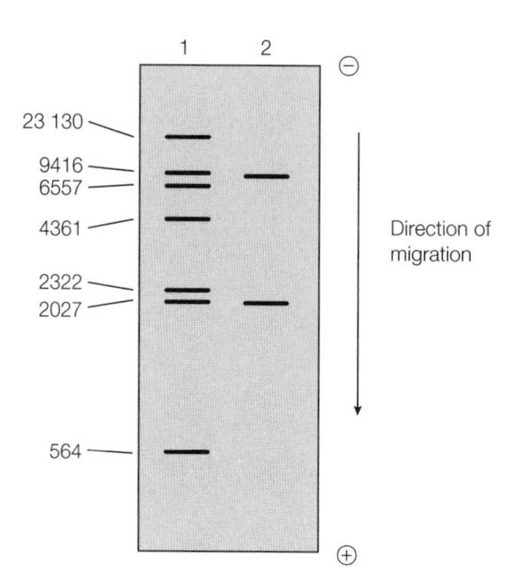


*Fig. 3.* Agarose gel electrophoresis of DNA fragments. DNA fragments of known size were electrophoresed in lane 1 (the sizes in bp are given on the left). A restriction digest of the sample DNA was electrophoresed in lane 2. By comparison with the migration positions of fragments in lane 1, it can be seen that the two sample DNA fragments have sizes of approximately 9000 bp and 2000 bp. The sizes could be determined more accurately by plotting the data from lane 1 as a standard curve of log DNA size vs. migration distance and then using this to estimate the size of the sample DNA fragments from their measured migration distances.

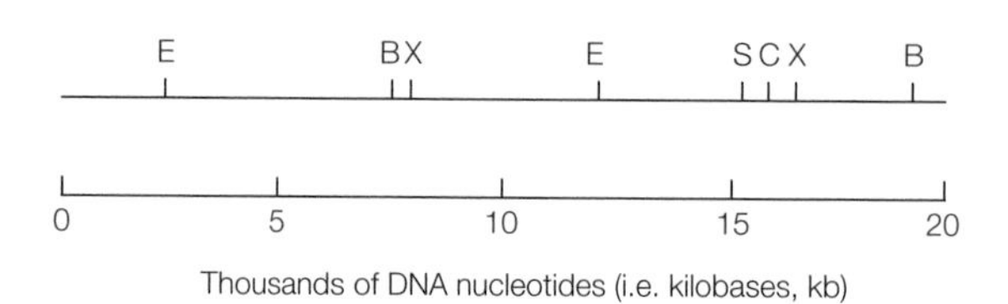


*Fig. 4.* A typical restriction map of a DNA molecule. The cleavage sites of different restriction enzymes, indicated by letters, are shown.

## Restriction fragment length polymorphisms

Analysis of human genomic DNA has revealed that there are many differences in DNA sequence between individuals that have no obvious effect, often because the changes lie in introns or between genes. Some of these changes are very common in individuals in a population and are called **polymorphisms**. Some polymorphisms affect the size of fragments generated by a particular restriction enzyme, for example by changing a nucleotide in the recognition sequence and so eliminating a cut site. Instead of two restriction fragments being generated from this region, a single large restriction fragment is now formed (*Fig. 5*). Alternatively the polymorphism may result from the insertion or deletion of sequences between two cut sites, so increasing or decreasing the size of that restriction fragment produced. This type of polymorphism that affects restriction fragment sizes is called **a restriction fragment length polymorphism (RFLP)**. Provided that a DNA probe (see Topic I2) exists for a sequence of DNA within the affected region, so that this sequence can be detected by hybridization, RFLPs can be detected by Southern blotting (see Topic I2).

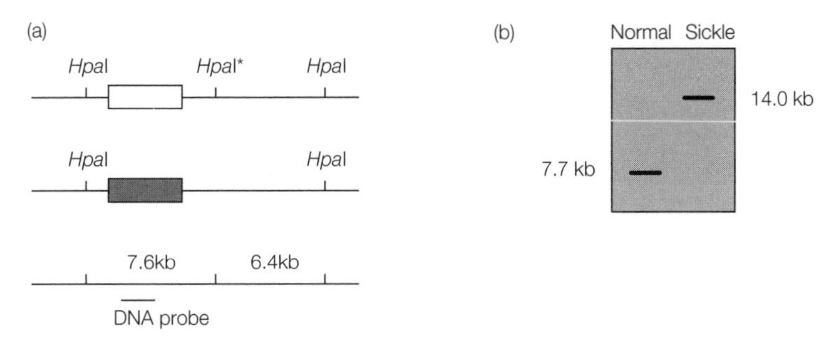


*Fig. 5. Analysis of human genetic disease using RFLPs. The analysis concerns two individuals from a family, one of whom has normal β-globin and one of whom has an abnormal β-globin gene that leads to sickle cell anemia. (a) The sickle β-globin is associated with a nucleotide change that results in the loss of the* HpaI *site marked with an asterisk. The presence or absence of this* HpaI* *site is detected by hybridization and Southern blotting (see Topic I2) using a DNA probe for the 7.6 kb fragment; (b) normal DNA with three* HpaI *sites yields a 7.6 kb fragment detected by the DNA probe but sickle DNA yields a 14.0 kb fragment due to loss of the* HpaI* *site.*

The value of RFLPs has been in the ability to use these as markers for particular human genetic diseases. Consider a polymorphism that happens to be very near to the site of a change in a key gene that results in a human genetic disease (*Fig. 5*). Because these two changes, the polymorphism and the genetic defect, lie close together on the same chromosome, they will tend to be co-inherited. Identifying such a **closely linked** RFLP has two major advantages. First, experiments can be directed to cloning DNA near the RFLP in the hope of identifying the gene itself which can then be sequenced and studied. Second, even in the absence of the gene, the RFLP acts as a screening marker for the disease; individuals who have the RFLP have a high probability of having the associated gene defect. Of course other RFLPs that are located a very long way from the gene, or even on a different chromosome, will essentially be **unlinked** (i.e. because of the high probability of cross-over events during meiosis to produce germ-line cells, the gene and RFLP will have only a 50:50 chance of being co-inherited). Thus a large amount of very painstaking work has to be carried out to identify a useful RFLP for a particular human genetic disease. Large numbers of individuals in family groups, some of whom suffer from the disease, need to be screened for a range of likely RFLPs to attempt to locate an RFLP that is routinely co-inherited with the gene defect.

As the genes themselves are identified and sequenced, so the need for RFLP markers declines since specific DNA probes (see Topic I2) for the most common types of gene defect can be employed. In addition, the use of the polymerase chain reaction (PCR; see Topic I6) in screening for human genetic disease is becoming the method of choice rather than RFLP analysis since it is much faster to perform and requires far less clinical material for analysis.

# I2 Nucleic acid hybridization

## Key Notes

**The hybridization reaction**

Double-stranded DNA denatures into single strands as the temperature rises but renatures into a double-stranded structure as the temperature falls. Any two single-stranded nucleic acid molecules can form double-stranded structures (hybridize) provided that they have sufficient complementary nucleotide sequence to make the resulting hybrid stable under the reaction conditions.

**Monitoring specific nucleic acid sequences**

The concentration of a specific nucleic acid sequence in a sample can be measured by hybridization with a suitable labeled DNA probe. After hybridization, nuclease is used to destroy unhybridized probe and the probe remaining is a measure of the concentration of the target sequence. The hybridization conditions can be altered to ensure that only identical sequences (high stringency conditions) or identical plus related sequences (low stringency conditions) will hybridize with the probe and hence be detected.

**Southern blotting**

Southern blotting involves electrophoresis of DNA molecules in an agarose gel and then blotting the separated DNA bands on to a nitrocellulose filter. The filter is then incubated with a labeled DNA probe to detect those separated DNA bands that contain sequences complementary to the probe.

**Northern blotting**

Northern blotting is analogous to Southern blotting except that the sample nucleic acid that is separated by gel electrophoresis is RNA rather than DNA.

***In situ* hybridization**

For *in situ* hybridization, a tissue sample is incubated with a labeled nucleic acid probe, excess probe is washed away and the location of hybridized probe is examined. The technique enables the spatial localization of gene expression to be determined as well as the location of individual genes on chromosomes.

**Related topics**

DNA structure (F1)
Restriction enzymes (I1)
DNA cloning (I3)

## The hybridization reaction

As double-stranded DNA is heated, a temperature is reached at which the two strands separate. This process is called **denaturation**. The temperature at which half of the DNA molecules have denatured is called the **melting temperature** or $T_m$ for that DNA. If the temperature is now lowered and falls below the $T_m$, the two complementary strands will form hydrogen bonds with each other once more to reform a double-stranded molecule. This process is called **renaturation** (or **reannealing**). In fact, double-stranded structures can form between any two single-stranded nucleic acid molecules (DNA–DNA, DNA–RNA, RNA–RNA) provided that they have sufficient complementary nucleotide sequence to make the double-stranded molecule stable under the conditions used. The general name given to this process is **hybridization** and the double-stranded nucleic acid product is called a **hybrid**.

### Monitoring specific nucleic acid sequences

The rate of formation of double-stranded hybrids depends on the concentration of the two single-stranded species. This can be used to measure the concentration of either specific DNA or RNA sequences in a complex mixture. The first task is to prepare a single-stranded **DNA probe** (i.e. a DNA fragment that is complementary to the nucleic acid being assayed). This can be one strand of a DNA restriction fragment, cloned DNA or a synthetic oligonucleotide. It must be labeled in order to be able to detect the formation of hybrids between it and the target nucleic acid. Whereas most labeling used to be via the incorporation of a radioisotope, nonradioactive chemical labels are now often used instead. For example, a DNA probe can be labeled with digoxygenin, a steroid, by using digoxygenin-labeled dUTP during DNA synthesis. Hybrids containing the digoxygenin-labeled DNA probe can then be detected using anti-digoxygenin antibody linked to a fluorescent dye. Irrespective of the method of probe labeling, the DNA probe is incubated with the nucleic acid sample and then nuclease is added to degrade any unhybridized single-stranded probe. The amount of labeled probe remaining indicates the concentration of the target nucleic acid in the sample.

The hybridization conditions (e.g. temperature, salt concentration) can be varied so as to govern the type of hybrids formed. The conditions may be arranged so that only perfectly matched hybrids are stable and hence assayed (conditions known as **high stringency**). Alternatively, the conditions may be such that even poorly complementary hybrids are stable and will be detected (**low stringency**). Thus by varying the reaction conditions it is possible to detect and quantify only those target sequences that are identical to the DNA probe or, alternatively, to detect and quantify related sequences also. Hybridization of nucleic acid probes with genomic DNA, for example, can be used to measure the copy number of particular DNA sequences in the genome. Hybridization of a DNA probe with cellular RNA as the target will indicate the concentration of the corresponding RNA transcript and hence give information about the level of gene expression. Variants of the methodology even allow determination of the transcriptional Start and Stop sites and the number and location of intron sequences in protein-coding genes.

### Southern blotting

Gel electrophoresis is widely used to separate and size DNA molecules during recombinant DNA experiments. After gel electrophoresis, there is often a need to detect one or more DNA fragments containing a specific nucleotide sequence. This is easily carried out by Southern blotting. After electrophoresis of the restriction fragments through an agarose gel, the gel is soaked in alkali to denature the DNA to single strands and the pH is then neutralized. The gel is placed in contact with a nitrocellulose or nylon membrane filter sheet arranged so that buffer flows through the gel and carries the DNA fragments to the membrane (*Fig. 1*). The membrane binds the single-stranded DNA and so the band pattern in the gel is now transferred to it. The membrane filter is peeled from the gel, baked at high temperature to fix the DNA to it, and then incubated with a radiolabeled DNA probe. After hybridization, the probe will have bound only to DNA fragments with complementary sequences. These can be visualized by washing away excess probe and then placing the filter against an X-ray film for autoradiography. The images produced on the autoradiogram indicate those bands that contain the probe sequence (*Fig. 1*).

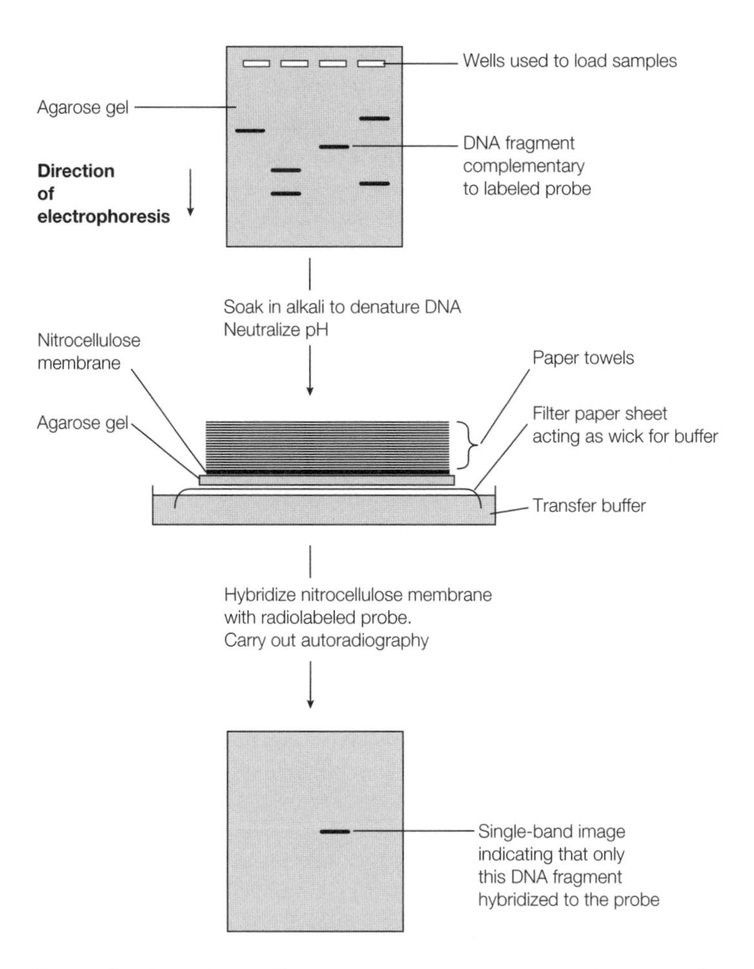


*Fig. 1. Southern blotting. The procedure shown is the original method of Southern using capillary action to blot the DNA bands from the gel to the nitrocellulose membrane. Electrolytic transfer is now often used instead.*

## Northern blotting

Northern blotting follows much the same procedure as Southern blotting except that the sample analyzed by gel electrophoresis and then bound to the filter is RNA not DNA. Therefore the technique detects RNA molecules that are complementary to the DNA probe. If cellular RNA is electrophoresed, for example, a DNA probe for a specific mRNA could be used to detect whether that mRNA was present in the sample. The migration distance of the RNA in the gel would also allow estimation of its size. Note that Southern blotting (for DNA) obtained its name after its inventor (E. Southern); the name Northern blotting (for RNA) was devised later and is a geographical pun!

## *In situ* hybridization

It is also possible to incubate radioactive or fluorescent nucleic acid probes with sections of tissues or even chromosomes, wash away excess probe and then detect where the probe has hybridized. This technique (***in situ* hybridization**) has proved to be very powerful in determining which cells in a complex tissue such as the mammalian brain express a particular gene and for locating specific genes on individual chromosomes.

# I3 DNA CLONING

## Key Notes

**The principle of DNA cloning**

Most foreign DNA fragments cannot self-replicate in a cell and must therefore be joined (ligated) to a vector (virus or plasmid DNA) that can replicate autonomously. Each vector typically will join with a single fragment of foreign DNA. If a complex mixture of DNA fragments is used, a population of recombinant DNA molecules is produced. This is then introduced into the host cells, each of which will typically contain only a single type of recombinant DNA. Identification of the cells that contain the DNA fragment of interest allows the purification of large amounts of that single recombinant DNA and hence the foreign DNA fragment.

**The basics of DNA cloning**

To clone into a plasmid vector, both the plasmid and the foreign DNA are cut with the same restriction enzyme and mixed together. The cohesive ends of each DNA reanneal and are ligated together. The resulting recombinant DNA molecules are introduced into bacterial host cells. If the vector contains an antibiotic resistance gene(s) and the host cells are sensitive to these antibiotic(s), plating on nutrient agar containing the relevant antibiotic will allow only those cells that have been transfected and contain plasmid DNA to grow.

**DNA libraries**

Genomic DNA libraries are made from the genomic DNA of an organism. A complete genomic DNA library contains all of the nuclear DNA sequences of that organism. A cDNA library is made using complementary DNA (cDNA) synthesized from mRNA by reverse transcriptase. It contains only those sequences that are expressed as mRNA in the tissue or organism of origin.

**Screening DNA libraries**

Genomic and cDNA libraries can be screened by hybridization using a labeled DNA probe complementary to part of the desired gene. The probe may be an isolated DNA fragment (e.g. restriction fragment) or a synthetic oligonucleotide designed to encode part of the gene as deduced from a knowledge of the amino acid sequence of part of the encoded protein. In addition, expression cDNA libraries may be screened using a labeled antibody to the protein encoded by the desired gene or by using any other ligand that binds to that protein.

**Related topics**

DNA structure (F1)
Restriction enzymes (I1)
Nucleic acid hybridization (I2)
Viruses (I4)

## The principle of DNA cloning

Consider an experimental goal which is to make large amounts of a particular DNA fragment in pure form from a mixture of DNA fragments. Although the DNA fragments can be introduced into bacterial cells, most or all will lack the ability for self-replication and will quickly be lost. However, two types of DNA molecule are known which can replicate autonomously in bacterial cells; bacteriophages (see Topic I4) and plasmids. Plasmids are small circular double-stranded DNA molecules that exist free inside bacterial cells, often carry particular genes that confer drug resistance, and are self-replicating. If a recombinant DNA molecule is made by joining a foreign DNA fragment and plasmid (or bacteriophage) DNA, then the foreign DNA is replicated when the plasmid or phage DNA is replicated. In this role, the plasmid or phage DNA is known as a **vector**. Now, a population of recombinant DNA molecules can be made, each recombinant molecule containing one of the foreign DNA fragments in the original mixture. This can then be introduced into a population of bacteria such that each bacterial cell contains, in general, a different type of recombinant DNA molecule. If we can identify the bacterial cell that contains the recombinant DNA bearing the foreign DNA fragment we want, it can be grown in culture and large amounts of the recombinant DNA isolated. The foreign DNA can then be recovered from this in pure form; it is then said to have been **cloned**. The vector that was used to achieve this cloning is called a **cloning vector**. Vectors are not limited to bacterial cells. Animal and plant viruses can also act as vectors.

## The basics of DNA cloning

There are a variety of different procedures for cloning DNA into either plasmid or viral vectors but the basic scheme of events is broadly the same. To clone into a plasmid vector, the circular plasmid DNA is cut with a restriction enzyme (see Topic I1) that has only a single recognition site in the plasmid. This creates a linear plasmid molecule with cohesive ends (*Fig. 1*). The simplest cloning strategy is now to cut the foreign donor DNA with the same restriction enzyme. Alternatively, different restriction enzymes can be used, provided that they leave the same cohesive ends (see Topic I1). The donor DNA and linear plasmid DNA are now mixed. The cohesive ends of the foreign DNA anneal with the ends of the plasmid DNA and are joined covalently by DNA ligase. The resulting **recombinant plasmid** DNA is introduced into bacterial host cells that have been treated to become permeable to DNA. This uptake of DNA by the bacterial cells is called **transfection**; the bacterial cells are said to have been **transfected** by the recombinant plasmid. The bacterial cells are now allowed to grow and divide, during which time the recombinant plasmids will replicate many times within the cells. One useful procedure is to use as cloning vector a plasmid that carries one or more antibiotic resistance genes plus a host that is sensitive to those antibiotics (*Fig. 1*). Then, after transfection, the cells are grown in the presence of the antibiotic(s). Only cells containing plasmid DNA will be resistant to the antibiotic(s) and can grow. If the cells are spread on an agar plate, each cell will multiply to form a bacterial colony where all the cells of that colony contain the same recombinant plasmid DNA bearing the same foreign DNA fragment. Thus all that is now needed is to identify the particular bacterial colony that contains the foreign DNA sequence of interest.

## DNA libraries

A DNA library is a collection of cloned DNA fragments in a cloning vector that can be searched for a DNA of interest. If the goal is to isolate particular gene sequences, two types of library are useful:

- **genomic DNA libraries**. A genomic DNA library is made from the genomic DNA of an organism. For example, a mouse genomic library could be made by digesting with a restriction nuclease to produce a large number of different DNA fragments but all with identical cohesive ends. The DNA fragments would then be ligated into the linearized plasmid vector molecules or into a suitable virus vector. This library would contain all of the nuclear DNA sequences of the mouse and could be searched for any particular mouse

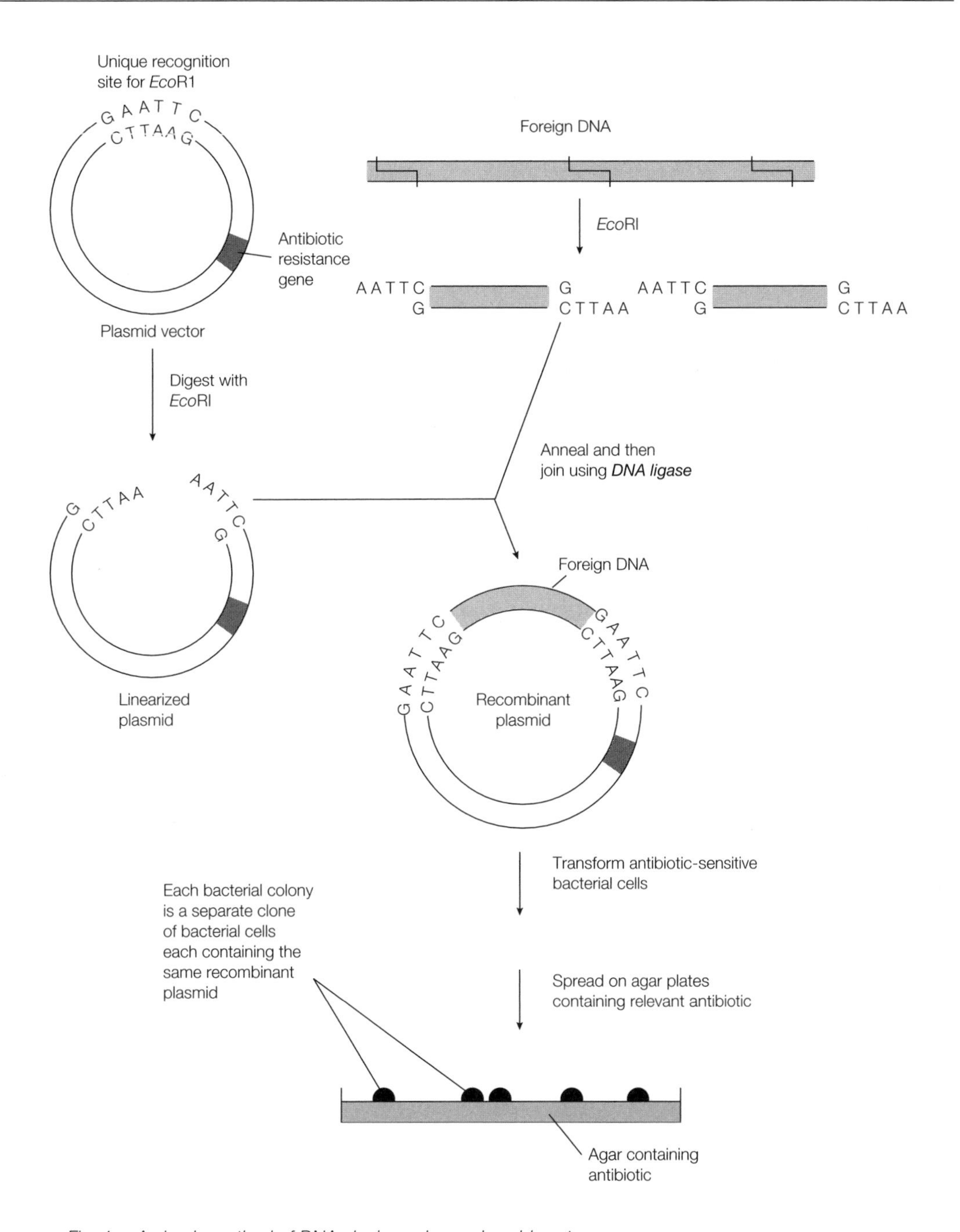


*Fig. 1. A simple method of DNA cloning using a plasmid vector.*

gene of interest. Each clone in the library is called a **genomic DNA clone**. Not every genomic DNA clone would contain a complete gene since in many cases the restriction enzyme will have cut at least once within the gene. Thus some clones will contain only a part of a gene.

- **cDNA libraries**. A cDNA library is made by using the reverse transcriptase of a retrovirus (see Topic I4) to synthesize **complementary DNA** (cDNA) copies of the total mRNA from a cell (or perhaps a subfraction of it). The single-stranded cDNA is converted into double-stranded DNA and inserted into the vector. Each clone in the library is called a **cDNA clone**. Unlike a complete genomic library that contains all of the nuclear DNA sequences of an organism, a cDNA library contains only sequences that are expressed as mRNA. Different tissues of an animal, that express some genes in common but also many different genes, will thus yield different cDNA libraries.

## Screening DNA libraries

Genomic libraries are screened by hybridization (see Topic I2) with a DNA probe that is complementary to part of the nucleotide sequence of the desired gene. The probe may be a DNA restriction fragment or perhaps part of a cDNA clone. Another approach is possible if some of the protein sequence for the desired gene is known. Using the genetic code, one can then deduce the DNA sequence of this part of the gene and synthesize an oligonucleotide with this sequence to act as the DNA probe.

When using a plasmid vector, a typical procedure for screening would be to take agar plates bearing bacterial colonies that make up the genomic library and overlay each plate with a nitrocellulose membrane (*Fig. 2*). This is peeled off and is a **replica** of the plate in that some of the colonies will have adhered to it and in the same pattern as the colonies on the plate. This filter is often called a '**colony lift**'. It is treated with alkali to lyse the bacterial cells and denature the DNA and then hybridized with a radiolabeled DNA probe. After washing away unreacted probe, autoradiography of the filter shows which colonies have hybridized with the probe and thus contain the desired sequences. These are then recovered from the agar plate.

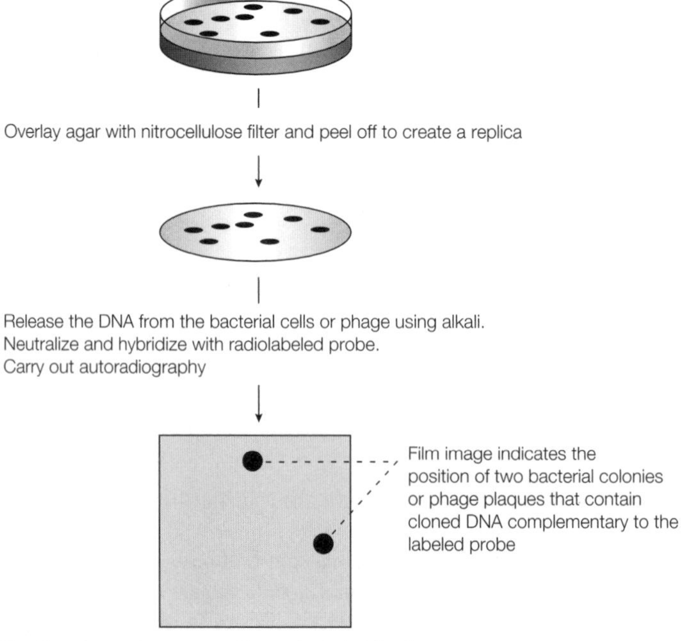


*Fig. 2. Screening a gene library by hybridization.*

When a bacteriophage is used as the cloning vector, the gene library is screened as an array of plaques in a bacterial lawn. A hybridization screening method is used similar to that described for plasmid screening above; in this case the replica filter is called a '**plaque lift**'.

For cDNA libraries, screening can similarly be carried out by hybridization. In addition, it is possible to make the cDNA library using a vector that will actually transcribe the inserted cDNA and then translate the resulting mRNA to form protein corresponding to the cloned gene. A library made with such an **expression vector** is an **expression cDNA library**. It can be screened using a labeled antibody that recognizes the specific protein and hence identifies those bacteria which contain the desired gene and are synthesizing the protein. Not just antibody but any ligand that binds to the target protein can be used as a probe. For example, labeled hormone may be used to identify clones synthesizing hormone receptor proteins.

# I4 VIRUSES

## Key Notes

**Overview**

A virus particle (virion) has a DNA or RNA genome packaged inside a protein capsid. Each virus can replicate only by infecting a limited range of host cells. Viral replication often leads to cell lysis to release the viral progeny but some viruses exit the host cell by budding through the plasma membrane without causing cell death. Viruses that exit this way have an outer lipid–glycoprotein envelope derived from the host cell plasma membrane.

**Bacteriophages**

Bacteriophages adsorb to a bacterial cell surface and inject the phage DNA through the cell wall into the cytosol. In the lytic cycle, this DNA then replicates inside the cell and is packaged within newly synthesized capsids, eventually being released by cell lysis. Some phages (temperate phages) can enter an alternative (lysogenic) cycle whereby the phage DNA becomes integrated into the bacterial genome. The integrated provirus replicates with the bacterial DNA unless the host is exposed to damaging UV light or ionizing radiation, whereupon the provirus enters the lytic cycle.

**Animal viruses**

Permissive cells infected with an animal DNA virus enter a lytic cycle, but in nonpermissive cells an animal virus may become integrated into the nuclear genome or become a plasmid. In either of the latter two states, viral oncogenes may be expressed that can lead to transformation of the host cell, uncontrolled proliferation and possible cancer. In this case the virus is known as a DNA tumor virus. RNA tumor viruses (members of the retrovirus family) become integrated into the host DNA following synthesis of a double-stranded DNA copy using a viral reverse transcriptase and also cause transformation by expressing various oncogenes. The human immunodeficiency virus (HIV) is a retrovirus that kills human helper T cells and so cripples the immune system, causing acquired immune deficiency syndrome (AIDS).

**Related topics**

Microscopy (A3)
Introduction to enzymes (C1)
DNA structure (F1)
RNA structure (G1)

## Overview

An intact virus, called a **virion**, consists of a small nucleic acid genome within a protein coat called a **capsid**. The capsid consists of regular arrays of a few viral proteins. Depending on the virus, the nucleic acid may be single- or double-stranded RNA, single- or double-stranded linear DNA, or single- or double-stranded circular DNA. Viral genomes range in size; some have 300 or so genes but others have far fewer. Each virus can multiply only in a suitable host cell since it does not possess the ability to replicate independently. The viral genome codes for the capsid proteins as well as enzymes that replicate the genome itself. The smaller viruses carry fewer genes and thus rely more heavily on subverting host replication enzymes to carry out replication of their genome. Thus viruses are **mobile genetic elements** that move from cell to cell, replicating as they do so.

Many types of virus exist, each with a limited **host range** in certain plants, animals or bacteria. Bacterial viruses are called **bacteriophages**. Their multiplication typically leads to lysis of the host bacterial cell to release the viral progeny. In contrast, some animal viruses leave the cell by passing through the plasma membrane, a process called **budding**. This does not kill the cell. During exit, the capsid becomes enclosed by a lipid and glycoprotein **envelope** derived from the plasma membrane.

## Bacteriophages

Bacteriophages (or phages) adsorb to the bacterial host cell and then inject the DNA genome into the cell, leaving the protein capsid outside. Two alternative modes of infection may follow; **lytic infection** or **lysogeny** (*Fig. 1*).

Bacteriophage lambda (λ) is a good example for considering phage infection. In lytic infection, the injected linear double-stranded λ DNA first circularizes. It is then transcribed to produce viral proteins needed for viral DNA replication and packaging as well as many molecules of viral capsid proteins. The viral DNA is replicated and the DNA copies are packaged into new phage

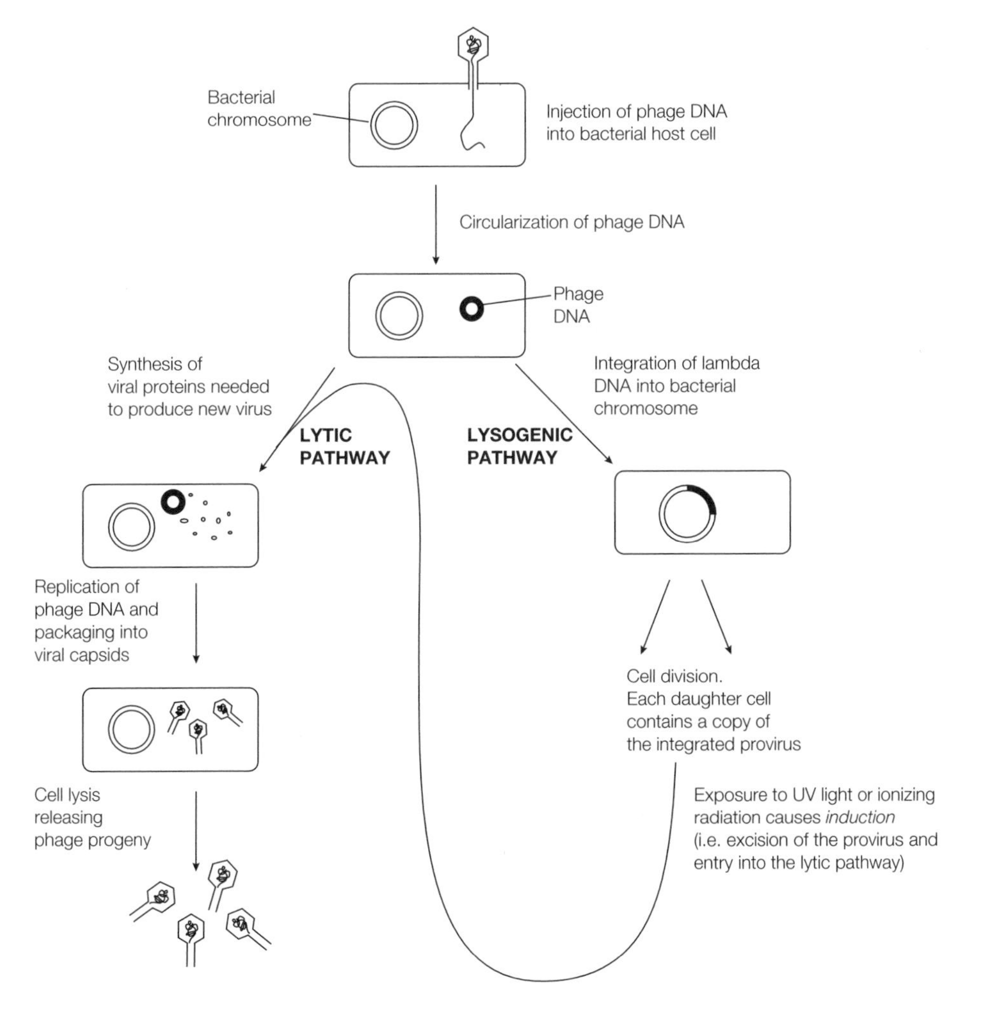


*Fig. 1. Lytic and lysogenic cycles of bacteriophage λ.*

particles. Cell lysis then leads to the release of several hundred viruses. Thus four events characterize lytic infection; adsorption, penetration, replication and release. In the lysogenic mode, after the initial infection by the virus, the circular λ DNA molecule integrates into the circular *E. coli* genome (*Fig. 1*). The integrated viral genome is called a **provirus**. Thus **lysogenic bacteria** carry dormant viral genomes integrated into the host genome. Not all bacteriophages can integrate into the host genome; those that can are called **temperate bacteriophages**. The integrated viral DNA is replicated along with the *E. coli* DNA during cell division unless the bacterium is subjected to an environmental stimulus such as UV light or ionizing radiation. If this occurs then the provirus is induced to excise from the host chromosome and enters the lytic cycle (*Fig. 1*).

## Animal viruses

When **permissive** animal cells are infected with a DNA virus, the virus enters the lytic cycle, multiplies and lyses the cell to release the viral progeny. If a **nonpermissive** animal cell is infected with the virus, however, the virus may not enter a lytic cycle but instead may become integrated into the host genome or may become a plasmid (a circular double-stranded DNA molecule) and replicate at a low level without killing the cell. However, some of the viral genes expressed may force the cell to start dividing uncontrollably, transforming the cell into a state which may lead to the formation of a tumor and hence cancer. Thus the viruses that can achieve this **neoplastic transformation** are called **DNA tumor viruses**. The viral genes that lead to transformation and cell proliferation are called **oncogenes**.

In addition to DNA tumor viruses, **RNA tumor viruses** also exist. These are members of a class of viruses called **retroviruses**, so called because they reverse the normal flow of information from DNA to RNA. Each retrovirus contains two molecules of its RNA genome, each with a cellular tRNA bound to it, and about 50 molecules of an enzyme called **reverse transcriptase**. On infection, the reverse transcriptase copies the infecting viral RNA to make a DNA copy. The DNA copy remains hydrogen-bonded to the RNA as a DNA–RNA hybrid (*Fig. 2*). The reverse transcriptase then degrades the RNA strand of the hybrid and synthesizes a second DNA strand, forming a double-stranded DNA molecule that now integrates into the host genome. This integrated copy acts as a template for the production of many RNA copies, each of which is packaged by viral capsid proteins to form a new retrovirus that escapes from the cell by budding from the plasma membrane. The RNA tumor virus also contains one or more oncogenes that are expressed by the integrated virus and transform the cell. Unlike the DNA tumor virus, however, the oncogenes in an RNA tumor virus are aberrant versions of normal cellular genes that control growth and have been acquired by the virus.

The **human immunodeficiency virus (HIV)** is a retrovirus. It infects (and normally kills) helper T lymphocytes that are a vital part of the immune system (see Topic D1). This greatly impairs the immune system and may lead to **acquired immune deficiency syndrome (AIDS)**. However, the HIV virus can also exist in a dormant state in T cells as a provirus until activated to enter the lytic cycle at a later time. The existence of this dormant state makes it difficult to design an effective strategy to overcome the HIV virus in an infected individual.

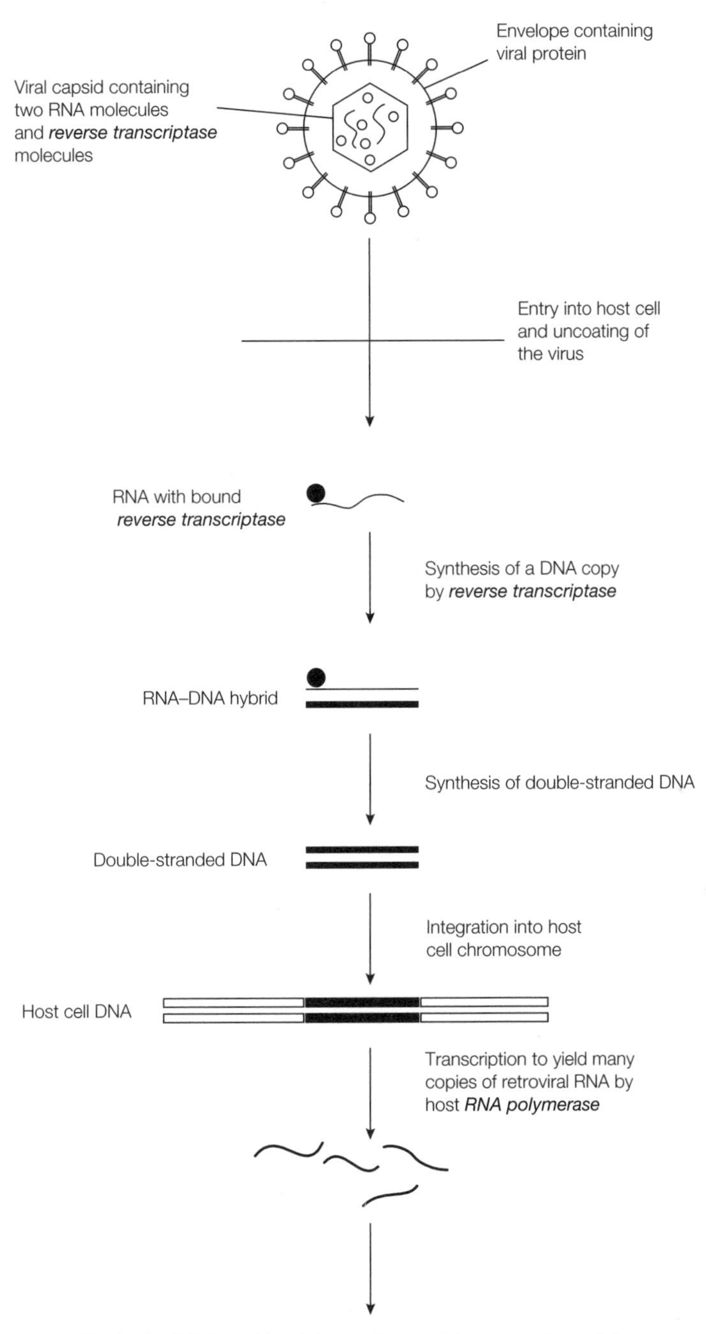


*Fig. 2. Life cycle of a retrovirus.*

# I5 DNA SEQUENCING

## Key Notes

**Two methods for DNA sequencing**

DNA can be sequenced by the chemical method or the chain termination procedure. The latter is now the method of choice in which the (single-stranded) DNA to be sequenced serves as the template for the synthesis of a complementary strand when supplied with a specific primer and *E. coli* DNA polymerase I.

**Chain termination method**

Four incubation mixtures are set up, each containing the DNA template, a specific DNA primer, *E. coli* DNA polymerase I and all four deoxyribonucleoside triphosphates (dNTPs). In addition, each mixture contains a different dideoxynucleoside triphosphate analog, ddATP, ddCTP ddGTP or ddTTP. Incorporation of a dideoxy analog prevents further elongation and so produces a chain termination extension product. The products are electrophoresed on a polyacrylamide gel and the DNA sequence read from the band pattern produced.

**Automated DNA sequencing**

Automated DNA sequencing uses the chain termination method but with an oligonucleotide primer labeled with a fluorescent dye. Each of the four reactions receives a primer labeled with a different dye. After incubation, the reaction mixtures are pooled and electrophoresed on one lane of a polyacrylamide gel. The order in which the different fluorescently labeled termination products elute from the gel gives the DNA sequence.

**Related topics**

DNA structure (F1)
DNA replication in bacteria (F3)
Nucleic acid hybridization (I2)

## Two methods for DNA sequencing

Two main methods have been devised to sequence DNA; the **chemical method** (also called the *Maxam–Gilbert method* after its inventors) and the **chain termination method** (also known as the *Sanger dideoxy method* after its inventor). The chain termination method is now the method usually used because of its speed and simplicity. In this procedure, the DNA to be sequenced is prepared as a single-stranded molecule so that it can act as a template for DNA synthesis (see Topic F3) in the sequencing reaction. ***E. coli* DNA polymerase I** is used to copy this DNA template. However, this enzyme needs a primer to start synthesis (see Topic F3). The primer used can be either a DNA restriction fragment complementary to the single-stranded template or it can be a short sequence of complementary DNA that has been synthesized chemically (a **synthetic oligonucleotide**).

## Chain termination method

An incubation mixture is set up containing the single-stranded DNA template, the primer, DNA polymerase I and all four deoxyribonucleoside triphosphates (dATP, dGTP, dCTP, dTTP), one of which is radioactively labeled, plus a single 2′3′ dideoxyribonucleoside triphosphate analog, say ddGTP. In this incubation, the DNA polymerase begins copying template molecules by extending the bound primer. As the new DNA strand is synthesized, every time that dGTP should be incorporated there is a chance that ddGTP will be incorporated instead. If this happens, no further chain elongation can occur because dideoxy analogs lack the 3′-OH group needed to make the next 3′5′ phosphodiester bond. Thus this particular chain stops at this point. In this first incubation mixture, a large population of templates is being copied and each new strand will stop randomly at positions where a G must be added to the newly synthesized strand. Thus, for every G in the complementary sequence there will be some new DNA strands that have terminated at that point (*Fig. 1*).

In fact, four incubation mixtures are set up, each containing the same components except that each contains a different dideoxy analog; one of ddATP, ddCTP, ddTTP or ddGTP (*Fig. 1*). This produces four sets of chain-terminated fragments corresponding to the positions of A, C, T and G in the sequence. After the incubation, all four reaction mixtures are electrophoresed in parallel lanes of a polyacrylamide gel and then subjected to autoradiography. The DNA sequence is determined simply by reading the band pattern on the autoradiogram from the bottom of the gel towards the top (i.e. reading the DNA sequence as it is synthesized from the primer; *Fig. 1*). The sequence read off the gel is the sequence of the synthesized DNA strand and hence is the complementary sequence to the original DNA template strand.

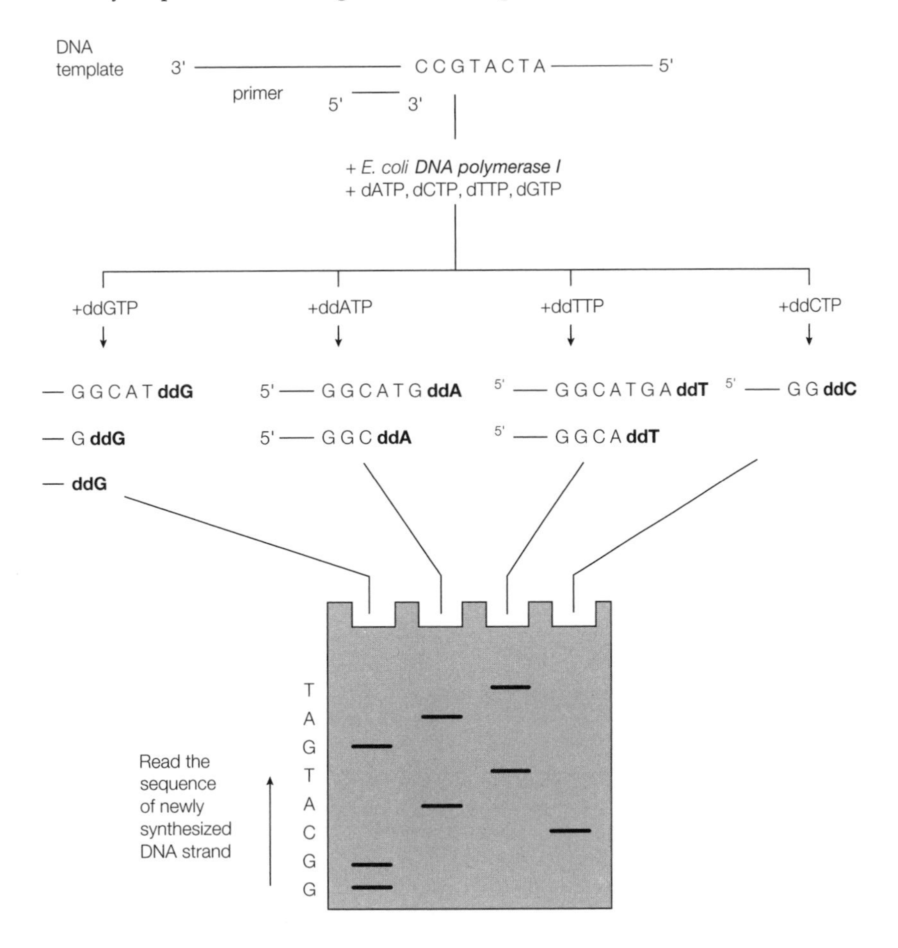


*Fig. 1. DNA sequencing by the chain termination (Sanger) method.*

**Automated DNA sequencing**

Automated DNA sequencing is now commonplace, based on the chain termination method but using a fluorescent dye attached to an oligonucleotide primer instead of using radioactive labeling. A different fluorescent dye is attached to the primer for each of the four sequencing reactions but, after incubation, all four mixtures are combined and electrophoresed on one gel lane. Laser detection systems then distinguish the identity of each termination product as it elutes from the gel. The sequence in which the different fluorescent products elutes from the gel gives the DNA sequence.

# I6 Polymerase chain reaction

## Key Notes

**Principles of PCR**

The polymerase chain reaction (PCR) allows an extremely large number of copies to be synthesized of any given DNA sequence provided that two oligonucleotide primers are available that hybridize to the flanking sequences on the complementary DNA strands. The reaction requires the target DNA, the two primers, all four deoxyribonucleoside triphosphates and a thermostable DNA polymerase such as *Taq* DNA polymerase. A PCR cycle consists of three steps; denaturation, primer annealing and elongation. This cycle is repeated for a set number of times depending on the degree of amplification required.

**Applications of PCR**

PCR has made a huge impact in molecular biology, with many applications in areas such as cloning, sequencing, the creation of specific mutations, medical diagnosis and forensic medicine.

**Related topics**

DNA structure (F1)
DNA replication in bacteria (F3)
Restriction enzymes (I1)
Nucleic acid hybridization (I2)
DNA cloning (I3)
DNA sequencing (I5)

## Principles of PCR

PCR (polymerase chain reaction) is an extremely simple yet immensely powerful technique. It allows enormous amplification of any specific sequence of DNA provided that short sequences either side of it are known. The technique is shown in *Fig. 1*. A PCR reaction contains the target double-stranded DNA, two primers that hybridize to flanking sequences on opposing strands of the target, all four deoxyribonucleoside triphosphates and a DNA polymerase. Because, as we shall see, the reaction periodically gets heated to high temperature, PCR depends upon using a heat-stable DNA polymerase. Many such heat-stable enzymes from thermophilic bacteria (bacteria that live in high temperature surroundings) are now available commercially. The first one used was *Taq* polymerase from the thermophilic bacterium *Thermus aquaticus*.

PCR consists of three steps:

- **Denaturation**. The reaction mixture is heated to 95°C for a short time period (about 15–30 sec) to denature the target DNA into single strands that can act as templates for DNA synthesis.
- **Primer annealing**. The mixture is rapidly cooled to a defined temperature which allows the two primers to bind to the sequences on each of the two strands flanking the target DNA. This **annealing temperature** is calculated carefully to ensure that the primers bind only to the desired DNA sequences. One primer binds to each strand (*Fig. 1*). The two parental strands do not reanneal with each other because the primers are in large excess over parental DNA.

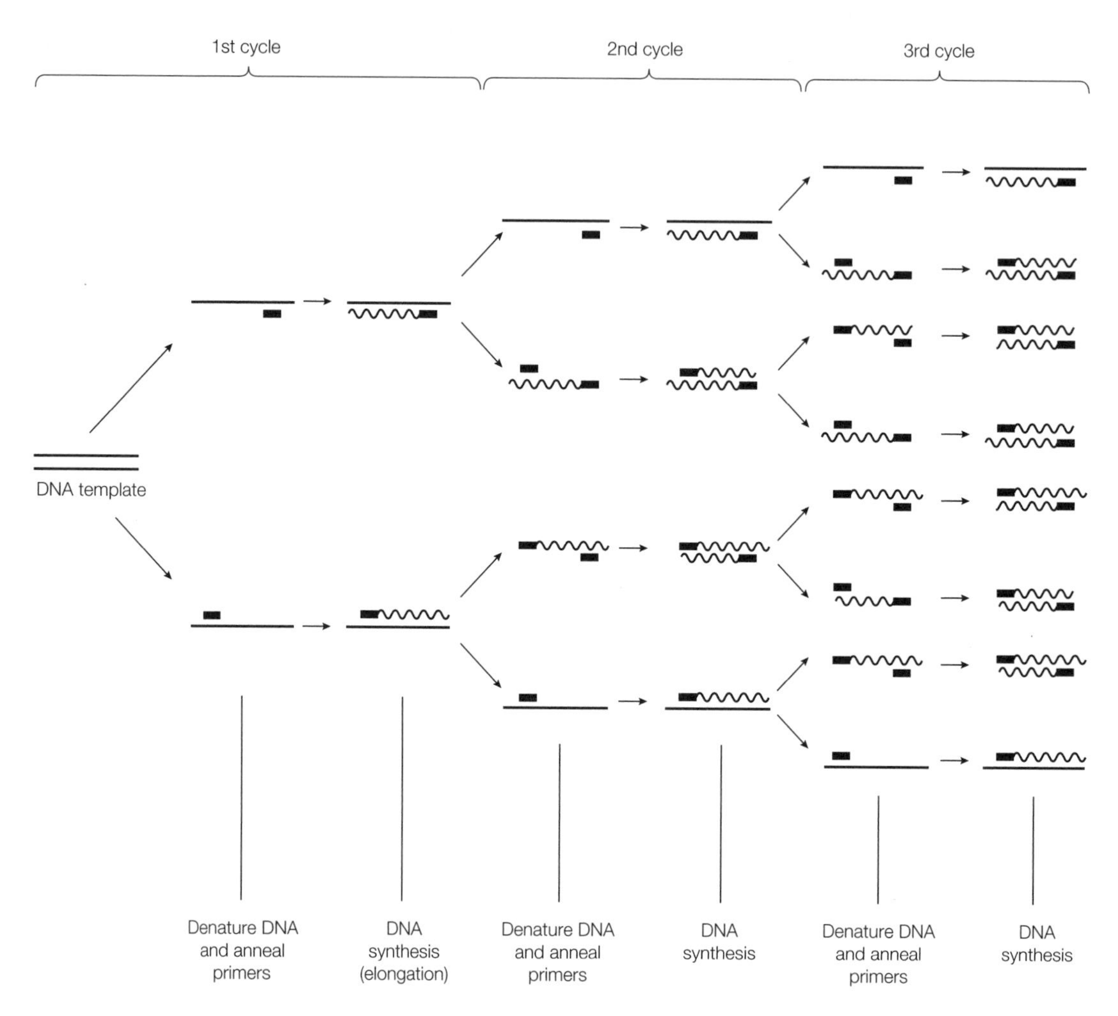


*Fig. 1. The polymerase chain reaction (PCR). Asterisks indicate the first PCR products that arise (in the third cycle) which consist of DNA sequence only between the two primer sites.*

- **Elongation**. The temperature of the mixture is raised to 72°C (usually) and kept at this temperature for a pre-set period of time to allow DNA polymerase to elongate each primer by copying the single-stranded templates. Thus at the end of this incubation, both single-stranded template strands have been made partially double stranded. The new strand of each double-stranded DNA extends for a variable distance downstream.

The three steps of the PCR cycle are repeated. Thus in the second cycle, the four strands denature, bind primers and are extended. No other reactants need to be added. The three steps are repeated once more for a third cycle (*Fig. 1*) and so on for a set number of additional cycles. By the third cycle, some of the PCR products (indicated by asterisks in *Fig. 1*) represent DNA sequence only between the two primer sites and the sequence does not extend beyond these sites. As more and more reaction cycles are carried out, this type of double-stranded DNA molecule becomes the majority species present. After 20 cycles, the original DNA has been amplified a million-fold and this rises to a billion-fold (1000) million after 30 cycles. At this point the vast majority of the products are identical in that the DNA amplified is only that between the two primer sites. Automated **thermocyclers** are now routinely used to cycle the reaction without manual interference so that a billion-fold amplification of the DNA sequence between the two primer sites (30 cycles) can take less than one hour!

## Applications of PCR

PCR already has very widespread applications, and new uses are being devised on a regular basis. Some (and certainly not all) of the applications are as follows:

- PCR can amplify a single DNA molecule from a complex mixture, largely avoiding the need to use DNA cloning to prepare that molecule. Variants of the technique can similarly amplify a specific single RNA molecule from a complex mixture.
- DNA sequencing has been greatly simplified using PCR, and this application is now common.
- By using suitable primers, it is possible to use PCR to create point mutations, deletions and insertions of target DNA which greatly facilitates the analysis of gene expression and function.
- PCR is exquisitely sensitive and can amplify vanishingly small amounts of DNA. Thus, using appropriate primers, very small amounts of specified bacteria and viruses can be detected in tissues, making PCR invaluable for medical diagnosis.
- PCR is now invaluable for characterizing medically important DNA samples. For example, in screening for human genetic diseases, it is rapidly replacing the use of RFLPs (see Topic I1). The PCR screen is based on the analysis of **microsatellites**. These are di-, tri- and tetranucleotide repeats in the DNA of the type $(CA)_n$ or $(CCA)_n$ where $n$ is a number from 10 to more than 30. The microsatellites can be used as markers in the same way that RFLPs were used in the past (see Topic I1). Thus two primers are chosen that bind to the DNA flanking the microsatellite. PCR is then carried out and the different sizes of microsatellite give different sizes of amplified DNA fragments that can then be used as screening markers. The method is very fast, reliable and uses very small amounts of clinical material.
- Because of its extreme sensitivity, PCR is now fundamentally important to forensic medicine. It is even possible to use PCR to amplify the DNA from a single human hair or a microscopic drop of blood left at the scene of a crime to allow detailed characterization.

# J1 MONOSACCHARIDES AND DISACCHARIDES

## Key Notes

**Aldoses and ketoses**

A monosaccharide has the general formula $(CH_2O)_n$ and contains either an aldehyde group (an aldose) or a ketone group (a ketose). The free aldehyde or ketone group can reduce cupric ions ($Cu^{2+}$) to cuprous ions ($Cu^+$) and hence such a monosaccharide is called a reducing sugar.

**Stereoisomers**

The D and L stereoisomers of sugars refer to the configuration of the asymmetric carbon atom furthest from the aldehyde or ketone group. The sugar is said to be a D isomer if the configuration of the atoms bonded to this carbon atom is the same as for the asymmetric carbon in D-glyceraldehyde.

**Ring structures**

Tetroses and larger sugars can cyclize by reaction of the aldehyde or ketone group with a hydroxyl group on another carbon atom of the sugar. Glucose cyclizes to form a six-membered pyranose ring whilst five-carbon sugars and six-carbon ketose sugars (ketohexoses) such as fructose form furanose rings. Two forms (anomers) of D-glucopyranose exist, depending on whether the hydroxyl group attached to the anomeric carbon atom (C-1) lies below the plane of the ring (the α form) or above the plane of the ring (β form). In solution the α and β forms interconvert via the open-chain form (mutarotation). The pyranose ring can exist in either boat or chair conformations but the chair form predominates since the side groups, which are usually OH groups, are less sterically hindered in this conformation.

**Disaccharides**

A disaccharide is formed when two monosaccharides become joined by a glycosidic bond. The bond may be an α- or β-bond depending on the configuration of the anomeric carbon atom involved in the bond. Usually the anomeric carbon atom of only one of the two monosaccharides is involved in the bond so that the disaccharide still has one free aldehyde or ketone group and is still reducing. However, in sucrose both anomeric carbon atoms are bonded together so that sucrose is a nonreducing disaccharide.

**Sugar derivatives**

The hydroxyl groups of sugars can be replaced by other groups to form a wide range of biologically important molecules including phosphorylated sugars, amino sugars and nucleotides.

**Nomenclature**

The names of simple sugars and sugar derivatives can all be abbreviated. This also allows an abbreviated description of the component sugars present in disaccharides, for example.

**Related topics**

Protein glycosylation (H5)
Polysaccharides and oligosaccharides (J2)
Glycolysis (J3)
Gluconeogenesis (J4)
Pentose phosphate pathway (J5)

## Aldoses and ketoses

A carbohydrate is composed of carbon (*carbo-*), and hydrogen and oxygen (*-hydrate*). The simplest carbohydrates are the **monosaccharides** that have the general formula $(CH_2O)_n$ where $n$ is 3 or more. A monosaccharide or simple **sugar**, consists of a carbon chain with a number of hydroxyl (OH) groups and either one **aldehyde group** ( $-C\overset{H}{\underset{O}{\lessgtr}}$ often written as –CHO) or one **ketone group** ( $>C=O$ ). A sugar that bears an aldehyde group is called an **aldose** whereas a sugar with a ketone group is a **ketose**. The smallest carbohydrates, for which $n = 3$, are called **trioses**. The terms can be combined. Thus glyceraldehyde (*Fig. 1*) is a triose that has an aldehyde group and so is an aldose. Thus it can also be called an **aldotriose**. Similarly, dihydroxyacetone (*Fig. 1*) is a **ketotriose**.

O=C–H
|
CHOH
|
$CH_2OH$

Glyceraldehyde
(an aldose)
$C_3H_6O_3$

$CH_2OH$
|
C=O
|
$CH_2OH$

Dihydroxyacetone
(a ketose)
$C_3H_6O_3$

*Fig. 1. Structures of glyceraldehyde and dihydroxyacetone.*

Sugars that contain a free aldehyde or ketone group in the open-chain configuration can reduce cupric ions ($Cu^{2+}$) to cuprous ions ($Cu^+$) and hence are called **reducing sugars**. This is the basis of the Fehling's and Benedict's tests for reducing sugars. The **reducing end** of such a sugar chain is thus the end that bears the aldehyde or ketone group.

Note that glyceraldehyde and dihydroxyacetone have the same chemical composition, $C_3H_6O_3$, but differ in structure (i.e. they are **structural isomers**).

## Stereoisomers

Glyceraldehyde has a single asymmetric carbon atom (the central one) and so two **stereoisomers** (also called **optical isomers**) are possible, that is two forms of glyceraldehyde, denoted as D- and **L-glyceraldehyde**, which are mirror images of each other (*Fig. 2*). Stereoisomers also exist for amino acids (see Topic B1).

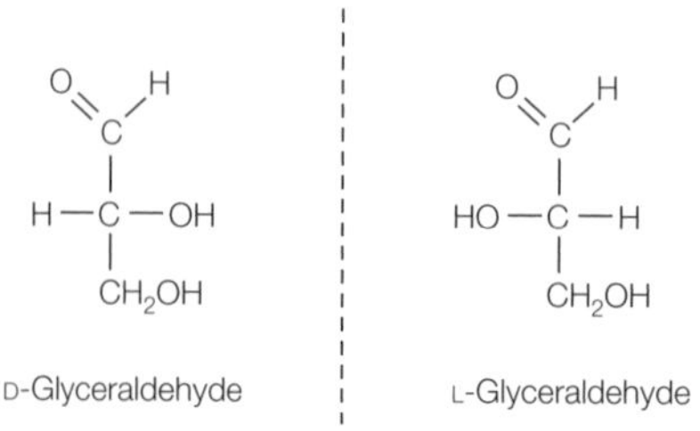

O=C–H
|
H—C—OH
|
$CH_2OH$

D-Glyceraldehyde

O=C–H
|
HO—C—H
|
$CH_2OH$

L-Glyceraldehyde

Mirror plane

*Fig. 2. D- and L-glyceraldehyde are mirror images of each other (stereoisomers or optical isomers).*

D-Glucose (an aldose) L-Glucose D-Fructose (a ketose) L-Fructose

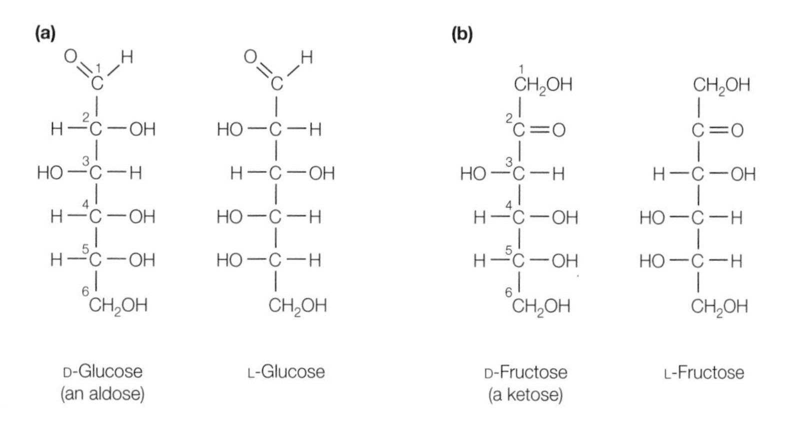

*Fig. 3. (a) D- and L-glucose; (b) D- and L-fructose.*

Sugars with four, five, six or seven carbons are called **tetroses, pentoses, hexoses** and **heptoses** respectively. In these cases the sugars may have more than one asymmetric carbon atom. The convention for numbering carbon atoms and naming configurations is as follows:

- the carbon atoms are numbered from the end of the carbon chain starting with the aldehyde or ketone group, which is carbon 1 (C-1);
- the symbols D and L refer to the configuration of the asymmetric carbon atom furthest from the aldehyde or ketone group.

Thus, for example, glucose, an aldohexose, exists as D and L forms (*Fig. 3a*). The furthest asymmetric carbon from the aldehyde group is C-5. D-Glucose (*Fig. 3a*) is called D because the configuration of the atoms bonded to C-5 is the same as for D-glyceraldehyde (*Fig. 2*). Similarly D-fructose (a ketohexose; *Fig. 3b*) is designated D because the configuration at C-5 matches that for D-glyceraldehyde. D sugars that differ in configuration at only a single asymmetric carbon atom are called **epimers**. Thus D-glucose and D-galactose are epimers, differing only in their configuration at C-4 (*Fig. 4*).

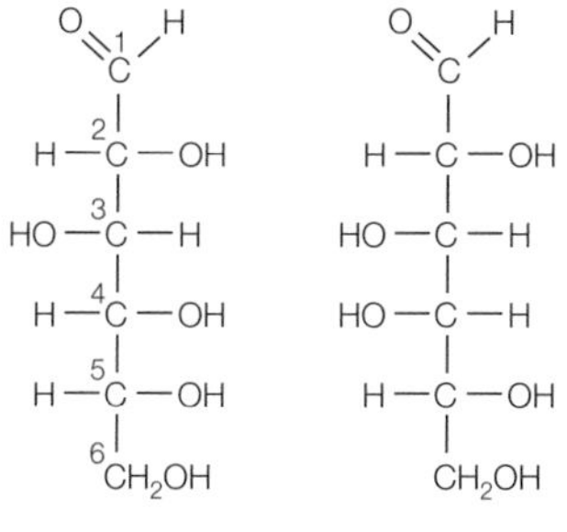

D-Glucose D-Galactose

*Fig. 4. The epimers D-glucose and D-galactose.*

## Ring structures

The aldehyde or ketone group can react with a hydroxyl group to form a covalent bond. Formally, the reaction between an aldehyde and the hydroxyl group of a sugar (an alcohol) creates a **hemiacetal** (Equation 1) whereas a ketone reacts with a hydroxyl group (alcohol) to form a **hemiketal** (Equation 2).

$$\underset{\text{Alcohol}}{R{-}OH} + \underset{\text{Aldehyde}}{R'{-}C(=O){-}H} \rightleftharpoons \underset{\text{Hemiacetal}}{R{-}O{-}C(OH)(R')(H)} \quad (1)$$

$$\underset{\text{Alcohol}}{R{-}OH} + \underset{\text{Ketone}}{R'{-}C(=O){-}R''} \rightleftharpoons \underset{\text{Hemiketal}}{R{-}O{-}C(OH)(R')(R'')} \quad (2)$$

For tetroses and larger sugars, the reaction can take place within the same molecule so that the straight-chain form of the sugar cyclizes. For example, *Fig. 5* shows the cyclization of D-glucose to form a six-carbon ring. The ring structures shown in *Fig. 5* are called **Haworth projections** in which the plane of the ring can be imagined as approximately perpendicular to the plane of the paper with the thick lines of the ring in the diagram pointing towards the reader.

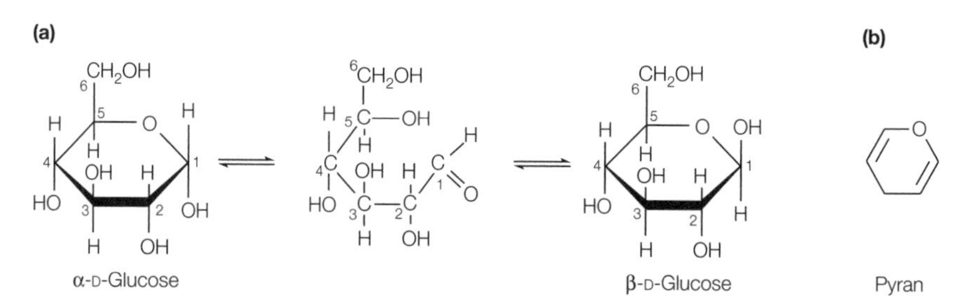


*Fig. 5. (a) Cyclization of the open-chain form of D-glucose; (b) the structure of pyran.*

Note that a new asymmetric center is formed during cyclization, at C-1. Thus **two isomers** of D-glucose exist (*Fig. 5a*), α-D-glucose (in which the OH group at C-1 lies below the plane of the ring) and β-D-glucose (in which the OH group at C-1 lies above the plane of the ring). The C-1 carbon is called the **anomeric carbon atom** and so the α and β forms are called **anomers**. In aqueous solution, the α and β forms rapidly interconvert via the open-chain structure, to give an equilibrium mixture (*Fig. 5a*). This process is called **mutarotation**. Because of its structural similarity to the ring compound called pyran (*Fig. 5b*), the six-membered ring structures of hexoses such as glucose are called **pyranoses**. Thus β-D-glucose may also be written as β-D-glucopyranose.

Five-carbon sugars, such as D-ribose (Topic G1) and D-deoxyribose (Topic F1), and six-carbon ketose sugars (ketohexoses), such as D-fructose, form rings called **furanoses** (*Fig. 6a*) by comparison with the compound furan (*Fig. 6b*). Again furanoses can exists in both α and β forms (*Fig. 6a*) except here the nomenclature refers to the hydroxyl group attached to C-2 which is the anomeric carbon atom.

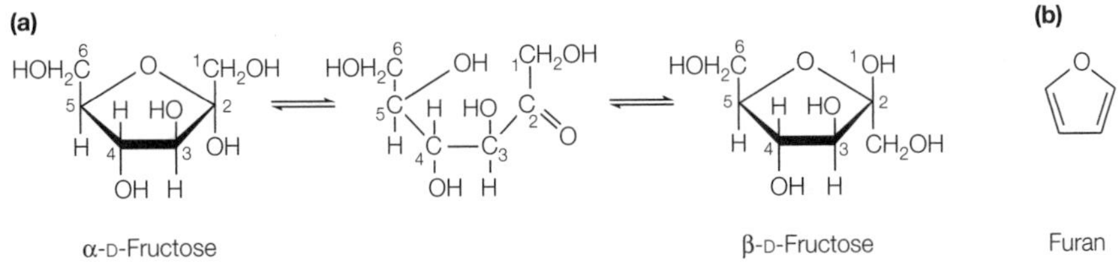

*Fig. 6.* *(a) Cyclization of the open-chain form of D-fructose; (b) the structure of furan.*

The pyranose ring of a six-carbon aldose sugar can exist in either a **boat** or a **chair** configuration (*Fig. 7*). The substituents attached to the ring carbons that extend parallel to the symmetry axis are said to be axial (a) whilst those that extend outward from this axis are said to be equatorial (e) (*Fig. 7*). In the boat form, there is considerable steric hindrance between the various groups attached to the carbon atoms of the ring and therefore this form is less favorable energetically. Hence the chair form predominates, as shown for β-D-glucose in *Fig. 7*, where all the axial positions are occupied by hydrogen atoms.

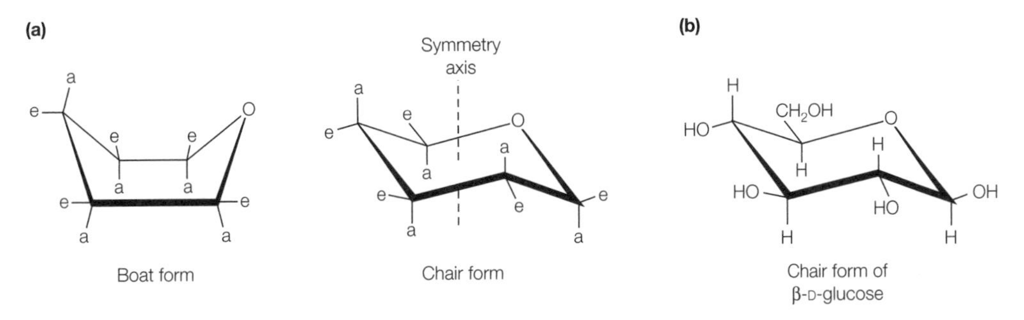


*Fig. 7.* *(a) Chair and boat conformations of pyranose rings; (b) stable chair form of β-D-glucose.*

## Disaccharides

The aldehyde or ketone group on the anomeric carbon atom of one monosaccharide can react with the hydroxyl group of a second monosaccharide to form a **disaccharide**. The covalent bond formed is called a **glycosidic bond**. Lactose (*Fig. 8a*) is a disaccharide formed between the anomeric carbon (C-1) of D-galactose and C-4 of D-glucose. Since the anomeric carbon of the galactose molecule is involved in the bond and is in the β configuration, this is called a **β(1→4) bond** which can be abbreviated as β1–4. Maltose (*Fig. 8b*) is a disaccharide formed between the C-1 and C-4 positions of two glucose units. However, here the configuration of the anomeric carbon atom involved is the α form and hence the bond is called an **α(1→4) bond** or abbreviated as α1–4. For lactose and maltose, one of the anomeric carbons has been used to form the bond, leaving the second anomeric carbon free. Thus both lactose and maltose have a **reducing end**. In contrast, sucrose (*Fig 8c*) is a disaccharide formed by bond formation between the anomeric C-1 of glucose and the anomeric C-2 of fructose so that sucrose lacks a free reducing group.

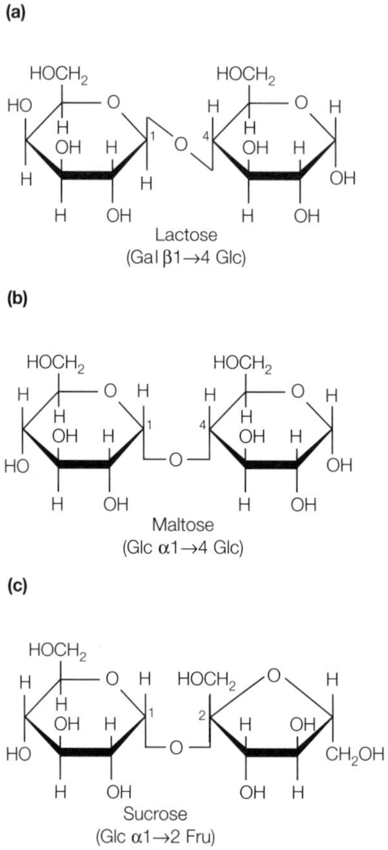

(a)

Lactose
(Gal β1→4 Glc)

(b)

Maltose
(Glc α1→4 Glc)

(c)

Sucrose
(Glc α1→2 Fru)

*Fig. 8. Structure of common disaccharides.*

## Sugar derivatives

The hydroxyl groups of simple monosaccharides can be replaced with other groups to form a range of sugar derivatives. For example, **phosphorylated sugars** such as glucose 6-phosphate (*Fig. 9*) are important metabolites in glycolysis (Topic J3). In **amino sugars**, one or more hydroxyl groups are each replaced

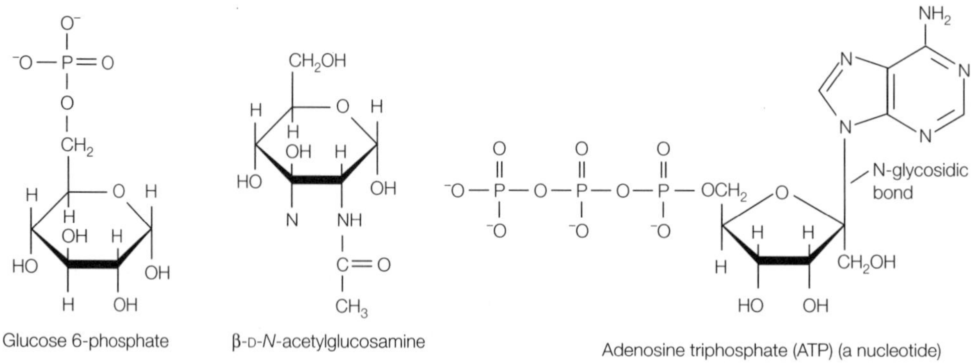

Glucose 6-phosphate

β-D-*N*-acetylglucosamine

Adenosine triphosphate (ATP) (a nucleotide)

N-glycosidic bond

*Fig. 9. Examples of sugar derivatives.*

by an amino group (which is often acetylated), for example acetyl β-D-*N*-acetyl-glucosamine (*Fig. 9*). This and other sugar derivatives are common components of many glycoproteins. Nucleotides (see Topic F1), such as ATP, each consist of a sugar in which the anomeric carbon atom has formed a covalent bond with a nitrogenous base (*Fig. 9*). Since the bond is between the anomeric carbon of the sugar and a nitrogen atom of the base, it is called an **N-glycosidic bond**.

**Nomenclature**

Simple sugars have three-letter abbreviations [e.g. Glc (glucose), Gal (galactose), Man (mannose), Fuc (fucose)]. Sugar derivatives can also be abbreviated, such as GlcNAc (*N*-acetylglucosamine), GalNAc (*N*-acetylgalactosamine). This also allows an abbreviated form of description for sugars that are bonded together and the nature of the covalent bonds. Thus, for example, lactose (*Fig. 8a*) can be represented as Galβ1→4Glc.

# J2 POLYSACCHARIDES AND OLIGOSACCHARIDES

## Key Notes

**Polysaccharides**

Long chains of monosaccharides joined together are collectively called polysaccharides. The major storage polysaccharides are glycogen (in animals), starch (in plants) and dextran (in yeast and bacteria). Cellulose is a structural polysaccharide found in plant cell walls.

**Glycogen**

Glycogen is a branched-chain polysaccharide containing glucose residues linked by α1–4 bonds with α1–6 branchpoints. The branched nature of glycogen makes it more accessible to glycogen phosphorylase during degradation, since this enzyme degrades the molecule by sequential removal of glucose residues from the nonreducing ends.

**Starch**

Starch is a mixture of unbranched amylose (glucose residues joined by α1–4 bonds) and branched amylopectin (glucose residues joined α1–4 but with some α1–6 branchpoints).

**Dextran**

Dextran consists of glucose residues linked mainly by α1–6 bonds but with occasional branchpoints that may be formed by α1–2, α1–3 or α1–4 bonds.

**Cellulose**

Cellulose is a straight-chain polymer of glucose units linked by β1–4 bonds. The polysaccharide chains are aligned to form fibrils that have great tensile strength. Cellulases, enzymes that degrade cellulose, are absent in mammals but are produced by some bacteria, fungi and protozoa.

**Oligosaccharides**

Short chains of monosaccharides linked by glycosidic bonds are called oligosaccharides. Oligosaccharides found in glycoproteins are either linked to a serine or threonine residue (O-linked oligosaccharide) or to an asparagine residue (N-linked oligosaccharide). All N-linked oligosaccharides have a common pentasaccharide core. High-mannose N-linked oligosaccharides have additional mannose residues linked to the core whilst the complex type N-linked oligosaccharides have branches comprising combinations of GlcNAc, Gal, sialic acid and L-fucose.

**Related topics**

Protein glycosylation (H5)
Monosaccharides and disaccharides (J1)
Glycogen metabolism (J6)

## Polysaccharides

Polysaccharides are large chains of sugar units joined together. Depending on the polysaccharide, the chains may be linear or branched. In animals, excess glucose is stored as a large branched polysaccharide called **glycogen** whereas in most plants the storage form of glucose is the polysaccharide called **starch**. Bacteria and yeasts store glucose as yet another type of polysaccharide called **dextran.** In each case these are nutritional reserves; when required, they are broken down and the monosaccharide products are metabolized to yield energy (see Topic J3). In contrast, **cellulose** is a structural polysaccharide used to make plant cell walls.

## Glycogen

A glycogen molecule consists entirely of glucose units, most of which are linked in long chains by α1–4 bonds. However, every 10 units or so, the chain is branched by the formation of an α1–6 glycosidic bond (*Fig. 1*). Each straight-chain segment of glycogen forms an open helix conformation which increases its accessibility to the enzymes of glycogen metabolism. Each chain terminates in a nonreducing end, that is an end with a free 4′ OH group. Since the enzyme that degrades glycogen (**glycogen phosphorylase**; see Topic J6) catalyzes the sequential removal of glycosyl units from the nonreducing end of a glycogen chain, the numerous branches, each with a nonreducing end, greatly increase the accessibility of the polysaccharide to degradation. This allows rapid mobilization of stored glycogen in times of need. The α1–6 branches themselves are removed by a **debranching enzyme** (see Topic J6).

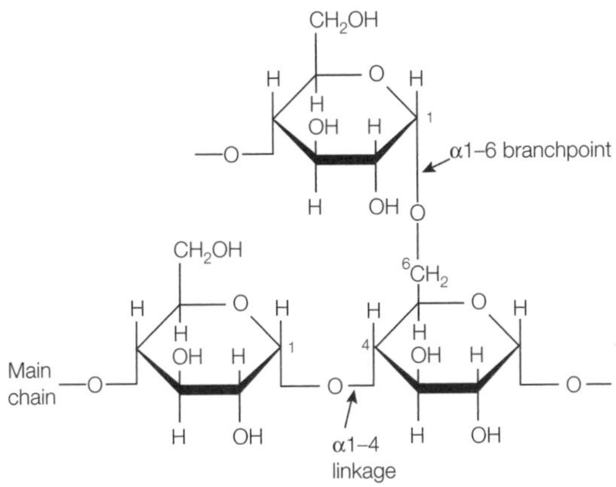

*Fig. 1. The α1–4 linkages in the straight chain and α1–6 branchpoint linkages in glycogen.*

## Starch

Starch exists in plants as insoluble starch granules in the cytoplasm. Each starch granule contains a mixture of two polysaccharide forms, **amylose** and **amylopectin**. Amylose is an unbranched polymer of glucose residues joined in α1–4 linkages. Amylopectin is the branched form; most of the constituent glucose residues are joined in α1–4 linkages but additional α1–6 bonds occur every 25–30 residues, creating the branchpoints.

## Dextran

Dextran is a glucose polymer where the glucose residues are mainly linked by α1–6 bonds. However, a few branches also occur. These are typically formed by α1–2, α1–3 or α1–4 bonds depending on the bacterial or yeast species that is the source of the dextran.

## Cellulose

Cellulose is an unbranched polysaccharide of glucose units linked by β1–4 bonds (*Fig. 2*). Unlike glycogen where the α1–4 linkages lead to a helical conformation of the polysaccharide, the β linkage between glucose residues in cellulose creates very long straight chains that are arranged together in fibrils.

In plant cell walls the cellulose fibrils are embedded in (and cross-linked to) a matrix of other polysaccharides. In wood, this matrix also contains **lignin**, a complex polymer of phenolic residues (see Topic A2). This composite has a very high tensile strength. Mammals, including humans, lack enzymes capable of digesting the β1–4 linkages of cellulose and so cannot digest plant cell walls. However, some bacteria produce **cellulases**, enzymes that degrade cellulose. Ruminant animals such as cattle have cellulase-producing bacteria in their digestive tracts and so can digest cellulose. In addition, some fungi and protozoa produce and secrete cellulase.

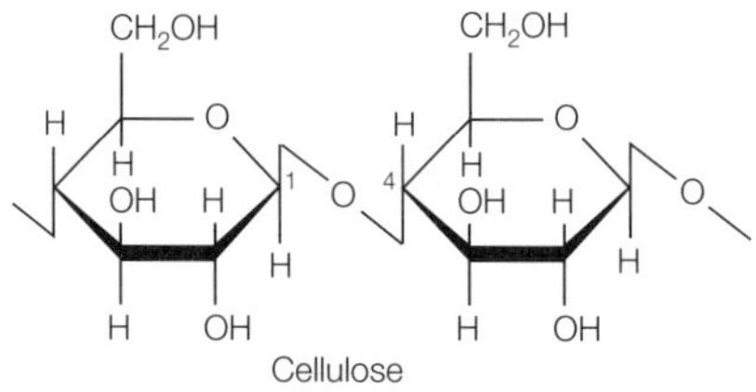

Cellulose

*Fig. 2. The repeating unit of cellulose showing the β1–4 linkage.*

## Oligosaccharides

**Oligosaccharides** are short chains of monosaccharides linked together by glycosidic bonds. In the case of oligosaccharides linked to proteins (glycoproteins) or lipids (glycolipids), the oligosaccharide is not a repeating unit but consists of a range of different monosaccharides joined by a variety of types of bonds. In glycoproteins, two main types of oligosaccharide linkages exist:

- **O-linked oligosaccharides** attached to the protein via **O-glycosidic bonds**, to the OH groups of serine or threonine side-chains (see Topic H5, *Fig. 3*).
- **N-linked oligosaccharides** attached to the protein via **N-glycosidic bonds**, to the $NH_2$ groups of asparagine side-chains (see Topic H5, *Fig. 3*). All N-linked oligosaccharides have a common pentasaccharide core of two GlcNAc and three Man residues but the nature of the side-chains differs (*Fig. 3*). In the **high mannose type** of N-linked oligosaccharide, typically two to six additional Man residues are joined to the pentasaccharide core (e.g. *Fig. 3a*). The **complex type** of N-linked oligosaccharide contain two to five outer branches attached to the Man of the polysaccharide core; these branches contain different combinations of GlcNAc, Gal, sialic acid (*N*-acetylneuraminic acid), mannose and L-fucose. *Fig. 3b* shows a complex oligosaccharide with two outer branches.

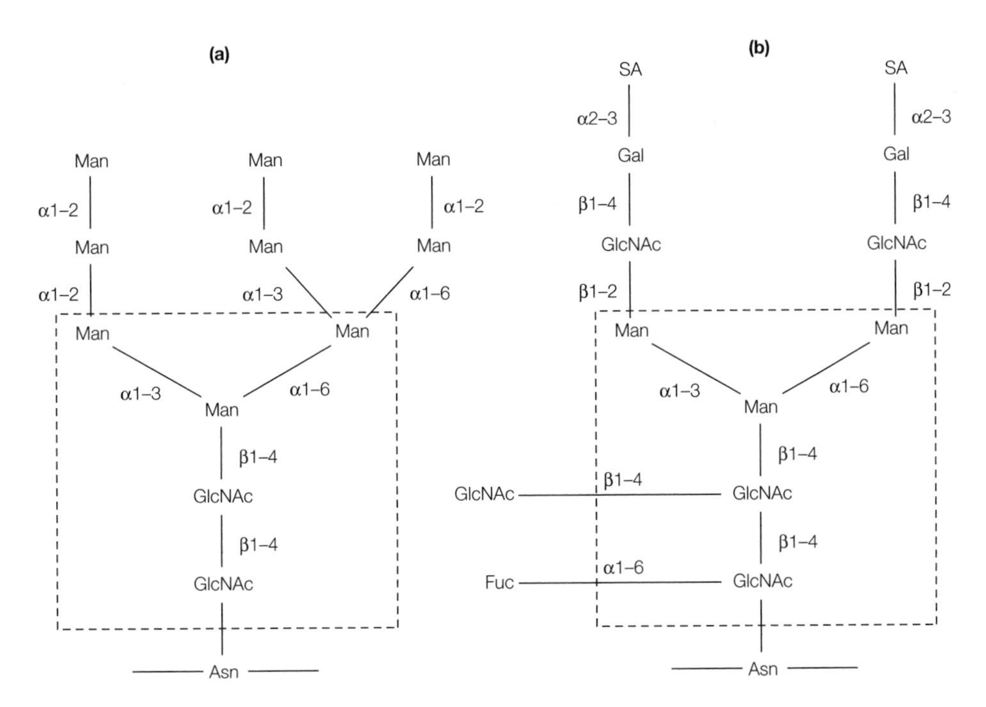


*Fig. 3. Examples of (a) high mannose type and (b) complex type oligosaccharides. In each case, the sugars that comprise the common pentasaccharide core are boxed. SA, sialic acid.*

# J3 GLYCOLYSIS

## Key Notes

**Overview**

Glycolysis is a set of reactions that take place in the cytoplasm of prokaryotes and eukaryotes. The roles of glycolysis are to produce energy (both directly and by supplying substrate for the citric acid cycle and oxidative phosphorylation) and to produce intermediates for biosynthetic pathways.

**The pathway**

Glucose is phosphorylated to glucose 6-phosphate (by hexokinase) which is converted to fructose 6-phosphate (by phosphoglucoisomerase) and then to fructose 1,6-bisphosphate (by phosphofructokinase, PFK). The fructose 1,6-bisphosphate is split into glyceraldehyde 3-phosphate and dihydroxyacetone phosphate (by aldolase) and these two trioses are interconverted by triose phosphate isomerase. Glyceraldehyde 3-phosphate is converted to 1,3-bisphosphoglycerate (by glyceraldehyde 3-phosphate dehydrogenase) which reacts with ADP to give 3-phosphoglycerate and ATP (catalyzed by phosphoglycerate kinase). The 3-phosphoglycerate is converted to 2-phosphoglycerate (by phosphoglycerate mutase) and then to phosphoenolpyruvate (PEP) by enolase. Finally, PEP and ADP react to form pyruvate and ATP (catalyzed by pyruvate kinase).

**Fates of pyruvate**

Under aerobic conditions, pyruvate can be converted by pyruvate dehydrogenase to acetyl coenzyme A (CoA) which can then enter the citric acid cycle. Under anaerobic conditions, pyruvate is converted to lactate by lactate dehydrogenase (LDH). The $NAD^+$ regenerated by this reaction allows glycolysis to continue, despite the lack of oxygen. When oxygen becomes available, the lactate is converted back to pyruvate. In anaerobic conditions, yeast and other organisms carry out alcoholic fermentation that converts pyruvate to acetaldehyde and then to ethanol, regenerating $NAD^+$ that allows glycolysis to continue.

**Energy yield**

Two ATPs are used in glycolysis and four ATPs are synthesized for each molecule of glucose so that the net yield is two ATPs per glucose. Under aerobic conditions, the two NADH molecules arising from glycolysis also yield energy via oxidative phosphorylation.

**Metabolism of fructose**

Fructose can be metabolized by two routes. In adipose tissue and muscle, hexokinase can phosphorylate fructose to fructose 6-phosphate that then enters glycolysis. In liver, most of the enzyme present is glucokinase not hexokinase and this does not phosphorylate fructose. In this tissue, fructose is metabolized instead by the fructose 1-phosphate pathway.

**Metabolism of galactose**

Galactose enters glycolysis via the galactose–glucose interconversion pathway, a four-step reaction sequence. The lack of the second enzyme in this pathway, galactose 1-phosphate uridylyl transferase, leads to the disease galactosemia through the accumulation of toxic products, including galactitol formed by the reduction of galactose.

**Regulation of glycolysis**

The main control step is that catalyzed by PFK but hexokinase and pyruvate kinase are additional control sites. PFK is allosterically inhibited by ATP, but this inhibition is relieved by AMP. Citrate also inhibits PFK. The build up of fructose 6-phosphate stimulates the formation of fructose 2,6-bisphosphate that in turn stimulates PFK. The enzyme that synthesizes fructose 2,6-bisphosphate (phosphofructokinase 2; PFK2) and the enzyme that hydrolyzes it back to fructose 6-phosphate (fructose bisphosphatase 2; FBPase2) are also regulated hormonally by glucagon that causes glycolysis to slow down when the blood glucose level falls. PFK is also inhibited by $H^+$ ions, thus preventing excessive formation of lactate under anaerobic conditions. Hexokinase is inhibited by glucose 6-phosphate which builds up after PFK is inhibited. Pyruvate kinase is activated by fructose 1,6-bisphosphate but allosterically inhibited by ATP and alanine. Like PFK, it is also regulated hormonally by glucagon.

**Related topics**

Monosaccharides and disaccharides (J1)
Polysaccharides and oligosaccharides (J2)
Gluconeogenesis (J4)
Pentose phosphate pathway (J5)
Glycogen metabolism (J6)
Citric acid cycle (L1)

## Overview

Glycolysis is a series of reactions (*Fig. 1*) that takes place in the cytoplasm of all prokaryotes and eukaryotes. Glycolysis converts one molecule of glucose into two molecules of pyruvate [which are then converted to acetyl coenzyme A (CoA) ready for entry into the citric acid cycle]. Two ATP molecules are needed for early reactions in the glycolytic pathway but four ATPs are generated later, giving a net yield of two ATPs per molecule of glucose degraded. Overall, glycolysis has a dual role. The first is to generate ATP. Although only two ATPs per glucose are made directly from the reactions of the glycolytic pathway, it also feeds substrates into the citric acid cycle and oxidative phosphorylation, where most ATP is made. The second role is to produce intermediates that act as precursors for a number of biosynthetic pathways. Thus acetyl CoA, for example, is the precursor for fatty acid synthesis (see Topic K3).

## The pathway

The individual steps in glycolysis are described below.

1. Glucose is phosphorylated by ATP to form glucose 6-phosphate and ADP. The reaction is catalyzed by the enzyme **hexokinase**.

$$\text{Glucose} + \text{ATP} \xrightarrow{\textit{Hexokinase}} \text{Glucose 6-phosphate} + \text{ADP} + H^+$$

CH$_2$OH, H, O, H, H, OH, H, HO, OH, H, OH — **Glucose**

CH$_2$OPO$_3^{2-}$, H, O, H, H, OH, H, HO, OH, H, OH — **Glucose 6-phosphate**

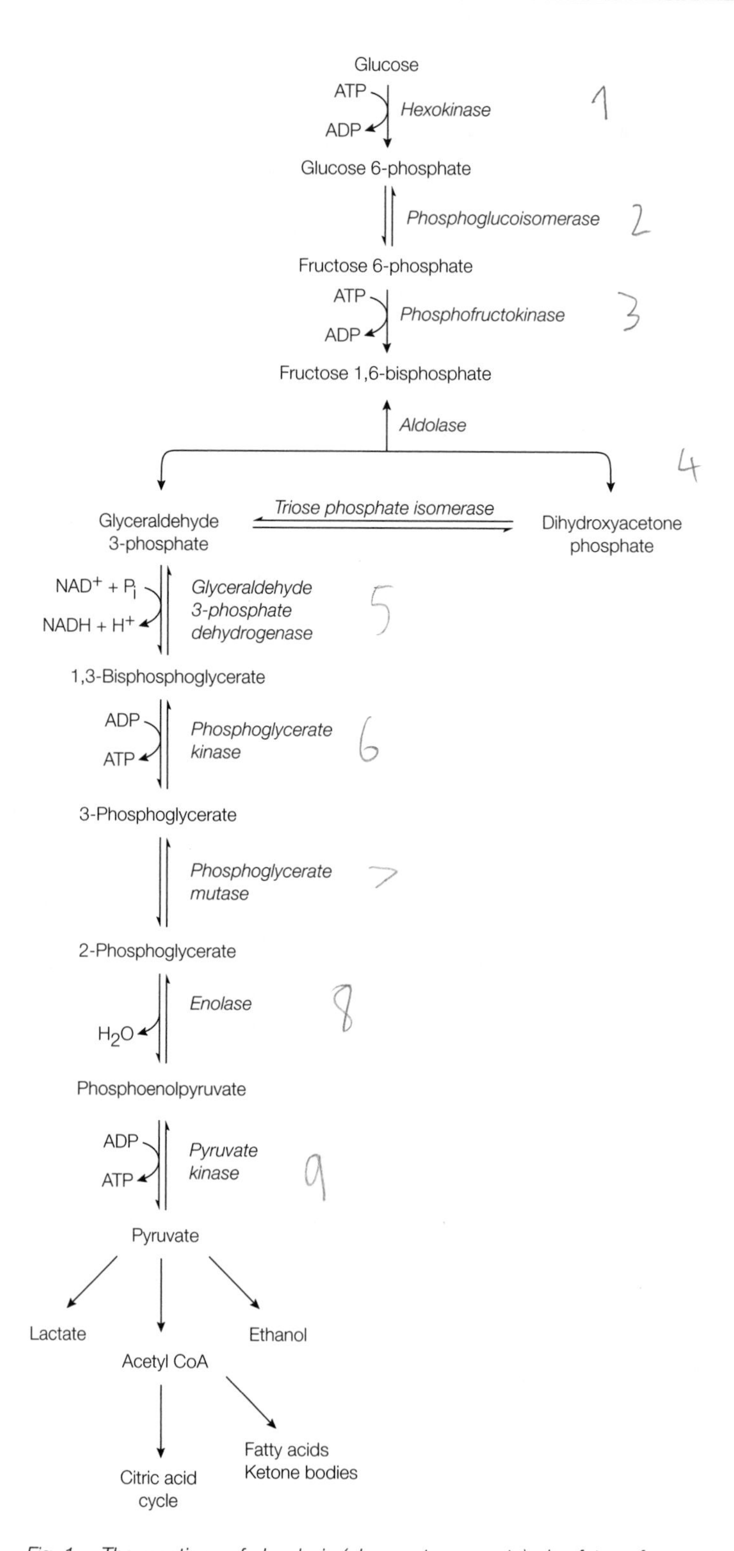


*Fig. 1. The reactions of glycolysis (glucose to pyruvate) plus fates of pyruvate.*

2. Glucose 6-phosphate is converted to fructose 6-phosphate by **phosphoglucoisomerase**.

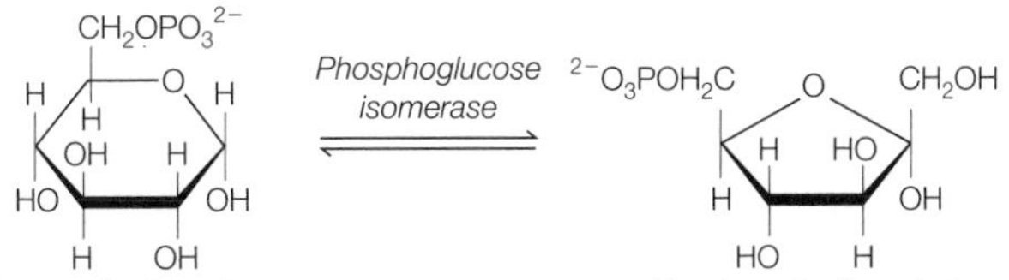

*Phosphoglucose isomerase*

**Glucose 6-phosphate** ⇌ **Fructose 6-phosphate**

This isomerization involves the conversion of an aldose to a ketose, a conversion that is better seen by viewing the open chain representations of these molecules.

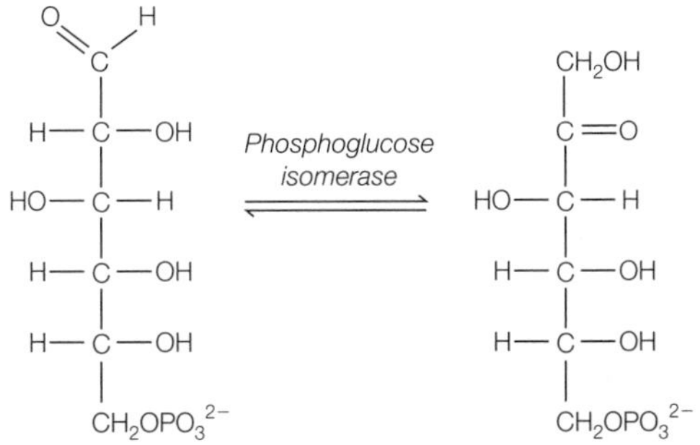

*Phosphoglucose isomerase*

**Glucose 6-phosphate** (An aldose) ⇌ **Fructose 6-phosphate** (A ketose)

3. Fructose 6-phosphate is phosphorylated by ATP to form fructose 1,6-bisphosphate and ADP. The enzyme catalyzing this step is **phosphofructokinase** (PFK).

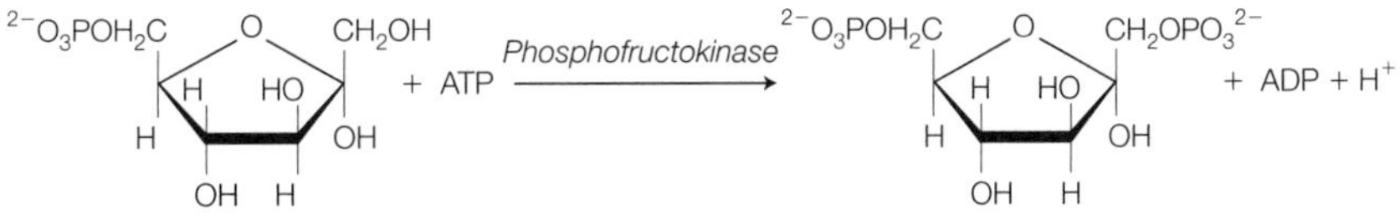

**Fructose 6-phosphate** + ATP → (*Phosphofructokinase*) **Fructose 1,6-bisphosphate** + ADP + $H^+$

4. **Aldolase** splits fructose 1,6-bisphosphate (a six-carbon molecule) into two three-carbon molecules, glyceraldehyde 3-phosphate and dihydroxyacetone phosphate.

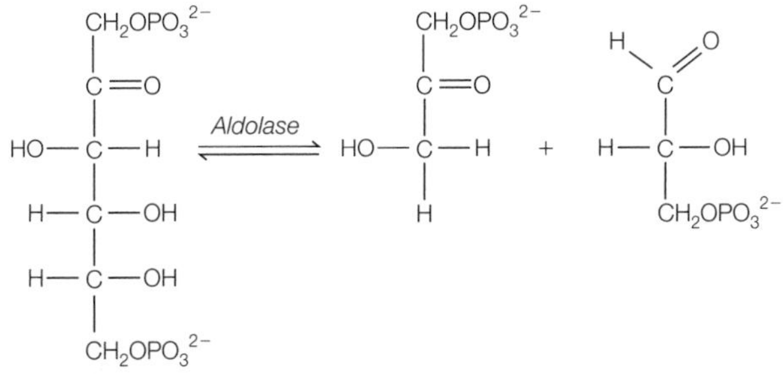

**Fructose 1,6-bisphosphate** ⇌ (*Aldolase*) **Dihydroxyacetone phosphate** + **Glyceraldehyde 3-phosphate**

5. Glyceraldehyde 3-phosphate is the only molecule that can be used for the rest of glycolysis. However, the dihydroxyacetone phosphate formed in the previous step can rapidly be converted to glyceraldehyde 3-phosphate by **triose phosphate isomerase**. This is an equilibrium reaction; as the glyceraldehyde 3-phosphate is used by the rest of glycolysis, more dihydroxyacetone phosphate is converted to glyceraldehyde 3-phosphate as replacement. Thus effectively, for each molecule of fructose 1,6-bisphosphate that is cleaved in step 4, two molecules of glyceraldehyde 3-phosphate continue down the pathway.

$CH_2OH$–C(=O)–$CH_2OPO_3^{2-}$ ⇌ (*Triose phosphate isomerase*) O=C(H)–H–C–OH–$CH_2OPO_3^{2-}$

**Dihydroxyacetone phosphate** (a ketose) **Glyceraldehyde 3-phosphate** (an aldose)

6. Glyceraldehyde 3-phosphate is converted to 1,3-bisphosphoglycerate. The reaction is catalyzed by **glyceraldehyde 3-phosphate dehydrogenase** and uses inorganic phosphate and $NAD^+$. The other product is NADH. The energy for creating this new high-energy phosphate bond comes from the oxidation of the aldehyde group of the glyceraldehyde 3-phosphate.

O=C(H)–H–C–OH–$CH_2OPO_3^{2-}$ + $NAD^+$ + $P_i$ ⇌ (*Glyceraldehyde 3-phosphate dehydrogenase*) O=C($OPO_3^{2-}$)–H–C–OH–$CH_2OPO_3^{2-}$ + NADH + $H^+$

**Glyceraldehyde 3-phosphate** **1,3-Bisphosphoglycerate (1,3-BPG)**

7. The newly created high-energy phosphate bond of 1,3-bisphosphoglycerate is now used to synthesize ATP. **Phosphoglycerate kinase** catalyzes the transfer of the phosphoryl group from the 1,3-bisphosphoglycerate to ADP, generating ATP and 3-phosphoglycerate.

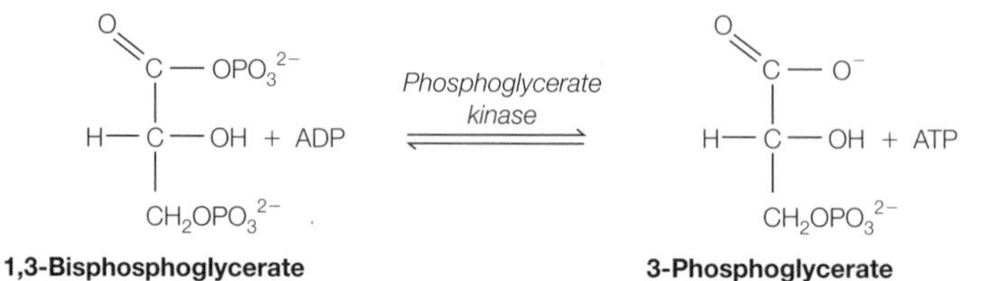


**1,3-Bisphosphoglycerate** **3-Phosphoglycerate**

8. 3-Phosphoglycerate is converted to 2-phosphoglycerate by **phosphoglycerate mutase**. Thus the reaction is a movement of the phosphate group to a different carbon atom within the same molecule.

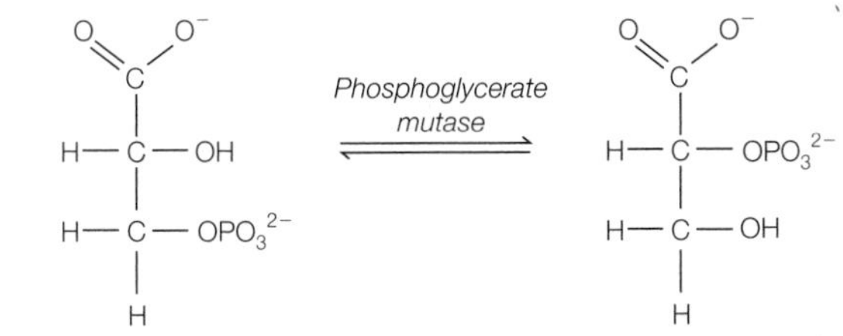

Phosphoglycerate mutase

**3-Phosphoglycerate** **2-Phosphoglycerate**

9. **Enolase** catalyzes the dehydration of 2-phosphoglycerate to form phosphoenolpyruvate (PEP). This reaction converts the low-energy phosphate ester bond of 2-phosphoglycerate into the high-energy phosphate bond of PEP.

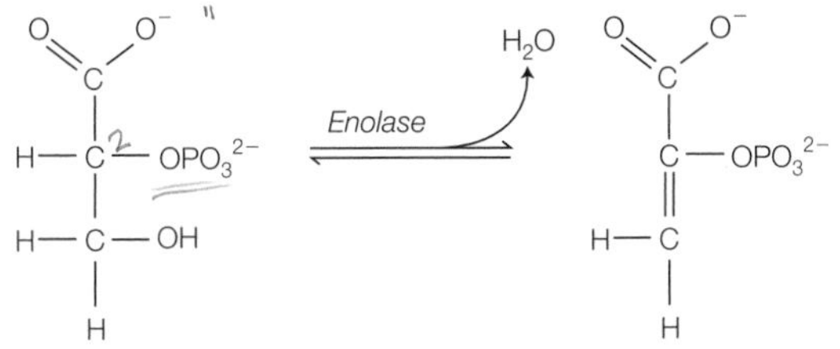

Enolase $H_2O$

**3-Phosphoglycerate** **Phosphoenolpyruvate**

10. In the last reaction, **pyruvate kinase** catalyzes the physiologically irreversible transfer of the phosphoryl group from PEP to ADP to form ATP and pyruvate.

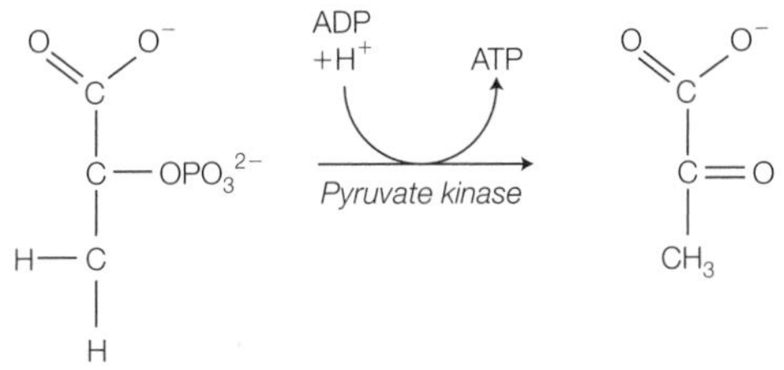

ADP $+H^+$ ATP Pyruvate kinase

**Phosphoenolpyruvate** **Pyruvate**

*Substrate-level phosphorylation*

There are two distinct methods by which cells synthesize ATP. In **oxidative phosphorylation**, involving the electron transport chain, the generation of ATP is linked to the oxidation of NADH and $FADH_2$ to $NAD^+$ and FAD respectively (see Topic L2), and occurs via the generation of a proton gradient across the inner mitochondrial membrane. In contrast, the two ATP synthetic reactions in glycolysis (catalyzed by phosphoglycerate kinase and pyruvate kinase) involve the direct transfer of a phosphate from a sugar–phosphate intermediate to ADP; these reactions are examples of **substrate-level phosphorylation**. A third example of substrate-level phosphorylation is the synthesis of GTP by succinate dehydrogenase in the citric acid cycle (see Topic L1). The GTP can be used to phosphorylate ADP to form ATP.

## Fates of pyruvate

- *Entry into the citric acid cycle.* Glycolysis releases relatively little of the energy present in a glucose molecule; much more is released by the subsequent operation of the citric acid cycle and oxidative phosphorylation.

Following this route under aerobic conditions, pyruvate is converted to acetyl CoA by the enzyme **pyruvate dehydrogenase** and the acetyl CoA then enters the citric acid cycle. The pyruvate dehydrogenase reaction is an **oxidative decarboxylation** (see Topic L1 for details):

$$\text{pyruvate} + \text{NAD}^+ + \text{CoA} \xrightarrow{\textit{Pyruvate dehydrogenase}} \text{acetyl CoA} + \text{CO}_2 + \text{NADH}$$

- *Conversion to fatty acid or ketone bodies.* When the cellular energy level is high (ATP in excess), the rate of the citric acid cycle (Topic L1) decreases and acetyl CoA begins to accumulate. Under these conditions, acetyl CoA can be used for fatty acid synthesis or the synthesis of ketone bodies (Topic K3).
- *Conversion to lactate.* The $NAD^+$ used during glycolysis (in the formation of 1,3-bisphosphoglycerate by glyceraldehyde 3-phosphate dehydrogenase; *Fig. 1*) must be regenerated if glycolysis is to continue. Under aerobic conditions, $NAD^+$ is regenerated by the reoxidation of NADH via the electron transport chain (see Topic L2). When oxygen is limiting, as in muscle during vigorous contraction, the reoxidation of NADH to $NAD^+$ by the electron transport chain becomes insufficient to maintain glycolysis. Under these conditions, $NAD^+$ is regenerated instead by conversion of the pyruvate to lactate by **lactate dehydrogenase**:

$$\text{pyruvate} + \text{NADH} + \text{H}^+ \xrightleftharpoons{\textit{Lactate dehydrogenase}} \text{lactate} + \text{NAD}^+$$

When sufficient oxygen becomes available once more, $NAD^+$ levels rise through operation of the electron transport chain. The lactate dehydrogenase reaction then reverses to regenerate pyruvate that is converted by pyruvate dehydrogenase to acetyl CoA which can enter the citric acid cycle (see above). Thus the operation of lactate dehydrogenase in mammals is a mechanism for the reoxidation of NADH to $NAD^+$ hence allowing glycolysis to continue, and ATP to be made, under anaerobic conditions. The process is even more sophisticated in the case of vigorously contracting skeletal muscle. Here the lactate produced is transported in the bloodstream to the liver where it is converted back to glucose and can return once again via the bloodstream to the skeletal muscle to be metabolized to yield energy. This is the Cori cycle and is described in Topic J4. Finally, in some microorganisms, lactate is the normal product from pyruvate.

- *Conversion to ethanol.* In yeast and some other microorganisms under anaerobic conditions, the $NAD^+$ required for the continuation of glycolysis is regenerated by a process called **alcoholic fermentation**. The pyruvate is converted to acetaldehyde (by **pyruvate decarboxylase**) and then to ethanol (by **alcohol dehydrogenase**), the latter reaction reoxidizing the NADH to $NAD^+$:

$$\text{pyruvate} + \text{H}^+ \xrightarrow{\textit{Pyruvate decarboxylase}} \text{acetaldehyde} + \text{CO}_2$$

$$\text{acetaldehyde} + \text{NADH} + \text{H}^+ \xrightleftharpoons{\textit{Alcohol dehydrogenase}} \text{ethanol} + \text{NAD}^+$$

## Energy yield

Early in glycolysis, two ATPs are required for the conversion of glucose to glucose 6-phosphate by hexokinase and for the conversion of fructose 6-phosphate to fructose 1,6-bisphosphate by PFK. However, fructose 1,6-bisphosphate then gives rise to two three-carbon units, each of which generates two ATPs in subsequent steps (catalyzed by phosphoglycerate kinase and pyruvate kinase) giving a net yield of two ATPs per original glucose molecule (*Fig. 1*). The overall reaction is:

$$\text{Glucose} + 2\,P_i + 2\,\text{ADP} + 2\,\text{NAD}^+ \rightarrow 2\,\text{pyruvate} + 2\,\text{ATP} + 2\,\text{NADH} + 2\,H^+ + 2\,H_2O$$

Note that, under aerobic conditions, the two NADH molecules that are synthesized are reoxidized via the electron transport chain generating ATP. Given the cytoplasmic location of these NADH molecules, each is reoxidized via the glycerol 3-phosphate shuttle (see Topic L2) and produces approximately two ATPs during oxidative phosphorylation or via the malate–aspartate shuttle (see Topic L2) and produces approximately three ATPs during oxidative phosphorylation.

## Metabolism of fructose

Fructose is an abundant sugar in the human diet; sucrose (table sugar) is a disaccharide which when hydrolyzed yields fructose and glucose (see Topic J1) and fructose is also a major sugar in fruits and honey. There are two pathways for the metabolism of fructose; one occurs in muscle and adipose tissue, the other in liver.

1. In muscle and adipose tissue, fructose can be phosphorylated by hexokinase (which is capable of phosphorylating both glucose and fructose) to form fructose 6-phosphate which then enters glycolysis.
2. In liver, the cells contain mainly glucokinase instead of hexokinase and this enzyme phosphorylates only glucose. Thus in liver, fructose is metabolized instead by the **fructose 1-phosphate pathway** (*Fig. 2*).

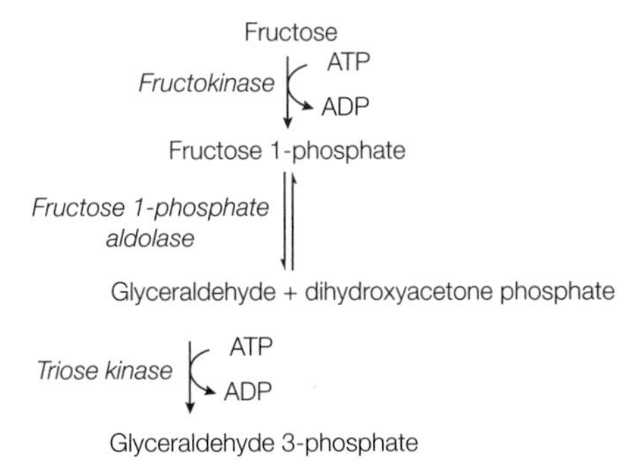


*Fig. 2. The fructose 1-phosphate pathway.*

- Fructose is converted to fructose 1-phosphate by **fructokinase.**
- Fructose 1-phosphate is then split into glyceraldehyde and dihydroxyacetone phosphate by **fructose 1-phosphate aldolase**. The dihydroxyacetone feeds into glycolysis at the triose phosphate isomerase step (*Fig. 1*).
- The glyceraldehyde is phosphorylated by **triose kinase** to glyceraldehyde 3-phosphate and so also enters glycolysis.

## Metabolism of galactose

The hydrolysis of the disaccharide lactose (in milk) yields galactose and glucose. Thus galactose is also a major dietary sugar for humans. Galactose and glucose are epimers that differ in their configuration at C-4 (Topic J1, *Fig. 4*). Thus the entry of galactose into glycolysis requires an epimerization reaction. This occurs via a four-step pathway called the **galactose–glucose interconversion pathway** (*Fig. 3*):

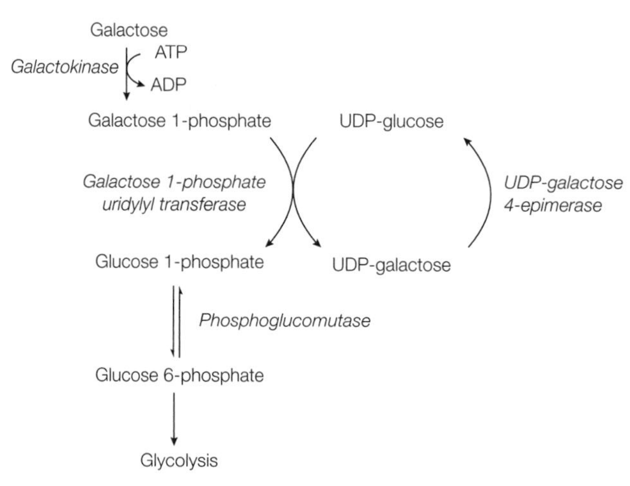


*Fig. 3. The galactose–glucose interconversion pathway.*

1. Galactose is phosphorylated by **galactokinase** to give galactose 1-phosphate.
2. **Galactose 1-phosphate uridylyl transferase** catalyzes the transfer of a uridyl group from UDP-glucose to galactose 1-phosphate to form UDP-galactose and glucose 1-phosphate.

**Galactose 1-phosphate** + **UDP-glucose** ⇌ **UDP-galactose** + **Glucose 1-phosphate**

3. The UDP-galactose is converted back to UDP-glucose by **UDP-galactose 4-epimerase**. Thus, overall, UDP-glucose is not consumed in the reaction pathway.
4. Finally the glucose 1-phosphate is converted to glucose 6-phosphate by **phosphoglucomutase**. The glucose 6-phosphate then enters glycolysis.

**Galactosemia** is a genetic disease caused by an inability to convert galactose to glucose. Toxic substances accumulate such as **galactitol**, formed by the reduction of galactose, and lead to dire consequences for the individual. Children who have the disease fail to thrive, may vomit or have diarrhea after drinking milk, and often have an enlarged liver and jaundice. The formation of cataracts in the eyes, mental retardation and an early death from liver damage are also possible. Most cases of galactosemia are due to a deficiency of the galactose 1-phosphate uridylyl transferase enzyme and hence these individuals cannot metabolize galactose. The disease is treated by prescribing a galactose-free diet which causes all the symptoms to regress except mental retardation which may be irreversible. Since such patients have normal levels of UDP-galactose 4-epimerase, they can still synthesize UDP-galactose from UDP-glucose and so can still synthesize, for example, oligosaccharides in glycoproteins that involve Gal residues.

## Regulation of glycolysis

*Phosphofructokinase*

The most important control step of glycolysis is the irreversible reaction catalyzed by phosphofructokinase (PFK). The enzyme is regulated in several ways:

- *ATP/AMP.* PFK is allosterically inhibited by ATP but this inhibition is reversed by AMP. This allows glycolysis to be responsive to the energy needs of the cell, speeding up when ATP is in short supply (and AMP is plentiful) so that more ATP can be made, and slowing down when sufficient ATP is already available.
- *Citrate.* PFK is also inhibited by citrate, the first product of the citric acid cycle proper (see Topic L1). A high level of citrate signals that there is a plentiful supply of citric acid cycle intermediates already and hence no additional breakdown of glucose via glycolysis is needed.
- *Fructose 2,6-bisphosphate.* Fructose 2,6-bisphosphate (F-2,6-BP) is synthesized (*Fig. 4*) from fructose 6-phosphate by an enzyme called **phosphofructokinase 2 (PFK2)**, a different enzyme from PFK. F-2,6-BP is hydrolyzed back to fructose 6-phosphate (*Fig. 4*) by **fructose bisphosphatase 2 (FBPase2)**. Amazingly, both PFK2 and FBPase2 are activities catalyzed by the same polypeptide; hence this is a **bi-functional enzyme**. Fructose 6-phosphate stimulates the synthesis of F-2,6-BP and inhibits its hydrolysis (*Fig. 4*). F-2,6-BP in turn strongly activates PFK and hence stimulates glycolysis. The overall effect is that when fructose 6-phosphate levels are high, PFK (and hence glycolysis) is stimulated. PFK2 and

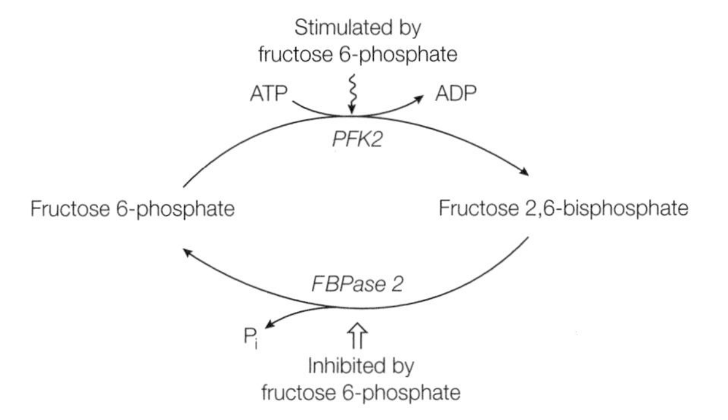


*Fig. 4. Synthesis and degradation of fructose 2,6-bisphosphate.*

FBPase2 are also controlled by covalent modification (see Topic C4). When blood glucose levels fall, the hormone glucagon is released into the bloodstream and triggers a cAMP cascade (see Topic J7) that leads to phosphorylation of the PFK2/FBPase2 polypeptide at a single serine residue. This activates FBPase2 and inhibits PFK2, lowering the level of F-2,6-BP and hence decreasing the rate of glycolysis. The reverse is true as glucose levels rise; the phosphate group is removed from the PFK2/FBPase2 polypeptide by a phosphatase, thus inhibiting FBPase2 and activating PFK2, raising the level of F-2,6-BP and hence increasing the rate of glycolysis.

F-2,6-BP is also important in preventing glycolysis (glucose degradation) and gluconeogenesis (glucose synthesis) operating simultaneously. This is called **reciprocal regulation** and is described in Topic J4.

- *$H^+$ ions.* PFK is inhibited by $H^+$ ions and hence the rate of glycolysis decreases when the pH falls significantly. This prevents the excessive formation of lactate (i.e. lactic acid) under anaerobic conditions (see above) and hence prevents the medical condition known as acidosis (a deleterious drop in blood pH).

### *Hexokinase*

Hexokinase, which catalyzes the first irreversible step of glycolysis, is inhibited by glucose 6-phosphate. Thus when PFK is inhibited, fructose 6-phosphate builds up and so does glucose 6-phosphate since these two metabolites are in equilibrium via phosphoglucoisomerase (see *Fig. 1*). The hexokinase inhibition then reinforces the inhibition at the PFK step. At first sight this seems unusual since it is usually the first irreversible step of a pathway (the committed step) that is the main control step. On this basis, it may appear that hexokinase should be the main control enzyme, not PFK. However, glucose 6-phosphate, the product of the hexokinase reaction, can also feed into glycogen synthesis (see Topic J6) or the pentose phosphate pathway (see Topic J5). Thus the first irreversible step that is unique to glycolysis is that catalyzed by PFK and hence this is the main control step.

### *Pyruvate kinase*

Pyruvate kinase catalyzes the third irreversible step in glycolysis. It is activated by fructose 1,6-bisphosphate. ATP and the amino acid alanine allosterically inhibit the enzyme so that glycolysis slows when supplies of ATP and biosynthetic precursors (indicated by the levels of Ala) are already sufficiently high. In addition, in a control similar to that for PFK (see above), when the blood glucose concentration is low, glucagon is released and stimulates phosphorylation of the enzyme via a cAMP cascade (see Topic J7). This covalent modification inhibits the enzyme so that glycolysis slows down in times of low blood glucose levels.

# J4 GLUCONEOGENESIS

## Key Notes

**Overview**

Gluconeogenesis synthesizes glucose from noncarbohydrate precursors and is important for the maintenance of blood glucose levels during starvation or during vigorous exercise. The brain and erythrocytes depend almost entirely on blood glucose as an energy source. Gluconeogenesis occurs mainly in the liver and to a lesser extent in the kidney. Most enzymes of gluconeogenesis are cytosolic, but pyruvate carboxylase and glucose 6-phosphatase are located in the mitochondrial matrix and bound to the smooth endoplasmic reticulum, respectively.

**The pathway**

Pyruvate is converted to oxaloacetate (by pyruvate carboxylase). The oxaloacetate is decarboxylated and phosphorylated to phosphoenolpyruvate (PEP) by phosphoenolpyruvate carboxykinase (PEP carboxykinase). PEP is converted to fructose 1,6-bisphosphate by a direct reversal of several reactions in glycolysis. Next, fructose 1,6-bisphosphate is dephosphorylated to fructose 6-phosphate (by fructose 1,6-bisphosphatase) and this is then converted to glucose 6-phosphate (by phosphoglucoisomerase). Finally, glucose 6-phosphate is dephosphorylated (by glucose 6-phosphatase) to yield glucose.

**Energy used**

The synthesis of one molecule of glucose from two molecules of pyruvate requires six molecules of ATP.

**Transport of oxaloacetate**

Oxaloacetate, the product of the first step in gluconeogenesis, must leave the mitochondrion and enter the cytosol where the subsequent enzyme steps take place. Since the inner mitochondrial membrane is impermeable to oxaloacetate, it is converted to malate by mitochondrial malate dehydrogenase. This leaves the mitochondrion and is converted back to oxaloacetate in the cytosol by cytoplasmic malate dehydrogenase.

**Pyruvate carboxylase activation**

Oxaloacetate, the product of the pyruvate carboxylase reaction, functions both as an important citric acid cycle intermediate in the oxidation of acetyl CoA and as a precursor for gluconeogenesis. The activity of pyruvate carboxylase depends on the presence of acetyl CoA so that more oxaloacetate is made when acetyl CoA levels rise.

**Reciprocal regulation of glycolysis and gluconeogenesis**

If glycolysis and gluconeogenesis operated simultaneously, the net effect would be a futile cycle resulting in the hydrolysis of two ATP and two GTP molecules. This is prevented by reciprocal regulation at the enzyme steps that are distinct in each pathway. AMP activates phosphofructokinase (PFK) (glycolysis) but inhibits fructose 1,6-bisphosphatase (gluconeogenesis). ATP and citrate inhibit PFK but citrate stimulates fructose 1,6-bisphosphatase. Glycolysis and gluconeogenesis are also responsive to starvation conditions via the concentration of fructose 2,6-bisphosphate (F-2,6-BP). During starvation, glucagon is released into the bloodstream and inhibits the synthesis of F-2,6-BP. In the fed state, insulin is released into the bloodstream and causes the accumulation of F-2,6-BP. Since F-2,6-BP activates PFK and inhibits fructose 1,6-bisphosphatase, glycolysis is stimulated and gluconeogenesis is inhibited in the fed animal and vice versa, during starvation.

In liver, ATP and alanine inhibit pyruvate kinase (glycolysis) whilst ADP inhibits pyruvate carboxylase and PEP carboxykinase (gluconeogenesis). Thus glycolysis is inhibited in times when ATP and biosynthetic intermediates are in excess whilst gluconeogenesis is inhibited in times when the ATP level is low (and ADP is high). Pyruvate kinase is also stimulated by fructose 1,6-bisphosphate so its rate increases when glycolysis is active. During starvation, glycogen secretion into the bloodstream activates a cAMP cascade that leads to the phosphorylation and inhibition of pyruvate kinase (glycolysis).

**The Cori cycle**

During vigorous exercise, pyruvate produced by glycolysis in muscle is converted to lactate by lactate dehydrogenase. The lactate diffuses into the bloodstream and is carried to the liver. Here it is converted to glucose by gluconeogenesis. The glucose is released into the bloodstream and becomes available for uptake by muscle (as well as other tissues, including brain). This cycle of reactions is called the Cori cycle.

**Related topics**

Protein glycosylation (H5)
Monosaccharides and disaccharides (J1)
Polysaccharides and oligosaccharides (J2)
Glycolysis (J3)
Pentose phosphate pathway (J5)
Glycogen metabolism (J6)
Citric acid cycle (L1)

## Overview

Gluconeogenesis synthesizes glucose from noncarbohydrate precursors, including lactate and pyruvate, citric acid cycle intermediates, the carbon skeletons of most amino acids and glycerol. This is extremely important since the brain and erythrocytes rely almost exclusively on glucose as their energy source under normal conditions. The liver's store of glycogen is sufficient to supply the brain with glucose for only about half a day during fasting. Thus gluconeogenesis is especially important in periods of starvation or vigorous exercise. During starvation, the formation of glucose via gluconeogenesis particularly uses amino acids from protein breakdown and glycerol from fat breakdown. During exercise, the blood glucose levels required for brain and skeletal muscle function are maintained by gluconeogenesis in the liver using lactate produced by the muscle.

The main site of gluconeogenesis is the liver, although it also occurs to a far lesser extent in the kidneys. Very little gluconeogenesis occurs in brain or muscle. Within liver cells, the first enzyme of gluconeogenesis, pyruvate carboxylase, is located in the mitochondrial matrix. The last enzyme, glucose 6-phosphatase is bound to the smooth endoplasmic reticulum. The other enzymes of the pathway are located in the cytosol.

## The pathway

In glycolysis (Topic J3), glucose is metabolized to pyruvate. In gluconeogenesis, pyruvate is metabolized to glucose. Thus, in principle, gluconeogenesis appears to be a reversal of glycolysis. Indeed, some of the reactions of glycolysis are reversible and so the two pathways have these steps in common. However,

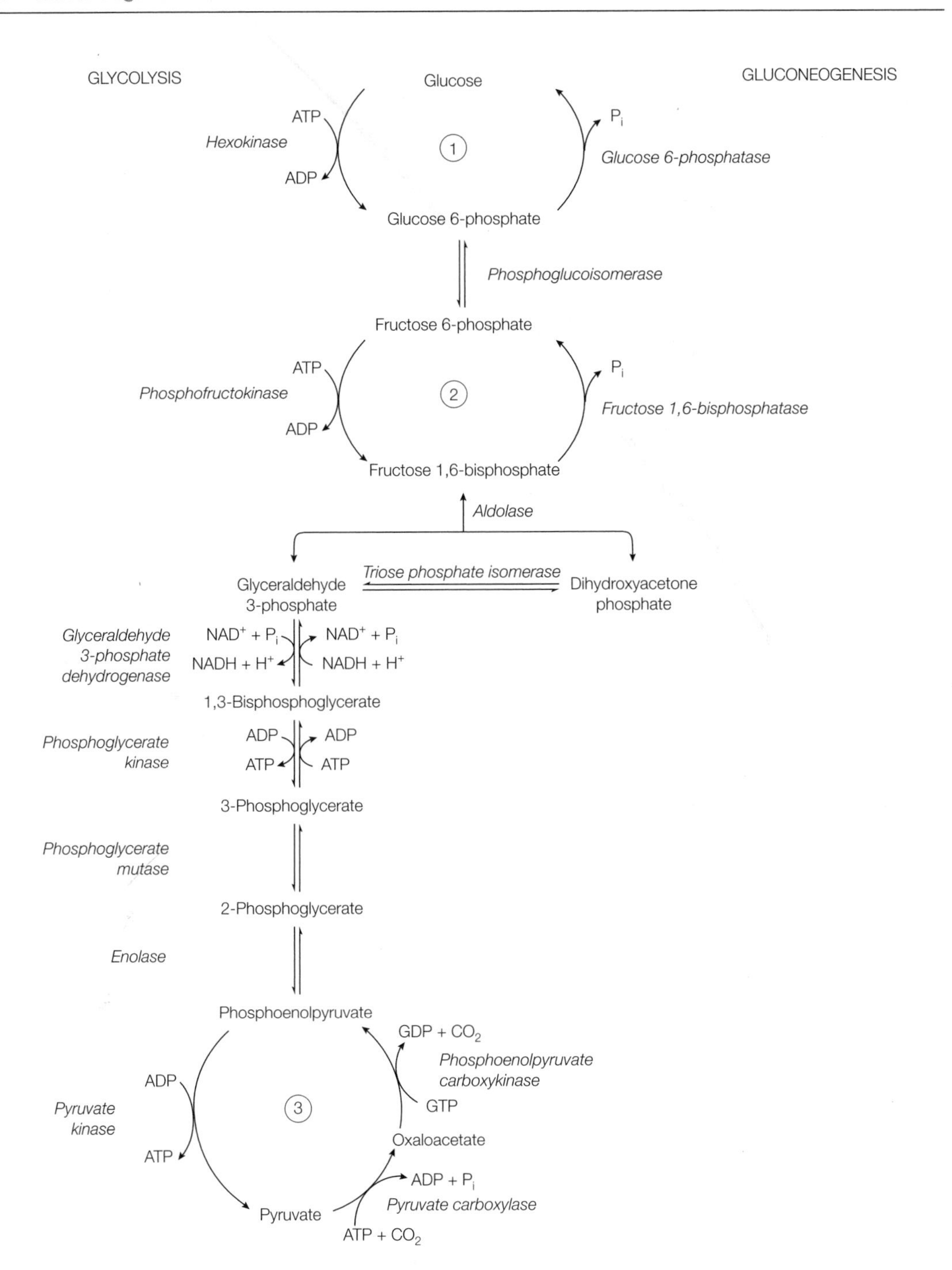


*Fig. 1. Comparison of gluconeogenesis and glycolysis. The three steps of glycolysis that are irreversible are numbered. (1) Hexokinase in glycolysis is reversed by glucose 6-phosphatase in gluconeogenesis; (2) PFK in glycolysis is reversed by fructose 1,6-bisphosphatase in gluconeogenesis; (3) pyruvate kinase in glycolysis is reversed by two sequential reactions in gluconeogenesis catalyzed by pyruvate carboxylase and PEP carboxykinase.*

three steps in glycolysis are essentially irreversible; those catalyzed by the enzymes hexokinase, phosphofructokinase (PFK) and pyruvate kinase (see Topic J3). Indeed it is the large negative free-energy change in these reactions that normally drives glycolysis forward towards pyruvate formation. Therefore, in gluconeogenesis, these three steps have to be reversed by using other reactions as shown in *Fig. 1*; gluconeogenesis is not a simple reversal of glycolysis.

*Precursors for gluconeogenesis*

**Glycerol** can act as a substrate for glucose synthesis by conversion to dihydroxyacetone phosphate, an intermediate in gluconeogenesis (*Fig. 1*). In order for **lactate, pyruvate, citric acid cycle intermediates** and the carbon skeletons of most **amino acids** to act as precursors for gluconeogenesis, these compounds must first be converted to oxaloacetate. Some of the carbon skeletons of the amino acids give rise to oxaloacetate directly. Others feed into the citric acid cycle as intermediates (see Topic L1 and M2) and the cycle then converts these molecules to oxaloacetate. Lactate is converted to pyruvate by the lactate dehydrogenase reaction (see Topic J3) and some amino acids also give rise to pyruvate (see Topic M2). Therefore, for these precursors, the first step in the gluconeogenic pathway is the conversion of pyruvate to oxaloacetate.

**The steps in gluconeogenesis** (see *Fig. 1*) are as follows:

1. Pyruvate is converted to oxaloacetate by carboxylation using the enzyme **pyruvate carboxylase** that is located in the mitochondrial matrix.

$$\underset{\textbf{Pyruvate}}{\begin{array}{l}COO^-\\ |\\ C{=}O\\ |\\ CH_3\end{array}} + CO_2 + ATP \xrightarrow{\text{Pyruvate carboxylase}} \underset{\textbf{Oxaloacetate}}{\begin{array}{l}COO^-\\ |\\ C{=}O\\ |\\ CH_2\\ |\\ COO^-\end{array}} + ADP + P_i$$

   This enzyme uses **biotin** as an **activated carrier of $CO_2$**, the reaction occurring in two stages:

$$\text{E-biotin} + ATP + HCO_3^- \longrightarrow \text{E-biotin-}CO_2 + ADP + P_i$$

$$\text{E-biotin-}CO_2 + \text{pyruvate} \longrightarrow \text{E-biotin} + \text{oxaloacetate}$$

2. The oxaloacetate is now acted on by **phosphoenolpyruvate carboxykinase** which simultaneously decarboxylates and phosphorylates it to form phosphoenolpyruvate (PEP), releasing $CO_2$ and using GTP in the process:

$$\underset{\textbf{Oxaloacetate}}{\begin{array}{l}COO^-\\ |\\ C{=}O\\ |\\ CH_2\\ |\\ COO^-\end{array}} + GTP \xrightleftharpoons{\text{Phosphoenolpyruvate carboxykinase}} \underset{\textbf{Phosphoenolpyruvate}}{\begin{array}{l}COO^-\quad O\\ |\qquad\quad \|\\ C-O-P-O^-\\ \|\qquad\quad |\\ CH_2\qquad O^-\end{array}} + GDP + CO_2$$

Thus, reversal of the glycolytic step from PEP to pyruvate requires two reactions in gluconeogenesis, pyruvate to oxaloacetate by pyruvate carboxylase and oxaloacetate to PEP by PEP carboxykinase. Given that the conversion of PEP to pyruvate in glycolysis synthesizes ATP, it is not surprising that the overall reversal of this step needs the input of a substantial amount of energy, one ATP for the pyruvate carboxylase step and one GTP for the PEP carboxykinase step.

3. PEP is converted to fructose 1,6-bisphosphate in a series of steps that are a direct reversal of those in glycolysis (see Topic J3), using the enzymes enolase, phosphoglycerate mutase, phosphoglycerate kinase, glyceraldehyde 3-phosphate dehydrogenase, triose phosphate isomerase and aldolase (see *Fig 1*). This sequence of reactions uses one ATP and one NADH for each PEP molecule metabolized.
4. Fructose 1,6-bisphosphate is dephosphorylated to form fructose 6-phosphate by the enzyme **fructose 1,6-bisphosphatase**, in the reaction:

$$\text{fructose 1,6-bisphosphate} + H_2O \longrightarrow \text{fructose 6-phosphate} + P_i$$

5. Fructose 6-phosphate is converted to glucose 6-phosphate by the glycolytic enzyme phosphoglucoisomerase.
6. Glucose 6-phosphate is converted to glucose by the enzyme **glucose 6-phosphatase**. This enzyme is bound to the smooth endoplasmic reticulum and catalyzes the reaction:

$$\text{glucose 6-phosphate} + H_2O \longrightarrow \text{glucose} + P_i$$

## Energy used

As would be expected, the synthesis of glucose by gluconeogenesis needs the input of energy. Two pyruvate molecules are required to synthesize one molecule of glucose. Energy is required at the following steps:

| | |
|---|---|
| pyruvate carboxylase | 1 ATP ( × 2) = 2 ATP |
| PEP carboxykinase | 1 GTP ( × 2) = 2 ATP |
| phosphoglycerate kinase | 1 ATP ( × 2) = 2 ATP |
| | Total = 6 ATP |

This compares with only two ATPs as the net ATP yield from glycolysis. Thus an extra four ATPs per glucose are required to reverse glycolysis.

In fact, the glyceraldehyde 3-phosphate dehydrogenase reaction also consumes NADH, equivalent to two molecules of NADH for each molecule of glucose synthesized. Since each cytosolic NADH would normally be used to generate approximately two ATP molecules via the glycerol 3-phosphate shuttle and oxidative phosphorylation (see Topic L2), this is equivalent to the input of another four ATPs per glucose synthesized.

## Transport of oxaloacetate

Pyruvate carboxylase is a mitochondrial matrix enzyme whereas the other enzymes of gluconeogenesis are located outside the mitochondrion. Thus oxaloacetate, produced by pyruvate carboxylase, needs to exit the mitochondrion. However the inner mitochondrial membrane is not permeable to this compound. Thus oxaloacetate is converted to malate inside the mitochondrion by **mitochondrial malate dehydrogenase**, the malate is transported through the mitochondrial membrane by a special transport protein and then the malate is converted back to oxaloacetate in the cytoplasm by a **cytoplasmic malate dehydrogenase** (*Fig. 2*).

## Pyruvate carboxylase activation

Oxaloacetate has two main roles. It is an intermediate that is consumed in gluconeogenesis and it is also a key intermediate in the citric acid cycle where it fuses with acetyl CoA to form citrate, eventually being regenerated by the cycle. Thus pyruvate carboxylase generates oxaloacetate for gluconeogenesis but also must maintain oxaloacetate levels for citric acid cycle function. For the latter reason, the activity of pyruvate carboxylase depends absolutely on the presence of acetyl CoA; the biotin prosthetic group of the enzyme cannot be carboxylated unless acetyl CoA is bound to the enzyme. This allosteric activation by acetyl CoA ensures that more oxaloacetate is made when excess acetyl CoA is present. In this role of maintaining the level of citric acid cycle intermediates, the pyruvate carboxylase reaction is said to be **anaplerotic**, that is 'filling up'.

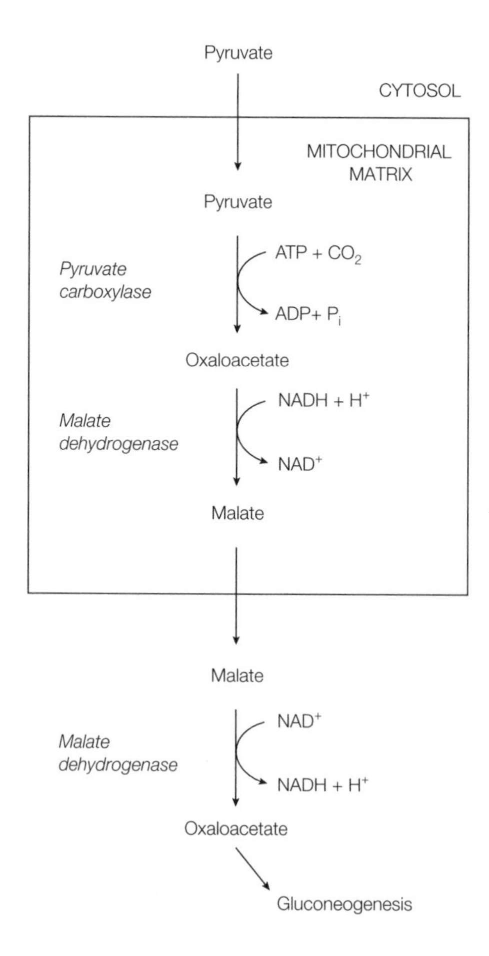


*Fig. 2. Transport of oxaloacetate out of the mitochondrion.*

**Reciprocal regulation of glycolysis and gluconeogenesis**

Glycolysis generates two ATPs net per glucose whereas gluconeogenesis uses four ATPs and two GTPs per glucose. Thus, if both glycolysis and gluconeogenesis were allowed to operate simultaneously, converting glucose to pyruvate and back again, the only net result would be the utilization of two ATPs and two GTPs, a so-called **futile cycle**. This is prevented by tight coordinate regulation of glycolysis and gluconeogenesis. Since many of the steps of the two pathways are common, the steps that are distinct in each pathway are the sites of this regulation, in particular the interconversions between fructose 6-phosphate and fructose 1,6-bisphosphate and between PEP and pyruvate. The situation is summarized in *Fig. 3* and described in detail below.

*Regulation of PFK and fructose 1,6-bisphosphatase*

When the level of AMP is high, this indicates the need for more ATP synthesis. AMP stimulates PFK, increasing the rate of glycolysis, and inhibits fructose 1,6-bisphosphatase, turning off gluconeogenesis. Conversely, when ATP and citrate levels are high, this signals that no more ATP need be made. ATP and citrate inhibit PFK, decreasing the rate of glycolysis, and citrate stimulates fructose 1,6-bisphosphatase, increasing the rate of gluconeogenesis.

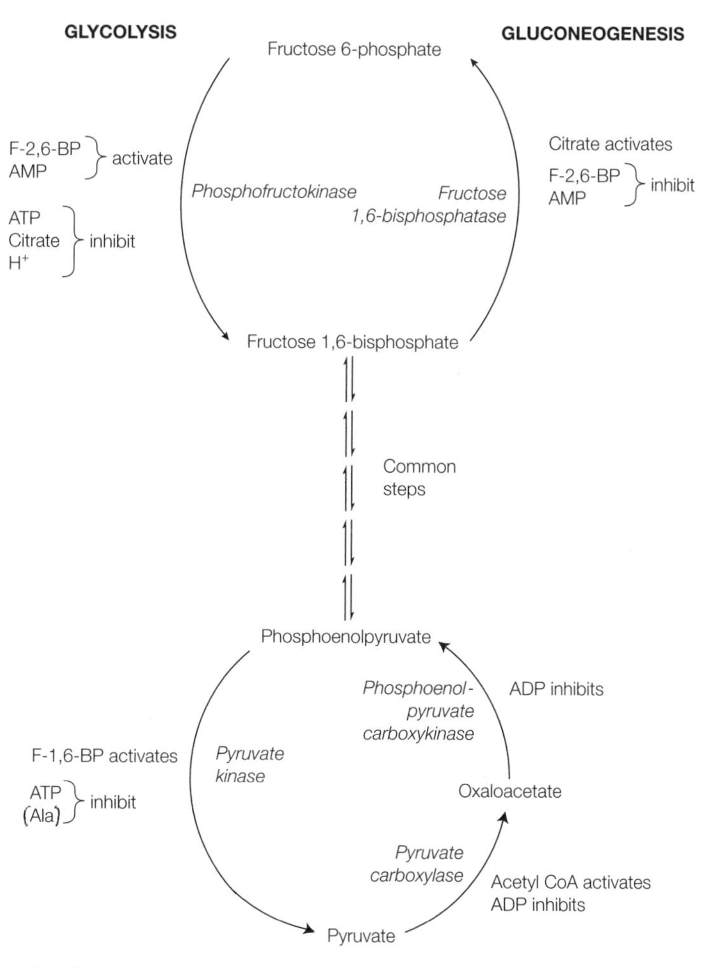


*Fig. 3. Reciprocal regulation of glycolysis and gluconeogenesis.*

Glycolysis and gluconeogenesis are made responsive to starvation by the level of the regulatory molecule **fructose 2,6-bisphosphate (F-2,6-BP)**. F-2,6-BP is synthesized from fructose 6-phosphate and hydrolyzed back to fructose 6-phosphate by a single polypeptide with two enzymatic activities (PFK2 and FBPase2; see Topic J3). The level of F-2,6-BP is under hormonal control. During starvation, when the level of blood glucose is low, the hormone glucagon is released into the bloodstream and triggers a cAMP cascade (Topic J7), eventually causing phosphorylation of the PFK2/FBPase2 polypeptide. This activates FBPase2 and inhibits PFK2, lowering the level of F-2,6-BP (see Topic J3). In the fed state, when blood glucose is at a high level, the hormone insulin is released and has the opposite effect, causing an elevation in the level of F-2,6-BP. Since F-2,6-BP strongly stimulates PFK and inhibits fructose 1,6-bisphosphatase (*Fig. 3*), glycolysis is stimulated and gluconeogenesis is inhibited in the fed animal. Conversely, during starvation, the low level of F-2,6-BP allows gluconeogenesis to predominate.

***Regulation of pyruvate kinase, pyruvate carboxylase and PEP carboxykinase***

- In liver, pyruvate kinase is inhibited by high levels of ATP and alanine so that glycolysis is inhibited when ATP and biosynthetic intermediates are already plentiful (see Topic J3). Acetyl CoA is also abundant under these conditions and activates pyruvate carboxylase, favoring gluconeogenesis. Conversely, when the energy status of the cell is low, the ADP concentration is high and this inhibits both pyruvate carboxylase and PEP carboxykinase, switching off gluconeogenesis. At this time, the ATP level will be low so pyruvate kinase is not inhibited and glycolysis will operate.
- Pyruvate kinase is also stimulated by fructose 1,6-bisphosphate (see Topic J3; feedforward activation) so that its activity rises when needed, as glycolysis speeds up.
- During starvation, the priority is to conserve blood glucose for the brain and muscle. Thus, under these conditions, pyruvate kinase in the liver is switched off. This occurs because the hormone glucagon is secreted into the bloodstream and activates a cAMP cascade (see Topic J7) that leads to the phosphorylation and inhibition of this enzyme.

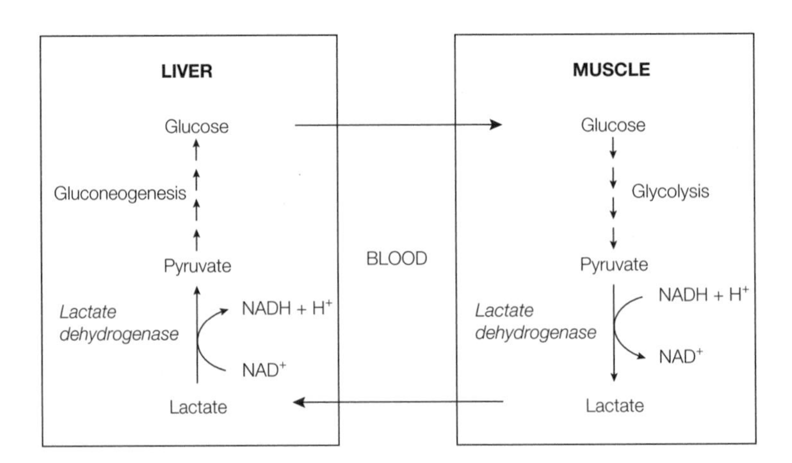


*Fig. 4. The Cori cycle.*

## The Cori cycle

Under the limiting oxygen conditions experienced during vigorous exercise, the formation of NADH by glycolysis exceeds the ability of the respiratory chain to oxidize it back to $NAD^+$. The pyruvate produced by glycolysis in muscle is then converted to lactate by lactate dehydrogenase, a reaction that regenerates $NAD^+$ and so allows glycolysis to continue to produce ATP (see Topic J3). However, lactate is a metabolic dead-end in that it cannot be metabolized further until it is converted back to pyruvate. Lactate diffuses out of the muscle and is carried in the bloodstream to the liver. Here it diffuses into liver cells and is converted back to pyruvate by lactate dehydrogenase. The pyruvate is then converted to glucose by gluconeogenesis and the glucose is released back into the bloodstream ready to be taken up by skeletal muscle (and brain). This cycle of reactions (*Fig. 4*) is called **the Cori cycle**.

# J5 PENTOSE PHOSPHATE PATHWAY

## Key Notes

**Overview**

The two major products of the pathway are nicotinamide adenine dinucleotide (reduced form; NADPH) and ribose 5-phosphate. Ribose 5-phosphate and its derivatives are components of important cellular molecules such as RNA, DNA, $NAD^+$, flavine adenine dinucleotide (FAD), ATP and coenzyme A (CoA). NADPH is required for many biosynthetic pathways and particularly for synthesis of fatty acids and steroids. Hence the pathway is very active in tissues such as adipose tissue, mammary gland and the adrenal cortex.

**Main reactions of the pathway**

The reactions of the pathway can be grouped into three stages. In the first stage, oxidative reactions convert glucose 6-phosphate into ribulose 5-phosphate, generating two NADPH molecules. In the second stage, ribulose 5-phosphate is converted to ribose 5-phosphate by isomerization. The third stage of reactions, catalyzed by transketolase and transaldolase, converts ribose 5-phosphate into fructose 6-phosphate and glyceraldehyde 3-phosphate and hence links the pentose phosphate pathway with glycolysis.

**Control of the pathway**

The transketolase and transaldolase reactions are reversible and so allow either the conversion of ribose 5-phosphate into glycolytic intermediates when it is not needed for other cellular reactions, or the generation of ribose 5-phosphate from glycolytic intermediates when more is required. The rate of the pentose phosphate pathway is controlled by $NADP^+$ regulation of the first step, catalyzed by glucose 6-phosphate dehydrogenase.

**Related topics**

Monosaccharides and disaccharides (J1)
Glycolysis (J3)
Fatty acid synthesis (K3)

## Overview

Reducing power is available in a cell both as NADH and NADPH but these have quite distinct roles. NADH is oxidized by the respiratory chain to generate ATP via oxidative phosphorylation (see Topic L2). NADPH is used for biosynthetic reactions that require reducing power. Despite their similar structures (see Topic C1), NADH and NADPH are not metabolically interchangeable and so the cell must carry out a set of reactions that specifically create NADPH. This set of reactions is the pentose phosphate pathway (also known as the **hexose monophosphate shunt** or the **phosphogluconate pathway**). It takes place in the cytosol and is particularly important in tissues such as adipose tissue, mammary gland and the adrenal cortex that synthesizes fatty acids and steroids from acetyl CoA (see Topic K3). The activity of the pathway is very low in skeletal muscle, for example, which does not synthesize fatty acids or steroids.

The core set of reactions of the pathway oxidize glucose 6-phosphate to ribose 5-phosphate and generate NADPH. Thus, as well as generating NADPH, the pathway has a second important role in converting hexoses into pentoses, in particular ribose 5-phosphate. Ribose 5-phosphate or derivatives of it are required for the synthesis of RNA, DNA, $NAD^+$, flavine adenine dinucleotide (FAD), ATP, coenzyme A (CoA) and other important molecules. Thus the two main products of the pathway are NADPH and ribose 5-phosphate.

## Main reactions of the pathway

The core reactions of the pathway can be summarized as:

$$\begin{matrix}\text{glucose} \\ \text{6-phosphate}\end{matrix} + 2\,NADP^+ + H_2O \rightarrow \begin{matrix}\text{ribose} \\ \text{5-phosphate}\end{matrix} + 2\,NADPH + 2\,H^+ + CO_2$$

The pathway has three stages:

**Stage 1 – Oxidative reactions that convert glucose 6-phosphate into ribulose 5-phosphate, generating two NADPH molecules**

Glucose 6-phosphate is oxidized by **glucose 6-phosphate dehydrogenase** to 6-phosphoglucono-δ-lactone (producing NADPH) and this is then hydrolyzed by **lactonase** to 6-phosphogluconate. The 6-phosphogluconate is subsequently converted by **6-phosphogluconate dehydrogenase** to ribulose 5-phosphate. This is an **oxidative decarboxylation** (i.e. the 6-phosphogluconate is oxidized and a carbon is removed as $CO_2$). These reactions are shown below:

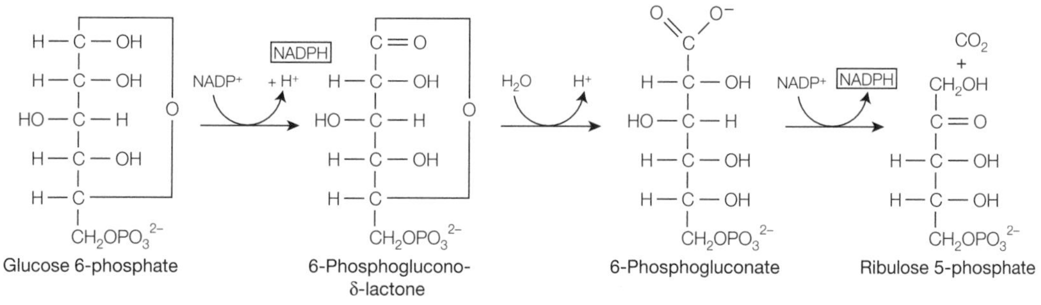

Glucose 6-phosphate → (NADP+ → NADPH + H+) → 6-Phosphoglucono-δ-lactone → (H2O → H+) → 6-Phosphogluconate → (NADP+ → NADPH) → Ribulose 5-phosphate + CO2

**Stage 2 – Isomerization of ribulose 5-phosphate to ribose 5-phosphate**

The ribulose 5-phosphate is now converted to ribose 5-phosphate by isomerization, a reaction catalyzed by **phosphopentose isomerase**:

Ribulose 5-phosphate ⇌ (Phosphopentose Isomerase) Ribose 5-phosphate

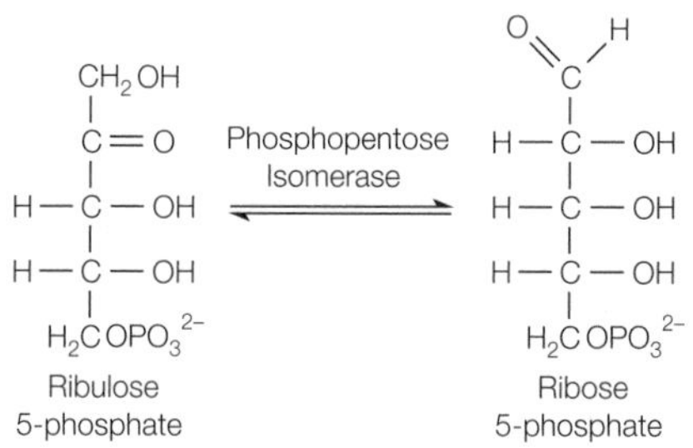

**Stage 3 – Linkage of the pentose phosphate pathway to glycolysis via transketolase and transaldolase**

If at any time only a little ribose 5-phosphate is required for nucleic acid synthesis and other synthetic reactions, it will tend to accumulate and is then converted to fructose 6-phosphate and glyceraldehyde 3-phosphate by the enzymes **transketolase** and **transaldolase**. These two products are intermediates of glycolysis. Therefore, these reactions provide a link between the pentose phosphate pathway and glycolysis. The outline reactions are shown below.

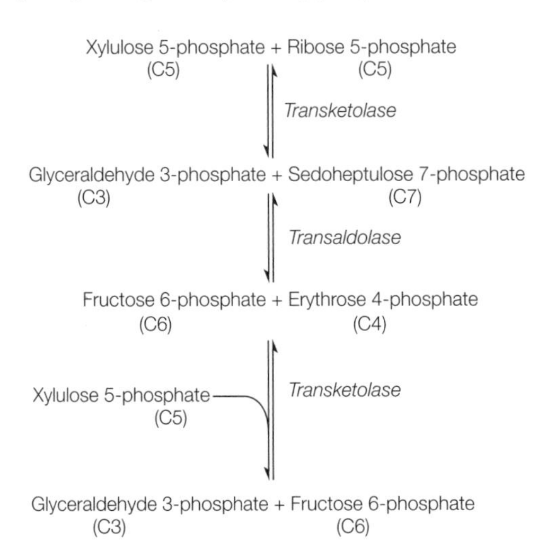


Details of these reactions, showing the structures of the molecules involved, are given in *Fig. 1*. These reactions require xylulose 5-phosphate as well as ribose-5-phosphate. Xylulose 5-phosphate is an epimer (see Topic J1) of ribulose 5-phosphate and is made by **phosphopentose epimerase**:

$CH_2OH$ / C=O / H—C—OH / H—C—OH / $CH_2OPO_3^{2-}$ ⇌ (*Phosphopentose epimerase*) $CH_2OH$ / C=O / HO—C—H / H—C—OH / $CH_2OPO_3^{2-}$

**Ribulose 5-phosphate** **Xylulose 5-phosphate**

Overall the reactions in this stage can be summarized as:

2 xylulose 5-phosphate + ribose 5-phosphate ⇌ 2 fructose 6-phosphate + glyceraldehyde 3-phosphate

## Control of the pathway

The transketolase and transaldolase reactions are reversible, so the final products of the pentose phosphate pathway can change depending on the metabolic needs of the cell. Thus when the cell needs NADPH but not ribose 5-phosphate, the latter is converted to glycolytic intermediates and enters glycolysis. At the other extreme, when the need for ribose 5-phosphate exceeds that for

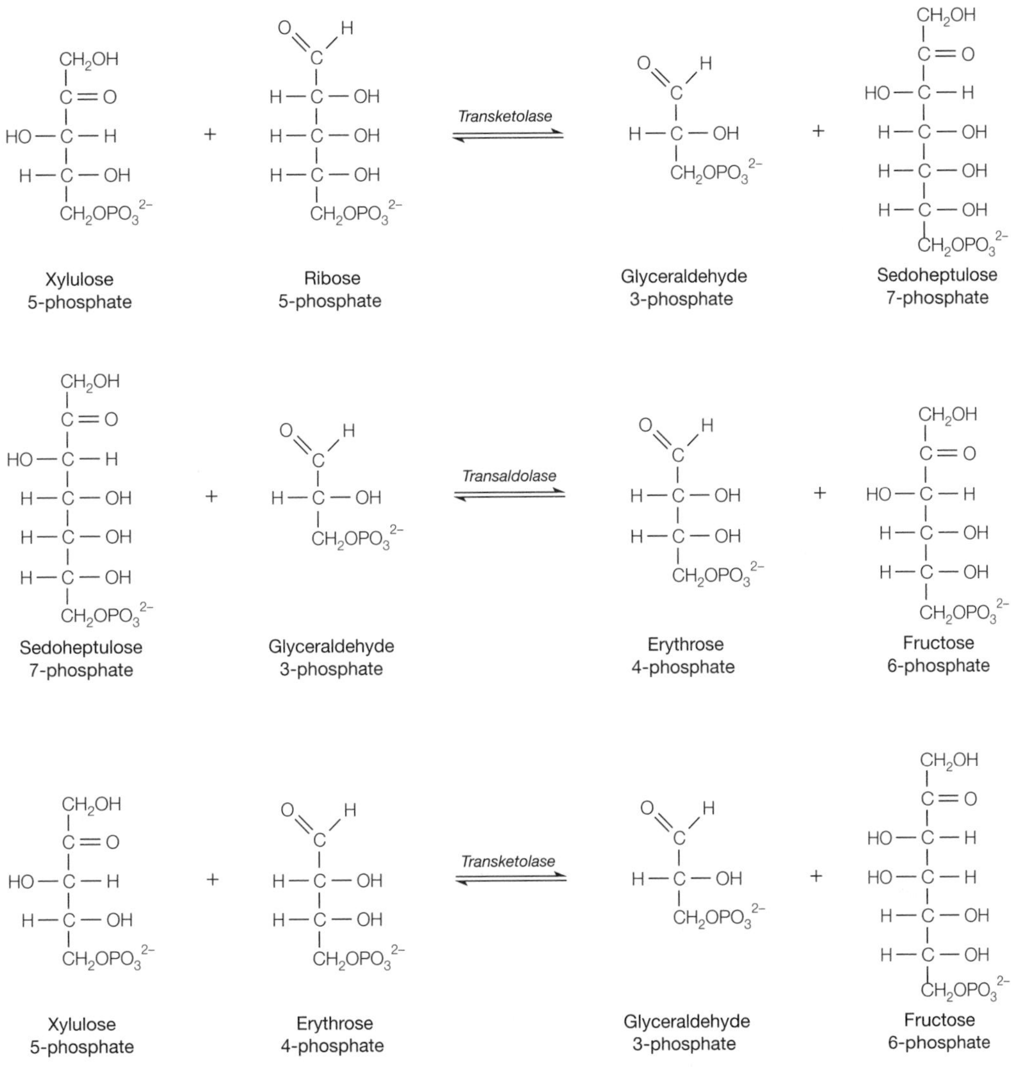

Xylulose 5-phosphate + Ribose 5-phosphate ⇌ (*Transketolase*) Glyceraldehyde 3-phosphate + Sedoheptulose 7-phosphate

Sedoheptulose 7-phosphate + Glyceraldehyde 3-phosphate ⇌ (*Transaldolase*) Erythrose 4-phosphate + Fructose 6-phosphate

Xylulose 5-phosphate + Erythrose 4-phosphate ⇌ (*Transketolase*) Glyceraldehyde 3-phosphate + Fructose 6-phosphate

*Fig. 1. Details of the transaldolase and transketolase reactions.*

NADPH, fructose 6-phosphate and glyceraldehyde 3-phosphate can be taken from glycolysis and converted into ribose 5-phosphate by reversal of the transketolase and transaldolase reactions.

The first reaction of the pathway, the oxidation of glucose 6-phosphate by glucose 6-phosphate dehydrogenase, is rate limiting and irreversible. The enzyme is regulated by $NADP^+$. As the cell uses NADPH, the concentration of $NADP^+$ rises, stimulating glucose 6-phosphate dehydrogenase and so increasing the rate of the pathway and NADPH regeneration.

# J6 Glycogen metabolism

## Key Notes

**Roles of glycogen metabolism**

Glycogen is stored mainly by the liver and skeletal muscle as an energy reserve. The role of stored glycogen in muscle is to provide a source of energy upon prolonged muscle contraction. In contrast, glycogen stored in the liver is used to maintain blood glucose levels.

**Glycogen degradation**

Glycogen degradation is carried out by glycogen phosphorylase and glycogen-debranching enzyme. Phosphorylase removes glucose units sequentially from the nonreducing ends of a glycogen molecule, producing glucose 1-phosphate as the product. It breaks only α1–4 glycosidic bonds and cannot break the α1–6 branchpoints. The glucose 1-phosphate is converted to glucose 6-phosphate by phosphoglucomutase. In liver this is further converted to glucose by glucose 6-phosphatase and the glucose enters the bloodstream. Muscle lacks glucose 6-phosphatase. Rather, here the glucose 6-phosphate enters glycolysis and is oxidized to yield energy for muscle contraction.

**Glycogen synthesis**

UDP-glucose is synthesized by UDP-glucose pyrophosphorylase from UTP and glucose 1-phosphate. Glycogen synthase then uses the UDP-glucose as a substrate to synthesize glycogen, adding one residue at a time to the nonreducing end of the glycogen molecule, forming α1–4 bonds between neighboring glucosyl residues. The enzyme can only extend chains and therefore requires a primer, called glycogenin, in order to begin synthesis. Glycogenin is a protein with eight glucose units joined by α1–4 bonds. The branches in glycogen are created by branching enzyme that breaks an α1–4 bond in the glycogen chain and moves about seven residues to an internal location, joining them to the main chain by an α1–6 bond.

**Related topics**

Monosaccharides and disaccharides (J1)
Glycolysis (J3)
Gluconeogenesis (J4)
Control of glycogen metabolism (J7)

## Roles of glycogen metabolism

Glycogen is a large polymer of glucose residues linked by α1–4 glycosidic bonds with branches every 10 residues or so via α1–6 glycosidic bonds (see Topic J2 for structure). Glycogen provides an important energy reserve for the body. The two main storage sites are the liver and skeletal muscle where the glycogen is stored as granules in the cytosol. The granules contain not only glycogen but also the enzymes and regulatory proteins that are required for glycogen degradation and synthesis. Glycogen metabolism is important because it enables the blood glucose level to be maintained between meals (via glycogen stores in the liver) and also provides an energy reserve for muscular activity. The maintenance of blood glucose is essential in order to supply tissues with an easily metabolizable energy source, particularly the brain which uses only glucose except after a long starvation period.

## Glycogen degradation

Glycogen degradation requires two enzymes; **glycogen phosphorylase** and **glycogen-debranching enzyme.**

Glycogen phosphorylase (often called simply **phosphorylase**) degrades glycogen by breaking α1–4 glycosidic bonds to release glucose units one at a time from the nonreducing end of the glycogen molecule (the end with a free 4′ OH group; see Topic J2) as glucose 1-phosphate. The other substrate required is inorganic phosphate ($P_i$). The reaction is an example of **phosphorolysis**, that is breakage of a covalent bond by the addition of a phosphate group. The (reversible) reaction is as follows:

$$\underset{(n \text{ residues})}{\text{glycogen}} + P_i \rightleftharpoons \underset{(n-1 \text{ residues})}{\text{glycogen}} + \text{glucose 1-phosphate}$$

However, glycogen phosphorylase can remove only those glucose residues that are more than five residues from a branchpoint. Glycogen-debranching enzyme removes the α1–6 branches and so allows phosphorylase to continue degrading the glycogen molecule. The glucose 1-phosphate produced is converted to glucose 6-phosphate by the enzyme **phosphoglucomutase**:

$$\text{glucose 1-phosphate} \rightleftharpoons \text{glucose 6-phosphate}$$

The fate of the glucose 6-phosphate depends on the tissue. Liver contains the enzyme glucose 6-phosphatase which converts the glucose 6-phosphate to glucose, which then diffuses out into the bloodstream and so maintains the blood glucose concentration:

$$\text{glucose 6-phosphate} + H_2O \longrightarrow \text{glucose} + P_i$$

During glycogen degradation in muscle, the main aim is to produce energy quickly and so the glucose 6-phosphate is metabolized immediately via glycolysis. This tissue does not contain glucose 6-phosphatase.

## Glycogen synthesis

Three enzymes are needed to synthesize glycogen:

1. **UDP-glucose pyrophosphorylase** catalyzes the synthesis of UDP-glucose (see *Fig. 1*) from UTP and glucose 1-phosphate:

   $$\text{UTP} + \text{glucose 1-phosphate} \longrightarrow \text{UDP-glucose} + PP_i$$

   The pyrophosphate ($PP_i$) is immediately hydrolyzed by inorganic pyrophosphatase, releasing energy. Thus the overall reaction is very exergonic and essentially irreversible.
2. **Glycogen synthase** now transfers the glucosyl residue from UDP-glucose to the C4 OH group at the nonreducing end of a glycogen molecule, forming an α1–4 glycosidic bond (see *Fig. 2*). Interestingly, glycogen synthase can only extend an existing chain. Thus it needs a primer; this is a protein called **glycogenin**. Glycogenin contains eight glucosyl units linked via α1–4 linkages, which are added to the protein by itself (i.e. autocatalysis). It is this molecule that glycogen synthase then extends. Each glycogen granule contains only a single glycogenin molecule at its core. The fact that glycogen synthase is fully active only when in contact with glycogenin limits the size of the glycogen granule.

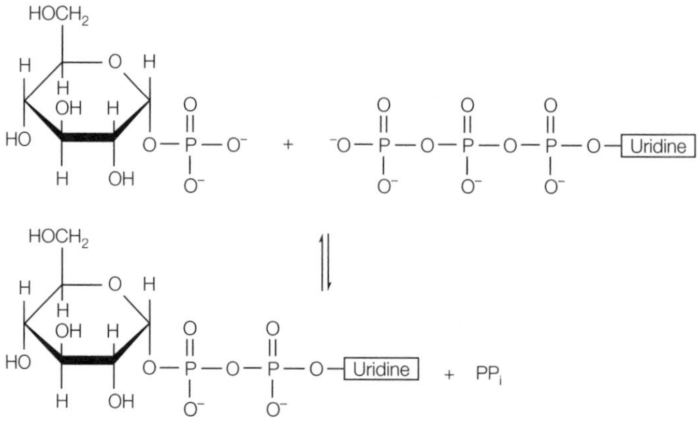

*Fig. 1. The UDP-glucose pyrophosphorylase reaction.*

3. **Branching enzyme [amylo-(1–4→1–6) transglycosylase]** is a different enzyme from glycogen-debranching enzyme. After a number of glucose units have been joined as a straight chain with α1–4 linkages, branching enzyme breaks one of the α1–4 bonds and transfers a block of residues (usually about seven) to a more interior site in the glycogen molecule, reattaching these by creating an α1–6 bond. The branches are important because the enzymes that degrade and synthesize glycogen (glycogen synthase and glycogen phosphorylase, respectively) work only at the ends of the glycogen molecule. Thus the existence of many termini allows a far more rapid rate of synthesis and degradation than would be possible with a nonbranched polymer.

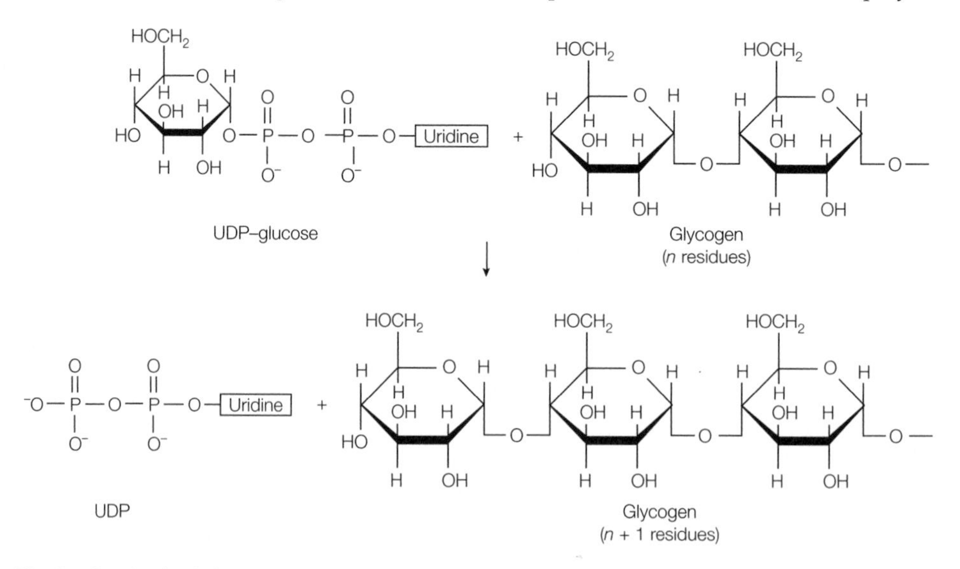


*Fig. 2. Synthesis of glycogen by glycogen synthase.*

# J7 Control of glycogen metabolism

## Key Notes

**Overview**

Glycogen degradation and glycogen synthesis are controlled both by allosteric regulation and by hormonal control.

**Allosteric control and covalent modification**

Phosphorylase exists in a phosphorylated active *a* form and a dephosphorylated normally inactive *b* form. The two forms are interconverted by phosphorylase kinase and protein phosphatase I. In muscle, phosphorylase *b* is activated by the high concentrations of AMP generated by strenuous exercise and thus degrades glycogen, but the AMP stimulation is opposed by high concentrations of ATP and glucose 6-phosphate and so the enzyme is inactive in resting muscle. In liver, phosphorylase *b* is not responsive to AMP but phosphorylase *a* is deactivated by glucose so that glycogen degradation to produce glucose occurs only when glucose levels are low. Glycogen synthase exists as a phosphorylated normally inactive *b* form and a dephosphorylated active *a* form. A high concentration of glucose 6-phosphate can activate synthase *b* in resting muscle, stimulating glycogen synthesis, but the enzyme is inactive in contracting muscle where the glucose 6-phosphate concentration is low.

**Hormonal control by epinephrine and glucagon**

Epinephrine (adrenaline) stimulates glycogen degradation in skeletal muscle. Epinephrine and glucagon stimulate glycogen degradation in liver. The hormone binds to a plasma membrane receptor and activates adenylate cyclase via a G protein. Adenylate cyclase synthesizes cAMP from ATP which in turn activates protein kinase A. Protein kinase A phosphorylates phosphorylase kinase which activates it. The phosphorylase kinase then converts inactive phosphorylase *b* to active phosphorylase *a* by phosphorylation. The same active protein kinase A inactivates glycogen synthase by phosphorylation, converting glycogen synthase *a* to glycogen synthase *b*. When hormone levels fall, stimulation of glycogen degradation is turned off by degradation of cAMP to 5′ AMP by phosphodiesterase and by dephosphorylation of the phosphorylated forms of phosphorylase and synthase by protein phosphatase I.

**Hormonal control by insulin**

Insulin is released into the bloodstream when the blood glucose concentration is high and it stimulates glycogen synthesis. It binds to and activates a receptor protein kinase in the plasma membrane of target cells. This leads to activation of an insulin-responsive protein kinase then activates protein phosphatase I by phosphorylation. Activated protein phosphatase I ensures that phosphorylase and glycogen synthase are dephosphorylated, thus inhibiting glycogen degradation and activating glycogen synthesis.

**Calcium control of glycogen metabolism**

During muscle contraction, $Ca^{2+}$ ions released from the sarcoplasmic reticulum partially activate dephosphorylated phosphorylase kinase and this in turn in turn activates phosphorylase, stimulating glycogen degradation.

**Related topics**

Signal transduction (E5)
Monosaccharides and disaccharides (J1)
Glycolysis (J3)
Gluconeogenesis (J4)
Glycogen metabolism (J6)
Muscle (N1)

## Overview

If glycogen synthesis and glycogen degradation were allowed to occur simultaneously, the net effect would be hydrolysis of UTP, a so-called **futile cycle** (*Fig. 1*). To avoid this, both pathways need to be tightly controlled. This control is carried out via allosteric regulation and covalent modification of both the glycogen synthase and phosphorylase. In addition, the covalent modification is under close hormonal control.

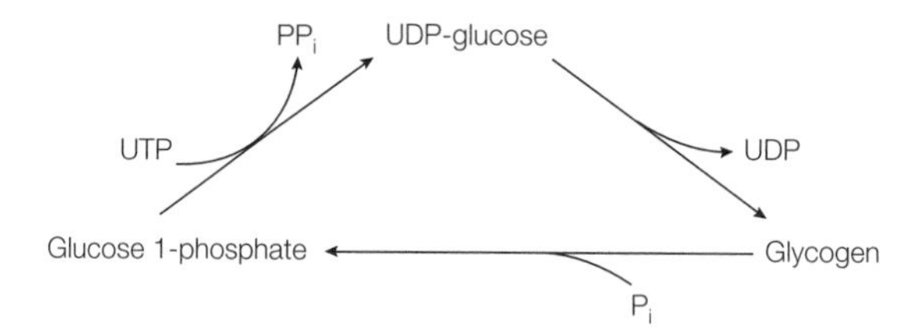


*Fig. 1. Simultaneous operation of glycogen synthesis and degradation resulting in net hydrolysis of UTP.*

## Allosteric control and covalent modification

Phosphorylase exists in two interchangeable forms; active **phosphorylase *a*** and a *normally* inactive **phosphorylase *b***. Phosphorylase *b* is a dimer and is converted into phosphorylase *a* by phosphorylation of a single serine residue on each subunit by the enzyme **phosphorylase kinase**. The process can be reversed and phosphorylase inactivated by removal of the phosphate group by **protein phosphatase I** (*Fig. 2a*) (see Topic C5).

In skeletal muscle, high concentrations of AMP can activate phosphorylase *b* (by acting at an **allosteric site**) but this is opposed by the concentrations of

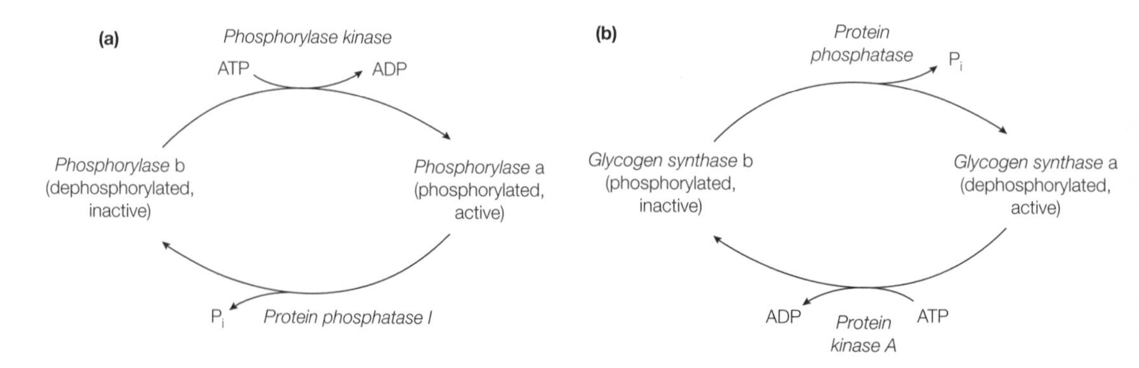


*Fig. 2. Regulation of (a) glycogen phosphorylase activity and (b) glycogen synthase activity by phosphorylation (covalent modification).*

ATP and glucose 6-phosphate found in resting muscle so that in this condition phosphorylase *b* is indeed inactive. Since most of the phosphorylase in resting muscle is phosphorylase *b*, significant glycogen degradation does not occur under these conditions. However, during exercise, the concentrations of ATP and glucose 6-phosphate fall and the concentration of AMP rises. Thus phosphorylase *b* becomes activated and this leads to the rapid degradation of glycogen to yield energy as required. Phosphorylase *a* is unaffected by ATP, AMP or glucose 6-phosphate and so remains active under all conditions.

In liver, phosphorylase *b* is not activated by AMP and is therefore always inactive. Unlike muscle, therefore, glycogen degradation in liver is not responsive to the energy status of the cell. Rather phosphorylase *a* is deactivated by glucose. This fits with the different role of glycogen storage in liver, namely to maintain blood levels of glucose. Thus as glucose levels rise, glycogen degradation by liver phosphorylase *a* is shut off and degradation starts again only as the glucose level falls.

Glycogen synthase is also regulated by covalent modification and allosteric interactions. The enzyme exists as an active glycogen synthase *a* and a *normally* inactive glycogen synthase *b*. However, in contrast to phosphorylase, it is the active form of glycogen synthase (synthase *a*) that is dephosphorylated whereas the inactive synthase *b* form is the phosphorylated form (*Fig. 2b*).

A high concentration of glucose 6-phosphate can activate glycogen synthase *b*. During muscle contraction, glucose 6-phosphate levels are low and therefore glycogen synthase *b* is inhibited. This is at the time when phosphorylase *b* is most active (see above). Thus glycogen degradation occurs and glycogen synthesis is inhibited, preventing the operation of a futile cycle. When the muscle returns to the resting state and ATP and glucose 6-phosphate levels rise, phosphorylase *b* is inhibited (see above), turning off glycogen degradation, whereas glycogen synthase is activated to rebuild glycogen reserves. The synthase *a* form is active irrespective of the concentration of glucose 6-phosphate.

## Hormonal control by epinephrine and glucagon

Glycogen metabolism is tightly controlled by hormones. When blood glucose levels fall, glucagon is secreted by the α cells of the pancreas and acts on the liver to stimulate glycogen breakdown to glucose which is then released into the bloodstream to boost blood glucose levels again. Muscular contraction or nervous stimulation (the 'fight or flight' response) causes the release of epinephrine (adrenaline) from the adrenal medulla and this acts on muscle to increase glycogen breakdown ready to supply the energy needs of the cells.

Consider first the activation of glycogen degradation by epinephrine in the liver. The hormone binds to a receptor, called the **β-adrenergic receptor**, in the plasma membrane of the target cell (*Fig. 3*). Binding of the hormone to the receptor causes a conformational change in the protein which in turn activates an enzyme called **adenylate cyclase**. The receptor does not activate adenylate cyclase directly but rather by activating a **G-protein** as an intermediary in the signaling process (see Topic E5 for details). Activated adenylate cyclase converts ATP to 3′5′ cyclic AMP (cAMP). The cAMP binds to **protein kinase A (PKA)**, also known as **cAMP-dependent protein kinase**. This enzyme consists of two regulatory subunits (R) and two catalytic subunits (C), making a complex, $R_2C_2$, that is normally inactive (*Fig. 3*). The binding of two molecules of cAMP to each of the regulatory subunits leads to dissociation of the complex into an $R_2$ complex and two free C subunits that are now catalytically active. The active protein kinase A

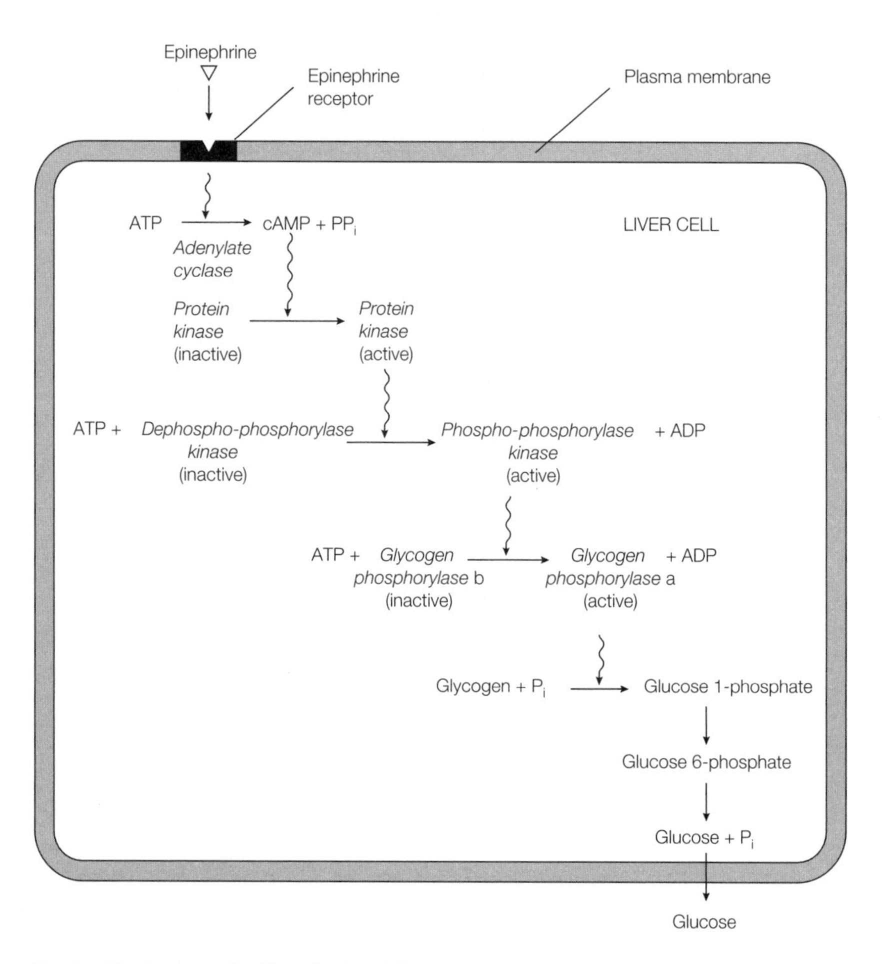


*Fig. 3. Mechanism of action of epinephrine.*

phosphorylates phosphorylase kinase which can exist as an inactive dephosphorylated form and an active phosphorylated form. Thus phosphorylase kinase is now also activated and in turn phosphorylates phosphorylase *b*, converting it to the active phosphorylase *a* that now carries out a rapid degradation of glycogen. This set of reactions is called a **cascade** and is organized so as to greatly amplify the original signal of a small number of hormone molecules. For example, each bound hormone causes the production of many cAMP molecules inside the cell; the activated protein kinase A in turn activates many molecules of phosphorylase kinase; each active phosphorylase kinase activates many molecules of phosphorylase. Thus a small hormonal signal can cause a major shift in cell metabolism.

To prevent the operation of a futile cycle, it is essential that glycogen synthesis is switched off during epinephrine or glucagon stimulation of glycogen breakdown. This is achieved by the activated protein kinase A that, as well as phosphorylating phosphorylase kinase, also phosphorylates glycogen synthase *a*, converting it to the inactive synthase *b* form (*Fig. 4*). Thus protein kinase A activates glycogen degradation and inhibits glycogen synthesis.

When epinephrine and glucagon levels in the bloodstream fall again, the hormone dissociates from the receptor, no more cAMP is made and existing

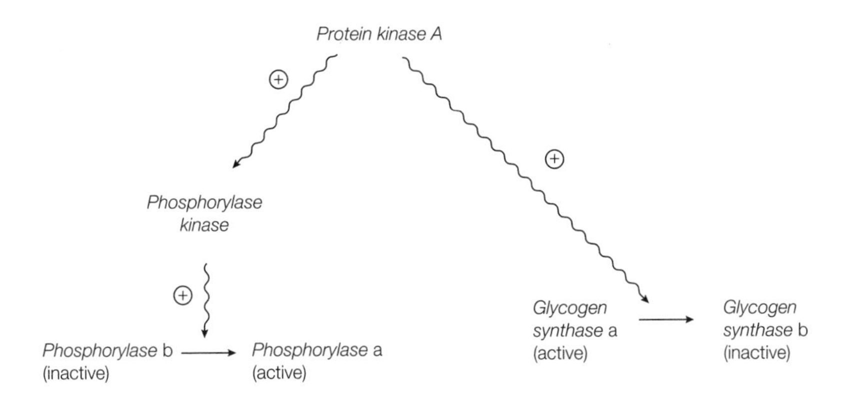


*Fig. 4. Dual control of glycogen degradation and glycogen synthesis by protein kinase A.*

cAMP is converted to 5′ AMP (i.e. 'normal' AMP not the cyclic form) by cAMP phosphodiesterase, an enzyme that is constantly active in the cell. This decline in the cAMP level shuts off the activation cascade. The enzymes that had been phosphorylated are now dephosphorylated by protein phosphatase I, restoring them to their original condition.

## Hormonal control by insulin

Insulin is released into the bloodstream by the β cells of the pancreas when blood glucose levels are high after feeding, and stimulates glycogen synthesis to store excess glucose as glycogen. This control is also achieved via phosphorylation events. Insulin binds to its receptor in the plasma membrane and activates it. This receptor has tyrosine kinase activity (i.e. it will phosphorylate selected tyrosine residues on target proteins; see Topic E5). Its activation leads to the activation of an insulin-responsive protein kinase that then phosphorylates protein phosphatase I, thus activating it. This enzyme now ensures that glycogen synthase is dephosphorylated (and hence active) and that phosphorylase kinase is also dephosphorylated (and hence inactive). The net effect is to stimulate glycogen synthesis.

## Calcium control of glycogen metabolism

As we have seen, during epinephrine or glucagon hormonal control, dephosphorylated phosphorylase kinase is activated by being phosphorylated by protein kinase. This then activates phosphorylase and stimulates glycogen degradation. However, there is also another way to activate dephosphorylated phosphorylase kinase, at least partially, and that is by a high concentration of $Ca^{2+}$ ions. This is important in muscle contraction which is triggered when $Ca^{2+}$ is released from the internal store in the sarcoplasmic reticulum (see Topic N1). Thus, as well as allosteric control and hormonal control during muscle contraction, both of which stimulate glycogen degradation, there is also calcium control.

# K1 STRUCTURES AND ROLES OF FATTY ACIDS

## Key Notes

**Structure and properties**

Fatty acids have a long hydrocarbon chain with a terminal carboxylic acid group. Most fatty acids have an even number of carbon atoms in an unbranched chain. Saturated fatty acids have no double bonds between the carbon atoms, whereas mono- and polyunsaturated fatty acids have one or more double bonds. The properties of a fatty acid depend on the chain length and the number of double bonds.

**Nomenclature**

Fatty acids are named according to the number of carbon atoms in the chain and the number and position of any double bonds. Some of the more common fatty acids are palmitate (C16:0), stearate (C18:0), oleate (C18:1), linoleate (C18:2), linolenate (C18:3) and arachidonate (C20:4). The double bonds in a fatty acid are usually in the *cis* configuration.

**Roles**

Fatty acids have four major biological roles:

1. They are components of membranes (glycerophospholipids and sphingolipids);
2. Several proteins are covalently modified by fatty acids;
3. They act as energy stores (triacylglycerols) and fuel molecules;
4. Fatty acid derivatives serve as hormones and intracellular second messengers.

**Prostaglandins**

Prostaglandins and the other eicosanoids (prostacyclins, thromboxanes and leukotrienes) are derived from arachidonate. These compounds all act as local hormones. Aspirin reduces inflammation by inhibiting prostaglandin synthase, the enzyme that catalyzes the first step in prostaglandin synthesis.

**Related topics**

Membrane lipids (E1)
Fatty acid breakdown (K2)
Fatty acid synthesis (K3)
Triacylglycerols (K4)
Cholesterol (K5)

## Structure and properties

A fatty acid consists of a **hydrocarbon chain** and a **terminal carboxylic acid group** (*Fig. 1*). Most fatty acids found in biology have an **even number of carbon atoms** arranged in an **unbranched chain**. Chain length usually ranges from 14 to 24 carbon atoms, with the most common fatty acids containing 16 or 18 carbon atoms. A **saturated fatty acid** has all of the carbon atoms in its chain saturated with hydrogen atoms (*Fig. 1a*). This gives the general formula $CH_3(CH_2)_nCOOH$, where *n* is an even number. **Mono-unsaturated fatty acids** have one double bond in their structure (*Fig. 1b* and *c*), while **polyunsaturated fatty acids** have two or more double bonds (*Fig. 1d*). The double bonds in polyunsaturated fatty acids are separated by at least one methylene group.

(a)

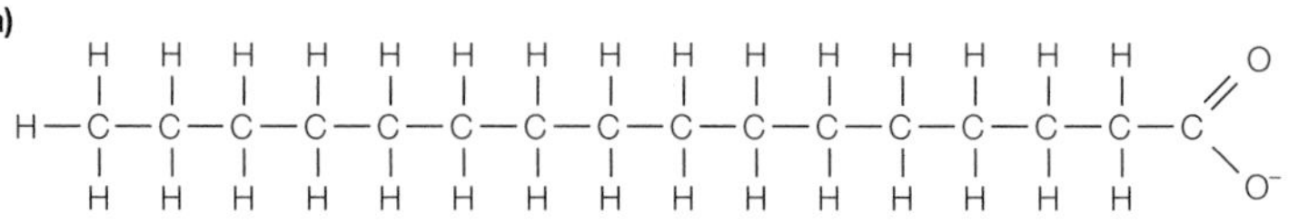

Palmitate (hexadecanoate) C16:0

(b)

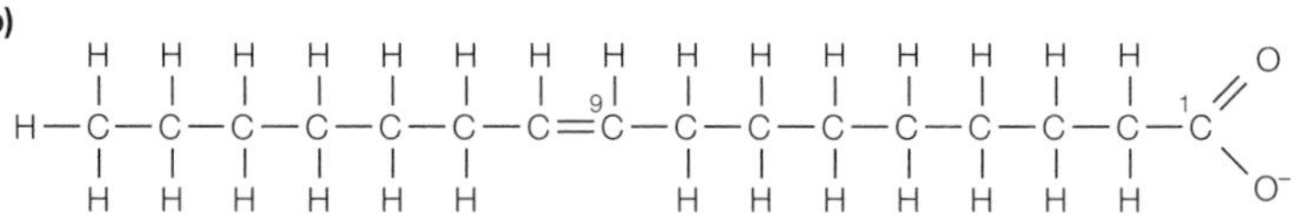

Palmitoleate (*cis*-$\Delta^9$ hexadecenoate) C16:1

(c)

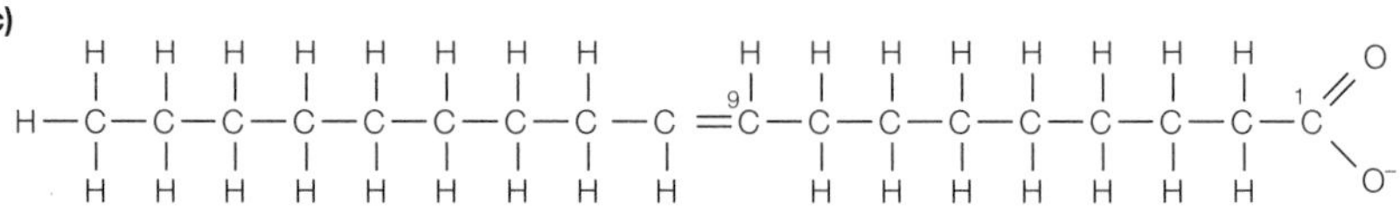

(*Trans*-$\Delta^9$-octadecenoate) C18:1

(d)

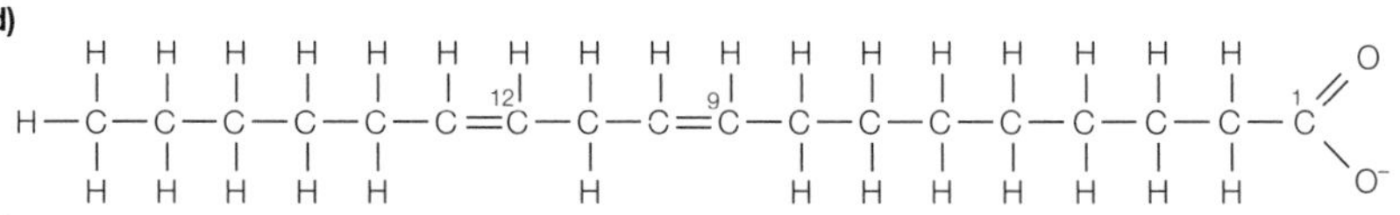

Linoleate (*cis*, *cis*-$\Delta^9$, $\Delta^{12}$-octadecadienoate) C18:2

*Fig. 1. Structures of (a) a saturated fatty acid (palmitate, C16:0); (b) a mono-unsaturated fatty acid with the double bond in the* cis *configuration (palmitoleate, C16:1); (c) a mono-unsaturated fatty acid with the double bond in the* trans *configuration (C18:1); and (d) a polyunsaturated fatty acid (linoleate, C18:2).*

The properties of fatty acids depend on their **chain length** and the **number of double bonds**. Shorter chain length fatty acids have lower melting temperatures than those with longer chains. Unsaturated fatty acids have lower melting temperatures than saturated fatty acids of the same chain length, whilst the corresponding polyunsaturated fatty acids have even lower melting temperatures (see Topic E1).

## Nomenclature

Fatty acids are named according to the total number of carbon atoms, and to the number and position of any double bonds. The systematic names for fatty acids are made by adding 'oic acid' on to the name of the parent hydrocarbon. However, as fatty acids are ionized at physiological pH they are usually written as $RCOO^-$, and have names ending in 'ate' rather than 'oic acid'. A C18 saturated fatty acid would be called octadecanoate, a C18 mono-unsaturated fatty acid octadecenoate, and a C18 fatty acid with two double bonds octadecadienoate (see *Fig. 1*). However, many nonsystematic names are still in use (*Table 1*).

There is also a shorthand notation to show the number of carbon atoms and the position of any double bonds in the structure. A fatty acid with 18 carbons and no double bonds is designated 18:0, while one with 18 carbons and two double bonds is 18:2. The carbon atoms in fatty acids are numbered from the carboxylic acid residue, and so the position of double bonds can be described using the number of the first carbon involved in the bond (e.g. $\Delta^9$ shows a

*Table 1. The names and formulae of some common fatty acids*

| Fatty acid | Formula | No. of double bonds | No. of carbon atoms |
|---|---|---|---|
| Palmitate | $CH_3(CH_2)_{14}COO^-$ | None | 16 |
| Stearate | $CH_3(CH_2)_{16}COO^-$ | None | 18 |
| Oleate | $CH_3(CH_2)_7CH{=}CH(CH_2)_7COO^-$ | 1 | 18 |
| Linoleate | $CH_3(CH_2)_4(CH{=}CHCH_2)_2(CH_2)_6COO^-$ | 2 | 18 |
| Linolenate | $CH_3CH_2(CH{=}CHCH_2)_3(CH_2)_6COO^-$ | 3 | 18 |
| Arachidonate | $CH_3(CH_2)_4(CH{=}CHCH_2)_4(CH_2)_2COO^-$ | 4 | 20 |

double bond between carbons 9 and 10 of the fatty acid chain; *Fig. 1b*). The configuration of the double bonds in most unsaturated fatty acids is ***cis***; so called because the two hydrogens on the carbon atoms either side of the double bond are on the same side of the molecule (*Fig. 1b*) (Latin, *cis* = on this side of). Thus, the full systematic name of linoleate (*Table 1*) is *cis, cis*-$\Delta^9$, $\Delta^{12}$-octadecadienoate (*Fig. 1d*). During the degradation of fatty acids (see Topic K2) some ***trans***-isomers are formed (*Fig. 1c*), where the hydrogens on the carbon atoms either side of the double bond are on opposite sides of the molecule (Latin, *trans* = across).

## Roles

Fatty acids have four major biological roles:

1. They are used to make **glycerophospholipids** and **sphingolipids** that are essential components of biological membranes (see Topic E1);
2. Numerous proteins are **covalently modified** by fatty acids (see Topic E2). Myristate (C14:0) and palmitate (C16:0) are directly attached to some proteins, while phosphatidylinositol is covalently linked to the C terminus of other proteins via a complex glycosylated structure;
3. Fatty acids act as **fuel molecules**, being stored as **triacylglycerols**, and broken down to generate energy (see Topics K2 and K4);
4. Derivatives of fatty acids serve as **hormones** (such as the prostaglandins) and **intracellular second messengers** (such as DAG and $IP_3$) (see Topic E5).

## Prostaglandins

**Prostaglandins**, and the structurally related molecules **prostacyclins, thromboxanes** and **leukotrienes**, are called **eicosanoids** because they contain 20 carbon atoms (Greek *eikosi* = 20). These hormones are relatively short-lived and hence act locally near to their site of synthesis in the body. They are derived from the common precursor **arachidonate** (*Fig. 2*). This polyunsaturated fatty

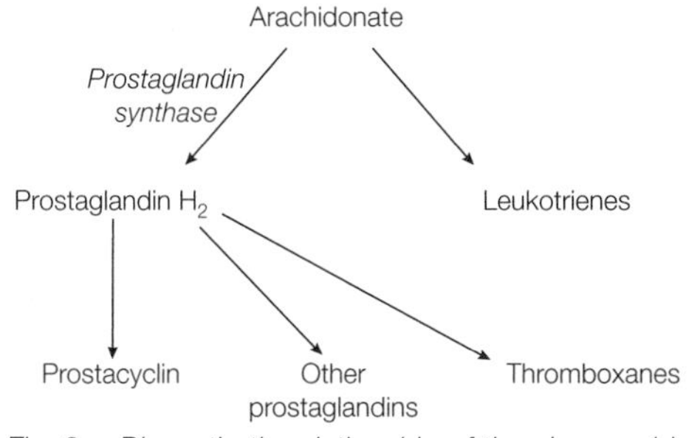


*Fig. 2. Biosynthetic relationship of the eicosanoids.*

acid is a derivative of linoleate (*Table 1*). Prostaglandins stimulate inflammation, modulate synaptic transmission between nerve cells, and induce sleep. Although **aspirin** (acetylsalicylic acid) has been used for centuries to decrease inflammation, pain and fever, it was not until 1974 that John Vane discovered how aspirin works. Aspirin inhibits the synthesis of prostaglandins by irreversibly inhibiting **prostaglandin synthase**. This enzyme catalyzes the first step in the synthesis of prostaglandins, prostacyclins and thromboxanes (*Fig. 2*).

# K2 FATTY ACID BREAKDOWN

## Key Notes

**Overview**

Fatty acid breakdown (also called β-oxidation) brings about the oxidation of long-chain fatty acids with the production of energy in the form of ATP. The fatty acids are converted into their acyl CoA derivatives and then metabolized by the removal of two-carbon acetyl CoA units from the end of the acyl chain.

**Activation**

Fatty acid breakdown occurs in the cytosol of prokaryotes and in the mitochondrial matrix of eukaryotes. The fatty acid is activated by forming a thioester link with CoA before entering the mitochondria.

**Transport into mitochondria**

The inner mitochondrial membrane is not permeable to long-chain acyl CoA derivatives and so these are transported into the mitochondria as carnitine derivatives by carnitine/acyl carnitine translocase.

**β-Oxidation pathway**

Fatty acid breakdown involves a repeating sequence of four reactions:

1. Oxidation of the acyl CoA by FAD to form a *trans*-$\Delta^2$-enoyl CoA;
2. Hydration to form 3-hydroxyacyl CoA;
3. Oxidation by $NAD^+$ to form 3-ketoacyl CoA;
4. Thiolysis by a second CoA molecule to form acetyl CoA and an acyl CoA shortened by two carbon atoms.

The $FADH_2$ and NADH produced feed directly into oxidative phosphorylation, while the acetyl CoA feeds into the citric acid cycle where further $FADH_2$ and NADH are produced. In animals the acetyl CoA produced in β-oxidation cannot be converted into pyruvate or oxaloacetate, and cannot therefore be used to make glucose. However, in plants two additional enzymes allow acetyl CoA to be converted into oxaloacetate via the glyoxylate pathway.

**Oxidation of unsaturated fatty acids**

Unsaturated fatty acids require the action of additional enzymes in order to be completely degraded by β-oxidation.

**Oxidation of odd-chain fatty acids**

Fatty acids having an odd number of carbon atoms give rise to acetyl CoA (two carbon atoms) and propionyl CoA (three carbon atoms) in the final round of fatty acid degradation.

**Regulation**

The rate of fatty acid degradation is controlled by the availability of free fatty acids in the blood which arise from the breakdown of triacylglycerols.

**Energy yield**

Complete degradation of palmitate (C16:0) in β-oxidation generates 35 ATP molecules from oxidation of the NADH and $FADH_2$ produced directly and 96 ATPs from the breakdown of the acetyl CoA molecules in the citric acid cycle. However, two ATP equivalents are required to activate the palmitate to its acyl CoA derivative prior to oxidation. Thus the net yield is 129 ATPs.

**Ketone bodies**

When in excess, acetyl CoA produced from the β-oxidation of fatty acids is converted into acetoacetate and D-3-hydroxybutyrate. Together with acetone, these compounds are collectively termed ketone bodies. Acetoacetate and D-3-hydroxybutyrate are produced in the liver and provide an alternative supply of fuel for the brain under starvation conditions or in diabetes.

**Related topics**

Structures and roles of fatty acids (K1)
Fatty acid synthesis (K3)
Triacylglycerols (K4)
Citric acid cycle (L1)

## Overview

**Fatty acid breakdown** brings about the oxidation of long-chain fatty acids. The fatty acids are first converted to their **acyl coenzyme A (CoA) derivatives** and then degraded by the successive removal of two-carbon units from the end of the fatty acid as **acetyl CoA**. The pathway produces $FADH_2$ and NADH directly. The acetyl CoA produced can also enter the citric acid cycle and produce further $FADH_2$ and NADH (see Topic L1). The $FADH_2$ and NADH are then oxidized by the respiratory electron transport chain to yield energy in the form of ATP (see Topic L2).

## Activation

Fatty acid breakdown occurs in the cytosol of prokaryotes and in the mitochondrial matrix of eukaryotes. Before entering the **mitochondrial matrix**, the fatty acid is **activated** by forming a **thioester link** with **CoA** (*Fig. 1*). This reaction is catalyzed by **acyl CoA synthase** (also called **fatty acid thiokinase**) which is present on the outer mitochondrial membrane, and uses a molecule of ATP. The overall reaction is irreversible due to the subsequent hydrolysis of $PP_i$ to two molecules of $P_i$.

## Transport into mitochondria

Small- and medium-chain acyl CoA molecules (up to 10 carbon atoms) are readily able to cross the **inner mitochondrial membrane** by diffusion. However, longer chain acyl CoAs do not readily cross the inner mitochondrial membrane, and require a specific transport mechanism. To achieve this, the longer chain acyl CoAs are conjugated to the polar **carnitine** molecule, that is found in both plants and animals. This reaction, catalyzed by an enzyme on the outer face of the inner mitochondrial membrane (**carnitine acyltransferase** I), removes the CoA group and substitutes it with a carnitine molecule (*Fig. 2*). The acylcarnitine is then transported across the inner mitochondrial membrane by a **carnitine/acylcarnitine translocase**. This integral membrane transport protein (see Topic E3) transports acylcarnitine molecules into the mitochondrial matrix and free carnitine molecules out. Once inside the mitochondrial matrix the acyl group is transferred back on to CoA, releasing free carnitine, by the enzyme **carnitine acyltransferase II** which is located on the matrix side of the inner mitochondrial membrane (*Fig. 2*).

$$R-C(=O)O^- + ATP + HS-CoA \xrightarrow[\text{Acyl CoA synthase}]{} R-C(=O)-S-CoA + AMP + PP_i$$

*Fig. 1. Activation of a fatty acid.*

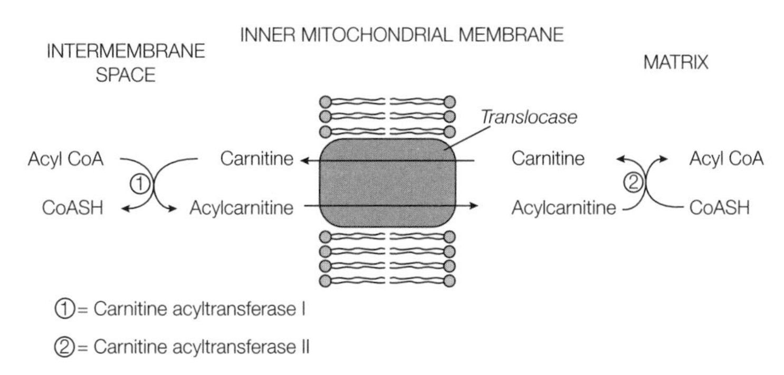


*Fig. 2. Transport of fatty acids across the inner mitochondrial membrane.*

## β-Oxidation pathway

The individual reactions involved in the degradation of fatty acids by β-oxidation are as follows (*see Fig. 3*):

1. **Oxidation** of the fatty acyl CoA to enoyl CoA forming a *trans* $\Delta^2$-double bond on the fatty acyl chain and producing $FADH_2$ (catalyzed by **acyl CoA dehydrogenase**).
2. **Hydration** of the *trans* $\Delta^2$-enoyl CoA to form 3-hydroxyacyl CoA (catalyzed by **enoyl CoA hydratase**).
3. **Oxidation** of 3-hydroxyacyl CoA to 3-ketoacyl CoA producing NADH (catalyzed by **hydroxyacyl CoA dehydrogenase**).
4. Cleavage, or **thiolysis**, of 3-ketoacyl CoA by a second CoA molecule, giving acetyl CoA and an acyl CoA shortened by two carbon atoms (catalyzed by **β-ketothiolase**).

Thus, the breakdown of individual fatty acids occurs as a repeating sequence of four reactions: **oxidation** (by FAD), **hydration**, **oxidation** (by $NAD^+$) and **thiolysis**. These four reactions form one 'round' of fatty acid degradation (*Fig. 3*) and their overall effect is to remove two-carbon units sequentially in the form of acetyl CoA from the fatty acid chain. The cleavage of the $\Delta^2$ (or β) bond of the fatty acyl chain (see *Fig. 3*, top structure, for nomenclature) gives fatty acid breakdown its alternative name, **β-oxidation**. The shortened acyl CoA then undergoes further cycles of β-oxidation until the last cycle, when the acyl CoA with four carbon atoms is split into two molecules of acetyl CoA. Thus a C16 saturated acyl CoA, such as palmitoyl CoA, would be completely degraded into eight molecules of acetyl CoA by seven rounds of degradation, leading to the overall equation:

$$\text{palmitoyl CoA} + 7\ \text{FAD} + 7\ \text{NAD}^+ + 7\ \text{CoA} + 7\ \text{H}_2\text{O} \rightarrow 8\ \text{acetyl CoA} + 7\ \text{FADH}_2 + 7\ \text{NADH} + 7\ \text{H}^+$$

Mitochondria contain three **acyl CoA dehydrogenases** which act on short-, medium- and long-chain acyl CoAs, respectively. In contrast, there is just one each of the enzymes enoyl CoA hydratase, hydroxyacyl CoA dehydrogenase and β-ketothiolase which all have a broad specificity with respect to the length of the acyl chain.

In animals the acetyl CoA produced from fatty acid degradation cannot be converted into pyruvate or oxaloacetate. Although the two carbon atoms from acetyl CoA enter the citric acid cycle, they are both oxidized to $CO_2$ in the reactions catalyzed by isocitrate dehydrogenase and α-ketoglutarate dehydrogenase (see

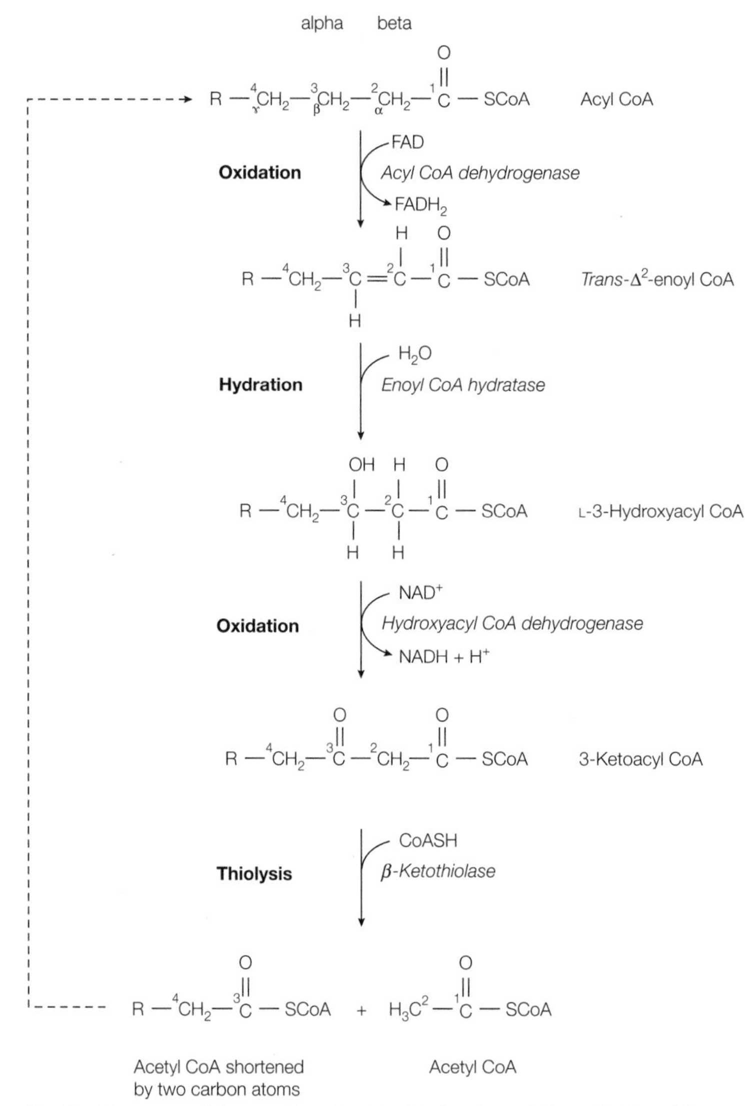


*Fig. 3. Summary of the reactions involved in the degradation of fatty acids.*

Topic L1). Thus, **animals cannot convert fatty acids into glucose**. In contrast, plants have two additional enzymes, isocitrate lyase and malate synthase, that enable them to convert the carbon atoms of acetyl CoA into oxaloacetate. This is accomplished via the **glyoxylate pathway**, a route involving enzymes of both the mitochondrion and the glyoxysome, a specialized membranous plant organelle.

## Oxidation of unsaturated fatty acids

**Unsaturated fatty acids** require some additional processing before they can be degraded completely by β-oxidation. Unsaturated fatty acyl CoAs with double bonds at odd-numbered carbon atoms (i.e. between, for example, C-9 and C-10 as in palmitoleate; see Topic K1, *Fig. 1b*) are acted on in the normal way by the degradation mechanism until the acyl CoA dehydrogenase encounters the *cis*-$\Delta^3$-enoyl CoA formed at the end of the third round. The presence of the double bond between C-3 and C-4 prevents the formation of another double bond between C-2 and C-3. To overcome this problem an **isomerase** converts the *cis*-$\Delta^3$ bond into a *trans*-$\Delta^2$ double bond, and the resulting *trans*-$\Delta^2$-enoyl CoA can then continue down the β-oxidation pathway (see *Fig. 4*).

Another enzyme, in addition to the isomerase, is required for the oxidation of **polyunsaturated fatty acids** which have a double bond at an even-numbered carbon atom. In this case the 2,4-dienoyl intermediate resulting from the action of acyl CoA dehydrogenase is acted on by **2,4-dienoyl CoA reductase** to form *cis*-$\Delta^3$-enoyl CoA (*Fig. 4*). This is then converted by the isomerase into the *trans* form which continues down the pathway. These reactions are important since over half the fatty acids of plant and animal lipids are unsaturated (and often polyunsaturated).

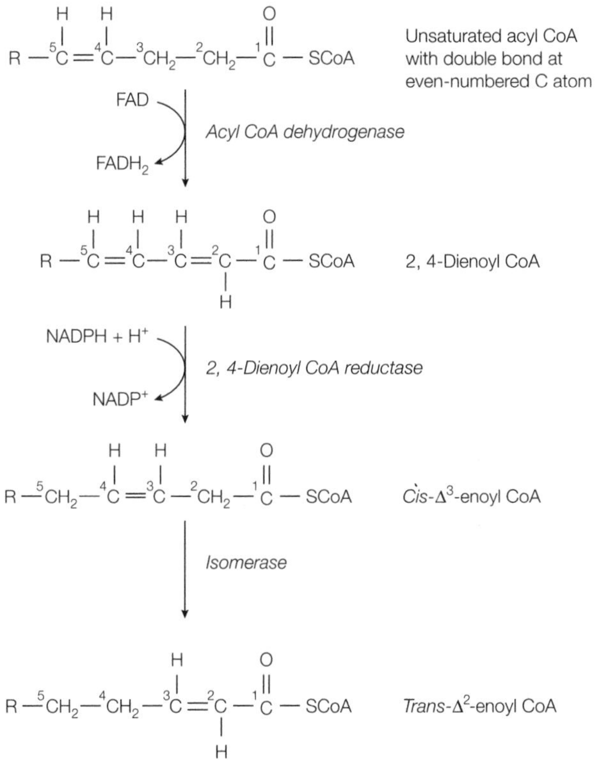

$R-{}^5CH={}^4CH-{}^3CH_2-{}^2CH_2-{}^1C(=O)-SCoA$ Unsaturated acyl CoA with double bond at even-numbered C atom

FAD → $FADH_2$ *Acyl CoA dehydrogenase*

$R-{}^5CH={}^4CH-{}^3CH={}^2CH-{}^1C(=O)-SCoA$ 2, 4-Dienoyl CoA

NADPH + $H^+$ → $NADP^+$ *2, 4-Dienoyl CoA reductase*

$R-{}^5CH_2-{}^4CH={}^3CH-{}^2CH_2-{}^1C(=O)-SCoA$ *Cis*-$\Delta^3$-enoyl CoA

*Isomerase*

$R-{}^5CH_2-{}^4CH_2-{}^3CH={}^2CH-{}^1C(=O)-SCoA$ *Trans*-$\Delta^2$-enoyl CoA

*Fig. 4. Accessory enzymes required for the metabolism of unsaturated fatty acids.*

## Oxidation of odd-chain fatty acids

Fatty acids having an **odd number of carbon atoms** (which are relatively rare in nature) are also degraded by the β-oxidation pathway in the same way as those with an even number of carbon atoms. The only difference is that in the final round the five carbon acyl CoA intermediate is cleaved into one molecule of the C3 **propionyl CoA** and one molecule of the C2 acetyl CoA. The propri-onyl CoA is then converted into succinyl CoA which enters the citric acid cycle (see Topic L1).

## Regulation

The major point of control of β-oxidation is the availability of fatty acids. The major source of free fatty acids in the blood is from the breakdown of **triacylglycerol** stores in adipose tissue which is regulated by the action of hormone-sensitive triacylglycerol lipase (see Topic K4). Fatty acid breakdown and fatty acid synthesis are coordinately controlled so as to prevent a futile cycle (see Topic K3).

## Energy yield

For each round of degradation, one $FADH_2$, one NADH and one acetyl CoA molecule are produced. Each NADH generates three ATP molecules, and each $FADH_2$ generates two ATPs during oxidative phosphorylation (see Topic L2). In addition, each acetyl CoA yields 12 ATPs on oxidation by the citric acid cycle (see Topic L1). The total yield for each round of fatty acid degradation is therefore 17 ATP molecules.

The complete degradation of palmitoyl CoA (C16:0) requires seven rounds of degradation and hence produces $7 \times 5 = 35$ ATP molecules. A total of eight acetyl CoA molecules are produced and hence another $8 \times 12 = 96$ ATP. Thus the total ATP yield per molecule of palmitate degraded is $35 + 96 = 131$ ATP. However, one ATP is hydrolyzed to AMP and $PP_i$ in the activation of palmitate to palmitoyl CoA, resulting in two high-energy bonds being cleaved. Thus the net yield is **129 ATPs** (*Table 1*).

The yield of ATP is reduced slightly for unsaturated fatty acids, since the additional metabolic reactions which enable them to be degraded by the β-oxidation pathway either involve using NADPH or bypass an $FADH_2$-producing reaction (see *Fig. 4*).

*Table 1. Calculation of the ATP yield from the complete oxidation of palmitate*

| Degradative step | ATP yield |
|---|---|
| 7 x 5 ATP for oxidation of NADH and $FADH_2$ produced by each round of degradation | 35 |
| 8 x 12 ATP for the breakdown of acetyl CoA by the citric acid cycle | 96 |
| –2 ATP equivalents for the activation of palmitate | –2 |
| | Total = 129 |

## Ketone bodies

When the level of acetyl CoA from β-oxidation increases in excess of that required for entry into the citric acid cycle, the acetyl CoA is converted into **acetoacetate** and **D-3-hydroxybutyrate** by a process known as **ketogenesis**. D-3-hydroxybutyrate, acetoacetate and its nonenzymic breakdown product **acetone** are referred to collectively as **ketone bodies** (*Fig. 5*).

Two molecules of acetyl CoA initially condense to form acetoacetyl CoA in a reaction which is essentially the reverse of the thiolysis step in β-oxidation. The acetoacetyl CoA reacts with another molecule of acetyl CoA to form **3-hydroxy-3-methylglutaryl CoA (HMG CoA)** (*Fig. 5*). This molecule is then cleaved to form acetoacetate and acetyl CoA. (HMG CoA is also the starting point for cholesterol biosynthesis; see Topic K5.) The acetoacetate is then either reduced to D-3-hydroxybutyrate in the mitochondrial matrix or undergoes a slow, spontaneous decarboxylation to acetone (*Fig. 5*). In diabetes, acetoacetate is produced faster than it can be metabolized. Hence untreated diabetics have high levels of ketone bodies in their blood, and the smell of acetone can often be detected on their breath.

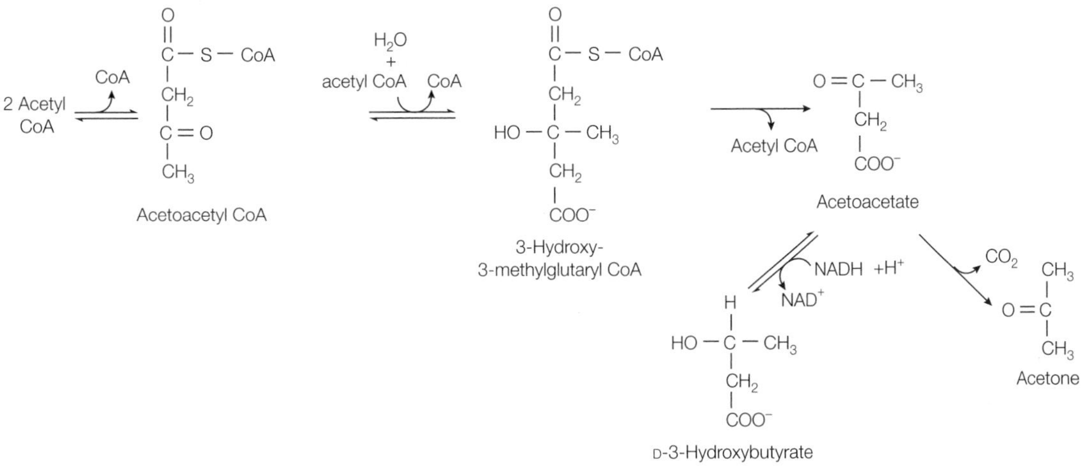

*Fig. 5. Conversion of acetyl CoA to the ketone bodies acetoacetate, acetone and D-3-hydroxybutyrate.*

Acetoacetate and D-3-hydroxybutyrate are produced mainly in the liver and are not just degradation products of little physiological value. They are used in preference to glucose as an energy source by certain tissues such as the heart muscle and kidney cortex. Although glucose is normally the major fuel for the brain, under conditions of starvation or diabetes this organ can switch to using predominantly acetoacetate.

# K3 FATTY ACID SYNTHESIS

## Key Notes

**Overview**

Fatty acid synthesis involves the condensation of two-carbon units, in the form of acetyl CoA, to form long hydrocarbon chains in a series of reactions. These reactions are carried out on the fatty acid synthase complex using NADPH as reductant. The fatty acids are covalently linked to acyl carrier protein (ACP) during their synthesis.

**Transport into the cytosol**

Since fatty acid synthesis takes place in the cytosol, the acetyl CoA produced from pyruvate has to be transported out of the mitochondria. However, the inner mitochondrial membrane is not permeable to this compound, so it is first combined with oxaloacetate to form citrate which readily crosses the membrane. In the cytosol the citrate is cleaved to regenerate the acetyl CoA.

**The pathway**

The first committed step in fatty acid biosynthesis is the carboxylation of acetyl CoA to form malonyl CoA which is catalyzed by the biotin-containing enzyme acetyl CoA carboxylase. Acetyl CoA and malonyl CoA are then converted into their ACP derivatives. The elongation cycle in fatty acid synthesis involves four reactions: condensation of acetyl-ACP and malonyl-ACP to form acetoacetyl-ACP releasing free ACP and $CO_2$, then reduction by NADPH to form D-3-hydroxybutyryl-ACP, followed by dehydration to crotonyl-ACP, and finally reduction by NADPH to form butyryl-ACP. Further rounds of elongation add more two-carbon units from malonyl-ACP on to the growing hydrocarbon chain, until the C16 palmitate is formed. Further elongation of fatty acids takes place on the cytosolic surface of the smooth endoplasmic reticulum (SER).

**Formation of double bonds**

The enzymes for introducing double bonds into the acyl chain are also present on the cytosolic surface of the SER. The polyunsaturated fatty acids linoleate and linolenate cannot be synthesized by mammals and are therefore termed essential fatty acids as they have to be ingested in the diet.

**Regulation**

The key control point of fatty acid synthesis is acetyl CoA carboxylase which catalyzes the formation of malonyl CoA. Acetyl CoA carboxylase is inactivated by phosphorylation by an AMP-activated protein kinase. Thus when the energy charge of the cell is low (high AMP, low ATP) acetyl CoA carboxylase is inactive. It is reactivated by dephosphorylation by protein phosphatase 2A. Glucagon and epinephrine inhibit fatty acid synthesis by inhibiting protein phosphatase 2A, whereas insulin stimulates fatty acid synthesis by activating the phosphatase. Acetyl CoA carboxylase is also allosterically regulated: citrate activates the enzyme, whereas palmitoyl CoA inhibits it.

**Related topics**

Structures and roles of fatty acids (K1)
Fatty acid breakdown (K2)
Triacylglycerols (K4)

## Overview

Fatty acids are synthesized by the condensation of two-carbon units. However, in terms of the enzymic steps involved, the process is not the reverse of β-oxidation (see Topic K2). **Fatty acid synthesis** involves a separate series of reactions to build up long-chain hydrocarbons from **acetyl CoA** units. The key differences between fatty acid synthesis and breakdown are:

- fatty acid synthesis occurs in the **cytosol** of both prokaryotes and eukaryotes whereas their degradation occurs in the mitochondria of eukaryotes;
- fatty acid synthesis uses **NADPH** as the reductant whereas NADH is produced in β-oxidation;
- during their synthesis, fatty acids are covalently linked to an **acyl carrier protein (ACP)** as opposed to CoA in their degradation;
- the enzyme activities of fatty acid synthesis in higher organisms are present in a single, multifunctional polypeptide chain (as a dimer) called **fatty acid synthase**, whereas in β-oxidation the individual activities are present on separate enzymes.

## Transport into the cytosol

Fatty acids are synthesized in the cytosol, but acetyl CoA is produced from pyruvate in the mitochondria (see Topic L1). Thus the acetyl CoA must be transferred from the mitochondria into the cytosol to allow fatty acid synthesis to occur. However, the **inner mitochondrial membrane** is not readily permeable to this molecule. This problem is overcome by the condensation of acetyl CoA with oxaloacetate to form **citrate** (*Fig. 1*). This is then transported into the cytosol where it is cleaved to regenerate acetyl CoA and oxaloacetate by **ATP-citrate lyase** in an energy-requiring process. The oxaloacetate, which also cannot cross the inner mitochondrial membrane, is returned to the mitochondrial matrix through conversion first to malate (catalyzed by **malate dehydrogenase**) and then to pyruvate (catalyzed by **NADP$^+$-linked malate enzyme**) (*Fig. 1*). This latter decarboxylation reaction generates NADPH which can be used in fatty acid synthesis. The remaining NADPH required for fatty acid synthesis is provided by the pentose phosphate pathway (see Topic J5). Once back in the mitochondrial matrix, pyruvate is carboxylated to form oxaloacetate by **pyruvate carboxylase** with the hydrolysis of a further molecule of ATP (*Fig. 1*).

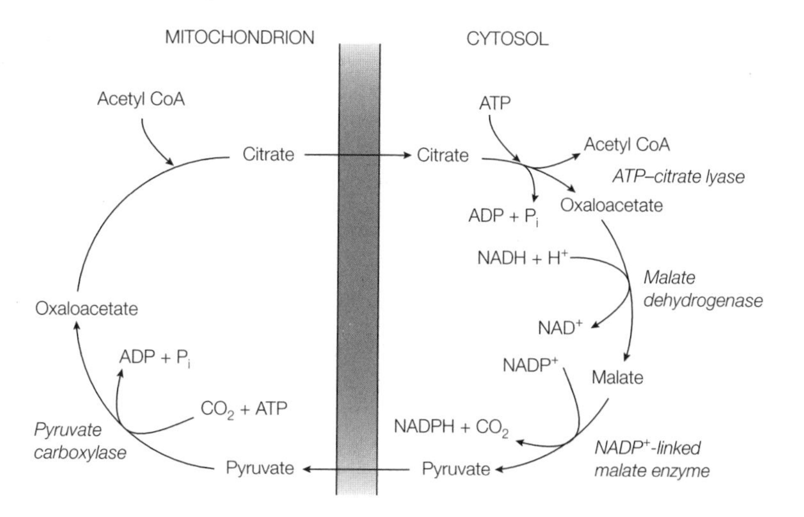


*Fig. 1. Transport of acetyl CoA from the mitochondrial matrix into the cytosol.*

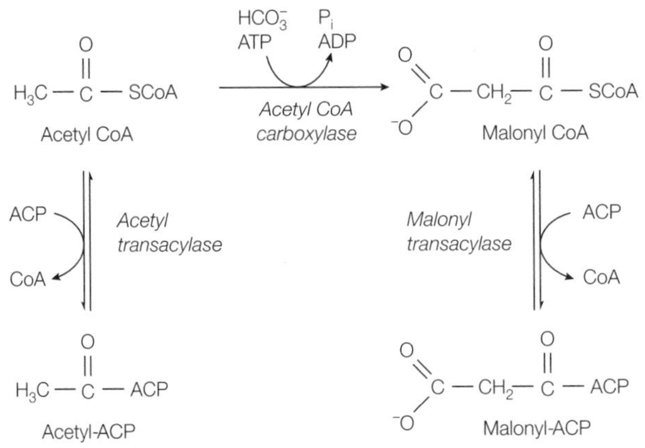

Fig. 2. Formation of acetyl- and malonyl-acyl carrier protein (ACP).

## The pathway

The **first committed step** in fatty acid biosynthesis is the carboxylation of acetyl CoA to form **malonyl CoA** using $CO_2$ in the form of bicarbonate $HCO_3^-$ (*Fig. 2*). This reaction is catalyzed by the enzyme **acetyl CoA carboxylase** which has **biotin** as a prosthetic group, a common feature in $CO_2$-binding enzymes. One molecule of ATP is hydrolyzed in the reaction, which is irreversible. The **elongation steps** of fatty acid synthesis all involve intermediates linked to the terminal sulfhydryl group of the **phosphopantetheine** reactive unit in **ACP**; phosphopantetheine is also the reactive unit in CoA. Therefore, the next steps are the formation of acetyl-ACP and malonyl-ACP by the enzymes **acetyl transacylase** and **malonyl transacylase**, respectively (*Fig. 2*). (For the synthesis of fatty acids with an odd number of carbon atoms the three-carbon propionyl-ACP is the starting point instead of malonyl-ACP.)

The elongation cycle of fatty acid synthesis has four stages for each round of synthesis (*Fig. 3*). For the first round of synthesis these are:

1. **Condensation** of acetyl-ACP and malonyl-ACP to form acetoacetyl-ACP, releasing free ACP and $CO_2$ (catalyzed by acyl-malonyl-ACP condensing enzyme).
2. **Reduction** of acetoacetyl-ACP to form D-3-hydroxybutyryl-ACP, using NADPH as reductant (catalyzed by β-ketoacyl-ACP reductase).
3. **Dehydration** of D-3-hydroxybutyryl-ACP to produce crotonyl-ACP (catalyzed by 3-hydroxyacyl-ACP dehydratase).
4. **Reduction** of crotonyl-ACP by a second NADPH molecule to give butyryl-ACP (catalyzed by enoyl-ACP reductase).

This first round of elongation produces the four-carbon butyryl-ACP. The cycle now repeats with malonyl-ACP adding two-carbon units in each cycle to the lengthening acyl-ACP chain. This continues until the 16-carbon **palmitoyl-ACP** is formed. This molecule is not accepted by the acyl-malonyl-ACP condensing enzyme, and so cannot be elongated further by this process. Instead it is hydrolyzed by a thioesterase to give palmitate and ACP.

The overall stoichiometry for the synthesis of palmitate is:

$$8\text{ acetyl CoA} + 7\text{ ATP} + 14\text{ NADPH} + 6\text{ H}^+ \rightarrow \text{palmitate} + 14\text{ NADP}^+ + 8\text{ CoA} + 6\text{ H}_2\text{O} + 7\text{ ADP} + 7\text{ P}_i$$

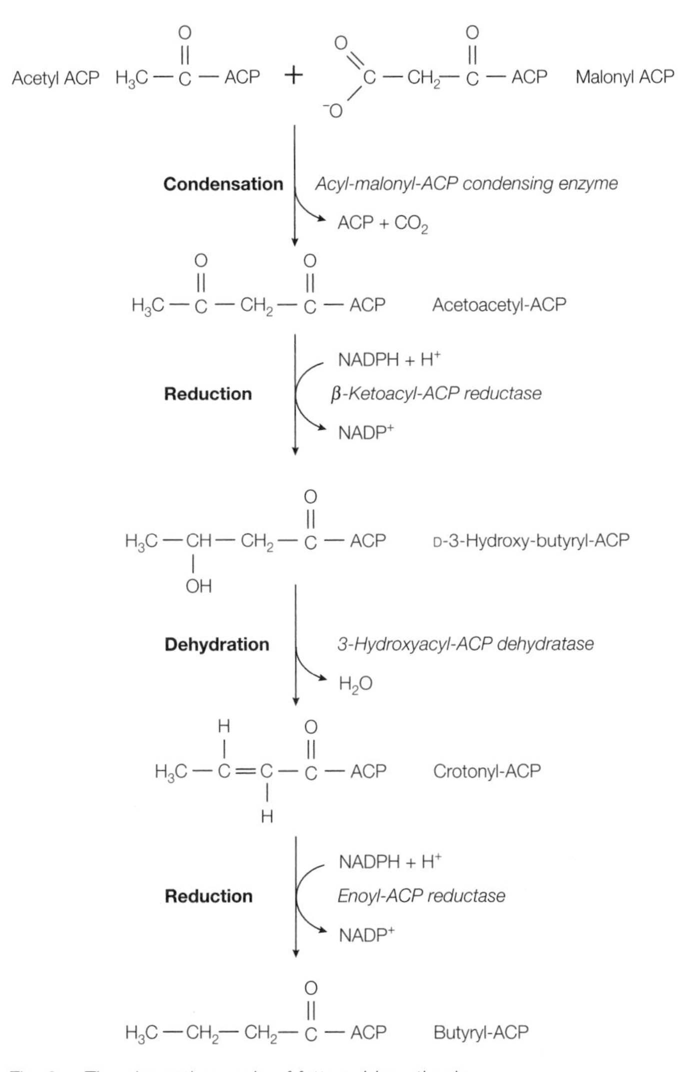


*Fig. 3. The elongation cycle of fatty acid synthesis.*

For each of the seven rounds of fatty acid elongation, one ATP is used in the synthesis of malonyl-CoA and two NADPH are used in the reduction reactions.

In eukaryotes the elongation of fatty acids beyond C16 palmitate is carried out by enzymes located on the cytosolic surface of the **smooth endoplasmic reticulum (SER)**. Malonyl CoA is used as the two-carbon donor, and the fatty acid is elongated as its CoA derivative rather than its ACP derivative.

In prokaryotes, each of the reactions of fatty acid synthesis is catalyzed by a separate enzyme. However, in eukaryotes, the enzymes of the fatty acid synthesis elongation cycle are present in a single polypeptide chain, multifunctional enzyme complex, called **fatty acid synthase**. The fatty acid synthase complex exists as a dimer, with the ACP moiety shuttling the fatty acyl chain between successive catalytic sites, and from one subunit of the dimer to the other. It is, in effect, a highly efficient production line for fatty acid biosynthesis.

## Formation of double bonds

In eukaryotes the SER has enzymes able to introduce **double bonds** into fatty acyl CoA molecules in an oxidation reaction that uses molecular oxygen. This reaction is catalyzed by a membrane-bound complex of three enzymes: NADH-cytochrome $b_5$ reductase, cytochrome $b_5$ and a desaturase. The overall reaction is:

$$\text{saturated fatty acyl CoA} + \text{NADH} + \text{H}^+ + \text{O}_2 \rightarrow \text{mono-unsaturated acyl CoA} + \text{NAD}^+ + 2\ \text{H}_2\text{O}$$

The reaction may be repeated to introduce more than one double bond into a fatty acid.

Mammals lack the enzymes to insert double bonds at carbon atoms beyond C-9 in the fatty acid chain. Thus they cannot synthesize **linoleate** and **linolenate**, both of which have double bonds later in the chain than C-9 (linoleate has *cis, cis* $\Delta^9$, $\Delta^{12}$ double bonds, and linolenate has all-*cis* $\Delta^9$, $\Delta^{12}$, $\Delta^{15}$ double bonds). Hence, in mammals linoleate and linolenate are called **essential fatty acids** since they have to be supplied in the diet. These two unsaturated fatty acids are also the starting points for the synthesis of other unsaturated fatty acids, such as **arachidonate**. This C20:4 fatty acid is the precursor of several biologically important molecules, including the prostaglandins, prostacyclins, thromboxanes and leukotrienes (see Topic K1).

## Regulation

The synthesis of fatty acids takes place when carbohydrate and energy are plentiful and when fatty acids are scarce. The key enzyme in the regulation of fatty acid synthesis is **acetyl CoA carboxylase** which synthesizes malonyl CoA. This is a good example of **control at the committed step** of a metabolic pathway. Acetyl CoA carboxylase is inactivated by the **phosphorylation** of a single serine residue by an **AMP-activated protein kinase** (*Fig. 4*) (see Topic C5). Unlike cAMP-dependent protein kinase (protein kinase A) (see Topic K4), this kinase is not affected by cAMP, but instead is stimulated by AMP and inhibited by ATP. Thus when the energy charge of the cell is low (i.e. there is a high AMP:ATP ratio) fatty acid synthesis is switched off. **Protein phosphatase 2A** removes the phosphate group from inactivated acetyl CoA carboxylase (*Fig. 4*), thereby reactivating it.

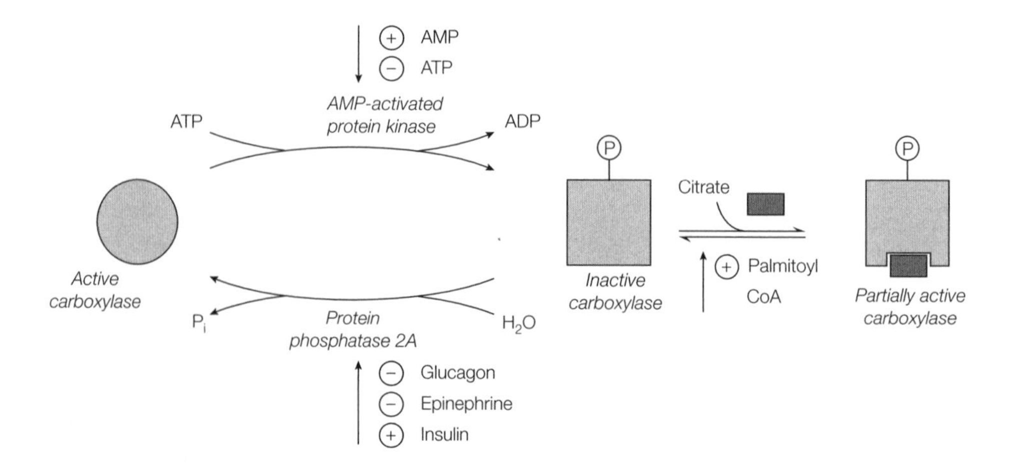


*Fig. 4. Summary of the control of acetyl CoA carboxylase by phosphorylation and allosteric regulation.*

Acetyl CoA carboxylase is also subject to hormonal regulation. When energy is required, **glucagon** and **epinephrine** inhibit protein phosphatase 2A, thus keeping acetyl CoA carboxylase in the inactive form (*Fig. 4*) and blocking fatty acid synthesis. In the well-fed state, when blood glucose levels are high, **insulin** stimulates acetyl CoA carboxylase, possibly by activating protein phosphatase 2A (*Fig. 4*), thereby leading to an increase in fatty acid synthesis.

As well as its control by phosphorylation/dephosphorylation, acetyl CoA carboxylase is also **allosterically regulated**. The citric acid cycle intermediate **citrate**, the level of which is high when both acetyl CoA and ATP are abundant, allosterically stimulates acetyl CoA carboxylase. This results in the conversion of the inactive phosphorylated form into a partially active form that is still phosphorylated (*Fig. 4*), thereby activating fatty acid synthesis so that the excess acetyl CoA is 'stored' as fatty acid residues within triacylglycerol in adipose tissue. In contrast, high levels of **palmitoyl CoA**, which is abundant when there is an excess of fatty acids, antagonize the effect of citrate on acetyl CoA carboxylase, reducing its activity (*Fig. 4*) and switching off further fatty acid synthesis.

# K4 TRIACYLGLYCEROLS

## Key Notes

**Structure and function**

Triacylglycerols (fats or triglycerides) consist of three fatty acid chains esterified to a glycerol backbone. Simple triacylglycerols have three identical fatty acids, mixed triacylglycerols have two or three different fatty acids. Triacylglycerols are the major energy store and the major dietary lipid in humans. They are insoluble in water and are stored in specialized adipose (fat) cells.

**Synthesis**

Triacylglycerols are synthesized from glycerol 3-phosphate, which is derived from the glycolytic intermediate dihydroxyacetone phosphate, and fatty acyl CoAs. Acyl CoA molecules are added on to glycerol 3-phosphate to form first lysophosphatidic acid and then phosphatidic acid. The phosphate group is then removed to form diacylglycerol (DAG), which is further acylated to triacylglycerol. The energy for the synthesis of triacylglycerols comes from the hydrolysis of the high-energy thioester bond in acyl CoA.

**Breakdown**

The fatty acids in triacylglycerols are released from the glycerol backbone by the action of lipases. The free fatty acids can then be degraded by β-oxidation to produce energy. The glycerol is converted into dihydroxyacetone phosphate which enters glycolysis.

**Regulation**

The concentration of free fatty acids in the blood is controlled by the rate at which hormone-sensitive triacylglycerol lipase hydrolyzes the triacylglycerols stored in adipose tissue. Glucagon, epinephrine and norepinephrine cause an increase in the intracellular level of cAMP which allosterically activates cAMP-dependent protein kinase. The kinase in turn phosphorylates hormone-sensitive lipase, activating it, and leading to the release of fatty acids into the blood. Insulin has the opposite effect; it decreases the level of cAMP which leads to the dephosphorylation and inactivation of hormone-sensitive lipase.

**Related topics**

Fatty acid breakdown (K2)
Fatty acid synthesis (K3)
Lipoproteins (K6)

## Structure and function

**Triacylglycerols** (also called **fats** or **triglycerides**) consist of three fatty acid chains esterified to a glycerol backbone. **Simple triacylglycerols** have three identical fatty acids esterified to the glycerol backbone, while **mixed triacylglycerols** have two or three different fatty acid chains (*Fig. 1*). Triacylglycerols constitute the **major fuel store** and the **major dietary lipid** in humans. Triacylglycerols are a highly concentrated **energy store**. The energy yield from the complete oxidation of fatty acids is about 39 kJ $g^{-1}$, compared with an energy yield of 13 kJ $g^{-1}$ of carbohydrate or protein. The hydrophobic properties of fats make them insoluble in water, and fats are stored in specialized cells called **adipose cells** (fat cells), which consist almost entirely of triacylglycerol. These

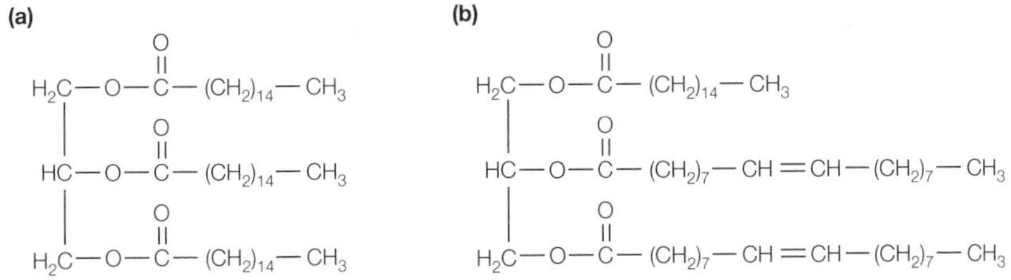

(a)

$H_2C-O-C(=O)-(CH_2)_{14}-CH_3$

$HC-O-C(=O)-(CH_2)_{14}-CH_3$

$H_2C-O-C(=O)-(CH_2)_{14}-CH_3$

(b)

$H_2C-O-C(=O)-(CH_2)_{14}-CH_3$

$HC-O-C(=O)-(CH_2)_7-CH=CH-(CH_2)_7-CH_3$

$H_2C-O-C(=O)-(CH_2)_7-CH=CH-(CH_2)_7-CH_3$

*Fig. 1. Structure of (a) a simple triacylglycerol (1,2,3-tripalmitoyl-glycerol) and (b) a mixed triacylglycerol (1-palmitoyl-2,3-dioleoyl-glycerol).*

cells are specialized for the synthesis and storage of triacylglycerols and for their mobilization into fuel molecules. Triacylglycerols are transported round the body in large lipid–protein particles called **lipoproteins** (see Topic K6).

## Synthesis

Triacylglycerols are synthesized from **fatty acyl CoAs** and **glycerol 3-phosphate** (*Fig. 2*). The glycolytic intermediate **dihydroxyacetone phosphate** is first reduced to glycerol 3-phosphate which is, in turn, acylated by glycerol-3-phosphate acyltransferase to form **lysophosphatidic acid**. This is then reacted with a further acyl CoA molecule to form **phosphatidic acid**. Removal of the phosphate group from phosphatidic acid generates **diacylglycerol (DAG)**, which is further acylated with a third acyl CoA molecule to form **triacylglycerol** (*Fig. 2*). ATP is not involved in the biosynthesis of triacylglycerols. Instead the reactions are driven by the cleavage of the high-energy thioester bond between the acyl moiety and CoA. Both phosphatidic acid (phosphatidate) and DAG are also used in the synthesis of membrane phospholipids (see Topic E1).

## Breakdown

The initial event in the utilization of both stored fat and dietary fat as energy sources is the hydrolysis of triacylglycerol by **lipases**. These enzymes release the three fatty acid chains from the glycerol backbone (*Fig. 3*). The fatty acids can then be broken down in **β-oxidation** to generate energy (see Topic K2). The glycerol backbone is also utilized, being transformed into dihydroxyacetone phosphate, an intermediate in glycolysis (*Fig. 4*). This requires two enzymes, glycerol kinase, which uses ATP to phosphorylate glycerol, producing L-glycerol 3-phosphate, and glycerol 3-phosphate dehydrogenase which produces dihydroxyacetone phosphate.

In the intestine, dietary fats are hydrolyzed by **pancreatic lipase** and the released fatty acids taken up into the **intestinal cells**. Both the digestion and uptake processes are aided by the detergent-like properties of the **bile salts** (see Topic K5).

## Regulation

The breakdown of fatty acids in β-oxidation (see Topic K2) is controlled mainly by the concentration of free fatty acids in the blood, which is, in turn, controlled by the hydrolysis rate of triacylglycerols in adipose tissue by **hormone-sensitive triacylglycerol lipase**. This enzyme is regulated by **phosphorylation** and **dephosphorylation** (*Fig. 5*) in response to hormonally controlled levels of the intracellular second messenger **cAMP** (see Topic E5). The catabolic hormones **glucagon**, **epinephrine** and **norepinephrine** bind to receptor proteins on the cell surface and increase the levels of cAMP in adipose cells through activation of **adenylate cyclase** (see Topic E5). The cAMP allosterically activates

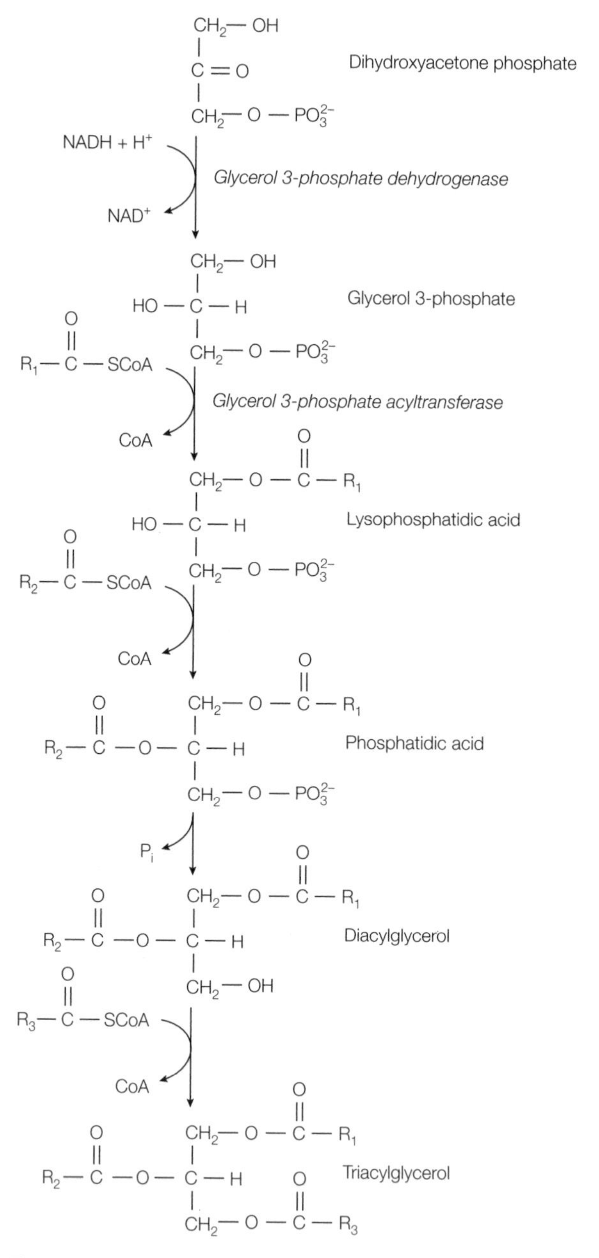


*Fig. 2. Synthesis of triacylglycerols.*

**cAMP-dependent protein kinase** (otherwise known as **protein kinase A**) which phosphorylates various intracellular enzymes including hormone-sensitive lipase. Phosphorylation of hormone-sensitive lipase activates it, thereby stimulating the hydrolysis of triacylglycerols, raising the levels of fatty acids in the blood, and subsequently activating β-oxidation in tissues such as muscle and liver. Glucagon and epinephrine also prevent the dephosphorylation, and therefore activation, of **acetyl CoA carboxylase**, so that fatty acid synthesis is inhibited (see Topic K3).

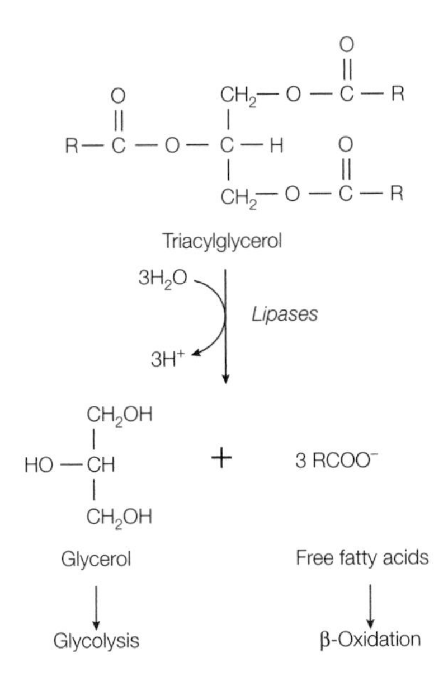


*Fig. 3. Breakdown of triacylglycerols.*

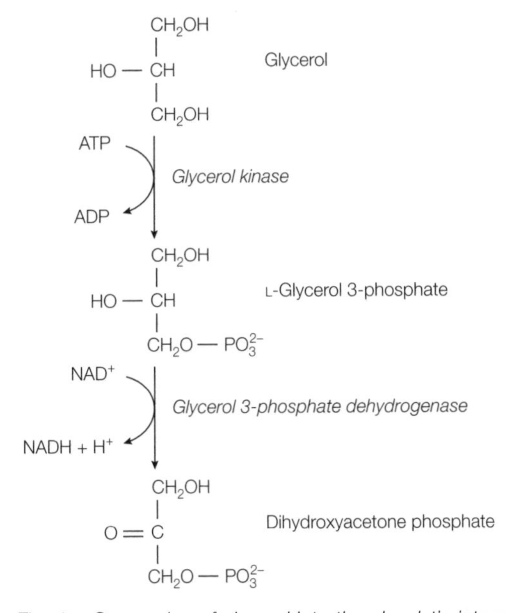


*Fig. 4. Conversion of glycerol into the glycolytic intermediate dihydroxyacetone phosphate.*

The anabolic hormone **insulin** has the opposite effect to glucagon and epinephrine. It stimulates the formation of triacylglycerols through decreasing the level of cAMP, which promotes the dephosphorylation and inactivation of hormone-sensitive lipase (*Fig. 5*). Insulin also stimulates the dephosphorylation of acetyl CoA carboxylase, thereby activating fatty acid synthesis (see Topic K3). Thus fatty acid synthesis and degradation are **coordinately controlled** so as to prevent a futile cycle.

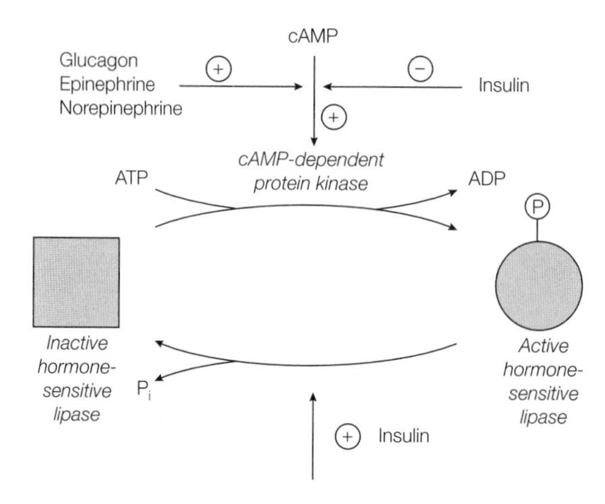


*Fig. 5. Summary of the control of hormone-sensitive triacylglycerol lipase.*

# K5 Cholesterol

## Key Notes

**Functions of cholesterol**

Cholesterol is a component of cell membranes and is the precursor of steroid hormones and the bile salts.

**Biosynthesis of cholesterol**

All 27 carbon atoms of cholesterol are derived from acetyl CoA. First acetyl CoA and acetoacetyl CoA combine to form HMG CoA which, in turn, is reduced to mevalonate by HMG CoA reductase. Mevalonate is converted into the five-carbon isoprene compounds 3-isopentenyl pyrophosphate and its isomer dimethylallyl pyrophosphate. These two compounds condense to form the C10 geranyl pyrophosphate, which is elongated to the C15 farnesyl pyrophosphate by the addition of another molecule of isopentenyl pyrophosphate. Two molecules of farnesyl pyrophosphate condense to form the C30 squalene, which is then converted via squalene epoxide and lanosterol to cholesterol.

**Regulation of cholesterol biosynthesis**

Cholesterol can either be obtained in the diet or synthesized in the liver. High levels of cholesterol and its metabolites decrease the amount and inhibit the activity of HMG CoA reductase, the enzyme that catalyzes the committed step in cholesterol biosynthesis. This enzyme can also be inhibited therapeutically by the compound lovastatin.

**Bile salts**

Bile salts (bile acids) are the major excretory form of cholesterol. These polar compounds are formed in the liver by converting cholesterol into the activated intermediate cholyl CoA and then combining this compound with either glycine, to form glycocholate, or taurine, to form taurocholate. The detergent-like bile salts are secreted into the intestine where they aid the digestion and uptake of dietary lipids.

**Vitamin D**

Vitamin D is derived via cholesterol in a series of reactions, one of which requires the action of UV light to break the bond between two carbon atoms. Deficiency of vitamin D causes rickets in children and osteomalacia in adults.

**Steroid hormones**

The steroid hormones are derived from cholesterol by a series of reactions that involve the heme-containing cytochrome P450 enzymes. These mono-oxygenases require both $O_2$ and NADPH to function. There are five classes of steroid hormones: (1) the progestagens (e.g. progesterone); (2) the androgens (e.g. testosterone); (3) the estrogens (e.g. estrone); (4) the glucocorticoids (e.g. cortisol); and (5) the mineralocorticoids (e.g. aldosterone).

**Related topics**

Membrane lipids (E1)    Lipoproteins (K6)

## Functions of cholesterol

**Cholesterol** is a **steroid**. It is an important constituent of **cell membranes**, where, in mammals, it modulates their fluidity (see Topic E1). Cholesterol is also the precursor of **steroid hormones** such as progesterone, testosterone and cortisol, and the **bile salts** (see below).

## Biosynthesis of cholesterol

Animals are able to synthesize cholesterol *de novo* by an elegant series of reactions in which all 27 carbon atoms of cholesterol are derived from **acetyl CoA**. The acetate units are first converted into **C5 isoprene units**, that are then condensed to form a linear precursor to the cyclic cholesterol.

The first stage in the synthesis of cholesterol is the formation of **isopentenyl pyrophosphate** (*Fig. 1*). Acetyl CoA and acetoacetyl CoA combine to form **3-hydroxy-3-methylglutaryl CoA (HMG CoA)**. This process takes place in the liver, where the HMG CoA in the mitochondria is used to form ketone bodies during starvation (see Topic K2), whereas that in the cytosol is used to synthesize cholesterol in the fed state (under the influence of cholesterol). HMG CoA is then reduced to **mevalonate** by **HMG CoA reductase** (*Fig. 1*). This is the committed step in cholesterol biosynthesis and is a key control point. Mevalonate is converted into **3-isopentenyl pyrophosphate** by three consecutive reactions each involving ATP, with $CO_2$ being released in the last reaction (*Fig. 1*).

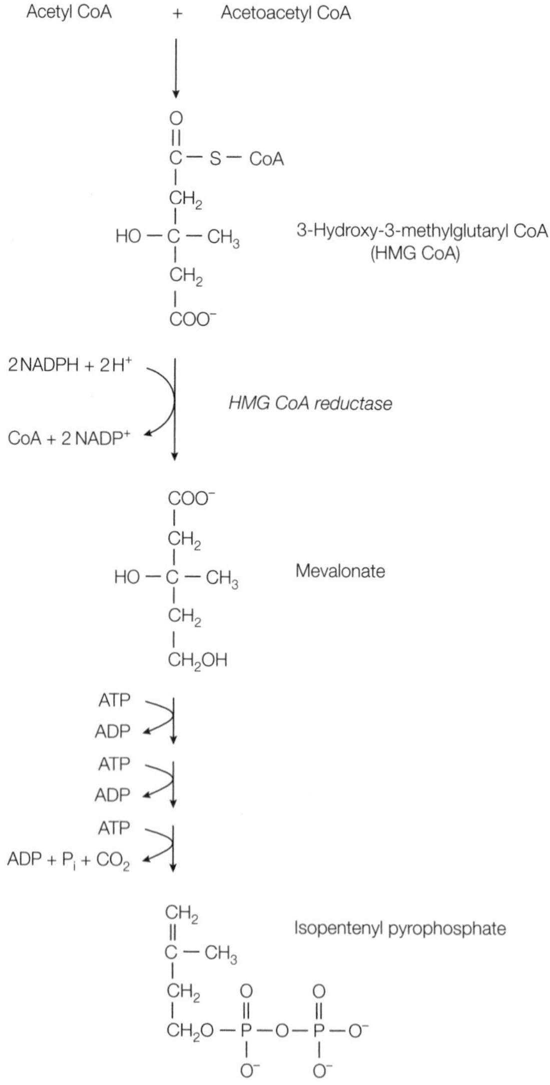


*Fig. 1. Synthesis of isopentenyl pyrophosphate.*

The C5 isoprene units in isopentenyl pyrophosphate are then condensed to form the C30 compound squalene (*Fig. 2*). First, isopentenyl pyrophosphate isomerizes to **dimethylallyl pyrophosphate** (*Fig. 2a*), which reacts with another molecule of isopentenyl pyrophosphate to form the C10 compound **geranyl pyrophosphate** (*Fig. 2b*). Another molecule of isopentenyl pyrophosphate then reacts with geranyl pyrophosphate to form the C15 compound **farnesyl pyrophosphate**. Next, two molecules of farnesyl pyrophosphate condense to form **squalene** (*Fig. 2b*).

Squalene is then converted into **squalene epoxide** in a reaction that uses $O_2$ and NADPH (*Fig. 2b*). The squalene epoxide cyclizes to form **lanosterol**, and finally cholesterol is formed from lanosterol by the removal of three methyl groups, the reduction of one double bond by NADPH, and the migration of the other double bond (*Fig. 2b*).

**Farnesyl pyrophosphate** and the C20 compound **geranylgeranyl pyrophosphate** (which is formed by the condensation of another isopentenyl pyrophosphate with farnesyl pyrophosphate) are covalently linked to cysteine residues in a number of proteins, promoting their association with membranes (see Topic E2). **Dolichol**, which contains some 20 isoprene units is used to carry the biosynthetic precursor of the N-linked oligosaccharides that are subsequently attached to proteins (see Topic H5).

## Regulation of cholesterol biosynthesis

Cholesterol can be obtained either from the diet or it can be synthesized *de novo*, mainly in the liver. Cholesterol is transported round the body in **lipoprotein** particles (see Topic K6). The rate of synthesis of cholesterol is dependent on the cellular level of cholesterol. High levels of cholesterol and its metabolites control cholesterol biosynthesis by:

- feedback-inhibiting the activity of **HMG CoA reductase**, the enzyme which catalyzes the committed step in cholesterol biosynthesis (see Topic C5);
- decreasing the amount of HMG CoA reductase by reducing the synthesis and translation of its mRNA;
- decreasing the amount of HMG CoA reductase by increasing its rate of degradation.

In addition, HMG CoA reductase, like acetyl CoA carboxylase in fatty acid synthesis (see Topic K3), is inactivated by phosphorylation by an AMP-activated protein kinase, retained in this form under the influence of glucagon during starvation.

HMG CoA reductase can be inhibited therapeutically by administering the drug **lovastatin**, based on the fungal products mevinolin and **compactin,** which competitively inhibit the enzyme and hence decrease the rate of cholesterol biosynthesis. Therefore, these compounds are routinely used for the treatment of **hypercholesterolemia** (high levels of blood cholesterol) (see Topic K6).

## Bile salts

**Bile salts** (or bile acids) are polar derivatives of cholesterol and constitute the major pathway for the excretion of cholesterol in mammals. In the liver, cholesterol is converted into the activated intermediate **cholyl CoA** which then reacts either with the amino group of glycine to form **glycocholate** (*Fig. 3a*), or with the amino group of taurine ($H_2N\text{-}CH_2\text{-}CH_2\text{-}SO_3^-$, a derivative of cysteine) to form **taurocholate** (*Fig. 3b*). After synthesis in the liver, the bile salts glycocholate and taurocholate are stored and concentrated in the gall bladder, before release into the small intestine. Since they contain both polar and nonpolar

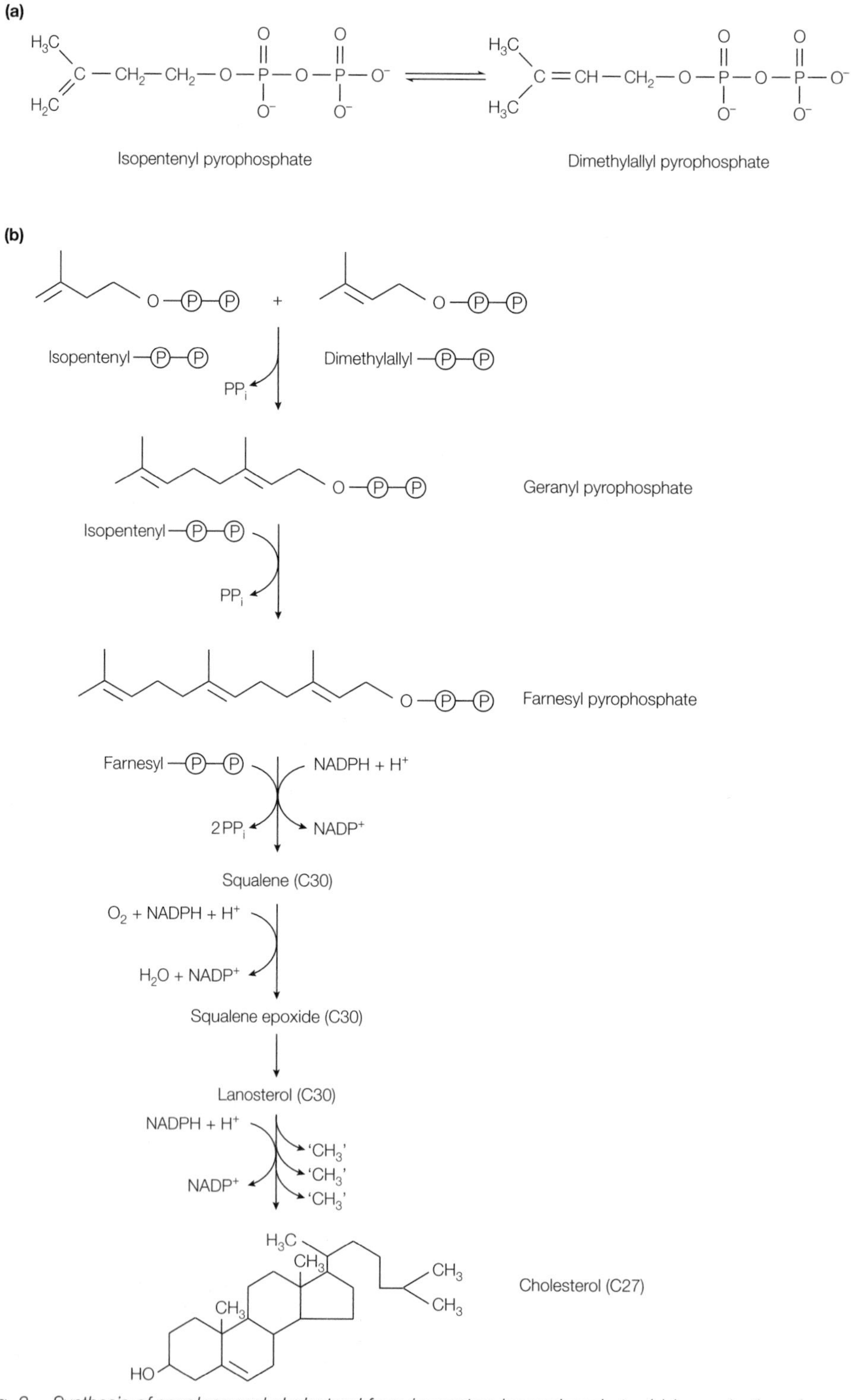

(a)

Isopentenyl pyrophosphate ⇌ Dimethylallyl pyrophosphate

(b)

Isopentenyl–P–P + Dimethylallyl–P–P → Geranyl pyrophosphate + $PP_i$

Geranyl pyrophosphate + Isopentenyl–P–P → Farnesyl pyrophosphate + $PP_i$

Farnesyl–P–P, NADPH + $H^+$ → Squalene (C30) + 2$PP_i$ + $NADP^+$

Squalene (C30), $O_2$ + NADPH + $H^+$ → Squalene epoxide (C30) + $H_2O$ + $NADP^+$

Squalene epoxide (C30) → Lanosterol (C30)

Lanosterol (C30), NADPH + $H^+$ → Cholesterol (C27) + $NADP^+$ + 'CH$_3$' + 'CH$_3$' + 'CH$_3$'

*Fig. 2. Synthesis of squalene and cholesterol from isopentenyl pyrophosphate. (a) Isomerization of isopentenyl pyrophosphate to dimethylallyl pyrophosphate; (b) synthesis of cholesterol.*

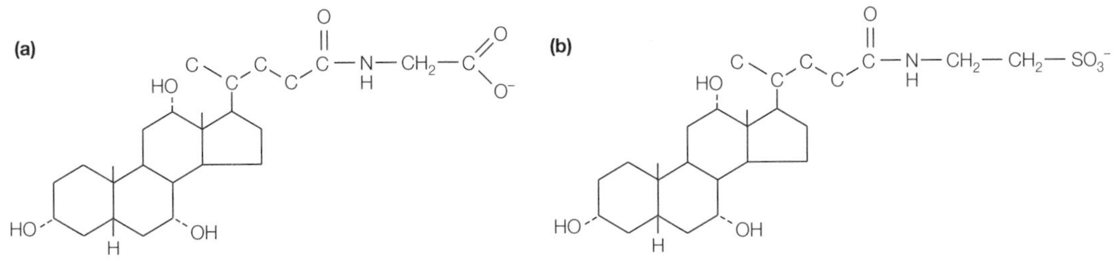

*Fig. 3. Structures of the bile salts (a) glycocholate and (b) taurocholate.*

regions (that is are amphipathic molecules), the bile salts are very effective detergents and act to solubilize dietary lipids. The resulting increase in the surface area of the lipids aids their hydrolysis by lipases and their uptake into intestinal cells (see Topic K4). The intestinal absorption of the **lipid-soluble vitamins** A, D, E and K also requires the action of the bile salts.

## Vitamin D

**Vitamin D** is derived from 7-dehydrocholesterol by the action of the **UV** component of sunlight on the skin. UV light brings about photolysis of 7-dehydrocholesterol between C-9 and -10, leading to a rearrangement of the double bonds of the molecule to form **previtamin $D_3$** (*Fig. 4*). This molecule spontaneously isomerizes to form **vitamin $D_3$ (cholecalciferol)**. Subsequent hydroxylation reactions take place in the liver and kidneys to produce 1,25-dihydroxycholecalcerol (1,25(OH)$D_3$), the active hormone (*Fig. 4*). **Rickets**, which is caused by a deficiency of vitamin D, was historically a common disease of childhood in Britain due to the low vitamin D content of the national diet, and lack of exposure to sunlight. Even today, people whose cultures require the body to be clothed so that no skin is exposed to sunlight have problems in maintaining an adequate vitamin D intake. In adults this takes the form of **osteomalacia** – the softening or weakening of the bones.

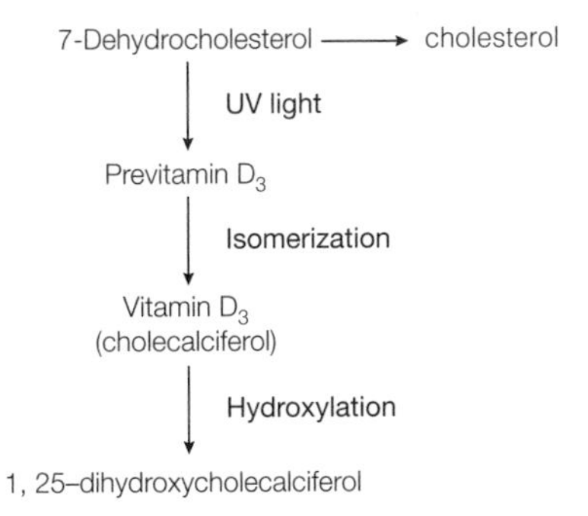


*Fig. 4. Formation of vitamin D.*

## Steroid hormones

Cholesterol is the precursor of the five major classes of **steroid hormones** (*Table 1*). The synthesis of steroid hormones is initiated by the removal of a six-carbon unit from carbon 20 of the cholesterol side chain to form **pregnenolone**, the common precursor of all steroid hormones (*Fig. 5*). A series of reactions catalyzed by **cytochrome P450** modify pregnenolone to give rise to the individual hormones (*Fig. 5*).

*Table 1. Classes of steroid hormone*

| Class | Site of synthesis | Hormone | Action |
|---|---|---|---|
| Progestagens | Corpus luteum | Progesterone | Prepares uterine lining for egg implantation; maintenance of pregnancy |
| Androgens | Testis | Testosterone | Development of male secondary sex characteristics |
| Estrogens | Ovary | Estrone | Development of female secondary sex characteristics |
| Glucocorticoids | Adrenal cortex | Cortisol | Promotes gluconeogenesis and glycogen formation; enhances fat and protein degradation |
| Mineralocorticoids | Adrenal cortex | Aldosterone | Increases reabsorption of $Na^+$ and excretion of $K^+$ and $H^+$ by kidney tubules |

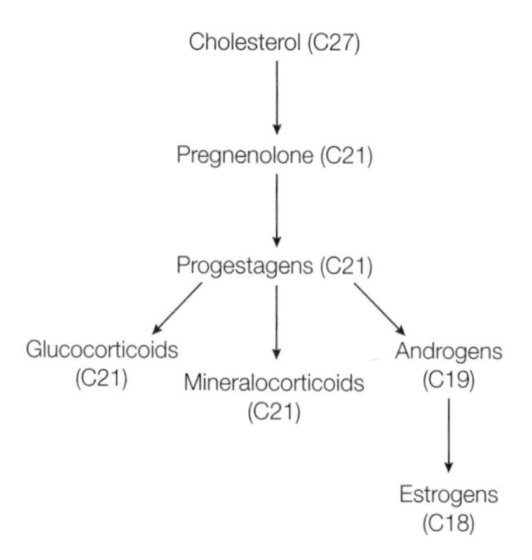


*Fig. 5. Biosynthetic pathway for the synthesis of the steroid hormones.*

The cytochrome P450s are a group of **heme-containing enzymes** (see Topic M4) that get their name from the wavelength maximum of their absorbance spectra when bound to carbon monoxide. They are present in both the mitochondria and the SER of many cells, and consist of a family of structurally related enzymes with different substrate specificities. The enzymes all catalyze so-called **mono-oxygenase** reactions, in which one oxygen atom from molecular oxygen is inserted into the substrate molecule, and the other oxygen atom forms water. The electrons required to bring about the reduction of oxygen to form water are supplied by specialized electron transport chains which are functionally linked to the P450 enzymes. These electron transport chains usually have NADPH as the ultimate electron donor, so a cytochrome P450-catalyzed reaction is often characterized by the involvement of both $O_2$ and NADPH.

# K6 Lipoproteins

## Key Notes

**Structure and function**

Lipoproteins are globular, micelle-like particles consisting of a hydrophobic core of triacylglycerols and cholesterol esters surrounded by an amphipathic coat of protein, phospholipid and cholesterol. The apolipoproteins (apoproteins) on the surface of the lipoproteins help to solubilize the lipids and target the lipoproteins to the correct tissues. There are five different types of lipoprotein, classified according to their functional and physical properties: chylomicrons, very low density lipoproteins (VLDLs), intermediate density lipoproteins (IDLs), low density lipoproteins (LDLs), and high density lipoproteins (HDLs). The major function of lipoproteins is to transport triacylglycerols, cholesterol and phospholipids around the body.

**Chylomicrons**

Chylomicrons are synthesized in the intestine and transport dietary triacylglycerols to skeletal muscle and adipose tissue, and dietary cholesterol to the liver. At these target tissues the triacylglycerols are hydrolyzed by lipoprotein lipase on the surface of the cells and the released fatty acids are taken up either for metabolism to generate energy or for storage. The resulting cholesterol-rich chylomicron remnants are transported in the blood to the liver where they are taken up by receptor-mediated endocytosis.

**VLDLs, IDLs and LDLs**

VLDLs are synthesized in the liver and transport triacylglycerols, cholesterol and phospholipids to other tissues, where lipoprotein lipase hydrolyzes the triacylglycerols and releases the fatty acids for uptake. The VLDL remnants are transformed first to IDLs and then to LDLs as all of their apoproteins other than apoB-100 are removed and their cholesterol esterified. The LDLs bind to the LDL receptor protein on the surface of target cells and are internalized by receptor-mediated endocytosis. The cholesterol, which is released from the lipoproteins by the action of lysosomal lipases, is either incorporated into the cell membrane or re-esterified for storage. High levels of intracellular cholesterol decrease the synthesis of the LDL receptor, reducing the rate of uptake of cholesterol, and inhibit HMG CoA reductase, preventing the cellular synthesis of cholesterol.

**HDLs**

HDLs are synthesized in the blood and extract cholesterol from cell membranes, converting it into cholesterol esters. Some of the cholesterol esters are then transferred to VLDLs. About half of the VLDLs and all of the HDLs are taken up into the liver cells by receptor-mediated endocytosis and the cholesterol disposed of in the form of bile salts.

**Atherosclerosis**

Atherosclerosis is characterized by cholesterol-rich arterial thickenings (atheromas) that narrow the arteries and cause blood clots to form. If these blood clots block the coronary arteries supplying the heart, the result is a myocardial infarction, or heart attack.

**Familial hypercholesterolemia**

This is an inherited disorder in which individuals have a lack of functional LDL receptors preventing cholesterol from being taken up by the tissues. The resulting high blood cholesterol level leads to an increase in the formation of atheromas and can cause death from myocardial infarction during childhood.

**Related topics**

Membrane transport: macromolecules (E4)
Triacylglycerols (K4)
Cholesterol (K5)

## Structure and function

**Triacylglycerols**, **phospholipids** and **cholesterol** are relatively insoluble in aqueous solution. Hence, they are transported around the body in the blood as components of **lipoproteins**. These globular, micelle-like particles consist of a hydrophobic core of triacylglycerols and cholesterol esters surrounded by an amphipathic coat of protein, phospholipid and cholesterol. The protein components of lipoproteins are called **apolipoproteins** (or **apoproteins**). At least 10 different apoproteins are found in the different human lipoproteins. Their functions are to help solubilize the hydrophobic lipids and to act as cellular targeting signals. Lipoproteins are classified into five groups on the basis of their physical and functional properties (*Table 1*):

- **chylomicrons** are the largest and least dense lipoproteins. They transport dietary (exogenous) triacylglycerols and cholesterol from the intestines to other tissues in the body.
- **Very low density liporoteins** (VLDLs), **intermediate density lipoproteins** (IDLs) and **low density lipoproteins** (LDLs) are a group of related lipoproteins that transport internally produced (endogenous) triacylglycerols and cholesterol from the liver to the tissues.
- **High density lipoproteins** (HDLs) transport endogenous cholesterol from the tissues to the liver.

## Chylomicrons

Chylomicrons, the largest of the lipoproteins, are synthesized in the intestine. They transport ingested triacylglycerols to other tissues, mainly skeletal muscle and adipose tissue, and transport ingested cholesterol to the liver (*Fig. 1*). At the target tissues the triacylglycerols are hydrolyzed by the action of **lipoprotein lipase**, an enzyme located on the outside of the cells that is activated by **apoC-II**, one of the apoproteins on the chylomicron surface. The released fatty acids and monoacylglycerols are taken up by the tissues, and either used for energy production or re-esterified to triacylglycerol for storage. As their triacylglycerol content is depleted, the chylomicrons shrink and form cholesterol-

*Table 1. Characteristics of the five classes of lipoproteins*

| Lipoprotein | Molecular mass (kDa) | Density (g $ml^{-1}$) | % Protein | Major lipids | Apoproteins |
|---|---|---|---|---|---|
| Chylomicrons | > 400 000 | < 0.95 | 1.5–2.5 | TG | A, B-48, C, E |
| VLDLs | 10 000–80 000 | <1.006 | 5–10 | TG, PL, CE | B-100, C, E |
| IDLs | 5000–10 000 | 1.006–1.019 | 15–20 | CE, TG, PL | B-100, C, E |
| LDLs | 2300 | 1.019–1.063 | 20–25 | CE, PL | B-100 |
| HDLs | 175–360 | 1.063–1.210 | 40–55 | PL, CE | A, C, D, E |

C, cholesterol; CE, cholesterol ester; TG, triglyceride; PL, phospholipid.

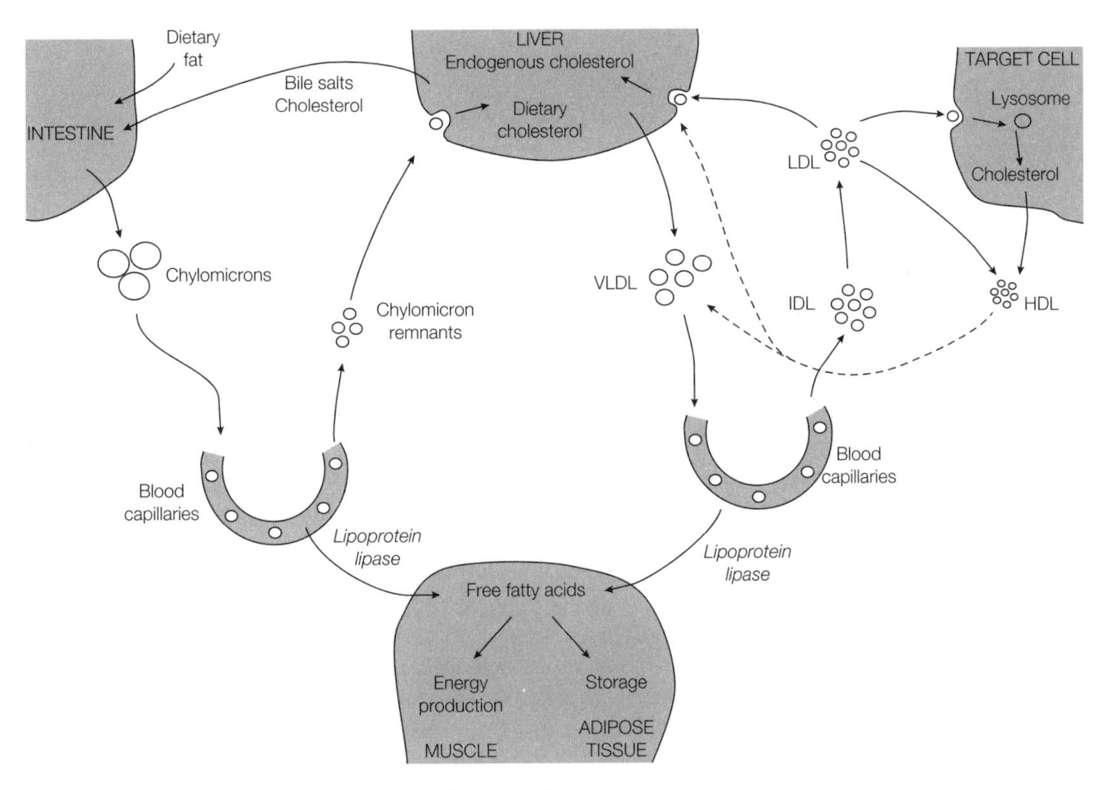


*Fig. 1. The transport of triacylglycerol and cholesterol by lipoproteins.*

rich **chylomicron remnants** which are transported in the blood to the liver (*Fig. 1*). Here they bind to a specific cell-surface remnant receptor and are taken up into the liver cells by **receptor-mediated endocytosis** (see Topic E4).

## VLDLs, IDLs and LDLs

VLDLs are synthesized in the liver and transport a variety of lipids (see *Table 1*) to other tissues, again mainly adipose tissue and skeletal muscle. As with chylomicrons, the triacylglycerols in VLDLs are acted on by **lipoprotein lipase** and the released fatty acids taken up by the tissues (*Fig. 1*). The VLDL remnants remain in the blood, first as IDLs and then as LDLs. In the transformation to LDLs, much of the cholesterol is esterified on its hydroxyl group on C-3 by the addition of a fatty acid chain from phosphatidylcholine (lecithin) by the enzyme **lecithin–cholesterol acyl transferase (LCAT)**. In addition, all of the apoproteins other than **apoB-100** are removed.

LDLs are then taken up by target cells through **receptor-mediated endocytosis** (see Topic E4). The **LDL receptor**, a transmembrane glycoprotein on the surface of the target cells, specifically binds apoB-100 in the LDL coat. The receptors then cluster into clathrin-coated pits and are internalized (see Topic E4, *Fig. 3*). Once in the lysosomes, the LDLs are digested by lysosomal enzymes, with the cholesterol esters being hydrolyzed by a lysosomal lipase to release the cholesterol (*Fig. 1*). This is then incorporated into the cell membrane and any excess is re-esterified for storage by **acyl CoA cholesterol acyltransferase** (ACAT).

To prevent the build up of cholesterol and its ester derivatives in the cell, high levels of cholesterol:

- decrease the synthesis of the LDL receptor, thereby reducing the rate of uptake of cholesterol by receptor-mediated endocytosis, and
- inhibit the cellular biosynthesis of cholesterol through inhibition of HMG CoA reductase (see Topic K5).

**HDLs**

HDLs have the opposite function to that of LDLs in that they remove cholesterol from the tissues. The HDLs are synthesized in the blood mainly from components derived from the degradation of other lipoproteins. HDLs then acquire their cholesterol by extracting it from cell membranes and converting it into cholesterol esters by the action of LCAT (*Fig. 1*). The HDLs are then either taken up directly by the liver or transfer their cholesterol esters to VLDLs, of which about half are taken up by the liver by receptor-mediated endocytosis (*Fig. 1*). The liver is the only organ that can dispose of significant quantities of cholesterol, primarily in the form of **bile salts** (see Topic K5).

**Atherosclerosis**

**Atherosclerosis**, the most common type of **hardening of the arteries**, is characterized by the presence of cholesterol-rich arterial thickenings (**atheromas**). This progressive disease begins with the intracellular deposition of lipids, mainly cholesterol esters, in the smooth muscle cells of the arterial wall. These lesions become fibrous, calcified plaques that narrow and can eventually block the arteries. **Blood clots** are also more likely to occur which may stop the blood flow and deprive the tissues of oxygen. If these blockages occur in the coronary arteries, those supplying the heart, the result is a **myocardial infarction** or **heart attack**, which is the most common cause of death in Western industrialized countries. Blood clots in cerebral arteries cause stroke, while those in peripheral blood vessels in the limbs can lead to possible gangrene and amputation.

**Familial hypercholesterolemia**

**Familial hypercholesterolemia** is an inherited disorder in which homozygotes have a markedly elevated level of cholesterol in their blood, whilst in heterozygotes the level is twice that of normal individuals. Not only does this result in the deposition of cholesterol in the skin as yellow nodules known as **xanthomas**, but also in the formation of **atheromas** that can cause death from **myocardial infarction** during childhood. The molecular defect in familial hypercholesterolemia is the lack of functional **LDL receptors**. Thus, the LDL cholesterol cannot be taken up by the tissues and results in a high concentration in the blood. Homozygotes can be treated by liver transplantation, while heterozygotes can be treated by inhibiting HMG CoA reductase with lovastatin (see Topic K5) and reducing the intestinal re-absorption of bile salts, thereby decreasing the blood cholesterol level.

# L1 CITRIC ACID CYCLE

## Key Notes

**Role**

The cycle oxidizes pyruvate (formed during the glycolytic breakdown of glucose) to $CO_2$ and $H_2O$, with the concomitant production of energy. Acetyl CoA from fatty acid breakdown and amino acid degradation products are also oxidized. In addition, the cycle has a role in producing precursors for biosynthetic pathways.

**Location**

The citric acid cycle occurs within the mitochondria of eukaryotes and the cytosol of prokaryotes.

**The cycle**

The citric acid cycle has eight stages:

1. Production of citrate from oxaloacetate and acetyl CoA (catalyzed by citrate synthase).
2. Isomerization of citrate to isocitrate (catalyzed by aconitase).
3. Oxidation of isocitrate to α-ketoglutarate (catalyzed by isocitrate dehydrogenase; the reaction requires $NAD^+$).
4. Oxidation of α-ketoglutarate to succinyl CoA (catalyzed by the α-ketoglutarate dehydrogenase complex; the reaction requires $NAD^+$).
5. Conversion of succinyl CoA to succinate [catalyzed by succinyl CoA synthetase; the reaction requires inorganic phosphate and GDP (or ADP)].
6. Oxidation of succinate to fumarate (catalyzed by succinate dehydrogenase; the reaction involves FAD).
7. Hydration of fumarate to malate (catalyzed by fumarase).
8. Oxidation of malate to oxaloacetate (catalyzed by malate dehydrogenase; the reaction requires $NAD^+$).

**Energy yield**

For each turn of the cycle, 12 ATP molecules are produced, one directly from the cycle and 11 from the re-oxidation of the three NADH and one $FADH_2$ molecules produced by the cycle by oxidative phosphorylation.

**Regulation**

The citric acid cycle is regulated at the steps catalyzed by citrate synthase, isocitrate dehydrogenase and α-ketoglutarate dehydrogenase via feedback inhibition by ATP, citrate, NADH and succinyl CoA, and stimulation of isocitrate dehydrogenase by ADP. Pyruvate dehydrogenase, which converts pyruvate to acetyl CoA to enter the cycle, is inhibited by acetyl CoA and NADH. In addition, this enzyme is inactivated by phosphorylation, a reaction catalyzed by pyruvate dehydrogenase kinase. A high ratio of NADH/$NAD^+$, acetyl CoA/CoA or ATP/ADP stimulates phosphorylation of pyruvate dehydrogenase and so inactivates this enzyme. Pyruvate inhibits the kinase. Removal of the phosphate group (dephosphorylation) by a phosphatase reactivates pyruvate dehydrogenase.

**Biosynthetic pathways**

Amino acids, purines and pyrimidines, porphyrins, fatty acids and glucose are all synthesized by pathways that use citric acid intermediates as precursors.

**Related topics**

Glycolysis (J3)
Gluconeogenesis (J4)
Fatty acid breakdown (K2)
Fatty acid synthesis (K3)
Electron transport and oxidative phosphorylation (L2)
Amino acid metabolism (M2)

## Role

The citric acid cycle, also known as the **TCA (tricarboxylic acid) cycle** or **Krebs cycle** (after its discoverer in 1937), is used to oxidize the pyruvate formed during the glycolytic breakdown of glucose into $CO_2$ and $H_2O$. It also oxidizes acetyl CoA arising from fatty acid degradation (Topic K2), and amino acid degradation products (Topic M2). In addition, the cycle provides precursors for many biosynthetic pathways.

## Location

The citric acid cycle operates in the mitochondria of eukaryotes and in the cytosol of prokaryotes. Succinate dehydrogenase, the only membrane-bound enzyme in the citric acid cycle, is embedded in the inner mitochondrial membrane in eukaryotes and in the plasma membrane in prokaryotes.

## The cycle

The cycle forms the central part of a three-step process which oxidizes organic fuel molecules into $CO_2$ with the concomitant production of ATP.

**Step 1 – Oxidation of fuel molecules to acetyl CoA**
A major source of energy is glucose which is converted by glycolysis (see Topic J3) into pyruvate. **Pyruvate dehydrogenase** (a complex of three enzymes and five coenzymes) then oxidizes the pyruvate (using $NAD^+$ which is reduced to NADH) to form acetyl CoA and $CO_2$. Since the reaction involves both an oxidation and a loss of $CO_2$, the process is called **oxidative decarboxylation**.

**Step 2 – The citric acid cycle**
The cycle carries out the oxidation of acetyl groups from acetyl CoA to $CO_2$ with the production of four pairs of electrons, stored initially in the reduced electron carriers NADH and $FADH_2$ (*Fig. 1*).

The cycle has eight stages:

1. Citrate (6C) is formed from the irreversible condensation of acetyl CoA (2C) and oxaloacetate (4C) – catalyzed by **citrate synthase**.
2. Citrate is converted to isocitrate (6C) by an isomerization catalyzed by **aconitase**. This is actually a two-step reaction during which *cis*-aconitate is formed as an intermediate. It is the *cis*-aconitate which gives the enzyme its name.
3. Isocitrate is oxidized to α-ketoglutarate (5C) and $CO_2$ by **isocitrate dehydrogenase**. This mitochondrial enzyme requires $NAD^+$, which is reduced to NADH.
4. α-Ketoglutarate is oxidized to succinyl CoA (4C) and $CO_2$ by the **α-ketoglutarate dehydrogenase complex**. Like pyruvate dehydrogenase, this is a complex of three enzymes and uses $NAD^+$ as a cofactor.
5. Succinyl CoA is converted to succinate (4C) by **succinyl CoA synthetase**. The reaction uses the energy released by cleavage of the succinyl–CoA bond to synthesize either GTP (mainly in animals) or ATP (exclusively in plants) from $P_i$ and, respectively, GDP or ADP.
6. Succinate is oxidized to fumarate (4C) by **succinate dehydrogenase**. FAD is tightly bound to the enzyme and is reduced to produce $FADH_2$.
7. Fumarate is converted to malate (4C) by **fumarase**; this is a hydration reaction requiring the addition of a water molecule.
8. Malate is oxidized to oxaloacetate (4C) by **malate dehydrogenase**. $NAD^+$ is again required by the enzyme as a cofactor to accept the free pair of electrons and produce NADH.

**Step 3 – Oxidation of NADH and $FADH_2$ produced by the citric acid cycle**
The NADH and $FADH_2$ produced by the citric acid cycle are reoxidized and the energy released is used to synthesize ATP **by oxidative phosphorylation** (see Topic L2).

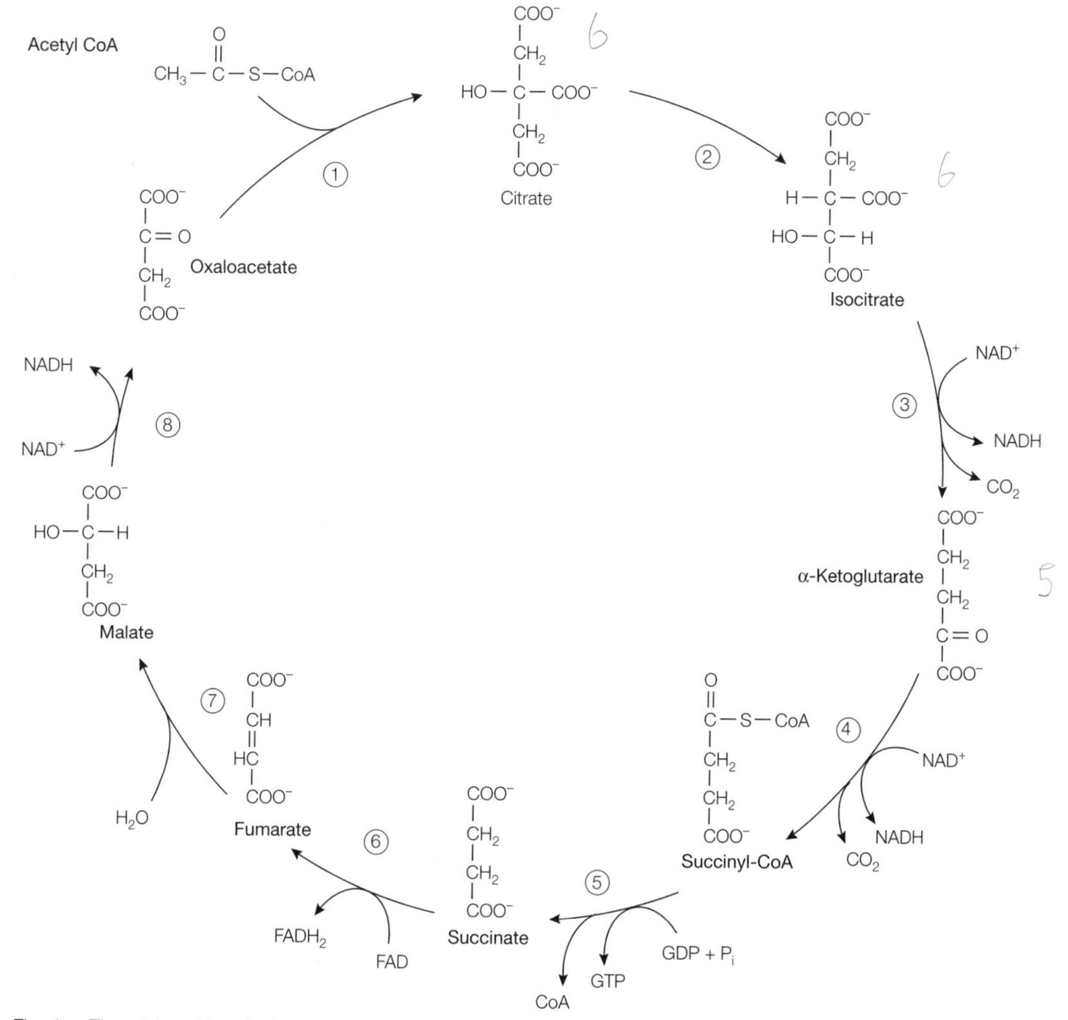


*Fig. 1. The citric acid cycle (reactions 1–8 are described in the text).*

**Energy yield**

Each of the three NADH molecules produced per turn of the cycle yields 3 ATPs and the single $FADH_2$ yields 2 ATPs by oxidative phosphorylation (although some measurements indicate that the quantities are 2.5 and 1.5 respectively – see p. 355). One GTP (or ATP) is synthesized directly during the conversion of succinyl CoA to succinate. Thus the oxidation of a single molecule of acetyl CoA via the citric acid cycle produces 12 ATP molecules.

**Regulation**

Regulation of the cycle is governed by substrate availability, inhibition by accumulating products, and allosteric feedback inhibition by subsequent intermediates in the cycle. Three enzymes in the cycle itself are regulated (citrate synthase, isocitrate dehydrogenase and α-ketoglutarate dehydrogenase) and so is the enzyme which converts pyruvate to acetyl CoA to enter the cycle, namely pyruvate dehydrogenase (*Fig. 2*):

- *citrate synthase* is inhibited by citrate and also by ATP (the $K_m$ for acetyl CoA is raised as the level of ATP rises);

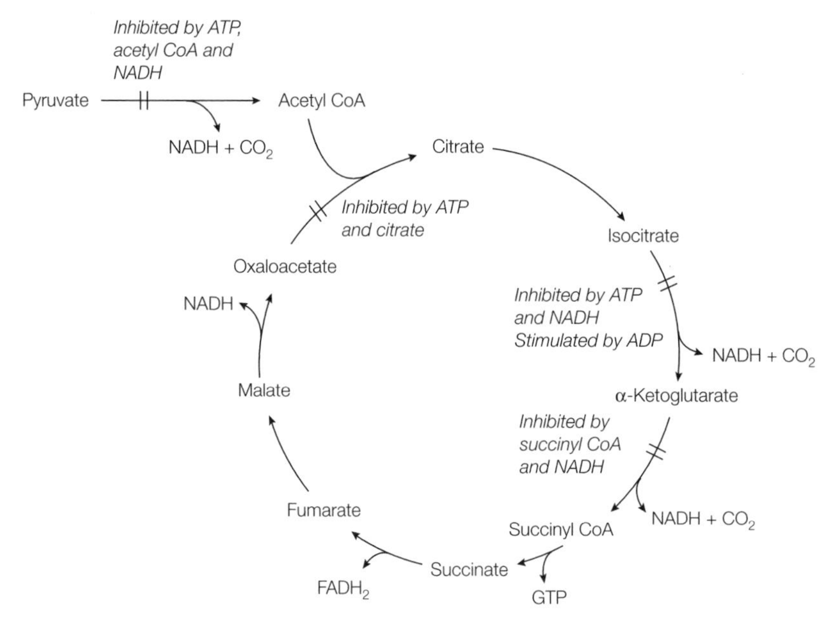


*Fig. 2. Regulation points of the citric acid cycle.*

- *isocitrate dehydrogenase* is inhibited by NADH and ATP but activated by ADP;
- *α-ketoglutarate dehydrogenase* is inhibited by NADH and succinyl CoA;
- *pyruvate dehydrogenase* is inhibited by NADH and acetyl CoA (i.e. product inhibition). However, in eukaryotes the enzyme is also controlled by phosphorylation/dephosphorylation via **pyruvate dehydrogenase kinase** and a **phosphatase**. The kinase catalyzes the phosphorylation of a specific Ser residue in pyruvate dehydrogenase, using ATP as the phosphate donor, and this inactivates the enzyme. Removal of the phosphate group by the phosphatase reactivates the enzyme. At any one time, the activity of pyruvate dehydrogenase is determined by the relative balance between the kinase and phosphatase reactions. Increasing the NADH/$NAD^+$, acetyl CoA/CoA or ATP/ADP ratio stimulates phosphorylation and hence inactivates pyruvate dehydrogenase. As pyruvate builds up, it inhibits the kinase and hence allows the phosphatase to reactivate pyruvate dehydrogenase, thus stimulating pyruvate conversion to acetyl CoA.

Overall, the cycle speeds up when cellular energy levels are low (high ADP concentration, low ATP and NADH) and slows down as ATP (and then $NADH_2$, succinyl CoA and citrate) accumulates.

## Biosynthetic pathways

The intermediates in the cycle provide precursors for many biosynthetic pathways. For example:

- synthesis of fatty acids from citrate (Topic K3);
- amino acid synthesis following transamination of α-ketoglutarate (Topic M2);
- synthesis of purine and pyrimidine nucleotides from α-ketoglutarate and oxaloacetate;
- oxaloacetate can be converted to glucose by gluconeogenesis (Topic J4);
- succinyl CoA is a central intermediate in the synthesis of the porphyrin ring of heme groups (Topic M4).

# L2 ELECTRON TRANSPORT AND OXIDATIVE PHOSPHORYLATION

## Key Notes

**Overview**

Electron transport and oxidative phosphorylation re-oxidize NADH and $FADH_2$ and trap the energy released as ATP. In eukaryotes, electron transport and oxidative phosphorylation occur in the inner membrane of mitochondria whereas in prokaryotes the process occurs in the plasma membrane.

**Redox potential**

The oxidation–reduction potential, $E$, (or redox potential) of a substance is a measure of its affinity for electrons. The standard redox potential ($E_0'$) is measured under standard conditions, at pH 7, and is expressed in volts. The standard free energy change of a reaction at pH 7, $\Delta G^{0\prime}$, can be calculated from the change in redox potential $\Delta E_0'$ of the substrates and products. A reaction with a positive $\Delta E_0'$ has a negative $\Delta G^{0\prime}$ (i.e. is exergonic).

**Electron transport from NADH**

Electrons are transferred from NADH to oxygen along the electron transport chain (also called the respiratory chain). NADH passes electrons to NADH dehydrogenase, a large protein complex that contains FMN and two types of iron–sulfur (FeS) clusters in iron–sulfur proteins. The electrons are accepted by the FMN to produce $FMNH_2$ and then passed to the iron atoms of the FeS clusters which accept and donate electrons by alternating between $Fe^{3+}$ and $Fe^{2+}$ states. Electrons from NADH dehydrogenase are passed to ubiquinone (coenzyme Q, CoQ), converting it to ubiquinol (or $CoQH_2$), and then to the cytochrome $bc_1$ complex. This contains cytochrome $b$ and cytochrome $c_1$, as well as an FeS protein. A cytochrome contains a heme group with a central iron atom which changes from the $Fe^{3+}$ state to the $Fe^{2+}$ state on accepting an electron. When the electron is donated to another component, the iron atom changes back to the $Fe^{3+}$ state. The cytochrome $bc_1$ complex passes the electrons to cytochrome $c$ which in turn passes them to cytochrome oxidase, a complex that contains two cytochromes (cytochrome $a$ and $a_3$) paired with copper atoms ($Cu_A$ and $Cu_B$ , respectively). During electron transfer, the copper atoms cycle between the $Cu^{2+}$ and $Cu^+$ states. Finally, cytochrome oxidase passes four electrons to molecular oxygen to form two molecules of water.

**Formation of an $H^+$ gradient**

The change in redox potential along the chain is a measure of the free energy change at each step. At the steps involving NADH dehydrogenase, the cytochrome $bc_1$ complex and cytochrome oxidase, the free energy change is large enough to pump $H^+$ ions across the inner mitochondrial membrane, from the mitochondrial matrix into the intermembrane space, to create an $H^+$ gradient. Therefore, each of these complexes is an $H^+$ pump driven by electron transport.

**Electron transport from $FADH_2$**

$FADH_2$ is reoxidized to FAD by donating two electrons to succinate-CoQ reductase (complex II), a protein complex that contains FeS clusters. It passes the electrons on to ubiquinone in the main electron transport chain where their further transport leads to the formation of an $H^+$ gradient and ATP synthesis. However succinate-CoQ reductase does not itself pump $H^+$ ions.

**Electron transport inhibitors**

Rotenone and amytal inhibit electron transport at NADH dehydrogenase, antimycin A inhibits the cytochrome $bc_1$ complex, and cyanide ($CN^-$), azide ($N_3^-$) and carbon monoxide (CO) all inhibit cytochrome oxidase.

**Oxidative phosphorylation**

Oxidative phosphorylation is ATP synthesis linked to the oxidation of NADH and $FADH_2$ by electron transport through the respiratory chain. This occurs via a mechanism originally proposed as the chemiosmotic hypothesis. Energy liberated by electron transport is used to pump $H^+$ ions out of the mitochondrion to create an electrochemical proton ($H^+$) gradient. The protons flow back into the mitochondrion through the ATP synthase located in the inner mitochondrial membrane, and this drives ATP synthesis. Approximately three ATP molecules are synthesized per NADH oxidized and approximately two ATPs are synthesized per $FADH_2$ oxidized.

**ATP synthase as a rotatory engine**

ATP synthase is located in the inner mitochondrial membrane. It consists of two major components, $F_1$ ATPase [seen as spheres under the electron microscope and with a subunit structure of $(\alpha\beta)_3\gamma\delta\varepsilon$] attached to component $F_0$ (coupling factor 0) which is a proton channel spanning this membrane. Hence, ATP synthase is also known as $F_0F_1$ ATPase. In mitochondria, this complete complex uses the energy released by electron transport to drive ATP synthesis but, in isolation, $F_1$ ATPase hydrolyzes ATP. During ATP hydrolysis, and presumably also during ATP synthesis, subunit $\gamma$ of $F_1$ ATPase rotates relative to $(\alpha\beta)_3$ and is the smallest rotatory engine known in nature.

**Coupling and respiratory control**

Electron transport is normally tightly coupled to ATP synthesis; electrons do not flow through the electron transport chain to oxygen unless ADP is simultaneously phosphorylated to ATP. If ADP is available, electron transport proceeds and ATP is made; as the ADP concentration falls, electron transport slows down. This process, called respiratory control, ensures that electron flow occurs only when ATP synthesis is required.

**Uncouplers**

Some chemicals (e.g. 2,4-dinitrophenol; DNP) are uncoupling agents; they allow electron transport to proceed without ATP synthesis. They uncouple mitochondria by carrying $H^+$ ions across the inner mitochondrial membrane and hence dissipate the proton gradient. The energy derived from uncoupled electron transport is released as heat. Uncoupling also occurs naturally in some tissues (e.g. the mitochondria of brown adipose tissue are uncoupled by a protein called thermogenin). The resulting production of heat (nonshivering thermogenesis) by the adipose tissue serves to protect sensitive body tissues in newborn animals and to maintain body temperature during hibernation.

**Reoxidation of cytosolic NADH**

Cytosolic NADH cannot cross the inner mitochondrial membrane and enter mitochondria to be reoxidized. However, it can be reoxidized via the glycerol 3-phosphate shuttle. Cytosolic glycerol 3-phosphate dehydrogenase oxidizes the NADH and reduces dihydroxyacetone phosphate to glycerol 3-phosphate.

The glycerol 3-phosphate enters the mitochondrion and is converted back to dihydroxyacetone phosphate by mitochondrial glycerol 3-phosphate dehydrogenase (an FAD-linked enzyme). The dihydroxyacetone phosphate diffuses back to the cytosol. The enzyme-linked $FADH_2$ is reoxidized by transferring its electrons to ubiquinone in the electron transport chain. Since the electrons enter the electron transport chain from $FADH_2$, only about two ATPs are synthesized per molecule of cytosolic NADH. In heart and liver, cytosolic NADH can be reoxidized via the malate–aspartate shuttle. Oxaloacetate in the cytosol is reduced to malate by NADH and enters the mitochondrion via a malate–α-ketoglutarate carrier. In the matrix, the malate is reoxidized to oxaloacetate by $NAD^+$ which is converted to NADH, resulting in a net transfer of electrons from cytosolic NADH to matrix NADH. The oxaloacetate is converted to aspartate by transamination, leaves the mitochondrion and is reconverted to oxaloacetate in the cytosol, again by transamination.

**Related topics**

Glycolysis (J3)
Citric acid cycle (L1)
Photosynthesis (L3)

## Overview

In eukaryotes, electron transport and oxidative phosphorylation occur in the inner membrane of mitochondria. These processes re-oxidize the NADH and $FADH_2$ that arise from the citric acid cycle (located in the mitochondrial matrix; Topic L2), glycolysis (located in the cytoplasm; Topic J3) and fatty acid oxidation (located in the mitochondrial matrix; Topic K2) and trap the energy released as ATP. Oxidative phosphorylation is by far the major source of ATP in the cell. In prokaryotes, the components of electron transport and oxidative phosphorylation are located in the plasma membrane (see Topic A1).

## Redox potential

The oxidation of a molecule involves the loss of electrons. The reduction of a molecule involves the gain of electrons. Since electrons are not created or destroyed in a chemical reaction, if one molecule is oxidized, another must be reduced (i.e. it is an **oxidation–reduction reaction**). Thus, by definition, oxidation–reduction reactions involve the transfer of electrons. In the oxidation–reduction reaction:

$$NADH + H^+ + \tfrac{1}{2}O_2 \rightleftharpoons NAD^+ + H_2O$$

when the NADH is oxidized to $NAD^+$, it loses electrons. When the molecular oxygen is reduced to water, it gains electrons.

The **oxidation–reduction potential**, $E$, (or **redox potential**) is a measure of the affinity of a substance for electrons and is measured relative to hydrogen. A positive redox potential means that the substance has a higher affinity for electrons than does hydrogen and so would accept electrons from hydrogen. A substance with a negative redox potential has a lower affinity for electrons than does hydrogen and would donate electrons to $H^+$, forming hydrogen. In the example above, NADH is a strong reducing agent with a negative redox potential and has a tendency to donate electrons. Oxygen is a strong oxidizing agent with a positive redox potential and has a tendency to accept electrons.

For biological systems, the **standard redox potential** for a substance ($E_0'$) is measured under standard conditions, at pH 7, and is expressed in volts. In an oxidation–reduction reaction, where electron transfer is occurring, the total voltage change of the reaction (change in electric potential, $\Delta E$) is the sum of the voltage changes of the individual oxidation–reduction steps. The standard free energy change of a reaction at pH 7, $\Delta G^{0\prime}$, can be readily calculated from the change in redox potential $\Delta E_0'$ of the substrates and products:

$$\Delta G^{0\prime} = -nF\,\Delta E_0'$$

where $n$ is the number of electrons transferred, $\Delta E_0'$ is in volts (V), $\Delta G^{0\prime}$ is in kilocalories per mole (kcal mol$^{-1}$) and $F$ is a constant called the Faraday (23.06 kcal V$^{-1}$ mol$^{-1}$). Note that a reaction with a **positive** $\Delta E_0'$ has a **negative** $\Delta G^{0\prime}$ (i.e. is exergonic).

Thus for the reaction:

$$NADH + H^+ + \tfrac{1}{2}O_2 \rightleftharpoons NAD^+ + H_2O$$

$$\Delta E_0' = +1.14\ V$$

$$\Delta G^{0\prime} = -52.6\ kcal\ mol^{-1}.$$

## Electron transport from NADH

Comparing the energetics of the oxidation of NADH:

$$NADH + H^+ + \tfrac{1}{2}O_2 \rightleftharpoons NAD^+ + H_2O \qquad \Delta G^{0\prime} = -52.6\ kcal\ mol^{-1}$$

and the synthesis of ATP:

$$ADP + P_i + H^+ \rightleftharpoons ATP + H_2O \qquad \Delta G^{0\prime} = +7.3\ kcal\ mol^{-1}$$

it is clear that the oxidation of NADH releases sufficient energy to drive the synthesis of several molecules of ATP. However, NADH oxidation and ATP synthesis do not occur in a single step. Electrons are not transferred from NADH to oxygen directly. Rather the electrons are transferred from NADH to oxygen along a chain of electron carriers collectively called the **electron transport chain** (also called the **respiratory chain**).

The main part of the electron transport chain consists of three large protein complexes embedded in the inner mitochondrial membrane, called **NADH dehydrogenase, the cytochrome *bc*$_1$ complex** and **cytochrome oxidase.** Electrons flow from NADH to oxygen through these three complexes as shown in *Fig. 1*. Each complex contains several electron carriers (see below) that work sequentially to carry electrons down the chain. Two small electron carriers are also needed to link these large complexes; **ubiquinone**, which is also called **coenzyme Q** (abbreviated here as **CoQ**), and **cytochrome c** (*Fig. 1*).

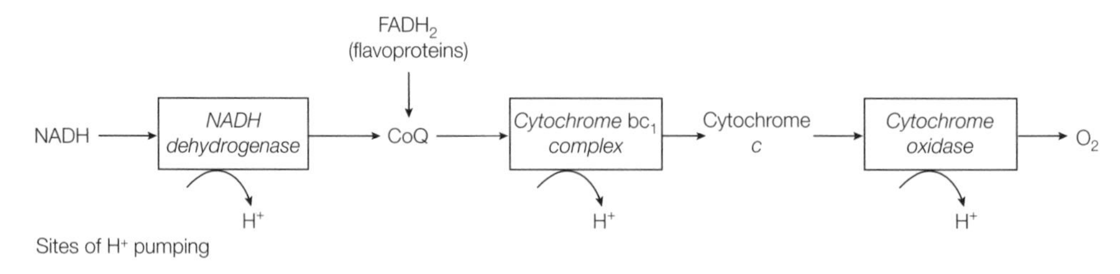


*Fig. 1. Overview of the electron transport chain (respiratory chain).*

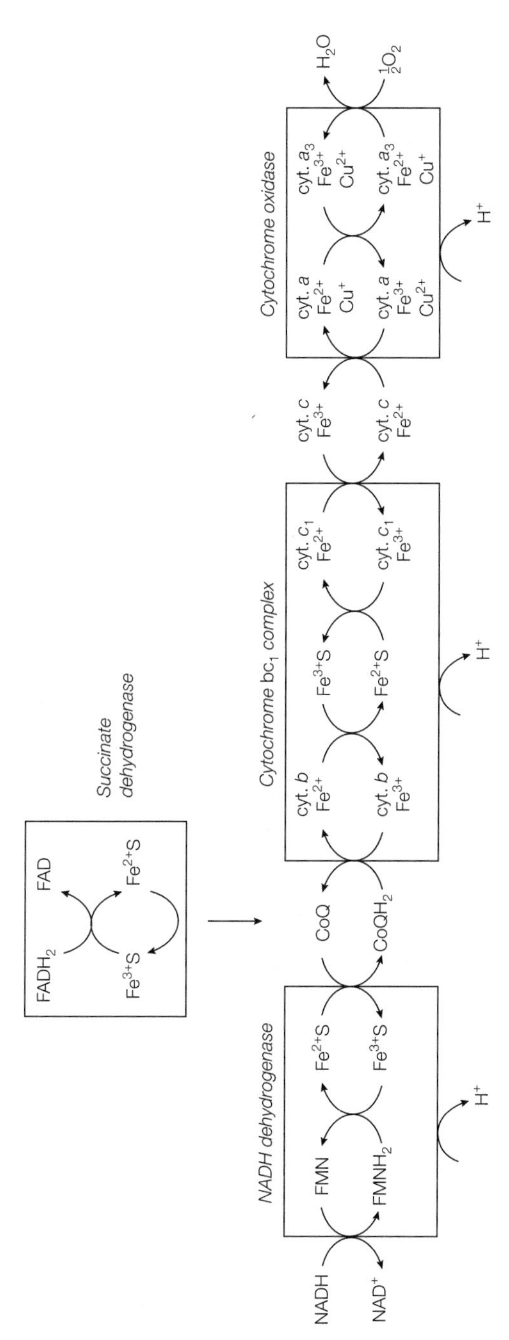


*Fig. 2. Details of electron transport.*

*NADH to NADH dehydrogenase*
**NADH dehydrogenase** (also called **NADH-Q reductase** or **complex I**) consists of at least 30 polypeptides. It binds the NADH and re-oxidizes it to $NAD^+$, passing the two electrons from NADH to a prosthetic group called **FMN** (**flavin mononucleotide**) (*Fig.* 2) to produce **$FMNH_2$** (see Topic C1 for structure of FMN). Each electron is accepted together with a hydrogen ion, $H^+$, such that two electrons and two $H^+$ are accepted in total. The electrons are then transferred, within the NADH dehydrogenase complex, to **iron–sulfur clusters** (FeS) in **iron–sulfur proteins** (also called **nonheme iron proteins**). Several types of FeS clusters exist but in each case the iron atoms are coordinated to inorganic sulfur atoms and the sulfur of cysteine side chains in the protein. Within an FeS cluster, an electron is carried by the iron atom which, on accepting the electron, changes from the $Fe^{3+}$ (ferric) state to the $Fe^{2+}$ (ferrous) state (*Fig.* 2). As the electron is passed to another electron carrier, the iron atom of the FeS cluster changes back again to the $Fe^{3+}$ state.

*NADH dehydrogenase to ubiquinone (CoQ)*
Electrons from the FeS clusters of NADH dehydrogenase are passed on to ubiquinone (CoQ), a small lipid-soluble molecule in the inner mitochondrial membrane. This molecule can act as an electron carrier by accepting up to two electrons and two $H^+$ ions. In so doing, ubiquinone (CoQ) is converted to ubiquinol ($CoQH_2$).

*Ubiquinol to cytochrome* bc$_1$ *complex*
When ubiquinol ($CoQH_2$) donates its two electrons to the next carrier in the chain, the **cytochrome *bc*$_1$ complex** (also called **cytochrome reductase** or **complex III**), the $H^+$ ions are released once more. The cytochrome *bc*$_1$ complex contains two types of cytochromes, **cytochrome *b*** and **cytochrome *c*$_1$**, as well as an FeS protein (*Fig.* 2). A cytochrome is a protein with a bound **heme group** that contains an iron atom (see Topic M4, *Fig. 1*). Different cytochromes have different heme groups, but all cytochromes have the ability to act as electron carriers. As the electron is accepted, the iron atom of the heme group changes from the $Fe^{3+}$ (ferric) state to the $Fe^{2+}$ (ferrous) state. *Figure 2* shows the electrons passing from ubiquinol ($QH_2$) through the cytochrome *b*, FeS and cytochrome *c*$_1$ components of the cytochrome *bc*$_1$ complex to the next electron carrier, cytochrome *c*. Since ubiquinol is a two-electron carrier whereas cytochromes are one-electron carriers, the pathway of electron transfer within the cytochrome *bc*$_1$ complex is complicated and involves ubiquinol ($CoQH_2$) releasing first one electron and an $H^+$ ion to become ubisemiquinone ($CoQH^{\bullet}$) and then the second electron and $H^+$ ion to become ubiquinone (CoQ).

*Cytochrome* bc$_1$ *complex to cytochrome* c *to cytochrome oxidase*
Cytochrome *c* is a peripheral membrane protein that is loosely bound to the outer surface of the inner mitochondrial membrane. It binds to the cytochrome *bc*$_1$ complex and accepts an electron via an $Fe^{3+}$ to $Fe^{2+}$ transition. Then it binds to cytochrome oxidase and donates the electron, with the iron atom of the heme of cytochrome *c* then reverting to the $Fe^{3+}$ state (*Fig.* 2).

*Cytochrome oxidase to oxygen*
**Cytochrome oxidase** (also called **complex IV**) contains two cytochromes (cytochrome *a* and *a*$_3$). Cytochrome *a* is paired with a copper atom, $Cu_A$, and cytochrome *a*$_3$ is paired with a different copper atom, $Cu_B$. During electron

transfer, the iron atoms of the cytochromes cycle between the $Fe^{3+}$ and $Fe^{2+}$ states whilst the copper atoms cycle between $Cu^{2+}$ and $Cu^{+}$. The cytochrome oxidase reaction is complex; it transfers four electrons from four cytochrome *c* molecules and four $H^+$ ions to molecular oxygen to form two molecules of water:

$$4\ \text{cyt.}\ c\ (Fe^{2+}) + 4\ H^+ + O_2 \xrightarrow{\textit{Cytochrome oxidase}} 4\ \text{cyt.}\ c\ (Fe^{3+}) + 2\ H_2O$$

**Formation of an $H^+$ gradient**

All of the electron carriers in the electron transport chain interact according to their redox potentials. Every time that an electron transfer occurs, the accepting carrier has a higher affinity for electrons than the donating carrier. Thus there is a net flow of electrons from NADH (most negative redox potential, least affinity for electrons) to oxygen (most positive redox potential, highest affinity for electrons). This ensures a unidirectional flow of electrons. However, note that each cytochrome, each FeS center and each copper atom can carry only one electron but each NADH donates two electrons. Furthermore, each molecule of oxygen ($O_2$) needs to accept four electrons to be reduced to a molecule of water, $H_2O$. The various components are arranged in such a manner as to allow their different electron-handling properties to work in harmony.

The change in redox potential along the chain is a measure of the free energy change occurring (see above). The potential falls (i.e. becomes more positive) throughout the chain but mainly in three large steps that correspond to the three main protein complexes: the NADH dehydrogenase complex, the cytochrome $bc_1$ complex and the cytochrome oxidase complex. The large free energy change at each of these three steps, and only these three steps, is large enough to pump $H^+$ ions from the mitochondrial matrix across the inner mitochondrial membrane and into the intermembrane space. Thus, each of these three complexes is an **$H^+$ pump** driven by electron transport (*Figs 1* and *2*). Overall, therefore, electron transport along the chain from NADH releases energy that is used to create an **$H^+$ gradient**.

**Electron transport from $FADH_2$**

Succinate dehydrogenase catalyzes the oxidation of succinate to fumarate in the citric acid cycle (Topic L1). The succinate dehydrogenase contains bound FAD that is reduced to $FADH_2$ in the reaction. The re-oxidation of the $FADH_2$ occurs via **succinate–coenzyme Q reductase** (also called **complex II**), an integral protein of the inner mitochondrial membrane. Succinate dehydrogenase is part of this complex but it also contains FeS clusters. During re-oxidation of $FADH_2$, the two electrons pass from the $FADH_2$ to the FeS clusters and are then passed on to ubiquinone (CoQ; see *Fig. 2*). They then enter the main electron transport chain and cause $H^+$ ions to be pumped out of the mitochondrion as for the oxidation of NADH. However, succinate–CoQ reductase itself is *not* an $H^+$ pump because the free energy change of the overall reaction is too small. The $FADH_2$ of other **flavoproteins,** such as mitochondrial glycerol 3-phosphate dehydrogenase in the glycerol 3-phosphate shuttle (see below) and fatty acyl CoA dehydrogenase in fatty acid oxidation (Topic K2), also feed their electrons into the electron transport chain at ubiquinone.

**Electron transport inhibitors**

Several inhibitors of specific electron carriers are known and were used in the original studies to determine the order of the components in the respiratory chain. For example:

- **rotenone** and **amytal** inhibit electron transport at NADH dehydrogenase and so prevent NADH oxidation but the oxidation of $FADH_2$ can still occur since this feeds electrons into the chain at CoQ (see *Fig. 1*) (i.e. past the point of inhibition);

- **antimycin A** inhibits electron transport at the cytochrome $bc_1$ complex;
- **cyanide** ($CN^-$), **azide** ($N_3^-$) and **carbon monoxide** (CO) all inhibit cytochrome oxidase.

## Oxidative phosphorylation

**Oxidative phosphorylation** is the name given to the synthesis of ATP (*phosphorylation*) that occurs when NADH and $FADH_2$ are oxidized (hence *oxidative*) by electron transport through the respiratory chain. Unlike substrate level phosphorylation (see Topics J3 and L1), it does not involve phosphorylated chemical intermediates. Rather, a very different mechanism was proposed by Peter Mitchell in 1961, the **chemiosmotic hypothesis**. This proposes that energy liberated by electron transport is used to create a proton gradient across the mitochondrial inner membrane and that it is this that is used to drive ATP synthesis. Thus the proton gradient couples electron transport and ATP synthesis, not a chemical intermediate. The evidence is overwhelming that this is indeed the way that oxidative phosphorylation works. The actual synthesis of ATP is carried out by an enzyme called **ATP synthase** located in the inner mitochondrial membrane (*Fig. 3*).

In summary, the process is as follows. Electron transport down the respiratory chain from NADH oxidation causes $H^+$ ions to be pumped out of the mitochondrial matrix across the inner mitochondrial membrane into the intermembrane space by the three $H^+$ pumps; NADH dehydrogenase, the cytochrome $bc_1$ complex and cytochrome oxidase (see above). [Because $FADH_2$ is reoxidized via ubiquinone (see *Figs 1* and *2*), its oxidation causes $H^+$ ions to be pumped out only by the cytochrome $bc_1$ complex and cytochrome oxidase and so the amount of ATP made from $FADH_2$ is less than from NADH.] The free energy change in transporting an electrically charged ion across a

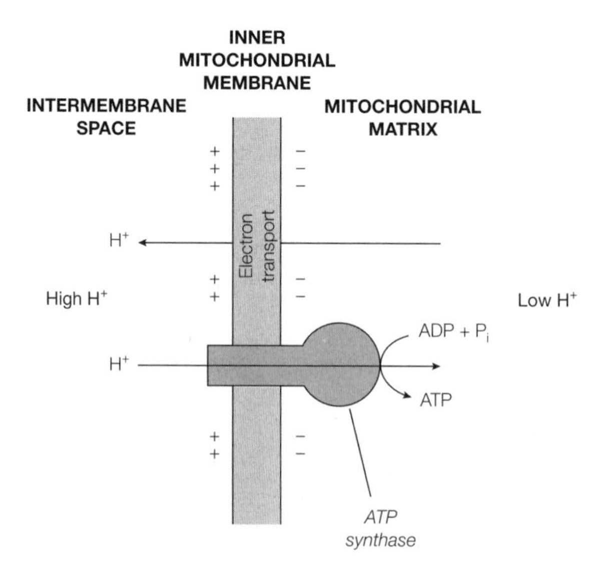


*Fig. 3. The mechanism of oxidative phosphorylation.*

membrane is related both to its electrical charge and the concentration of the species. The pumping out of the $H^+$ ions generates a higher concentration of $H^+$ ions in the intermembrane space and an electrical potential, with the side of the inner mitochondrial membrane facing the intermembrane space being positive (*Fig. 3*). Thus, overall, an **electrochemical proton gradient** is formed. The protons flow back into the mitochondrial matrix through the ATP synthase and this drives ATP synthesis. The ATP synthase is driven by **proton-motive force**, which is the sum of the pH gradient (i.e. the chemical gradient of $H^+$ ions) and the membrane potential (i.e. the electrical charge potential across the inner mitochondrial membrane). There is some debate over the exact stoichiometry of ATP production; in past years it was believed that 3 ATP were generated per NADH and 2 ATP per $FADH_2$ but some recent measurements have indicated that the numbers of ATP molecules generated may be 2.5 and 1.5, respectively.

## ATP synthase as a rotatory engine

The ATP synthase can be seen as spherical projections from the inner membrane (*Fig. 4a*). If mitochondria are subjected to sonic disruption,

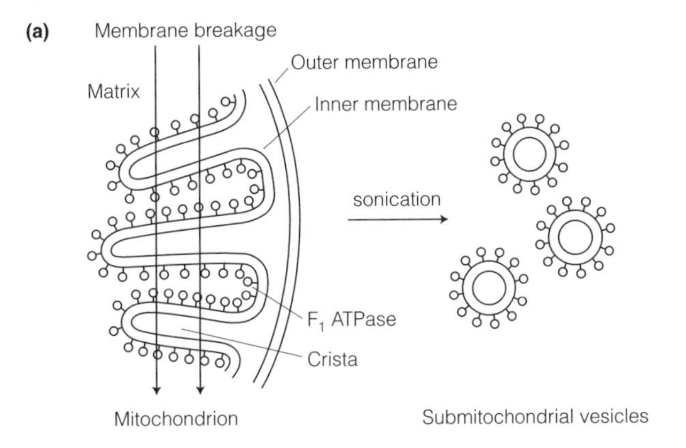


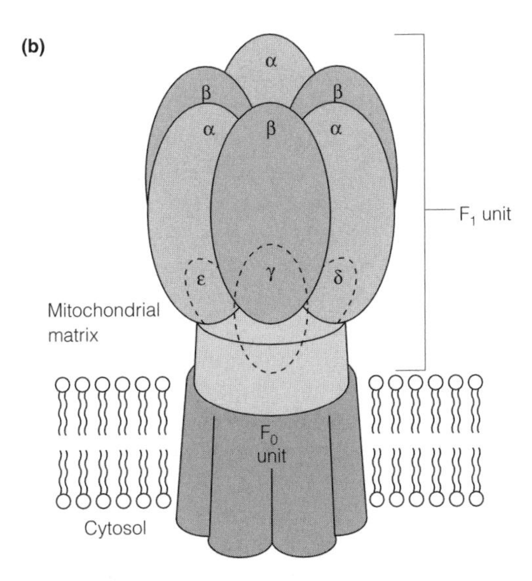


*Fig. 4. (a) Sonic disruption (sonication) of mitochondria produces submitochondrial vesicles, (b) schematic representation of the ATP synthase complex.*

submitochondrial vesicles are formed in which the spheres of the ATP synthase point outward (*Fig. 4a*). In 1960, Racker showed that the spheres can be removed and that the isolated spheres *hydrolyze* ATP, that is, the spheres have ATPase activity (called **$F_1$ ATPase**; *Fig. 4b*). $F_1$ ATPase contains five types of polypeptide in the ratio $(\alpha\beta)_3\gamma\delta\varepsilon$. The stripped submitochondrial vesicles, devoid of the $F_1$ ATPase, can still transport electrons along the electron transport chain but cannot synthesize ATP. These stripped submitochondrial vesicles contain the other major part of the ATP synthase, called **$F_0$** (coupling factor 0) which is a proton channel that spans the inner mitochondrial membrane (*Fig. 4b*). Since it is composed of these two major component parts, ATP synthase is also known as **$F_0F_1$ ATPase**. The stalk between $F_0$ and $F_1$ (*Fig. 4b*) contains several additional polypeptides. The complete complex harnesses the energy released by electron transport to drive ATP synthesis whereas alone, without coupling to electron transport, the $F_1$ component hydrolyzes ATP.

Amazingly it has recently been shown that the $F_1$ portion of ATP synthase behaves as a rotatory engine; during ATP hydrolysis (and presumably also during ATP synthesis) subunit $\gamma$ of the $F_1$ ATPase rotates relative to $(\alpha\beta)_3$. In fact, this is the smallest rotatory engine so far discovered in nature!

## Coupling and respiratory control

Electron transport is normally tightly **coupled** to ATP synthesis (i.e. electrons do not flow through the electron transport chain to oxygen unless ADP is simultaneously phosphorylated to ATP). Clearly, it also follows that ATP is not synthesized unless electron transport is occurring to provide the proton gradient. Thus oxidative phosphorylation needs NADH or $FADH_2$, oxygen, ADP and inorganic phosphate. The actual rate of oxidative phosphorylation is set by the availability of ADP. If ADP is added to mitochondria, the rate of oxygen consumption rises as electrons flow down the chain and then the rate of oxygen utilization falls when all the ADP has been phosphorylated to ATP; a process called **respiratory control**. This mechanism ensures that electrons flow down the chain only when ATP synthesis is needed. If the level of ATP is high and the ADP level is low, no electron transport occurs, NADH and $FADH_2$ build up, as does excess citrate, and the citric acid cycle (Topic L1) and glycolysis (Topic J3) are inhibited.

## Uncouplers

Some chemicals, such as 2,4-dinitrophenol (DNP), act as **uncoupling agents**, that is, when added to cells, they stop ATP synthesis but electron transport still continues and so oxygen is still consumed. The reason is that DNP and other uncoupling agents are lipid-soluble small molecules that can bind $H^+$ ions and transport them across membranes (i.e. they are **$H^+$ ionophores**). Electron transport occurs and pumps out $H^+$ ions across the inner mitochondrial membrane but DNP in the same membrane carries the $H^+$ ions back into the mitochondrion, preventing formation of a proton gradient. Since no proton gradient forms, no ATP can be made by oxidative phosphorylation. Rather the energy derived from electron transport is released as heat.

The production of heat by uncoupling is called nonshivering **thermogenesis.** It is important in certain biological situations. For example, uncoupling occurs naturally in brown adipose tissue. This tissue is rich in mitochondria, the inner mitochondrial membranes of which contain a protein called **thermogenin** (or **uncoupling protein**). Thermogenin allows $H^+$ ions to flow back into mitochondria without having to enter via the ATP synthase and so uncouples electron transport and oxidative phosphorylation, generating heat. The importance of this natural phenomenon is that brown adipose tissue is found in sensitive body areas of some newborn animals (including humans) where the heat production provides protection from cold conditions. In addition, thermogenesis by brown adipose tissue plays a role in maintaining body temperature in hibernating animals.

## Reoxidation of cytosolic NADH

The inner mitochondrial membrane is impermeable to NADH. Therefore NADH produced in the cytoplasm during glycolysis must be reoxidized via a **membrane shuttle**, a combination of enzyme reactions that bypass this impermeability barrier. *Figure 5* shows the **glycerol 3-phosphate shuttle**. Dihydroxyacetone phosphate in the cytosol is reduced to glycerol 3-phosphate, and NADH reoxidized to $NAD^+$, by cytosolic glycerol 3-phosphate dehydrogenase. The glycerol 3-phosphate diffuses across the inner mitochondrial membrane where it is converted back to dihydroxyacetone phosphate by mitochondrial glycerol 3-phosphate dehydrogenase, a transmembrane protein of the inner mitochondrial membrane. The dihydroxyacetone phosphate then diffuses back to the cytosol. The mitochondrial glycerol 3-phosphate dehydrogenase does not use $NAD^+$ but instead uses FAD. The enzyme-linked $FADH_2$ (E.$FADH_2$) is then reoxidized by transferring its electrons to ubiquinone in the same inner mitochondrial membrane (see above). Note that the shuttle does not allow cytoplasmic NADH to enter the mitochondrion but its operation effectively transports the two electrons from the NADH into the mitochondrion and feeds them into the electron transport chain. Since the electrons from cytoplasmic NADH actually enter the electron transport chain from $FADH_2$, only about two ATPs are synthesized instead of approximately three ATPs from each NADH that arises inside the mitochondrion from the citric acid cycle (Topic L1) and fatty acid oxidation (Topic K2).

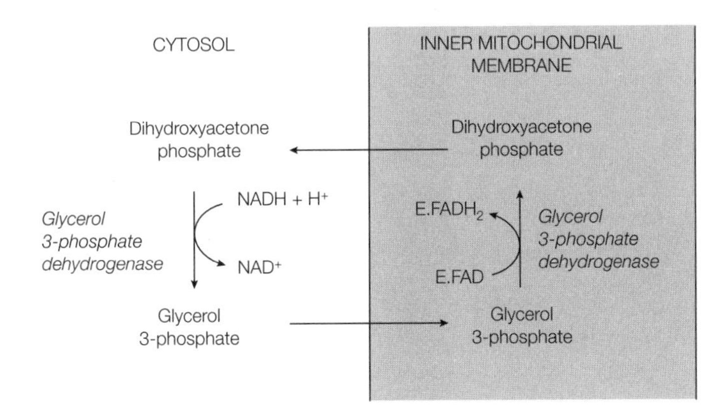


*Fig. 5. The glycerol 3-phosphate shuttle.*

A similar shuttle, the **malate–aspartate shuttle**, operates in heart and liver (*Fig. 6*). Oxaloacetate in the cytosol is converted to malate by cytoplasmic malate dehydrogenase, reoxidizing NADH to $NAD^+$ in the process. The malate enters the mitochondrion via a **malate–α-ketoglutarate carrier** in the inner mitochondrial membrane. In the matrix the malate is reoxidized to oxaloacetate by $NAD^+$ to form NADH. Oxaloacetate does not easily cross the inner mitochondrial membrane and so is transaminated to form aspartate which then exits from the mitochondrion

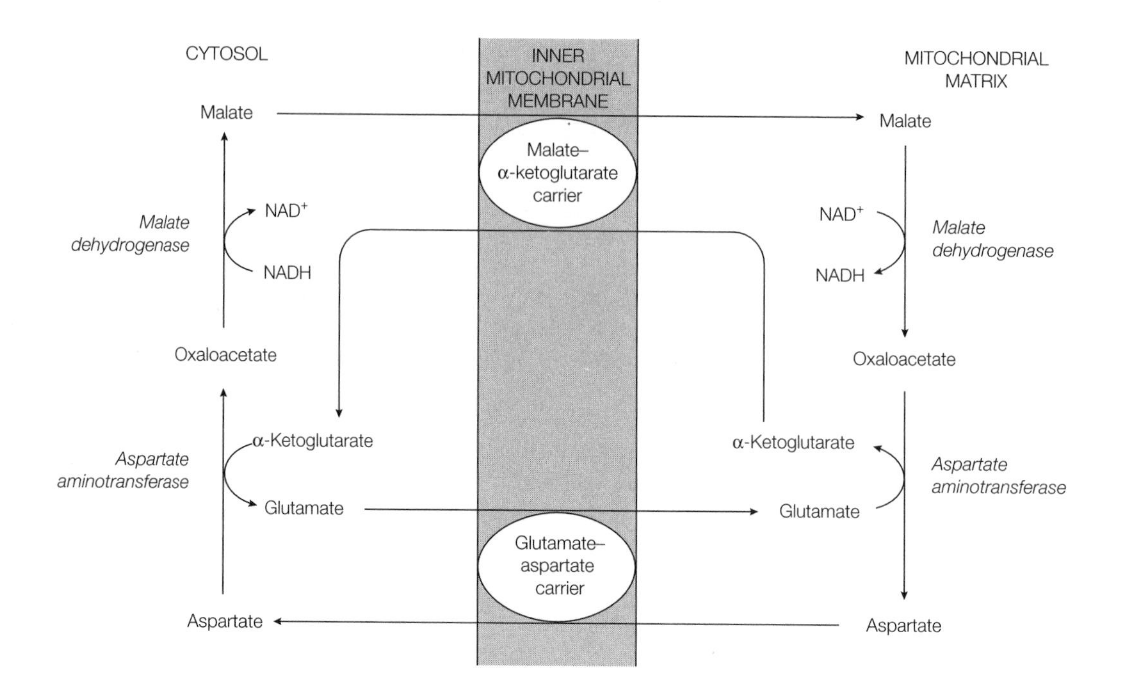


*Fig. 6. The malate–aspartate shuttle.*

and is reconverted to oxaloacetate in the cytosol, again by transamination. The net result of this cycle of reactions is to transfer the electrons from NADH in the cytosol to NADH in the mitochondrial matrix which is then reoxidized by the electron transport chain.

# L3 PHOTOSYNTHESIS

## Key Notes

**Overview**

Photosynthesis uses solar energy to synthesize carbohydrate from carbon dioxide and water. In the light reactions, the light energy drives the synthesis of NADPH and ATP. In the dark reactions (carbon-fixation reactions), the NADPH and ATP are used to synthesize carbohydrate from $CO_2$ and $H_2O$.

**Location**

In green plants and algae, photosynthesis takes place in chloroplasts. The light reactions occur in the thylakoid membranes and the dark reactions take place in the stroma. In photosynthetic bacteria the light reactions take place in the bacterial plasma membrane, or in invaginations of it (chromatophores).

**Light harvesting in green plants**

Sunlight is absorbed by chlorophyll molecules, each of which is a magnesium porphyrin. Accessory pigments, such as carotenoids, absorb light at other wavelengths so maximizing light absorption. The pigments are arranged as photosystems, each photosystem consisting of an antenna complex and a photosynthetic reaction center. An antenna complex has several hundred chlorophyll molecules and accessory pigments clustered together in the thylakoid membrane. The absorption of a photon of light by a chlorophyll molecule raises an electron to a higher energy orbital. The excited chlorophyll can pass its extra energy on to another chlorophyll molecule in the complex by exciton transfer. The energy is channeled to two special chlorophyll molecules in the photosynthetic reaction center.

**Photosystems I and II**

Green plants and algae use two types of photosystem, photosystem I with chlorophyll P700 in its reaction center and photosystem II with P680 in its reaction center. The two photosystems are linked by a chain of electron carriers. When arranged in order of their redox potentials, the components form the so-called Z scheme. Light excites P680 of photosystem II to P680*. The excited $P680^*$ passes a high-energy electron to pheophytin, and is oxidized to $P680^+$. The $P680^+$ accepts an electron from water and returns to the ground state. Overall, the removal of four electrons from two molecules of water generates four $H^+$ ions and one molecule of $O_2$. The high-energy electrons accepted by pheophytin are passed in order to plastoquinone (PQ), the cytochrome *bf* complex (also called cytochrome $b_6f$ complex) and plastocyanin. Light excites P700 of photosystem I to P700*. The excited $P700^+$ passes a high-energy electron to ferredoxin, and becomes oxidized to $P700^+$. The $P700^+$ accepts an electron from plastocyanin and returns to the ground state. Finally, two electrons from two molecules of reduced ferredoxin are transferred to $NADP^+$ to form NADPH.

**Noncyclic photophosphorylation**

The cytochrome *bf* complex is a proton pump and, during electron transport, pumps $H^+$ ions from the stroma into the thylakoid space, creating an $H^+$ gradient. $H^+$ ions are also released into the thylakoid space when photosystem II oxidizes water to produce oxygen whilst the $H^+$ ions used to reduce $NADP^+$ to NADPH are taken up from the stroma. Both effects contribute to the $H^+$ gradient.

The proton gradient drives ATP synthesis via an ATP synthase located in the thylakoid membrane (photophosphorylation). Since the electron transport involves a linear array of electron carriers, the system is called noncyclic photophosphorylation.

**Cyclic photophosphorylation**

When little $NADP^+$ is available to accept electrons, an alternative electron transport pathway is used. The high-energy electron donated by photosystem I passes to ferredoxin, then the cytochrome *bf* complex, then plastocyanin and back to the P700 of photosystem I. The resulting proton gradient generated by the cytochrome *bf* complex drives ATP synthesis (cyclic photophosphorylation) but no NADPH is made and no $O_2$ is produced.

**Bacterial photosynthesis**

Cyanobacteria use two photosystems as in green plants. The purple photosynthetic bacterium, *Rhodospirillum rubrum*, has only a single photosystem reaction center. This can carry out cyclic electron transport, synthesizing ATP (cyclic photophosphorylation). Alternatively, noncyclic electron transport can be used, producing NADH. Hydrogen sulfide ($H_2S$) can act as electron donor, generating sulfur (S). Hydrogen gas ($H_2$) and a variety of organic compounds can also be used as electron donors. Water is *not* used as electron donor and so no oxygen is produced.

**The dark reactions**

The dark reactions (carbon-fixation reactions) use the ATP and NADPH produced by the light reactions to 'fix' carbon dioxide as carbohydrate; sucrose and starch. The reactions form a cycle (the Calvin cycle) in which the enzyme ribulose bisphosphate carboxylase (rubisco), located in the stroma, condenses a $CO_2$ molecule with ribulose 1,5-bisphosphate to produce two molecules of 3-phosphoglycerate. Other reactions then regenerate the ribulose 1,5-bisphosphate. The fixation of three molecules of $CO_2$ requires six NADPH and nine ATP and leads to the net production of one molecule of glyceraldehyde 3-phosphate. For the synthesis of sucrose, glyceraldehyde 3-phosphate exits to the cytosol and is converted to fructose 6-phosphate and glucose 1-phosphate. The latter is then converted to UDP-glucose and reacts with fructose 6-phosphate to form sucrose 6-phosphate. Hydrolysis of the sucrose 6-phosphate yields sucrose. The glyceraldehyde 3-phosphate from the Calvin cycle is also used to synthesize glucose 1-phosphate which generates ADP-glucose, CDP-glucose or GDP-glucose as precursors for starch synthesis.

**The C4 pathway**

When the $CO_2$ concentration is low, rubisco can add $O_2$ to ribulose 1,5-bisphosphate (oxygenase activity) instead of $CO_2$ (carboxylase activity) producing phosphoglycolate and 3-phosphoglycerate. Metabolism of phosphoglycolate releases $CO_2$ and $NH_4^+$ and wastes energy. This consumption of $O_2$ and release of $CO_2$ is called photorespiration. Plants in hot climates close their stomata to reduce water loss. This causes a drop in the $CO_2$ concentration in the leaf, favoring photorespiration. To avoid this problem, these plants carry out the Calvin cycle only in bundle-sheath cells that are protected from the $O_2$ in air by mesophyll cells. The $CO_2$ is transported from the air via the mesophyll cells to the bundle-sheath cells by combining with three-carbon molecules (C3) to produce four-carbon molecules (C4). This C4 pathway ensures a high $CO_2$ concentration for carbon fixation by rubisco in the bundle-sheath cells.

**Related topics**

Eukaryotes (A2)
Electron transport and oxidative phosphorylation (L2)
Hemes and chlorophylls (M4)

## Overview

Photosynthesis occurs in green plants, algae and photosynthetic bacteria. Its role is to trap solar energy and use this to drive the synthesis of carbohydrate from carbon dioxide and water. Using ($CH_2O$) to represent carbohydrate, the overall reaction is:

$$H_2O + CO_2 \xrightarrow{Light} (CH_2O) + O_2$$

The reactions of photosynthesis occur in two distinct phases:

- **the light reactions**: which use light energy to synthesize NADPH and ATP;
- **the dark reactions**: that use the NADPH and ATP to synthesize carbohydrate from $CO_2$ and $H_2O$. In fact, the term 'dark reactions' is a misnomer; these **carbon-fixation reactions** should really be called light-independent reactions.

## Location

In green plants and algae, photosynthesis takes place in chloroplasts (see Topic A2). Similar to a mitochondrion, a chloroplast has a highly permeable **outer membrane** and an **inner membrane** that is impermeable to most molecules and ions. Within each chloroplast lies the **stroma**, containing soluble enzymes (analogous to the matrix of a mitochondrion). However, whereas the inner membrane of a mitochondrion contains the electron transport chain and ATP synthase (see Topic L2), in a chloroplast these are located, together with photosynthetic light-absorbing systems, in stacks of flattened membranes within the stroma called **thylakoids** (see Topic A2). Thus the primary events of trapping solar energy in photosynthesis, the light reactions, occur in the thylakoid membranes. The dark reactions take place in the stroma. In photosynthetic bacteria the light reactions take place in the bacterial plasma membrane, or in invaginations of it called **chromatophores**.

## Light harvesting in green plants

Sunlight is absorbed by **chlorophyll** molecules. **Chlorophyll** is a porphyrin in which nitrogen atoms are coordinated to a magnesium ion (see Topic M4, *Fig. 1*) (i.e. it is a **magnesium porphyrin**). This contrasts with a heme in which the nitrogen atoms are coordinated to an iron atom to form an iron porphyrin; see Topics L2 and M4). Green plants contain two types of chlorophyll molecules, **chlorophyll *a*** and **chlorophyll *b***, that differ slightly in structure (see Topic M4, *Fig. 1*) and in the wavelength of light they can absorb. Although light is trapped by **chlorophyll** molecules directly, several **accessory pigments** also exist that absorb light and pass the excitation energy on to chlorophyll molecules. Thus the **carotenoids** are important accessory pigments in green plants whilst **phycobilins** are accessory pigments in photosynthetic bacteria. These pigments absorb light at wavelengths different from that of chlorophyll and so act together to maximize the light harvested.

When a chlorophyll molecule is excited by a quantum of light (a **photon**), an electron is excited to a higher energy orbital. The excited chlorophyll can pass on its extra energy to a neighboring chlorophyll molecule by **exciton transfer** (also called **resonance energy transfer**) and so return to the unexcited state. Alternatively, the high-energy electron itself may be passed on, with the chlorophyll taking up a low-energy electron from another source.

The capture of solar energy occurs in **photosystems**. Each photosystem consists of an **antenna complex** and a **photosynthetic reaction center**. The antenna complex is composed of several hundred chlorophyll molecules and accessory pigments clustered together in the thylakoid membrane. When a chlorophyll molecule in the antenna complex absorbs light and is excited, the energy is passed by exciton transfer, from molecule to molecule, and is finally channeled to two special chlorophyll molecules in the photosynthetic reaction center. The reaction center passes on the energy as a high-energy electron to a chain of electron carriers in the thylakoid membrane (see below).

## Photosystems I and II

Green plants and algae use two types of photosystem called **photosystem I (PSI)** and **photosystem II (PSII)**. The chlorophyll in the reaction center of PSI has an absorption maximum at 700 nm and so is called **P700** (P for pigment) and that in the reaction center of PSII has an absorption maximum at 680 nm and so is called **P680**. The two photosystems are linked by other electron carriers. When arranged according to their redox potentials (see Topic L2) the various components form the so-called **Z scheme** (*Fig. 1*) because the overall shape of the redox diagram looks like a Z.

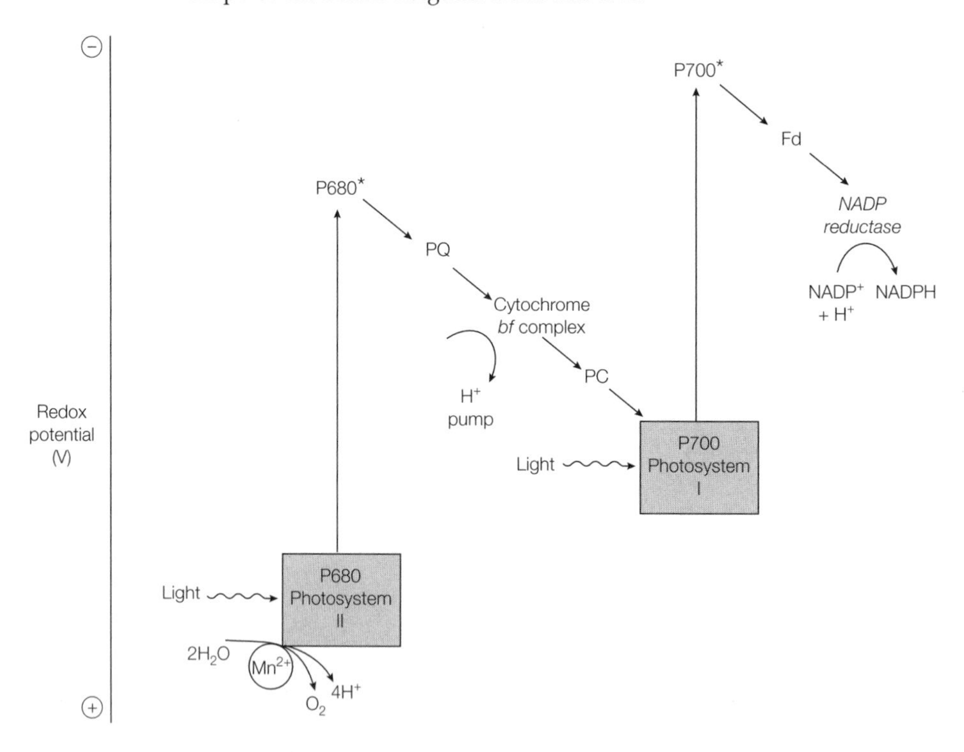


*Fig. 1. The Z scheme of noncyclic photophosphorylation in green plants.*

The sequence of reactions during light absorption (*Fig. 1*) is as follows:

1. Light is harvested by the antenna complex chlorophylls of PSII and the energy is channeled towards the reaction center at which P680 is located.
2. The excited P680 (P680*) emits a high-energy electron that passes to **plastoquinone** (PQ), a mobile quinone in the thylakoid membrane. This leaves P680 as the **P680$^+$ cation**. Plastoquinone accepts a total of two electrons and two $H^+$ ions to form $PQH_2$.

3. $P680^+$ extracts an electron from water, returning to its unexcited state. The removal of four electrons from two molecules of water requires four quanta of light to fall on PSII and leads to the production of four $H^+$ ions and one molecule of $O_2$:

$$2\ H_2O \xrightarrow{\textit{4 photons}} 4\ e^- + 4\ H^+ + O_2$$

   This reaction is mediated by a cluster of four manganese ions ($Mn^{2+}$) in PSII.

4. The electrons are now passed from $PQH_2$ via the **cytochrome *bf* complex** (also called **cytochrome $b_6f$ complex**) to **plastocyanin** (PC). PC is a copper-containing protein that accepts electrons by the copper cycling between $Cu^{2+}$ and $Cu^+$ states:

$$PQH_2 + 2\ PC\ (Cu^{2+}) \xrightarrow{\textit{Cytochrome}\ b_6f\ \textit{complex}} PQ + 2\ PC\ (Cu^+) + 2\ H^+$$

5. Light energy falling onto the antenna complex of PSI is funneled to the reaction center. Here P700 is excited (to P700*) and emits a high-energy electron to **ferredoxin**, a protein that contains at least one FeS cluster (see Topic L2), becoming the **$P700^+$ cation**. The $P700^+$ receives the electron from PC (see step 4 above) and so returns to the unexcited state.
6. Two high-energy electrons from two molecules of reduced ferredoxin are now transferred to $NADP^+$ to form NADPH. The reaction is carried out by **NADP reductase**.

$$NADP^+ + 2\ e^- + H^+ \xrightarrow{\textit{NADP reductase}} NADPH$$

   Taking account of the entire sequence of electron transport, the reaction can be written as:

$$2\ H_2O + 2\ NADP^+ \xrightarrow{\textit{Light}} 2\ NADPH + 2\ H^+ + O_2$$

   showing that electrons flow from $H_2O$ to $NADP^+$, reducing it to NADPH.

## Noncyclic photophosphorylation

During operation of the Z scheme, high-energy electrons are created by energy input via the two photosystems and the electrons then travel along a chain of carriers that decrease in redox potential (*Fig. 1*). This is analogous to the passage of electrons along the respiratory chain in mitochondria (Topic L2). In a further analogy, the cytochrome *bf* complex is a **proton pump** (*Fig. 1*) and pumps $H^+$ ions from the stroma into the thylakoid space (*Fig. 2*). Thus an $H^+$ gradient is formed during electron transport. Because of the orientation of the various electron transport components in the thylakoid membrane (*Fig. 2*), the $H^+$ ions released when PSII oxidizes water to produce oxygen are released into the thylakoid space whilst the $H^+$ used to reduce $NADP^+$ to NADPH by NADP reductase are taken up from the stroma. Thus these two reactions also contribute to the proton gradient. The proton gradient drives ATP synthesis via an ATP synthase located in the thylakoid membrane (*Fig. 2*). This is called **photophosphorylation** and is analogous to ATP synthesis via a proton gradient during oxidative phosphorylation in mitochondria (see Topic L2). The difference is that protons are pumped *out* of mitochondria but *into* a subcompartment, the

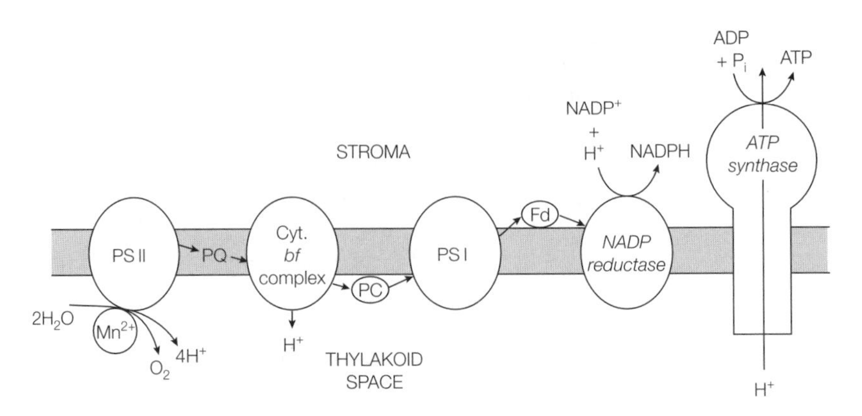


*Fig. 2. Formation of the proton gradient and ATP synthesis.*

thylakoid space, in chloroplasts. Because of the alternative ('cyclic') pathway for electron transport and ATP synthesis (see below), the formation of ATP via the joint operation of PSI and PSII (*Fig. 1*; the Z scheme) is called **noncyclic photophosphorylation**.

## Cyclic photophosphorylation

When the NADPH/$NADP^+$ ratio is high and little $NADP^+$ is available to accept electrons, an alternative electron transport pathway is used that involves only PSI and a few electron carriers (*Fig. 3*). Here the high-energy electron is passed by ferredoxin to the cytochrome *bf* complex instead of to $NADP^+$. It then flows

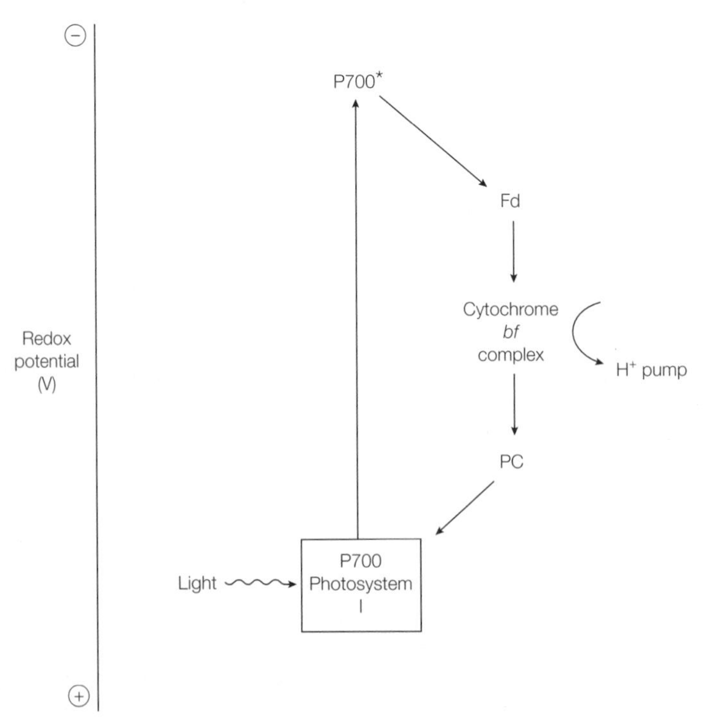


*Fig. 3. Cyclic photophosphorylation in green plants.*

to plastocyanin and back to the P700 of PSI. The resulting proton gradient generated from the $H^+$ pump, cytochrome *bf* complex, then drives ATP synthesis. During this **cyclic photophosphorylation**, ATP is formed but no NADPH is made. Furthermore, since PSII is not involved, no $O_2$ is produced.

In summary, when electron transport is operating in noncyclic mode, via PSI and PSII, the products are NADPH and ATP. In cyclic electron transport, on the other hand, the sole product is ATP.

## Bacterial photosynthesis

Cyanobacteria carry out photosynthesis using two photosystems as in green plants. However, other photosynthetic bacteria, such as the **purple photosynthetic bacterium** *Rhodospirillum rubrum*, have only a single photosystem reaction center. This can carry out cyclic electron transport, generating a proton gradient and hence synthesizing ATP (cyclic photophosphorylation). Alternatively, a noncyclic pattern of electron transport can be carried out in which the electrons from the cytochromes pass to $NAD^+$ (rather than $NADP^+$ as in green plants) to produce NADH. The electron donor is, for example, hydrogen sulfide ($H_2S$), which generates sulfur (S). Hydrogen gas ($H_2$) and a variety of organic compounds can also be used as electron donors by certain photosynthetic bacteria. Since $H_2O$ is *not* used as electron donor, no oxygen is produced.

## The dark reactions

The dark reactions (also called the **carbon-fixation reactions**) use the ATP and NADPH produced by the light reactions to convert carbon dioxide into carbohydrate. The final products are sucrose and starch.

The key carbon fixation reaction is catalyzed by a large enzyme called **ribulose bisphosphate carboxylase** (often abbreviated to **rubisco**) that is located in the stroma. The reaction condenses a $CO_2$ molecule with **ribulose 1,5-bisphosphate** (a five-carbon molecule) to produce a transient six-carbon intermediate that rapidly hydrolyzes to two molecules of 3-phosphoglycerate (*Fig. 4*):

$$CO_2 + \begin{array}{c} CH_2O\textcircled{P} \\ | \\ C=O \\ | \\ CHOH \\ | \\ CHOH \\ | \\ CH_2O\textcircled{P} \end{array} \xrightarrow{Rubisco} \begin{array}{c} CH_2O\textcircled{P} \\ | \\ CHOH \\ | \\ COO^- \\ + \\ COO^- \\ | \\ CHOH \\ | \\ CH_2O\textcircled{P} \end{array}$$

Ribulose 1,5-bisphosphate — Two molecules of 3-phosphoglycerate

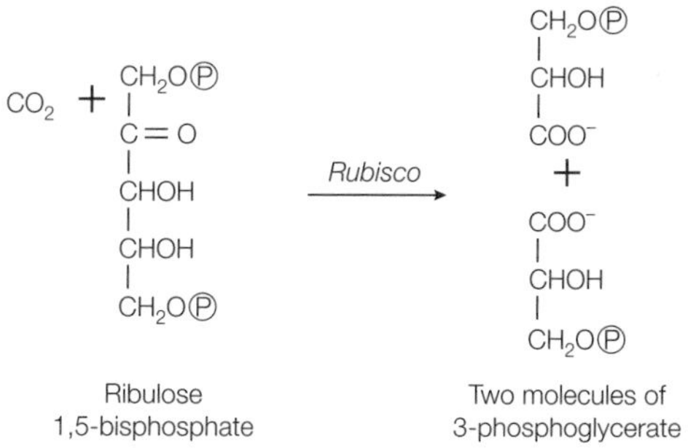

*Fig. 4. The rubisco reaction.*

Rubisco is a very slow enzyme, fixing only three molecules of its substrate every second and hence a large amount of this enzyme is needed by each plant. Typically, rubisco accounts for 50% or so of the total protein in a chloroplast. Indeed, it is probably the most abundant protein on earth!

The rubisco reaction forms part of a cycle of reactions, called the **Calvin cycle**, that leads to the regeneration of ribulose 1,5-bisphosphate (ready to fix another $CO_2$) and the net production of glyceraldehyde 3-phosphate for the synthesis

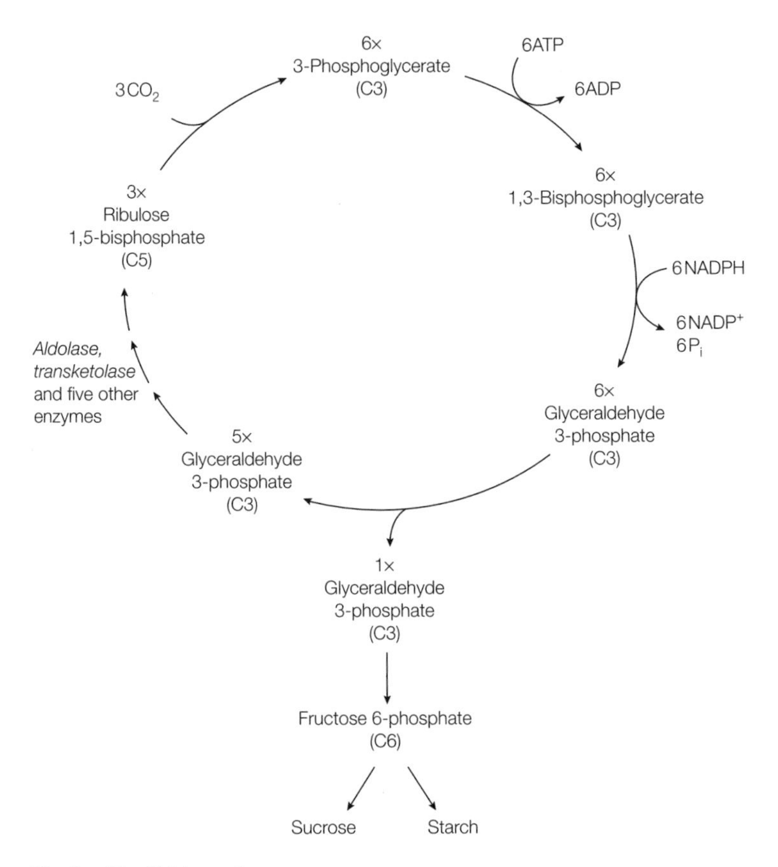


*Fig. 5. The Calvin cycle.*

of sucrose and starch. Three molecules of $CO_2$ must be fixed to generate one molecule of glyceraldehyde 3-phosphate (a three-carbon molecule). This is shown in *Fig. 5*. The conversion of glyceraldehyde 3-phosphate to ribulose 5-phosphate in the cycle requires seven enzymes including transketolase and aldolase.

Since three ATP and two NADPH are required for each $CO_2$ converted to carbohydrate, the overall reaction for the synthesis of one molecule of glyceraldehyde 3-phosphate is:

$$3\ CO_2 + 6\ NADPH + 9\ ATP \rightarrow \begin{matrix}\text{glyceraldehyde}\\ \text{3-phosphate}\end{matrix} + 6\ NADP^+ + 9\ ADP + 8\ P_i$$

### *Synthesis of sucrose*

Much of the glyceraldehyde 3-phosphate produced by the Calvin cycle in chloroplasts is exported to the cytosol and used to produce the disaccharide, sucrose. First the glyceraldehyde 3-phosphate is converted to fructose 6-phosphate and glucose 1-phosphate. The chemical reactions involved are essentially a reversal of glycolysis (see Topic J3). The glucose 1-phosphate is then converted to UDP-glucose and this reacts with fructose 6-phosphate to synthesize sucrose 6-phosphate:

$$\text{UDP-glucose} + \text{fructose 6-phosphate} \rightarrow \text{sucrose 6-phosphate} + \text{UDP}$$

Hydrolysis of the sucrose 6-phosphate yields sucrose. This is the major sugar that is transported between plant cells, analogous to the supply of glucose via the bloodstream to animal tissues (see Topic J4).

*Synthesis of starch*
Whereas animals store excess carbohydrate as glycogen (see Topics J2 and J6), plants do so in the form of starch (Topic J2). Starch is produced in the stroma of chloroplasts and stored there as starch grains. Starch synthesis occurs from ADP-glucose, CDP-glucose or GDP-glucose (but *not* UDP-glucose). The pathway involves the conversion of glyceraldehyde 3-phosphate (from the Calvin cycle) to glucose 1-phosphate which in turn is used to synthesize the nucleotide sugar derivatives.

## C4 pathway

Under normal atmospheric conditions, rubisco adds $CO_2$ to ribulose 1,5-bisphosphate. However, when the $CO_2$ concentration is low, it can add $O_2$ instead. This produces phosphoglycolate and 3-phosphoglycerate. The phosphoglycolate can be salvaged and used for biosynthetic reactions but the pathway for achieving this releases $CO_2$ and $NH_4^+$ and wastes metabolic energy. Because the net result of this process is to consume $O_2$ and release $CO_2$, it is known as **photorespiration**. This is a major problem for plants in hot climates. The plants close the gas exchange pores in their leaves (stomata) to conserve water but this leads to a drop in the $CO_2$ concentration within the leaf, favoring photorespiration. In addition, as temperature rises, the oxygenase activity of rubisco (using $O_2$) increases more rapidly than the carboxylase activity (using $CO_2$), again favoring photorespiration. To avoid these problems, some plants adapted to live in hot climates, such as corn and sugar cane, have evolved a mechanism to maximize the carboxylase activity of rubisco. In these plants, carbon fixation using the Calvin cycle takes place only in **bundle-sheath cells** that are protected from the air by **mesophyll cells**. Since the bundle-sheath cells are not exposed to air, the $O_2$ concentration is low. The $CO_2$ is transported from the air via the mesophyll cells to the bundle-sheath cells by combining with three-carbon molecules (C3) to produce four-carbon molecules (C4). These enter the bundle-sheath cells where they are broken down to C3 compounds, releasing $CO_2$. The C3 molecules return to the mesophyll cell to accept more $CO_2$. This cycle ensures a high $CO_2$ concentration for the carboxylase activity of rubisco action in the bundle-sheath cells. Since it relies on $CO_2$ transport via four-carbon molecules, it is called the **C4 pathway** and plants that use this mechanism are called **C4 plants**. All other plants are called **C3 plants** since they trap $CO_2$ directly as the three-carbon compound 3-phosphoglycerate (*Fig. 4*).

Details of the C4 pathway are shown in *Fig. 6*. The steps involved are as follows:

- in the mesophyll cell, phosphoenolpyruvate (C3) accepts $CO_2$ to form oxaloacetate (C4); a reaction catalyzed by **phosphoenolpyruvate carboxylase**
- oxaloacetate is converted to malate (C4) by **NADP$^+$-linked malate dehydrogenase**
- malate enters the bundle-sheath cell and releases $CO_2$, forming pyruvate (C3); catalyzed by **NADP$^+$-linked malate enzyme**
- pyruvate returns to the mesophyll cell and is used to regenerate phosphoenolpyruvate. This reaction, catalyzed by **pyruvate-$P_i$ dikinase**, is unusual in that it requires ATP and $P_i$ and breaks a high-energy bond to generate AMP and pyrophosphate.

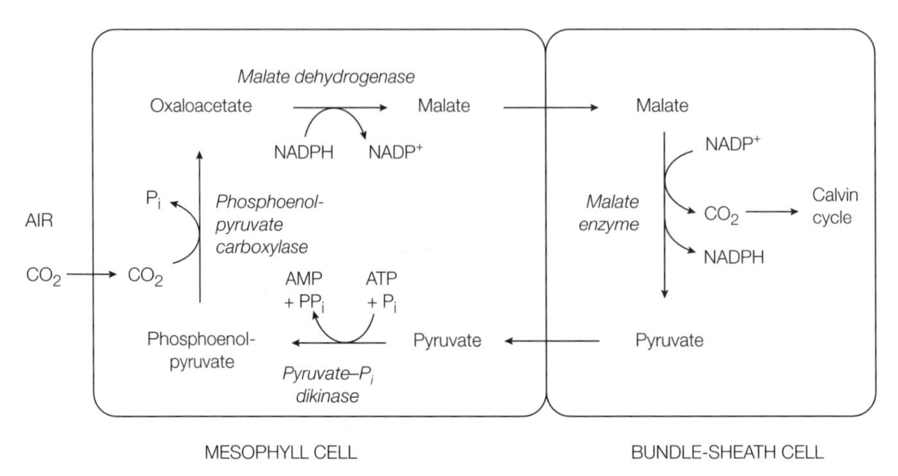


*Fig. 6. The C4 pathway.*

The pyrophosphate from the pyruvate-$P_i$ dikinase is rapidly degraded so that, overall, the net price the plant pays for operation of this $CO_2$ pump is the hydrolysis of two high-energy phosphate bonds for every molecule of $CO_2$ transported:

$$CO_2 \text{ (in air)} + ATP \rightarrow CO_2 \text{ (bundle-sheath cell)} + AMP + 2\ P_i$$

# M1 NITROGEN FIXATION AND ASSIMILATION

## Key Notes

**The nitrogen cycle**

The nitrogen cycle is the movement of nitrogen through the food chain from simple inorganic compounds, mainly ammonia, to complex organic compounds.

**Nitrogen fixation**

Nitrogen fixation is the conversion of $N_2$ gas into ammonia, a process carried out by some soil bacteria, cyanobacteria and the symbiotic bacteria *Rhizobium* that invade the root nodules of leguminous plants. This process is carried out by the nitrogenase complex, which consists of a reductase and an iron–molybdenum-containing nitrogenase. At least 16 ATP molecules are hydrolyzed to form two molecules of ammonia. Leghemoglobin is used to protect the nitrogenase in the *Rhizobium* from inactivation by $O_2$.

**Nitrogen assimilation**

Ammonia is assimilated by all organisms into organic nitrogen-containing compounds (amino acids, nucleotides, etc.) by the action of glutamate dehydrogenase (to form glutamate) and glutamine synthetase (to form glutamine).

**Related topics**

Amino acid metabolism (M2) — Hemes and chlorophylls (M4)

## The nitrogen cycle

The nitrogen cycle refers to the movement of nitrogen through the food chain of living organisms (*Fig. 1*). This complex cycle involves bacteria, plants and animals. All organisms can convert **ammonia** ($NH_3$) to organic nitrogen compounds, that is compounds containing C–N bonds. However, only a few microorganisms can synthesize ammonia from nitrogen gas ($N_2$). Although $N_2$ gas makes up about 80% of the earth's atmosphere, it is a chemically unreactive compound. The first stage in the nitrogen cycle is the reduction of $N_2$ gas to ammonia, a process called **nitrogen fixation**. Ammonia can also be obtained by reduction of nitrate ion ($NO_3^-$) that is present in the soil. **Nitrate reduction** can be carried out by most plants and microorganisms. The ammonia resulting from these two processes can then be assimilated by all organisms. Within the biosphere there is a balance between total inorganic and total organic forms of nitrogen. The conversion of organic to inorganic nitrogen comes about through catabolism, denitrification and decay (*Fig. 1*).

## Nitrogen fixation

The process of converting atmospheric $N_2$ gas into ammonia (nitrogen fixation) is carried out by only a few microorganisms, termed **diazatrophs**. These are some free-living soil bacteria such as *Klebsiella* and *Azotobacter*, cyanobacteria (blue–green algae), and the **symbiotic bacteria** (mainly *Rhizobium*). The symbiotic *Rhizobium* bacteria invade the roots of leguminous green plants (plants belonging to the pea family, e.g. beans, clover, alfalfa) and form **root nodules** where nitrogen fixation takes place. The amount of $N_2$ fixed by these

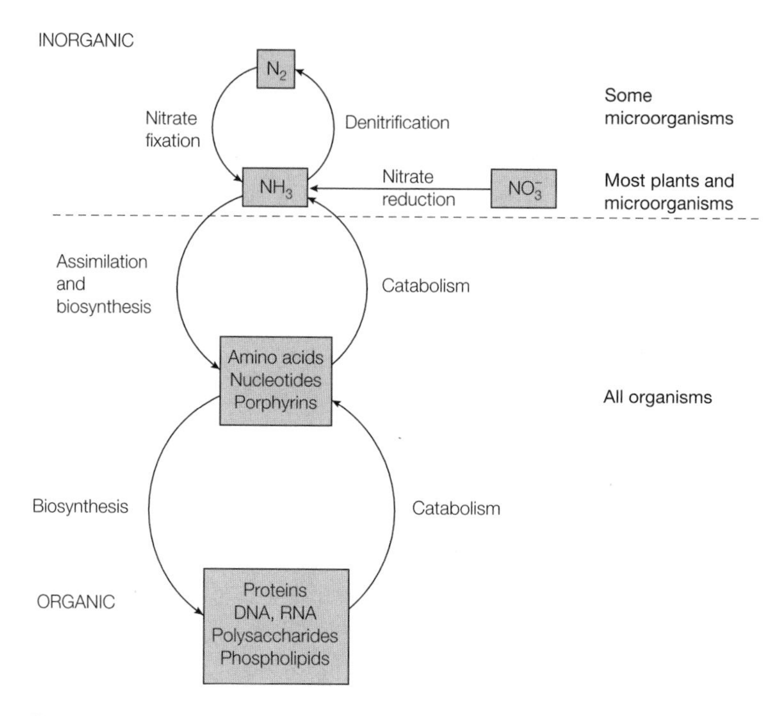


*Fig. 1. The interrelationships between inorganic and organic nitrogen metabolism.*

diazatrophic microorganisms has been estimated to be in the order of $10^{11}$ kg per year, about 60% of the earth's newly fixed nitrogen. Lightning and ultraviolet radiation fix another 15%, with the remainder coming from industrial processes.

The chemical unreactivity of the N≡N bond is clearly seen when one considers the industrial process of nitrogen fixation. This process, devised by Fritz Haber in 1910 and still used today in fertilizer factories, involves the reduction of $N_2$ in the presence of $H_2$ gas over an iron catalyst at a temperature of 500°C and a pressure of 300 atmospheres.

$$N_2 + 3\ H_2 \rightleftharpoons 2\ NH_3$$

### *Nitrogenase complex*

Biological nitrogen fixation is carried out by the **nitrogenase complex** which consists of two proteins: a **reductase**, which provides electrons with high reducing power, and a **nitrogenase**, which uses these electrons to reduce $N_2$ to $NH_3$ (*Fig. 2*). The reductase is a 64 kDa dimer of identical subunits that contains one iron–sulfur cluster and two ATP binding sites. The nitrogenase is a larger protein of 220 kDa that consists of two α- and two β-subunits ($\alpha_2\beta_2$) and contains an iron–molybdenum complex. The transfer of electrons from the reductase to the nitrogenase protein is coupled to the hydrolysis of ATP by the reductase. Although the reduction of $N_2$ to $NH_3$ is only a six-electron process:

$$N_2 + 6\ e^- + 6\ H^+ \rightarrow 2\ NH_3$$

the reductase is imperfect and $H_2$ is also formed. Thus two additional electrons are also required:

$$N_2 + 8\ e^- + 8\ H^+ \rightarrow 2\ NH_3 + H_2$$

The eight high-potential electrons come from reduced **ferredoxin** that is produced either in chloroplasts by the action of photosystem I or in oxidative electron transport (*Fig. 2*) (see Topics L2 and L3). The overall reaction of biological nitrogen fixation:

$$N_2 + 8\ e^- + 16\ ATP + 16\ H_2O \rightarrow 2\ NH_3 + H_2 + 16\ ADP + 16\ P_i + 8\ H^+$$

highlights that it is energetically very costly, with at least 16 ATP molecules being hydrolyzed.

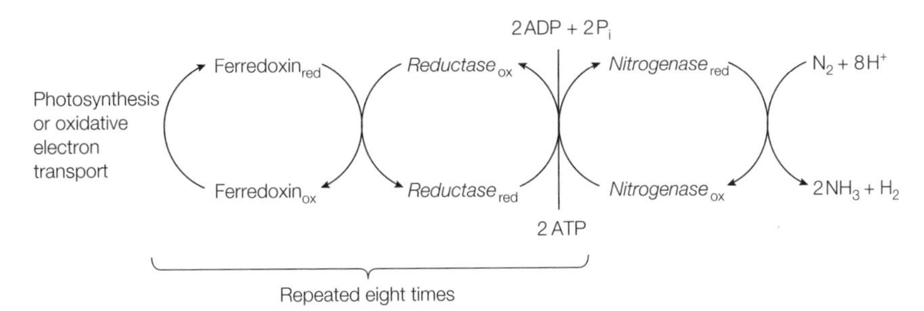


*Fig. 2. The flow of electrons in the nitrogenase-catalyzed reduction of $N_2$.*

### *Leghemoglobin*

The nitrogenase complex is extremely sensitive to inactivation by $O_2$, so the enzyme must be protected from this reactive substance. In the root nodules of leguminous plants, protection is afforded by the symbiotic synthesis of **leghemoglobin**. The globin part of this monomeric oxygen-binding protein (see Topic B4) is synthesized by the plant, whereas the **heme** group (see Topic M4) is synthesized by the *Rhizobium*. The leghemoglobin has a very high affinity for $O_2$, so maintaining a low enough concentration to protect the nitrogenase.

## Nitrogen assimilation

The next step in the nitrogen cycle is the assimilation of inorganic nitrogen, in the form of ammonia, into organic nitrogen-containing compounds. All organisms assimilate ammonia via two main reactions catalyzed by **glutamate dehydrogenase** and **glutamine synthetase** giving rise to the amino acids glutamate (Glu) and glutamine (Gln), respectively. The amino nitrogen in Glu and the amide nitrogen in Gln are then used in further biosynthetic reactions to give rise to other compounds.

### *Glutamate dehydrogenase*

Glutamate dehydrogenase catalyzes the reductive amination of the citric acid cycle intermediate α-ketoglutarate (*Fig. 3a*) (see Topic L1). Although the reaction is reversible, the reductant used in the biosynthetic reaction is NADPH. This enzyme is also involved in the catabolism of amino acids (see Topic M2).

### *Glutamine synthetase*

Glutamine synthetase catalyzes the incorporation of ammonia into glutamine, deriving energy from the hydrolysis of ATP (*Fig. 3b*). This enzyme is named a **synthetase**, rather than a synthase, because the reaction couples bond formation with the hydrolysis of ATP. In contrast, a **synthase** does not require ATP.

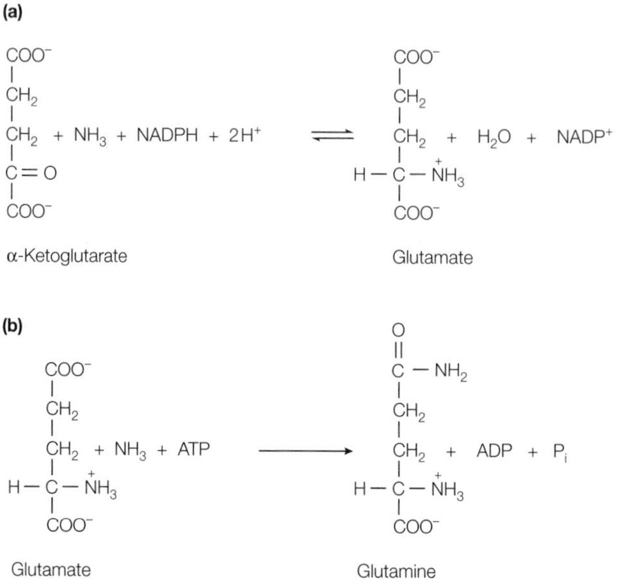

**(a)**

$$^{-}OOC{-}CH_2{-}CH_2{-}C(=O){-}COO^{-} + NH_3 + NADPH + 2H^+ \rightleftharpoons {}^{-}OOC{-}CH_2{-}CH_2{-}CH(\overset{+}{N}H_3){-}COO^{-} + H_2O + NADP^+$$

α-Ketoglutarate Glutamate

**(b)**

$$^{-}OOC{-}CH_2{-}CH_2{-}CH(\overset{+}{N}H_3){-}COO^{-} + NH_3 + ATP \longrightarrow H_2N{-}C(=O){-}CH_2{-}CH_2{-}CH(\overset{+}{N}H_3){-}COO^{-} + ADP + P_i$$

Glutamate Glutamine

*Fig. 3. Assimilation of ammonia by (a) glutamate dehydrogenase and (b) glutamine synthetase.*

# M2 Amino acid metabolism

## Key Notes

**Biosynthesis of amino acids**

Some organisms can synthesize all of the 20 standard amino acids, others cannot. Nonessential amino acids are those that can be synthesized, essential amino acids have to be taken in the diet. The 20 standard amino acids can be grouped into six biosynthetic families depending on the metabolic intermediate from which their carbon skeleton is derived.

**Amino acid degradation**

Amino acids are degraded by the removal of the α-amino group and the conversion of the resulting carbon skeleton into one or more metabolic intermediates. Amino acids are termed glucogenic if their carbon skeletons can give rise to the net synthesis of glucose, and ketogenic if they can give rise to ketone bodies. Some amino acids give rise to more than one intermediate and these lead to the synthesis of glucose as well as ketone bodies. Thus these amino acids are both glucogenic and ketogenic.

**Transamination**

The α-amino groups are removed from amino acids by a process called transamination. The acceptor for this reaction is usually the α-keto acid called α-ketoglutarate which results in the formation of glutamate and the corresponding α-keto acid. The coenzyme of all transaminases is pyridoxal phosphate which is derived from vitamin $B_6$ and which is transiently converted during transamination into pyridoxamine phosphate.

**Oxidative deamination of glutamate**

The glutamate produced by transamination is oxidatively deaminated by glutamate dehydrogenase to produce ammonia. This enzyme is unusual in being able to use either $NAD^+$ or $NADP^+$, and is subject to allosteric regulation. GTP and ATP are allosteric inhibitors, whereas GDP and ADP are allosteric activators.

**Amino acid oxidases**

Small amounts of amino acids are degraded by L- and D-amino acid oxidases that utilize flavin mononucleotide (FMN) or flavin adenine dinucleotide (FAD) as coenzyme, respectively.

**Metabolism of phenylalanine**

Phenylalanine is first converted to tyrosine by the monooxygenase phenylalanine hydroxylase; a reaction involving the coenzyme tetrahydrobiopterin. The tyrosine is then converted first by transamination and then by a dioxygenase reaction to homogentisate, which in turn is further metabolized to fumarate and acetoacetate.

**Inborn errors of metabolism**

Inborn errors of metabolism are inherited metabolic disorders caused by the absence of an enzyme in a metabolic pathway. Alkaptonuria is caused by the lack of homogentisate oxidase and is harmless, whereas phenylketonuria, which is due to a lack of phenylalanine hydroxylase, can cause severe mental retardation.

**Related topics**

Nitrogen fixation and assimilation (M1) The urea cycle (M3)

## Biosynthesis of amino acids

Plants and microorganisms can synthesize all of the 20 standard amino acids. Mammals, however, cannot synthesize all 20 and must obtain some of them in their diet. Those amino acids that are supplied in the diet are referred to as **essential**, whereas the remainder that can be synthesized by the organism are termed **nonessential**. This designation refers to the needs of an organism under a particular set of conditions. In humans the nonessential amino acids are Ala, Arg, Asn, Asp, Cys, Glu, Gln, Gly, Pro, Ser and Tyr, while the essential ones are His, Ile, Leu, Lys, Met, Phe, Thr, Trp and Val. The pathways for the biosynthesis of amino acids are diverse and often vary from one organism to another. However, they all have an important feature in common: their carbon skeletons come from key intermediates in central metabolic pathways (glycolysis, Topic J3; the citric acid cycle, Topic L1; or the pentose phosphate pathway, Topic J5) (*Fig. 1*). The amino acids can be grouped together into six biosynthetic pathways depending on the intermediate from which they are derived (*Fig. 1*). The primary amino group usually comes from transamination of glutamate.

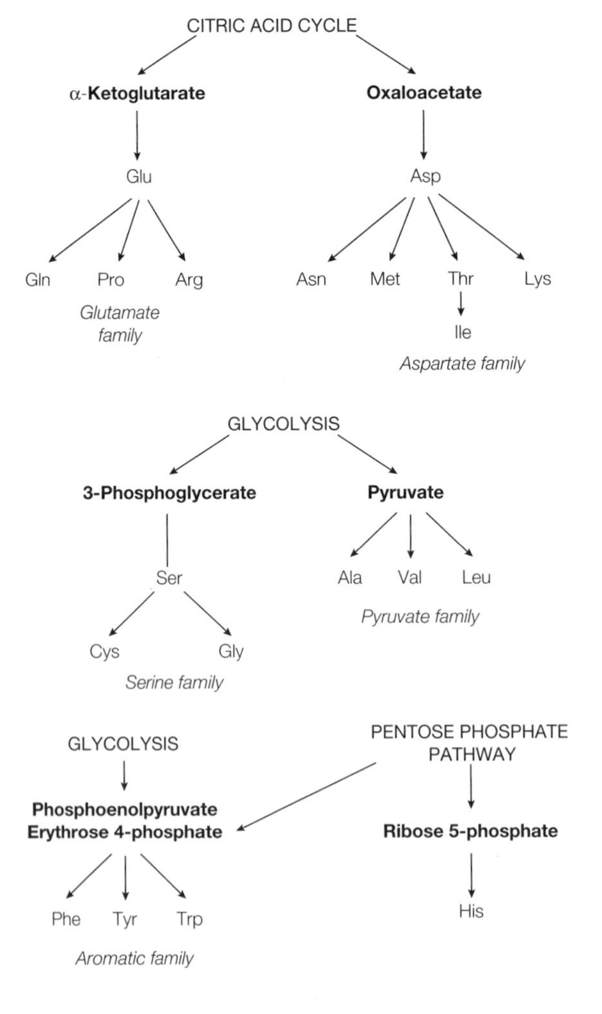


*Fig. 1. Biosynthetic families of amino acids.*

## Amino acid degradation

As there is no store for excess amino acids, and as proteins are constantly being turned over, amino acids have to be continually degraded. The α-amino group is removed first and the resulting carbon skeleton is converted into one or more major metabolic intermediates and used as metabolic fuel. The carbon skeletons of the 20 standard amino acids are funneled into only seven molecules: pyruvate, acetyl CoA, acetoacetyl CoA, α-ketoglutarate, succinyl CoA, fumarate and oxaloacetate (*Fig. 2*). Amino acids that are degraded to pyruvate, α-ketoglutarate, succinyl CoA, fumarate and oxaloacetate are termed **glucogenic** as they can give rise to the net synthesis of glucose. This is because the citric acid cycle intermediates and pyruvate can be converted into phosphoenolpyruvate and then into glucose via gluconeogenesis (see Topics J4 and L1). In contrast, amino acids that are degraded to acetyl CoA or acetoacetyl CoA are termed **ketogenic** because they give rise to ketone bodies (see Topic K2); the acetyl CoA or acetoacetyl CoA can also be used to synthesize lipids (see Topic K3). Of the standard set of 20 amino acids, only Leu and Lys are solely ketogenic. Ile, Phe, Trp and Tyr are both ketogenic and glucogenic as some of their carbon atoms end up in acetyl CoA or acetoacetyl CoA, whereas others end up in precursors of glucose. The remaining 14 amino acids are classified as solely glucogenic.

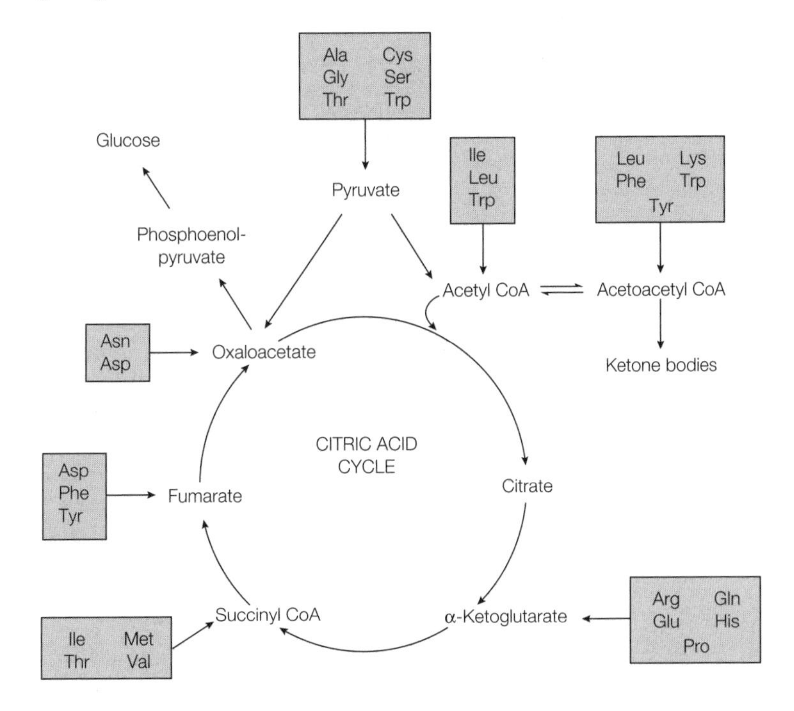


*Fig. 2. Fates of the amino acid carbon skeletons.*

## Transamination

Prior to the metabolism of their carbon skeletons into a major metabolic intermediate, the α-amino group of the amino acid has first to be removed by a process known as **transamination**. In this process the α-amino group of most amino acids is transferred to **α-ketoglutarate** to form **glutamate** and the corresponding α-keto acid:

$$\alpha\text{-amino acid} + \alpha\text{-ketoglutarate} \rightleftharpoons \alpha\text{-keto acid} + \text{glutamate}$$

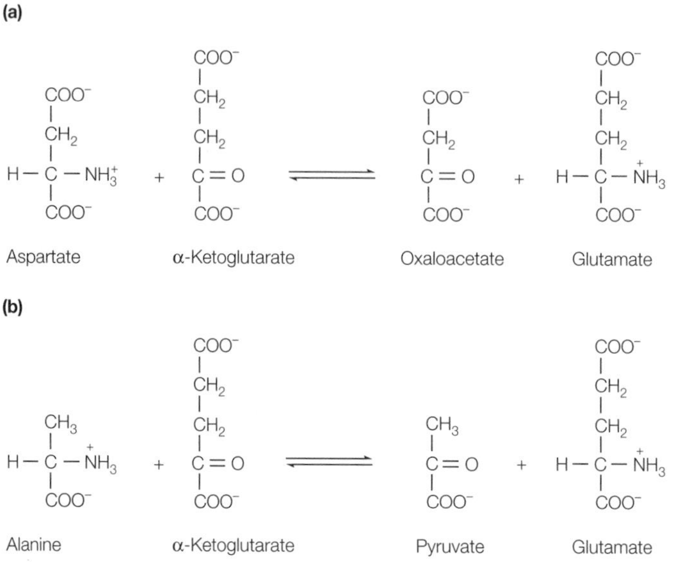

(a) Aspartate + α-Ketoglutarate ⇌ Oxaloacetate + Glutamate

(b) Alanine + α-Ketoglutarate ⇌ Pyruvate + Glutamate

*Fig. 3. Reactions catalyzed by (a) aspartate transaminase and (b) alanine transaminase.*

The enzymes that catalyze these reactions are called **transaminases** (aminotransferases) and in mammals are found predominantly in the liver. For example, aspartate transaminase catalyzes the transfer of the amino group of aspartate to α-ketoglutarate (*Fig. 3a*), while alanine transaminase catalyzes the transfer of the amino group of alanine to α-ketoglutarate (*Fig. 3b*).

### *Pyridoxal phosphate*

The coenzyme (or prosthetic group) of all transaminases is **pyridoxal phosphate**, which is derived from pyridoxine (**vitamin $B_6$**), and which is transiently converted into **pyridoxamine phosphate** during transamination (*Fig. 4*). In the absence of substrate, the aldehyde group of pyridoxal phosphate forms a covalent **Schiff base linkage** (imine bond) with the amino group in the side-chain of a specific lysine residue in the active site of the enzyme. On addition of substrate, the α-amino group of the incoming amino acid displaces the amino group of the active site lysine and a new Schiff base linkage is formed with the amino acid substrate. The resulting amino acid–pyridoxal phosphate–Schiff base that is formed remains tightly bound to the enzyme by multiple noncovalent interactions.

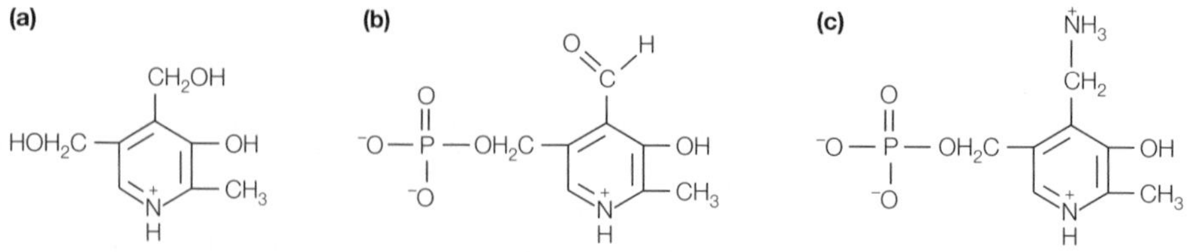

*Fig. 4. Structures of (a) pyridoxine (vitamin $B_6$), (b) pyridoxal phosphate and (c) pyridoxamine phosphate.*

The amino acid is then hydrolyzed to form an α-keto acid and pyridoxamine phosphate, the α-amino group having been temporarily transferred from the amino acid substrate on to pyridoxal phosphate (*Fig. 5*). These steps constitute one half of the overall transamination reaction. The second half occurs by a reversal of the above reactions with a second α-keto acid reacting with the pyridoxamine phosphate to yield a second amino acid and regenerate the enzyme–pyridoxal phosphate complex (*Fig. 5*).

The reactions catalyzed by transaminases are **anergonic** as they do not require an input of metabolic energy. They are also freely reversible, the direction of the reaction being determined by the relative concentrations of the amino acid–keto acid pairs. Pyridoxal phosphate is not just used as the coenzyme in transamination reactions, but is also the coenzyme for several other reactions involving amino acids including decarboxylations, deaminations, racemizations and aldol cleavages.

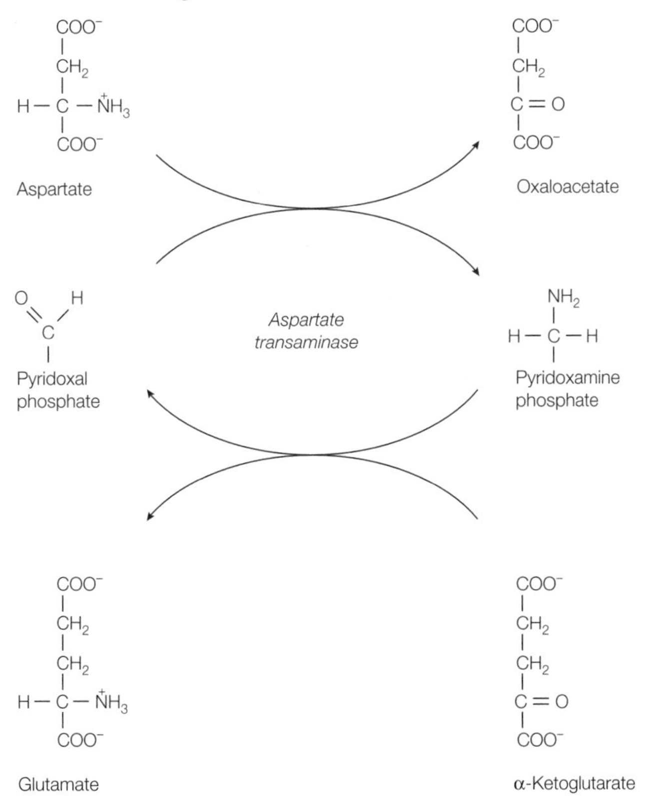


*Fig. 5. The overall reaction of transamination.*

## Oxidative deamination of glutamate

The α-amino groups that have been funneled into glutamate from the other amino acids are then converted into **ammonia** by the action of **glutamate dehydrogenase** (*Fig. 6*). This enzyme is unusual in being able to utilize either $NAD^+$ or $NADP^+$. In the biosynthesis of glutamate, the $NADP^+$ form of the coenzyme is used (see Topic M1), whereas $NAD^+$ is used in its degradation. Glutamate dehydrogenase consists of six identical subunits and is subject to **allosteric regulation** (see Topic C5). GTP and ATP are **allosteric inhibitors**, whereas GDP and ADP are **allosteric activators**. Hence, when the energy charge of the cell is low (i.e. there is more ADP and GDP than their triphosphate forms) glutamate dehydrogenase is activated and the oxidation of amino acids increases. The resulting carbon skeletons are then utilized as metabolic fuel, feeding into the citric acid cycle (see Topic L1) and ultimately giving rise to energy through oxidative phosphorylation (see Topic L2).

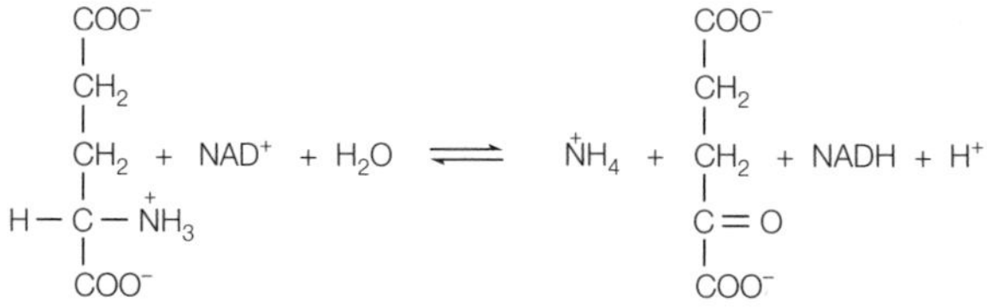

$$\text{Glutamate} + NAD^+ + H_2O \rightleftharpoons \overset{+}{N}H_4 + \alpha\text{-Ketoglutarate} + NADH + H^+$$

Glutamate: $^-OOC{-}CH_2{-}CH_2{-}CH(\overset{+}{N}H_3){-}COO^-$

α-Ketoglutarate: $^-OOC{-}CH_2{-}CH_2{-}C(=O){-}COO^-$

*Fig. 6. Oxidative deamination of glutamate by glutamate dehydrogenase.*

## Amino acid oxidases

The major route for the deamination of amino acids is transamination followed by the oxidative deamination of glutamate. However, a minor route also exists that involves direct oxidation of the amino acid by **L-amino acid oxidase**. This enzyme utilizes **flavin mononucleotide (FMN)** as its coenzyme (see Topic C1), with the resulting $FMNH_2$ being reoxidized by molecular $O_2$, a process that also generates the toxic $H_2O_2$ (*Fig. 7*). The $H_2O_2$ is rendered harmless by the action of **catalase** (see Topic A2). Kidney and liver are also rich in the **FAD**-containing **D-amino acid oxidase**. However, the function of this enzyme in animals is unclear, since the D-isomers of amino acids are rare in nature, only occurring in bacterial cell walls (see Topic A1).

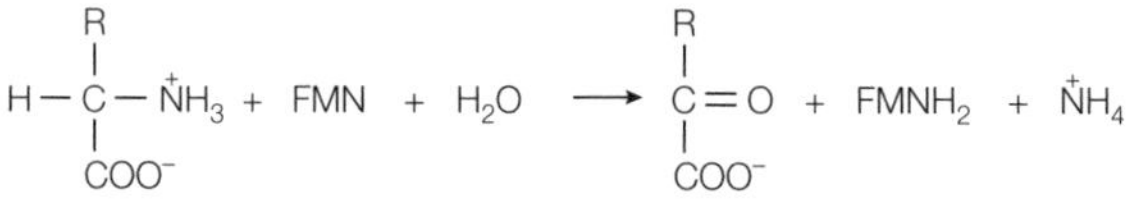

$$\underset{\text{Amino acid}}{R{-}CH(\overset{+}{N}H_3){-}COO^-} + FMN + H_2O \longrightarrow \underset{\alpha\text{-Keto acid}}{R{-}C(=O){-}COO^-} + FMNH_2 + \overset{+}{N}H_4$$

$$FMNH_2 + O_2 \longrightarrow FMN + H_2O_2$$

*Fig. 7. Action of L-amino acid oxidase.*

## Metabolism of phenylalanine

The metabolism of phenylalanine will now be considered in some detail, as two inborn errors of metabolism are known that affect this pathway. **Phenylalanine** is first hydroxylated by **phenylalanine hydroxylase** to form another aromatic amino acid **tyrosine** (*Fig. 8*). The coenzyme for this reaction is the reductant **tetrahydrobiopterin** which is oxidized to **dihydrobiopterin**. Phenylalanine hydroxylase is classified as a **monooxygenase** as one of the atoms of $O_2$ appears in the product and the other in $H_2O$. The tyrosine is then transaminated to *p*-hydroxyphenylpyruvate, which is in turn converted into **homogentisate** by *p*-hydroxyphenylpyruvate hydroxylase. This hydroxylase is an example of a **dioxygenase**, as both atoms of $O_2$ become incorporated into the product (*Fig. 8*). The homogentisate is then cleaved by homogentisate oxidase, another dioxygenase, before fumarate and acetoacetate are produced

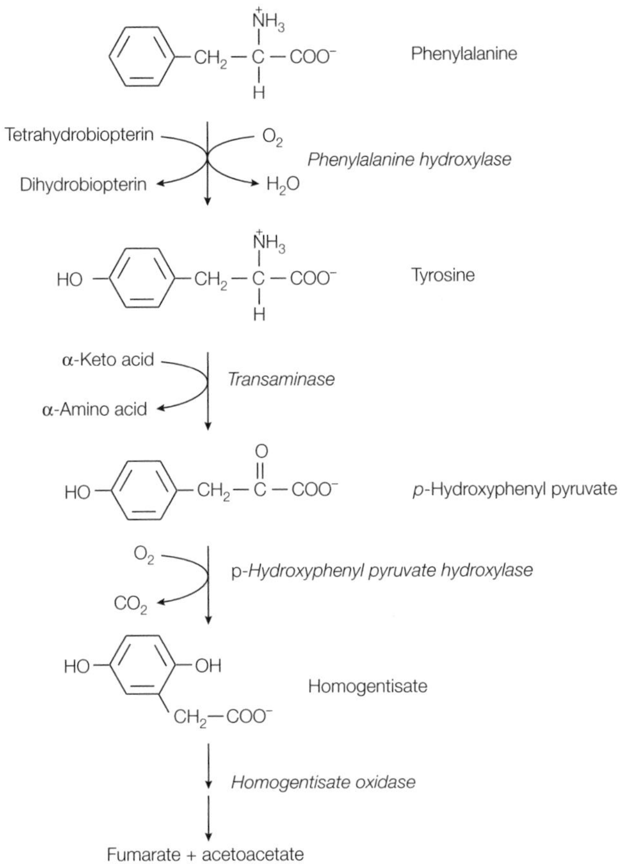

Phenylalanine

Tetrahydrobiopterin → Dihydrobiopterin; $O_2$ → $H_2O$

*Phenylalanine hydroxylase*

Tyrosine

α-Keto acid → α-Amino acid

*Transaminase*

*p*-Hydroxyphenyl pyruvate

$O_2$ → $CO_2$

*p-Hydroxyphenyl pyruvate hydroxylase*

Homogentisate

*Homogentisate oxidase*

Fumarate + acetoacetate

*Fig. 8. The metabolism of phenylalanine.*

in a final reaction. The fumarate can feed into the citric acid cycle (see Topic L1), whereas acetoacetate can be used to form ketone bodies (see Topic K2). Thus phenylalanine and tyrosine are each both glucogenic and ketogenic.

## Inborn errors of metabolism

Two inborn errors of metabolism are known that affect phenylalanine metabolism. These are **inherited metabolic disorders** caused by the absence of one of the enzymes in the pathway. One of these disorders, **alkaptonuria**, is caused by the absence of **homogentisate oxidase**. This results in the accumulation of homogentisate that is subsequently excreted in the urine, and which oxidizes to a black color on standing. This defect is harmless, in contrast with the other disorder, **phenylketonuria**. In phenylketonuria there is a block in the hydroxylation of phenylalanine to tyrosine caused by an absence or deficiency of **phenylalanine hydroxylase** or, more rarely, of its tetrahydrobiopterin coenzyme. The result is a 20-fold increase in the levels of phenylalanine in the blood with its subsequent **transamination** to phenylpyruvate. If untreated, severe mental retardation occurs, with a life expectancy of on average 20 years. With an incidence of 1:20 000 this condition is now screened for at birth, with treatment being to restrict the intake of phenylalanine (by putting the individual on a **low phenylalanine diet**) and thus minimize the need to metabolize the excess. However, enough phenylalanine must be provided to meet the needs for growth and replacement.

# M3 THE UREA CYCLE

## Key Notes

**Ammonia excretion**

Excess nitrogen is excreted as ammonia. Ammonotelic organisms excrete ammonia directly, uricotelic organisms excrete it as uric acid, and ureotelic organisms excrete it as urea.

**The urea cycle**

In the urea cycle ammonia is first combined with $CO_2$ to form carbamoyl phosphate. This then combines with ornithine to form citrulline. Citrulline then condenses with aspartate, the source of the second nitrogen atom in urea, to form argininosuccinate. This compound is in turn split to arginine and fumarate, and the arginine then splits to form urea and regenerate ornithine The first two reactions take place in the mitochondria of liver cells, the remaining three in the cytosol.

**Link to the citric acid cycle**

The fumarate produced in the urea cycle can enter directly into the citric acid cycle and be converted into oxaloacetate. Oxaloacetate can then be either transaminated to aspartate which feeds back into the urea cycle, or be converted into citrate, pyruvate or glucose.

**Hyperammonemia**

Hyperammonemia is an increase in the levels of ammonia in the blood caused by a defect in an enzyme of the urea cycle. The excess ammonia is channeled into glutamate and glutamine with a deleterious effect on brain function.

**Formation of creatine phosphate**

The urea cycle intermediate arginine can be condensed with glycine to form guanidinoacetate, which in turn is methylated by the methyl donor *S*-adenosyl methionine to creatine. The creatine is then phosphorylated to form creatine phosphate, a high-energy store found in muscle.

**The activated methyl cycle**

*S*-Adenosyl methionine is the major methyl donor in biological reactions. It is regenerated via the intermediates *S*-adenosyl homocysteine, homocysteine and methionine in the activated methyl cycle.

**Uric acid**

Uric acid, the major nitrogenous waste product of uricotelic organisms, is also formed in other organisms from the breakdown of purine bases. Gout is caused by the deposition of excess uric acid crystals in the joints.

**Related topics**

The citric acid cycle (L1)
Nitrogen fixation and assimilation (M1)
Amino acid metabolism (M2)

## Ammonia excretion

There is no store for nitrogen-containing compounds as there is for carbohydrate (glycogen) or lipids (triacylglycerol) (see Topics J6 and K4). Thus nitrogen ingested in excess of what is required by the organism has to be excreted. The excess nitrogen is first converted into **ammonia** and is then excreted from living organisms in one of three ways. Many aquatic animals simply excrete the ammonia itself directly into the surrounding water. Birds and terrestrial reptiles excrete the ammonia in the form of **uric acid**, while most terrestrial vertebrates convert the ammonia into **urea** before excretion. These three classes of organisms are called: **ammonotelic**, **uricotelic** and **ureotelic**, respectively.

## The urea cycle

Urea is synthesized in the liver by the **urea cycle**. It is then secreted into the bloodstream and taken up by the kidneys for excretion in the urine. The urea cycle was the first cyclic metabolic pathway to be discovered by Hans Krebs and Kurt Henseleit in 1932, 5 years before Krebs discovered the citric acid cycle (see Topic L1). The overall reaction of the pathway is:

$$NH_4^+ + HCO_3^- + H_2O + 3\ ATP + aspartate \rightarrow$$
$$urea + 2\ ADP + AMP + 2\ P_i + PP_i + fumarate$$

One of the nitrogen atoms of urea comes from ammonia, the other is transferred from the amino acid aspartate, while the carbon atom comes from $CO_2$. **Ornithine**, an amino acid that is not in the standard set of 20 amino acids and is not found in proteins, is the carrier of these nitrogen and carbon atoms. Five enzymatic reactions are involved in the urea cycle (*Fig. 1*), the first two of which take place in mitochondria, the other three in the cytosol:

1. **Carbamoyl phosphate synthetase**, which is technically not a member of the urea cycle, catalyzes the condensation and activation of ammonia (from the oxidative deamination of glutamate by glutamate dehydrogenase; Topic M2) and $CO_2$ (in the form of bicarbonate, $HCO_3^-$) to form **carbamoyl phosphate**. The hydrolysis of two ATP molecules makes this reaction essentially irreversible.
2. The second reaction also occurs in the mitochondria and involves the transfer of the carbamoyl group from carbamoyl phosphate to **ornithine** by **ornithine transcarbamoylase**. This reaction forms another nonstandard amino acid **citrulline** which then has to be transported out of the mitochondrion into the cytosol where the remaining reactions of the cycle take place.
3. The citrulline is then condensed with aspartate, the source of the second nitrogen atom in urea, by the enzyme **argininosuccinate synthetase** to form **argininosuccinate**. This reaction is driven by the hydrolysis of ATP to AMP and $PP_i$, with subsequent hydrolysis of the pyrophosphate. Thus both of the high-energy bonds in ATP are ultimately cleaved.
4. **Argininosuccinase** then removes the carbon skeleton of aspartate from argininosuccinate in the form of **fumarate**, leaving the nitrogen atom on the other product **arginine**. As the urea cycle also produces arginine, this amino acid is classified as nonessential in ureotelic organisms. Arginine is the immediate precursor of urea.
5. The **urea** is then formed from arginine by the action of **arginase** with the regeneration of ornithine. The ornithine is then transported back into the mitochondrion ready to be combined with another molecule of carbamoyl phosphate.

## Link to the citric acid cycle

The synthesis of fumarate by argininosuccinase links the urea cycle to the citric acid cycle (*Fig. 2*). Fumarate is an intermediate of this latter cycle which is then hydrated to malate, which in turn is oxidized to **oxaloacetate** (see Topic L1).

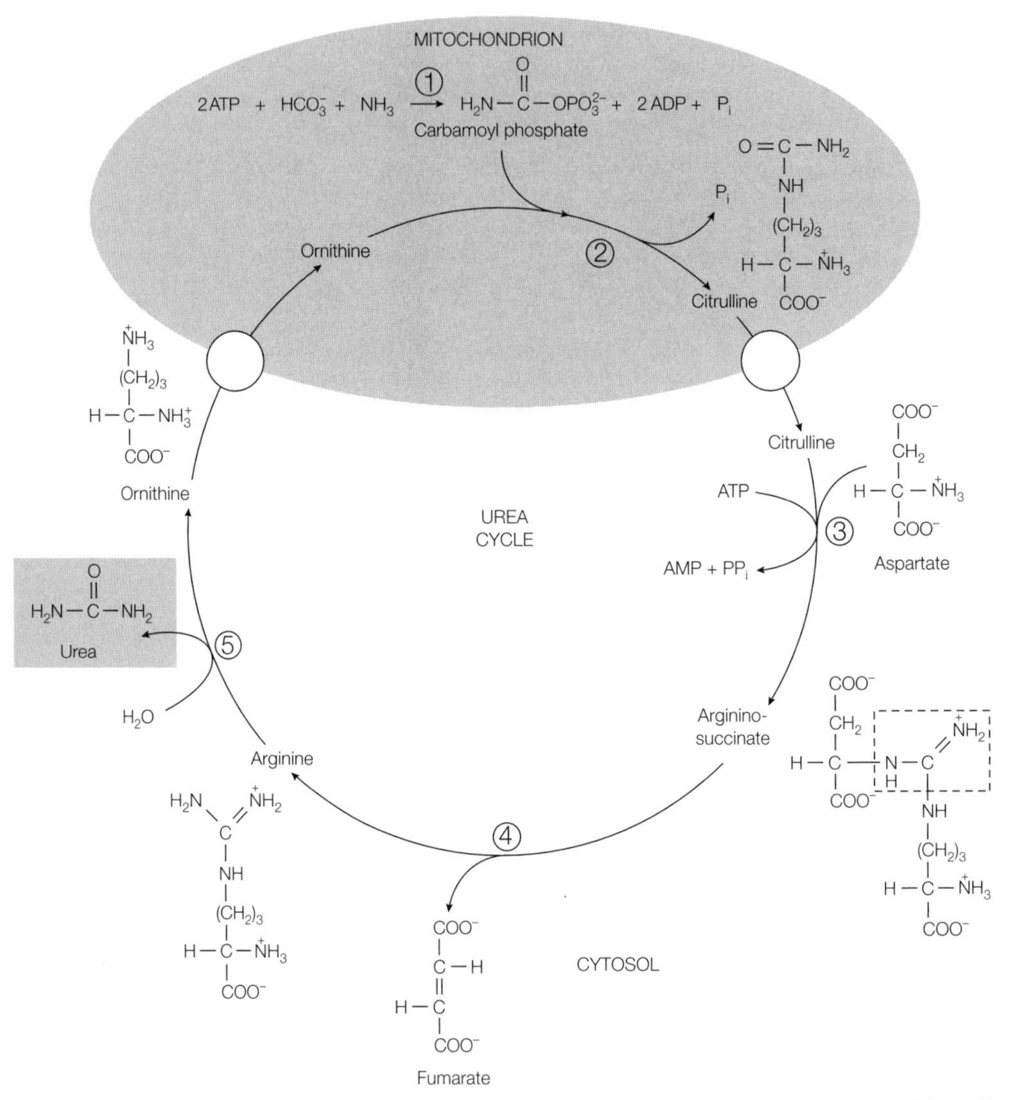


*Fig. 1. The urea cycle. The enzymes involved in this cycle are: (1) carbamoyl phosphate synthetase; (2) ornithine transcarbamoylase; (3) argininosuccinate synthetase; (4) arginosuccinase; and (5) arginase.*

Oxaloacetate has several possible fates:

- **transamination** to aspartate (see Topic M2) which can then feed back into the urea cycle;
- condensation with acetyl CoA to form citrate which then continues on round the citric acid cycle (see Topic L1);
- conversion into glucose via gluconeogenesis (see Topic J4);
- conversion into pyruvate.

## Hyperammonemia

Why do organisms need to detoxify ammonia in the first place? The answer to this question is obvious when one considers what happens if there is a block in the

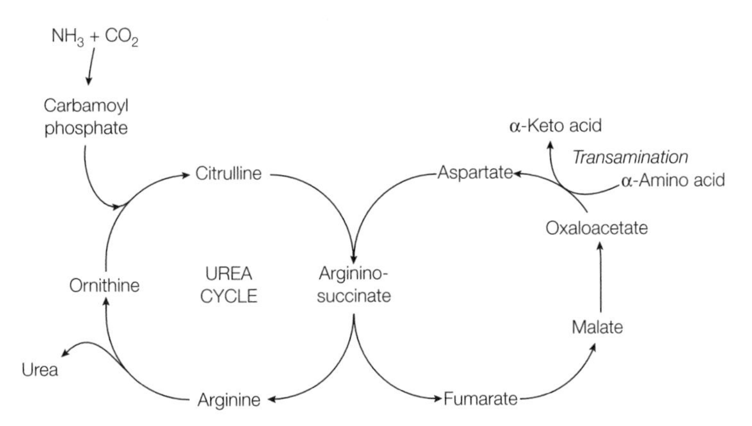


*Fig. 2. The urea cycle and the citric acid cycle are linked by fumarate and the transamination of oxaloacetate to aspartate.*

urea cycle due to a defective enzyme. A block in any of the urea cycle enzymes leads to an increase in the amount of ammonia in the blood, so-called **hyperammonemia**. The most common cause of such a block is a genetic defect that becomes apparent soon after birth, when the afflicted baby becomes lethargic and vomits periodically. If left untreated, coma and irreversible brain damage will follow. The reasons for this are not entirely clear but may be because the excess ammonia leads to the increased formation of **glutamate** and **glutamine** (*Fig. 3*) (see Topic M1). These reactions result via depletion of the citric acid cycle intermediate α-ketoglutarate which may then compromise energy production, especially in the brain. It also leads to an increase in the acidic amino acids glutamate and glutamine which may directly cause damage to the brain.

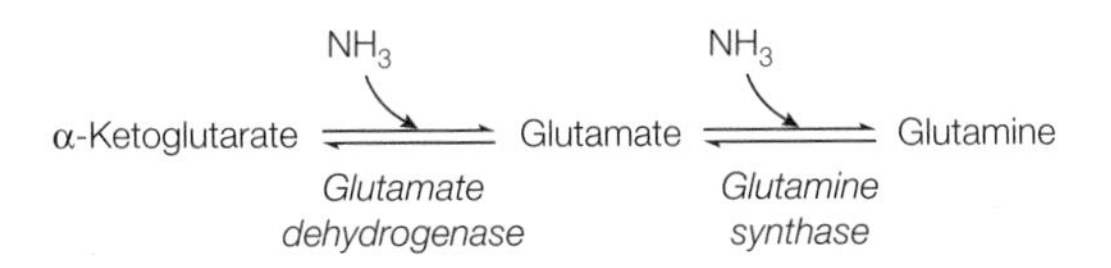


*Fig. 3. Excess ammonia leads to the formation of glutamate and glutamine.*

## Formation of creatine phosphate

The urea cycle is also the starting point for the synthesis of another important metabolite **creatine phosphate**. This **phosphagen** provides a reservoir of high-energy phosphate in muscle cells as the energy released upon its hydrolysis is greater than that released upon the hydrolysis of ATP ($\Delta G$ for creatine phosphate hydrolysis = $-10.3$ kcal mol$^{-1}$ compared with $-7.3$ kcal mol$^{-1}$ for ATP hydrolysis) (see Topic C2). The first step in the formation of creatine phosphate is the condensation of arginine and glycine to form guanidinoacetate (*Fig. 4*). Ornithine is released in this reaction and can then be re-utilized by the urea cycle. The guanidinoacetate is then methylated by the methyl group donor ***S*-adenosyl methionine** to form creatine, which is in turn phosphorylated by **creatine kinase** to form creatine phosphate (*Fig. 4*).

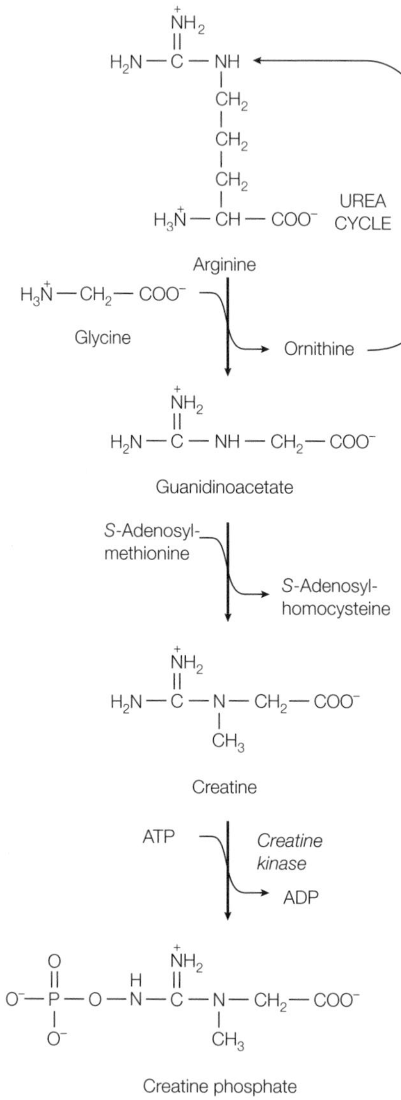

$H_2N-C(=\overset{+}{N}H_2)-NH-CH_2-CH_2-CH_2-CH(\overset{+}{N}H_3)-COO^-$

Arginine

UREA CYCLE

$H_3\overset{+}{N}-CH_2-COO^-$

Glycine

Ornithine

$H_2N-C(=\overset{+}{N}H_2)-NH-CH_2-COO^-$

Guanidinoacetate

*S*-Adenosyl-methionine

*S*-Adenosyl-homocysteine

$H_2N-C(=\overset{+}{N}H_2)-N(CH_3)-CH_2-COO^-$

Creatine

ATP

*Creatine kinase*

ADP

$O^--P(=O)(O^-)-O-N(H)-C(=\overset{+}{N}H_2)-N(CH_3)-CH_2-COO^-$

Creatine phosphate

Fig. 4. Formation of creatine phosphate.

## The activated methyl cycle

**S-Adenosyl methionine** serves as a donor of **methyl groups** in numerous biological reactions [e.g. in the formation of creatine phosphate (see above) and in the synthesis of nucleic acids]. It is formed through the action of the activated methyl cycle (*Fig. 5*). During donation of its methyl group to another compound, *S*-adenosyl methionine is converted into *S*-adenosyl homocysteine. To regenerate *S*-adenosyl methionine, the adenosyl group is removed from the *S*-adenosyl homocysteine to form homocysteine. This is then methylated by the enzyme homocysteine methyltransferase, one of only two **vitamin $B_{12}$**-containing enzymes found in eukaryotes, to form **methionine**. The resulting methionine is then activated to *S*-adenosyl methionine with the release of all three of the phosphates from ATP.

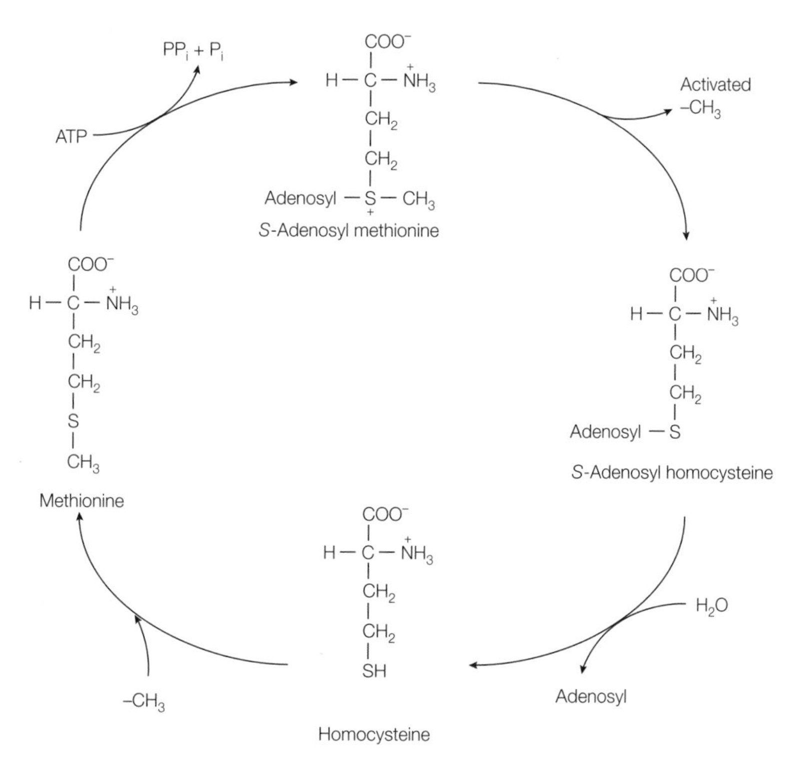


*Fig. 5. The activated methyl cycle.*

## Uric acid

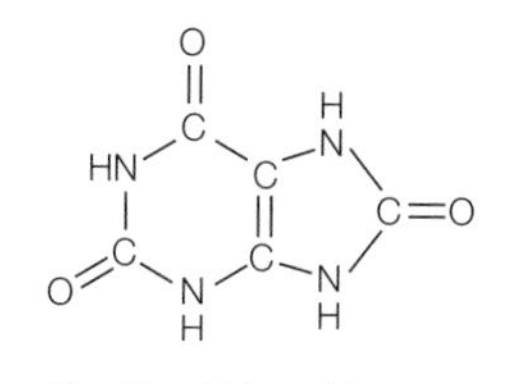


*Fig. 6. Uric acid.*

**Uric acid** (*Fig. 6*) is the main nitrogenous waste product of uricotelic organisms (reptiles, birds and insects), but is also formed in ureotelic organisms from the breakdown of the **purine bases** from DNA and RNA (see Topics F1 and G1). Some individuals have a high serum level of sodium urate (the predominant form of uric acid at neutral pH) which can lead to crystals of this compound being deposited in the joints and kidneys, a condition known as **gout**, a type of arthritis characterized by extremely painful joints.

# M4 HEMES AND CHLOROPHYLLS

## Key Notes

**Tetrapyrroles**

The tetrapyrroles are a family of pigments based on a common chemical structure that includes the hemes and chlorophylls. Hemes are cyclic tetrapyrroles that contain iron and are commonly found as the prosthetic group of hemoglobin, myoglobin and the cytochromes. The chlorophylls are modified tetrapyrroles containing magnesium that occur as light-harvesting and reaction center pigments of photosynthesis in plants, algae and photosynthetic bacteria.

**Biosynthesis of hemes and chlorophylls**

The starting point for heme and chlorophyll synthesis is aminolaevulinic acid (ALA) which is made in animals from glycine and succinyl CoA by the enzyme ALA synthase. This pyridoxal phosphate-requiring enzyme is feedback-regulated by heme. Two molecules of ALA then condense to form porphobilinogen in a reaction catalyzed by ALA dehydratase Porphobilinogen deaminase catalyzes the condensation of four porphobilinogens to form a linear tetrapyrrole. This compound then cyclizes to form uroporphyrinogen III, the precursor of hemes, chlorophylls and vitamin $B_{12}$. Further modifications take place to form protoporphyrin IX. The biosynthetic pathway then branches, and either iron is inserted to form heme, or magnesium is inserted to begin a series of conversions to form chlorophyll.

**Heme degradation**

Heme is broken down by heme oxygenase to the linear tetrapyrrole biliverdin This green pigment is then converted to the red-orange bilirubin by biliverdin reductase. The lipophilic bilirubin is carried in the blood bound to serum albumin, and is then converted into a more water-soluble compound in the liver by conjugation to glucuronic acid. The resulting bilirubin diglucuronide is secreted into the bile, and finally excreted in the feces. Jaundice is due to a build up of insoluble bilirubin in the skin and whites of the eyes. In higher plants heme is broken down to the phycobiliprotein phytochrome which is involved in coordinating light responses, while in algae it is metabolized to the light-harvesting pigments phycocyanin and phycoerythrin.

**Related topics**

Myoglobin and hemoglobin (B4)
Electron transport and oxidative phosphorylation (L2)
Photosynthesis (L3)
Amino acid metabolism (M2)

## Tetrapyrroles

The red **heme** and green **chlorophyll** pigments, so important in the energy-producing mechanisms of respiration and photosynthesis, are both members of the family of pigments called **tetrapyrroles**. They share similar structures (*Fig. 1*), and have some common steps in their synthesis and degradation. The basic structure of a tetrapyrrole is four pyrrole rings surrounding a central metal atom.

Hemes (*Fig. 1a*) are a diverse group of tetrapyrrole pigments, being present as the prosthetic group of both the **globins** (hemoglobin and myoglobin; Topic

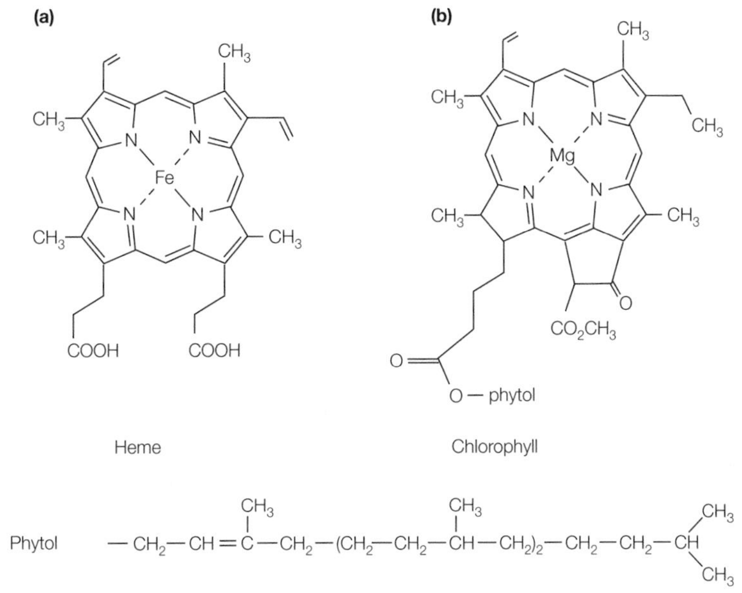

(a) Heme

(b) Chlorophyll

Phytol $-CH_2-CH=C(CH_3)-CH_2-(CH_2-CH_2-CH(CH_3)-CH_2)_2-CH_2-CH_2-CH(CH_3)_2$

*Fig. 1. Structure of (a) heme and (b) chlorophyll.*

B4) and the **cytochromes** (including those involved in respiratory and photosynthetic electron transport; Topic L2 and L3) and the cytochrome P450s that are used in detoxification reactions (see Topic A2). Some enzymes, including the catalases and peroxidases, contain heme. In all these hemoproteins the function of the heme is either to bind and release a ligand to its central **iron atom**, or for the iron atom to undergo a change in oxidation state, releasing or accepting an electron for participation in a redox reaction.

The chlorophylls are also a diverse family of pigments, existing in different forms in photosynthetic bacteria, algae and higher plants. They share a common function in all of these organisms to act as light-harvesting and reaction center pigments in **photosynthesis** (see Topic L3). This function is achieved by a number of modifications to the basic tetrapyrrole structure. These include: the insertion of **magnesium** as the central metal ion, the addition of a fifth ring to the tetrapyrrole structure, loss of a double bond from one or more of the pyrrole rings, and binding of one specific side-chain to a long fat-like molecule called **phytol** (*Fig. 1b*). These changes give chlorophylls and bacteriochlorophylls a number of useful properties. For example, chlorophylls are membrane bound, absorb light at longer wavelengths than heme, and are able to respond to excitation by light. In this way, chlorophylls can accept and release light energy and drive photosynthetic electron transport (see Topic L3).

## Biosynthesis of hemes and chlorophylls

In animals, fungi and some bacteria, the first step in tetrapyrrole synthesis is the condensation of the amino acid **glycine** with **succinyl CoA** (an intermediate of the citric acid cycle; Topic L1) to form **aminolaevulinic acid** (ALA). This reaction is catalyzed by the enzyme **ALA synthase** (*Fig. 2a*) which requires the coenzyme **pyridoxal phosphate** (see Topic M2) and is located in the mitochondria of eukaryotes. This committed step in the pathway is subject to regulation. The synthesis of ALA synthase is feedback-inhibited by heme.

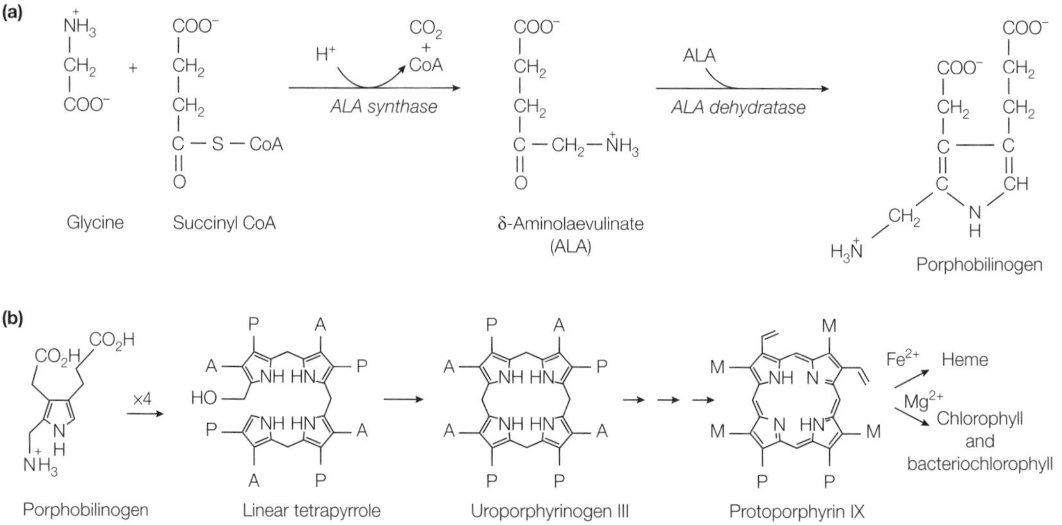

*Fig. 2. Pathway of the synthesis of heme and chlorophyll. (a) Synthesis of porphobilinogen from glycine and succinyl CoA; (b) synthesis of protoporphyrin IX from porphobilinogen. A = $CH_2COOH$, M = $CH_3$, P = $CH_2CH_2COOH$.*

In plants, algae and many bacteria there is an alternative route for ALA synthesis that involves the conversion of the intact five-carbon skeleton of glutamate in a series of three steps to yield ALA. In all organisms, two molecules of ALA then condense to form **porphobilinogen** in a reaction catalyzed by **ALA dehydratase** (also called **porphobilinogen synthase**) (*Fig. 2a*). Inhibition of this enzyme by lead is one of the major manifestations of acute **lead poisoning**.

Four porphobilinogens then condense head-to-tail in a reaction catalyzed by **porphobilinogen deaminase** to form a **linear tetrapyrrole** (*Fig. 2b*). This enzyme-bound linear tetrapyrrole then cyclizes to form **uroporphyrinogen III**, which has an asymmetric arrangement of side-chains (*Fig. 2b*). Uroporphyrinogen III is the common precursor of all hemes and chlorophylls, as well as of **vitamin $B_{12}$**. The pathway continues with a number of modifications to groups attached to the outside of the ring structure, finally forming **protoporphyrin IX** (*Fig. 2b*). At this point either iron or magnesium is inserted into the central cavity, committing the porphyrin to either heme or chlorophyll synthesis, respectively. From here further modifications occur, and finally the specialized porphyrin prosthetic groups are attached to their respective **apoproteins** (the form of the protein consisting of just the polypeptide chain) to form the biologically functional **holoprotein**.

Heme biosynthesis takes place primarily in immature erythrocytes (85% of the body's heme groups), with the remainder occurring in the liver. Several genetic defects in heme biosynthesis have been identified that give rise to the disorders called **porphyrias**.

## Heme degradation

**Bile pigments** exist in both the plant and animal kingdoms, and are formed by breakdown of the cyclic tetrapyrrole structure of heme. In animals this pathway is an excretory system by which the heme from the hemoglobin of aging red blood cells, and other hemoproteins, is removed from the body. In the plant kingdom, however, heme is broken down to form bile pigments which have major roles to play in coordinating light responses in higher plants (the phycobiliprotein **phytochrome**), and in light harvesting in algae (the phycobiliproteins **phycocyanin** and **phycoerythrin**).

In all organisms, the degradation of heme begins with a reaction carried out by a single common enzyme. This enzyme, **heme oxygenase**, is present mainly in the spleen and liver of vertebrates, and carries out the oxidative ring opening of heme to produce the green bile pigment **biliverdin**, a linear tetrapyrrole (*Fig. 3*). Heme oxygenase is a member of the cytochrome P450 family of enzymes, and requires NADPH and $O_2$. In birds, reptiles and amphibians this water-soluble pigment is the final product of heme degradation and is excreted directly. In mammals, however, a further conversion to the red-orange **bilirubin** takes place; a reaction catalyzed by **biliverdin reductase** (*Fig. 3*). The changing colour of a bruise is a visible indicator of these degradative reactions. The bilirubin, like other lipophilic molecules such as free fatty acids, is then transported in the blood bound to **serum albumin**. In the liver, its water solubility is increased by conjugation to two molecules of **glucuronic acid**, a sugar residue that differs from glucose in having a $COO^-$ group at C-6 rather than a $CH_2OH$ group. The resulting **bilirubin diglucuronide** is secreted into the bile and then into the intestine, where it is further metabolized by bacterial enzymes and finally excreted in the feces.

When the blood contains excessive amounts of the insoluble bilirubin, it is deposited in the skin and the whites of the eyes, resulting in a yellow discoloration. This condition, called **jaundice**, is indicative either of impaired liver function, obstruction of the bile duct, or excessive breakdown of erythrocytes.

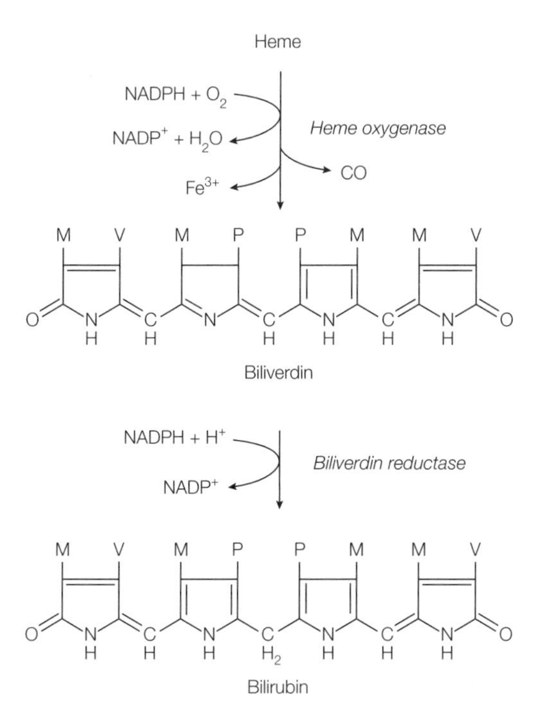


*Fig. 3. Degradation of heme to the bile pigments biliverdin and bilirubin. M = methyl ($CH_3$), V = vinyl ($CH = CH_2$), P = propionyl ($CH_2CH_2CH_2OH$).*

# N1 Muscle

## Key Notes

**Muscle structure**

Each cell within vertebrate striated muscle contains within its sarcoplasm many parallel myofibrils which in turn are made up of repeating sarcomere units. Within the sarcomere are the alternating dark A band and light I band, in the middle of which are the H zone and Z line, respectively. A myofibril contains two types of filaments: the thick filaments consisting of myosin which are present only in the A band, and the thin filaments consisting of actin, tropomyosin and troponin. When muscle contracts, the thick and thin filaments slide over one another, shortening the length of the sarcomere.

**Myosin**

The protein myosin consists of two heavy polypeptide chains and two pairs of light chains arranged as a double-headed globular region attached to a two-stranded α-helical coiled-coil. Myosin molecules spontaneously assemble into filaments, hydrolyze ATP and bind actin. Limited proteolysis can fragment myosin into smaller functional units: light meromyosin that can form filaments, and heavy meromyosin which can be further broken down into S1 and S2 subfragments that are linked by flexible hinge regions. The S1 heads bind actin and hydrolyze ATP.

**Actin**

Actin, the major constituent of the thin filaments, can exist as monomeric globular G-actin or as polymerized fibrous F-actin. The actin filaments are connected to the thick filaments by cross-bridges formed by the S1 heads of myosin.

**The generation of force in muscle**

The cyclic formation and dissociation of complexes between the actin filaments and the S1 heads of myosin leads to contraction of the muscle. On binding to actin, myosin releases its bound $P_i$ and ADP. This causes a conformational change to occur in the protein which moves the actin filament along the thick filament. ATP then binds to myosin, displacing the actin. Hydrolysis of the ATP returns the S1 head to its original conformation.

**Troponin and tropomyosin**

Tropomyosin is an elongated protein that lies along the thin filament and prevents the association of myosin with actin in the resting state. Troponin is a complex of three polypeptide chains: TnC, TnI and TnT. $Ca^{2+}$ ions released into the sarcoplasm from the sarcoplasmic reticulum in response to a nerve stimulation bind to TnC and cause a conformational change in the protein. This movement is transmitted by an allosteric mechanism through TnI and TnT to tropomyosin, causing the latter to move out of the way and allowing the actin and myosin to associate.

**Actin and myosin in nonmuscle cells**

Both actin and myosin are found in nonmuscle cells where they are involved in cell movement, the movement of organelles around the cell, and cell division.

**Related topics**

Eukaryotes (A2)
Microscopy (A3)

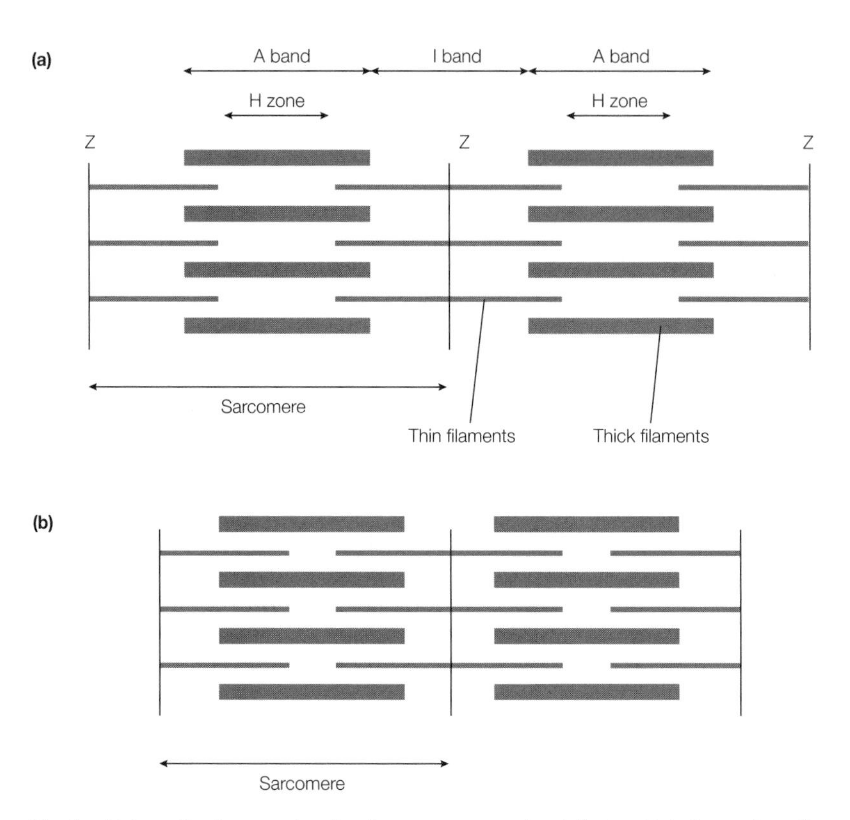


*Fig. 1. Schematic diagram showing the appearance of vertebrate striated muscle as it appears under phase-contrast microscopy. (a) Relaxed, (b) contracted.*

## Muscle structure

The best understood force-generating process in biological systems is the contraction of **vertebrate striated muscle**, so named because it appears striated (striped) under phase-contrast microscopy (see Topic A3). This muscle is composed of numerous multinucleate cells that are bounded by an electrically excitable plasma membrane. Each cell contains within its **sarcoplasm** (cytosol) many parallel **myofibrils**, each approximately 1 μm in diameter. The sarcoplasm is also rich in ATP, creatine phosphate (see Topic M3) and glycolytic enzymes (see Topic J3). The functional unit of the myofibril is the **sarcomere** which repeats every 2.3 μm along the fibril axis (*Fig. 1a*). A dark **A band** and a light **I band** alternate regularly along the length of the myofibril. The central region of the A band, the **H zone**, is less dense than the rest of the band. Within the middle of the I band is a very dense narrow **Z line**. A cross-section of a myofibril reveals that there are two types of interacting filaments. The **thick filaments** of diameter approximately 15 nm are found only in the A band (*Fig. 1a*) and consist primarily of the protein **myosin**, while the **thin filaments** of approximately 9 nm diameter contain **actin**, **tropomyosin** and the **troponin complex**.

When muscle contracts it can shorten by as much as a third of its original length. In the 1950s, information obtained from X-ray crystallographic, and light- and electron-microscopic studies led to the proposal of the **sliding filament model** to explain muscle contraction. The thick and thin filaments were seen not to change in length during muscle contraction, but the length of the **sarcomere** was observed to decrease as the thick and thin filaments slide past each other (*Fig. 1*). Thus, as muscle contracts, the sizes of the H zone and the I band are seen to decrease. The force of the contraction is generated by a process that actively moves one type of filament past neighboring filaments of the other type.

## Myosin

Myosin is a large protein of 520 kDa consisting of six polypeptide chains: two heavy chains of 220 kDa each, and two pairs of light chains (20 kDa each). This large protein has three biological activities:

1. Myosin molecules spontaneously assemble into **filaments** in solutions of physiological ionic strength and pH;
2. Myosin is an **ATPase**, hydrolyzing ATP to ADP and $P_i$;
3. Myosin binds the **polymerized form of actin**.

Electron micrographs revealed that myosin consists of a double-headed globular region joined to a long rod. The rod is a two-stranded α-helical coiled-coil formed by the two heavy chains, while the globular heads are also part of each heavy chain with the light chains attached (*Fig. 2a*). Limited **proteolysis** of myosin with **trypsin** results in its dissection into two fragments: **light meromyosin** (LMM) and **heavy meromyosin** (HMM) (*Fig. 2b*). Functional studies of these two fragments reveal that LMM can still form filaments but lacks ATPase activity, whereas HMM does not form filaments but possesses ATPase activity and can bind to actin. HMM can be further split into two identical **globular subfragments (S1)** and one **rod-shaped subfragment (S2)** by another protease **papain** (*Fig. 2b*). The S1 subfragment, whose structure has

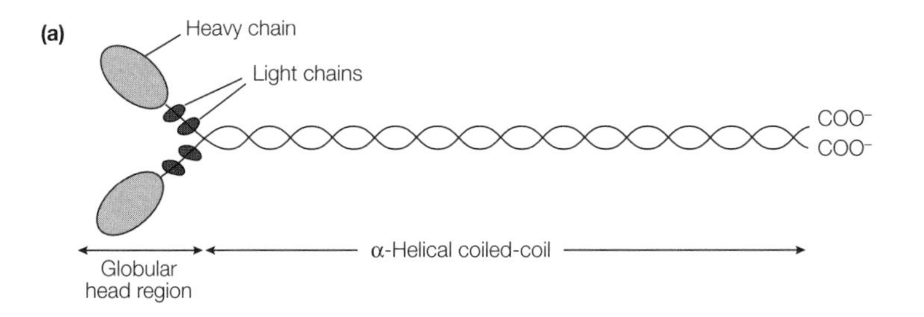


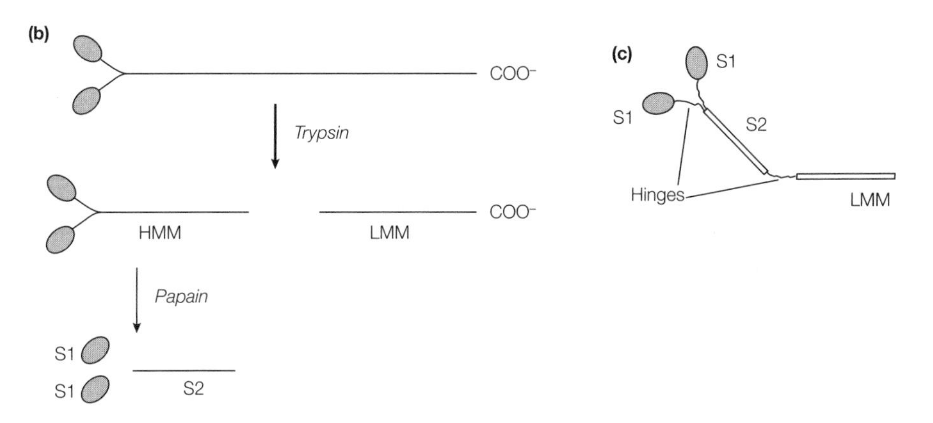


*Fig. 2. Structure of myosin: (a) showing the association of the two heavy and two pairs of light chains; (b) showing the proteolytic fragmentation of myosin; and (c) showing the hinge regions between domains.*

recently been determined by X-ray crystallography, contains an ATPase site, an actin-binding site and two light chain-binding sites. The proteolytic cleavage of myosin occurs at **flexible hinge regions** within the protein that separate the globular S1 domains from the rod-like S2 and LMM domains (*Fig. 2c*). These hinges have a crucial role to play in the contraction of muscle.

## Actin

Actin, the major constituent of the thin filaments, exists in two forms. In solutions of low ionic strength it exists as a 42 kDa monomer, termed **G-actin** because of its globular shape. As the ionic strength of the solution rises to that at the physiological level, G-actin polymerizes into a fibrous form, **F-actin**, that resembles the thin filaments found in muscle. Although actin, like myosin, is an **ATPase**, the hydrolysis of ATP is not involved in the contraction–relaxation cycle of muscle but rather in the assembly and disassembly of the actin filament.

On the thick filaments, **cross-bridges** emerge from the filament axis in a regular helical array towards either end, whereas there is a bare region in the middle that is devoid of cross-bridges (*Fig. 3*). In muscle depleted of ATP, the myosin cross-bridges interact with the surrounding actin filaments. The absolute direction of the actin and myosin molecules reverses halfway between the Z lines. Thus, as the two thin filaments that bind the cross-bridges at either end of a thick filament move towards each other, sliding over the thick filament, the distance between the Z lines shortens and the muscle contracts (*Fig. 3*).

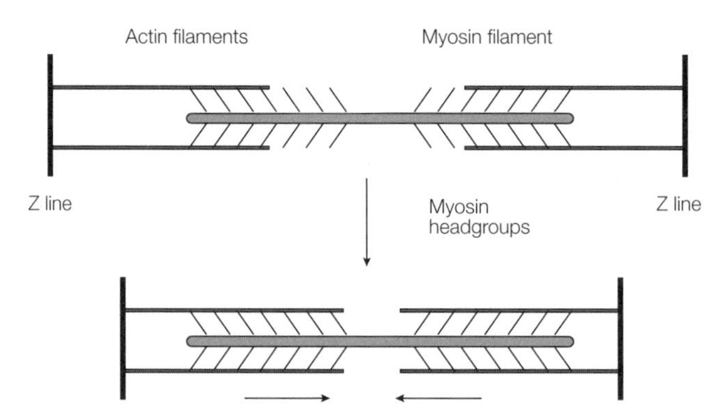


*Fig. 3. Schematic diagram showing the interaction of the myosin thick filaments and the actin thin filaments during skeletal muscle contraction.*

## The generation of force in muscle

The cyclic formation and dissociation of cross-bridges between actin and the S1 heads of myosin leads to contraction of the muscle because of **conformational changes** that take place in the myosin S1 head. In resting muscle, the S1 heads are unable to interact with the actin in the thin filaments because of steric interference by the regulatory protein **tropomyosin** (*Fig. 4a*). The myosin has bound to it ADP and $P_i$. When the muscle is stimulated, the tropomyosin moves out of the way, allowing the S1 heads projecting out from the thick filament to attach to the actin in the thin filament (*Fig. 4b*). On binding of myosin–ADP–$P_i$ to actin, first the $P_i$ and then the ADP are released. As the ADP is released the S1 head undergoes a **conformational change in the hinge region** between the S1 and S2 domains that alters its orientation relative to the actin molecule in the thin filament (*Fig. 4c*). This constitutes the **power stroke** of muscle contraction and results in the thin filament moving a distance of approximately 10 nm relative to the thick filament towards the center of the sarcomere. ATP then binds to the S1 head which leads to the rapid release of the actin [i.e. dissociation of the thin and thick filaments (*Fig. 4d*)]. The ATP is then hydrolyzed to ADP and $P_i$ by the free S1 head, which is returned to its original conformation ready for another round of attachment (*Fig. 4e*), conformational change and release.

## Troponin and tropomyosin

Troponin and tropomyosin mediate the regulation of muscle contraction in response to $\mathbf{Ca^{2+}}$. These two proteins are present in the thin filament, alongside the actin, and constitute about a third of its mass. **Tropomyosin** is an

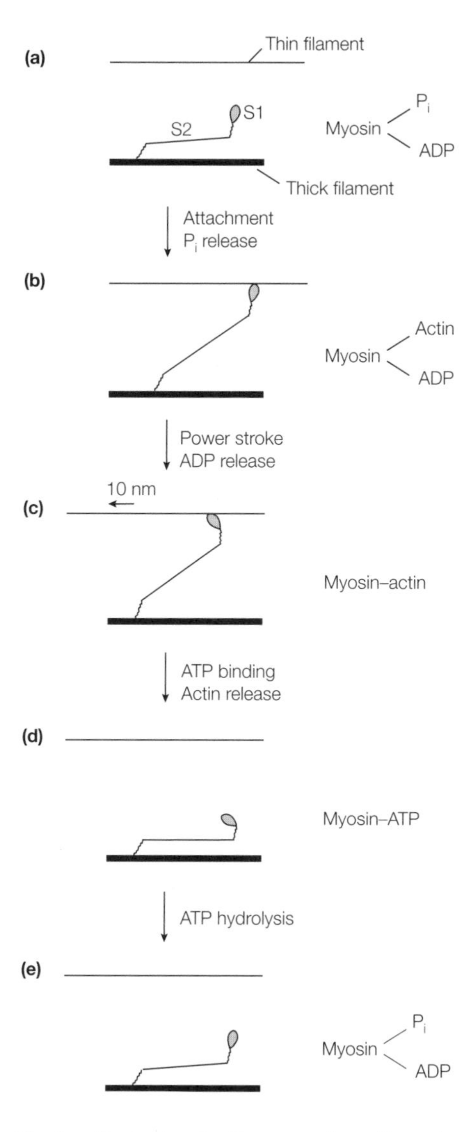


*Fig. 4. Mechanism for the generation of force in muscle as an S1 head of a myosin thick filament interacts with an actin thin filament.*

elongated protein of 70 kDa that forms a two-stranded α-helical rod which lies nearly parallel to the long axis of the thin filament. **Troponin** is a complex of three polypeptide chains: **TnC** (18 kDa) which binds $Ca^{2+}$, **TnI** (24 kDa) which binds to actin and **TnT** (37 kDa) which binds to tropomyosin. On muscle stimulation by a nerve impulse, $Ca^{2+}$ ions are released from the **sarcoplasmic reticulum** (a specialized form of the ER found in muscle cells; see Topic A2) into the cytosol, raising the cytosolic $Ca^{2+}$ concentration from the resting concentration of less than 1 μM to about 10 μM. The $Ca^{2+}$ binds to sites on TnC, causing a **conformational change** in this polypeptide that is transmitted through the other components of the troponin complex to the tropomyosin. The tropomyosin then moves out of the way, allowing the S1 head of myosin to interact with the actin and initiate a cycle of contraction. Thus, $Ca^{2+}$ controls muscle contraction by an **allosteric mechanism** (see Topic C5) involving troponin, tropomyosin, actin and myosin.

## Actin and myosin in nonmuscle cells

Both actin and myosin are present in nonmuscle cells. Actin is a highly conserved protein in organisms ranging from cellular slime molds to humans, whereas myosin is less well conserved. Within nonmuscle cells these two proteins are involved in such processes as cell movement, the transport of membrane-bound subcellular organelles around the cell on actin tracks, and cell division (see Topic A2).

# N2 CILIA AND FLAGELLA

## Key Notes

**Cilia**

Eukaryotic cilia are hair-like protrusions on the surface of the cell that consist mainly of microtubules. The microtubule fibers in a cilium are bundled together in a characteristic 9 + 2 arrangement within the axoneme. The outer nine microtubule doublets look like a figure eight with a smaller circle, subfiber A, and a larger circle, subfiber B. Two dynein arms protrude from subfiber A which, upon hydrolysis of ATP, move along adjacent B subfibers. Due to extensible nexin links between the doublets, this sliding motion is converted into a local bending of the cilium. In immotile-cilia syndrome, the cilia are unable to move due to a genetic defect in one or other of the proteins within the axoneme.

**Bacterial flagella**

Bacteria move through their surrounding media in response to chemicals (chemotaxis) by rotation of their tail-like flagella. Bacterial flagella are made of the protein flagellin that forms a long filament which is attached to the flagellar motor by the flagellar hook.

**Related topics**

Eukaryotes (A2)
Membrane protein and carbohydrate (E2)
Muscle (N1)

## Cilia

The hair-like protrusions or cilia on the surfaces of certain eukaryotic cells, such as those lining the respiratory passages, consist mainly of **microtubules**. Cilia are involved in moving a stream of liquid over the surface of the cell. Free cells such as protozoa and sperm from various species can be propelled by either cilia or a flagellum. In eukaryotic cells, flagella differ from cilia only in being much longer. Electron microscopic studies have shown that virtually all eukaryotic cilia and flagella have the same basic design: a bundle of fibers called an **axoneme** surrounded by a membrane that is continuous with the plasma membrane (*Fig. 1*). The microtubule fibers in an axoneme are in a characteristic **9 + 2 array**, with a peripheral group of nine pairs of microtubules surrounding two singlet microtubules (*Fig. 1*). Each of the nine outer doublets appears like a figure eight, the smaller circle is termed **subfiber A**, the larger circle, **subfiber B**. Subfiber A is joined to a central sheath by radial spokes, while neighboring microtubule doublets are held together by **nexin** links. Two **dynein** arms emerge from each subfiber A, with all the arms in a cilium pointing in the same direction (*Fig. 1*).

### *Dynein*

Dynein is a very large protein of 1000–2000 kDa consisting of one, two or three heads depending on the source. Like the heads of myosin (see Topic N1), the heads of dynein form **cross-bridges**, in this case with the B subfibers, and

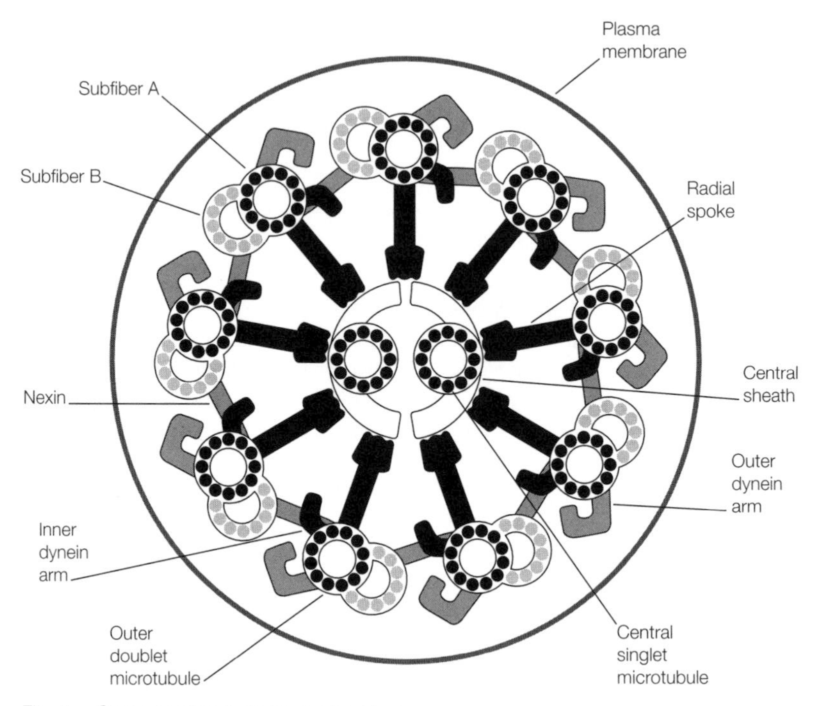


*Fig. 1. Cross-sectional diagram of a cilium.*

possess **ATPase activity**. The binding of ATP to dynein causes it to dissociate from the B subfiber. On hydrolysis of the ATP to ADP and $P_i$, the dynein binds again with the B subfiber with the subsequent release of the $P_i$ and ADP (a cycle very similar to that which occurs with the binding of the S1 heads of myosin to ATP; see Topic N1). This ATPase cycle leads to the movement of the cilium as the outer doublets of the axoneme slide past each other. The force between adjacent doublets is generated by the dynein cross-bridges. Thus, the dynein arms on subfiber A of one doublet walk along subfiber B of the adjacent doublet. Unlike in muscle, where the myosin and actin filaments slide past each other, in a cilium the radial spokes resist the sliding motion, which instead is converted into a local bending. The highly extensible protein, **nexin**, keeps adjacent doublets together during this process.

A defect or absence in any one of the proteins within the axoneme (e.g. dynein, nexin, etc.) results in cilia that are immotile, so called **immotile-cilia syndrome**. Patients suffering from this disease have chronic pulmonary disorders due to the cilia in the respiratory tract being unable to sweep out bacteria and other foreign particles. In addition, males with this genetic defect are infertile because their sperm are unable to move due to flagella inactivity.

## Bacterial flagella

Bacterial flagella are different from eukaryotic cilia and flagella in two ways: (1) each bacterial flagellum is made of the protein **flagellin** (53 kDa subunit) as opposed to tubulin; and (2) it **rotates** rather than bends. Bacteria can move through the extracellular medium towards attractants and away from repellents, so called **chemotaxis**, by rotating their flagella. An *E. coli* bacterium has about six flagella that emerge from random positions on the surface of the cell. Flagella are thin helical filaments, 15 nm in diameter and 10 μm long. Electron microscopy has revealed that the **flagellar filament** contains 11 subunits in two helical turns which, when viewed end-on, has the appearance of an 11-bladed propeller with a hollow central core. Flagella grow by the addition of new flagellin subunits to the end away from the cell, with the new subunits diffusing through the central core. Between the flagellar filament and the cell membrane is the **flagellar hook** composed of subunits of the 42 kDa hook protein that forms a short, curved structure. Situated in the plasma membrane is the basal body or **flagellar motor**, an intricate assembly of proteins. The flexible hook is attached to a series of protein rings which are embedded in the inner and outer membranes. The rotation of the flagella is driven by a flow of protons through an outer ring of proteins, called the stator. A similar proton-driven motor is found in the $F_1F_0$-ATPase that synthesizes ATP (see Topic L2).

# N3 NERVE

## Key Notes

**Nerve cells**

Nerve cells, or neurons, consist of a cell body from which the dendrites and axon extend. The dendrites receive information from other cells; the axon passes this information on to another cell, the post-synaptic cell. The axon is covered in a myelin membranous sheath except at the nodes of Ranvier. The axon ends at the nerve terminal where chemical neurotransmitters are stored in synaptic vesicles for release into the synaptic cleft.

**The action potential**

An electric membrane potential exists across the plasma membrane due to the unequal distribution of $Na^+$ and $K^+$ ions which is generated by the $Na^+/K^+$-ATPase. Upon stimulation, neurons depolarize their membrane potential from the resting state (–60 mV) to +40 mV, generating an action potential. The action potential is caused by $Na^+$ ions flowing into the cell through voltage-sensitive $Na^+$ channels. The resting membrane potential is restored by $K^+$ ions flowing out of the cell through voltage-sensitive $K^+$ channels. The poison tetrodotoxin acts by blocking the $Na^+$ channel.

**Neurotransmitters**

Chemical neurotransmitters, such as acetylcholine, the biogenic amines and small peptides, are stored in the pre-synaptic nerve terminal in synaptic vesicles. When the action potential reaches the nerve terminal it causes the synaptic vesicles to fuse with the plasma membrane in a $Ca^{2+}$-dependent manner and to release their contents by exocytosis. The neurotransmitter then diffuses across the synaptic cleft, binds to specific receptors on the post-synaptic cell membrane and initiates a response in that cell.

**Related topics**

Membrane lipids (E1)
Membrane protein and carbohydrate (E2)
Membrane transport: small molecules (E3)

## Nerve cells

In eukaryotes, probably the most rapid and complex signaling is mediated by **nerve impulses**. Nerve cells (**neurons**) consist of a cell body with numerous projections of the plasma membrane, called **dendrites** (*Fig. 1*). These interact with other cells and receive information from them in the form of nerve impulses. The cell body then assimilates the information derived from a number of dendritic contacts and passes on the information as another nerve impulse down the large **axon** (*Fig. 1*). The axon ends at the **synapse** where it makes contact with the **post-synaptic** (target) **cell**. The axon is covered in places by a membranous **myelin** sheath, made up mainly of the lipid sphingomyelin (see Topic E1), which acts as an electrical insulator, enabling the nerve impulses to be transmitted over long distances, sometimes more than 1 m in larger animals. Every millimeter or so along the axon the myelin sheath is interrupted by unmyelinated regions called the **nodes of Ranvier** (*Fig. 1*). The end of the axon, the nerve terminal, is full of **synaptic vesicles** that store the chemical **neurotransmitters**, such as acetylcholine. When a nerve impulse reaches the nerve terminal the synaptic vesicles release their contents into the **synaptic cleft**, the space between the pre- and post-synaptic cells. The neurotransmitter then

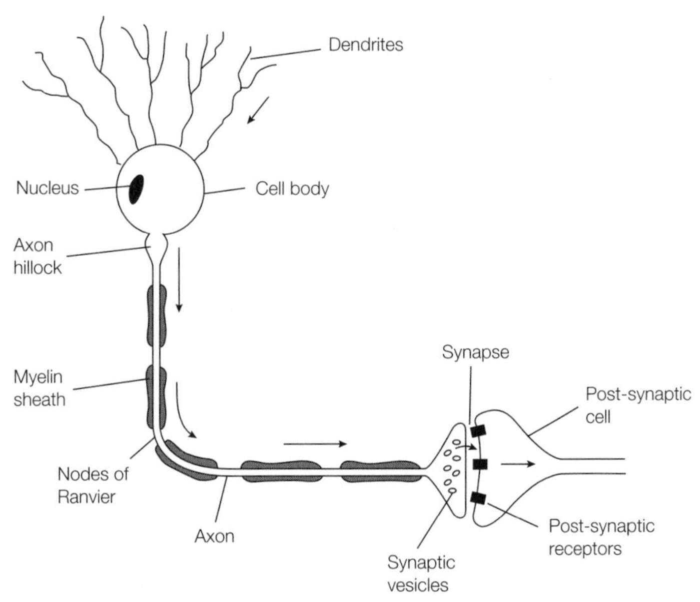


*Fig. 1. Schematic diagram of a typical nerve cell.*

diffuses across the space and interacts with receptors on the surface of the post-synaptic cell, causing a signal to be transduced in that cell.

## The action potential

An electric potential (the **membrane potential**) exists across the plasma membrane of all cells. Most cells are electrically inactive as this membrane potential does not vary with time. However, neurons and muscle cells are **electrically active** as their membrane potential *can* vary with time. In all cells the membrane potential is generated through the action of the **$Na^+/K^+$-ATPase** (see Topic E3), with a high concentration of $K^+$ inside the cell and a high concentration of $Na^+$ outside. Neurons vary their electric potential by controlled changes in the **permeability** of the plasma membrane to $Na^+$ and $K^+$ ions. Upon stimulation, the membrane potential of a neuron rises rapidly from the **resting potential** of −60 mV (millivolts) to approximately +40 mV (*Fig. 2a*); the membrane is said to **depolarize** and an **action potential** is generated. In order for this to occur, the membrane potential has to be depolarized beyond a critical **threshold level** (approximately −40 mV). With time, the membrane

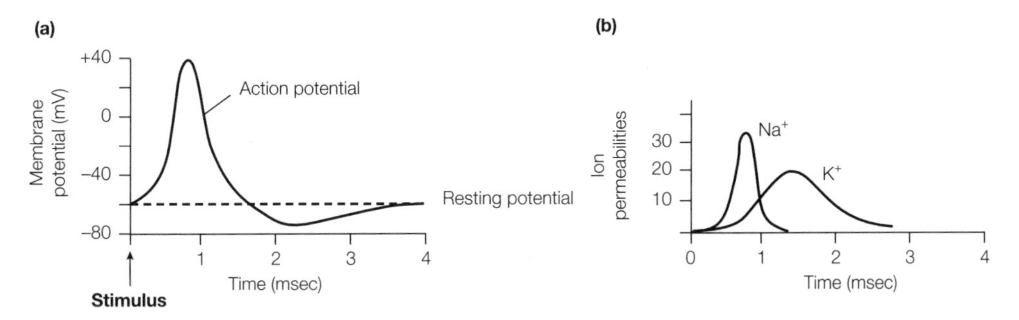


*Fig. 2. The action potential. (a) Depolarization of the membrane potential; (b) changes in the permeability of the plasma membrane to $Na^+$ and $K^+$.*

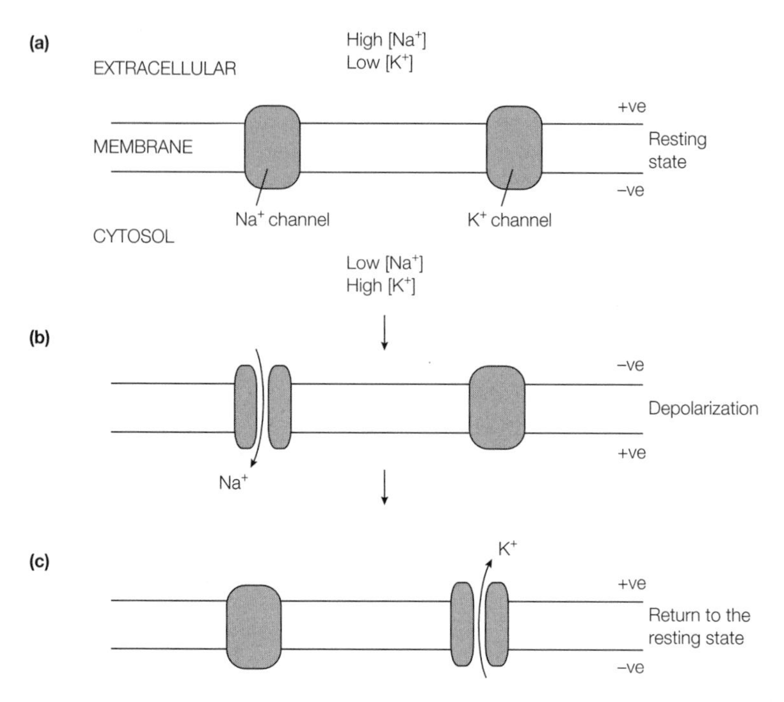


*Fig. 3. Mechanism of depolarization of the nerve membrane by the opening and closing of selective $Na^+$ and $K^+$ ion channels.*

potential returns to the resting potential. The action potential is propagated along the axon starting from the axon hillock (*Fig. 1*).

The action potential arises from large, transient changes in the permeability of the plasma membrane of the neuron to $Na^+$ and $K^+$ ions. Two types of **voltage-sensitive ion channels** are present in the membrane: one is selectively permeable to $Na^+$ ions, the other to $K^+$ ions (*Fig. 3a*). These **integral membrane proteins** (see Topic E2) are sensitive to the membrane potential, undergoing **conformational changes** as the potential alters. First, the conductance of the membrane to $Na^+$ changes. Depolarization of the membrane beyond the threshold level causes a conformational change in the $Na^+$ channel, allowing $Na^+$ ions to flow down their concentration gradient from the outside of the cell into the interior (*Fig. 3b*). The entry of $Na^+$ further depolarizes the membrane, causing more $Na^+$ channels to open, resulting in a rapid influx of $Na^+$ and a change in the membrane potential from –60 mV to +40 mV in a millisecond. The $Na^+$ channels then spontaneously close, and the $K^+$ channels open, allowing $K^+$ ions to flow out of the cell and restore the negative resting potential within a few milliseconds (*Fig. 3c*). The wave of depolarization is propagated along the axon by the opening of $Na^+$ channels on the nerve terminal side of the initial depolarized region (*Fig. 4*). The action potential can only move in that direction as the $Na^+$ channels have a **refractory period** when they are insensitive to further stimulation. Only approximately one in a million of the $Na^+$ and $K^+$ ions in a neuron flow across the plasma membrane during the action potential. Thus, this is a very efficient way of signaling over long distances.

The neurotoxin, **tetrodotoxin**, a highly potent poison from the puffer fish, blocks the conduction of nerve impulses along axons and so leads to respiratory paralysis by binding very tightly to the $Na^+$ channel and blocking its action.

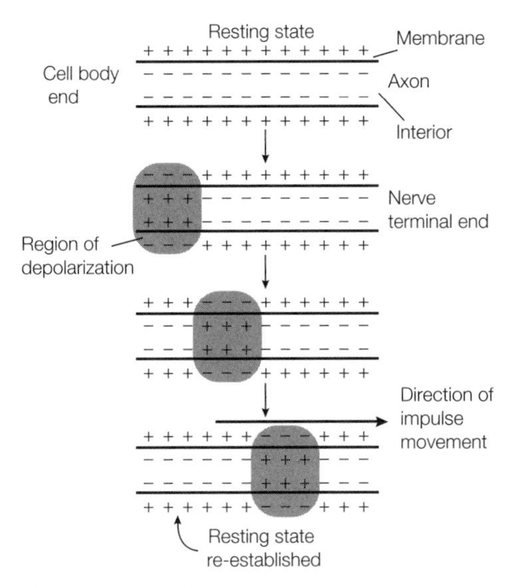


*Fig. 4. Propagation of the action potential along an axon.*

## Neurotransmitters

When the action potential reaches the nerve terminal it causes the release of a **chemical neurotransmitter** from the synaptic vesicles. The mammalian nervous system employs over 30 substances as neurotransmitters. These include the **amino acids** glutamate and glycine, **acetylcholine**, the **biogenic amines** such as **epinephrine** and **dopamine**, and a variety of **small peptides** such as the **enkephalins** (see Topic E5). For example, acetylcholine is stored in **synaptic vesicles**, a specialized form of secretory vesicle, and is released into the **synaptic cleft** by **exocytosis** (see Topic E4) in a $Ca^{2+}$-dependent manner (*Fig. 5*). The acetylcholine molecules then diffuse across to the plasma membrane of the **post-synaptic cell** where they bind to specific **receptors**. The acetylcholine receptor is a 250 kDa complex of four polypeptide chains that forms a **gated channel** through the membrane. On binding of two acetylcholine molecules, the channel opens, allowing $Na^+$ and $K^+$ ions to flow in and out of the cell, respectively. The resulting depolarization of the post-synaptic membrane initiates a new action potential in that cell.

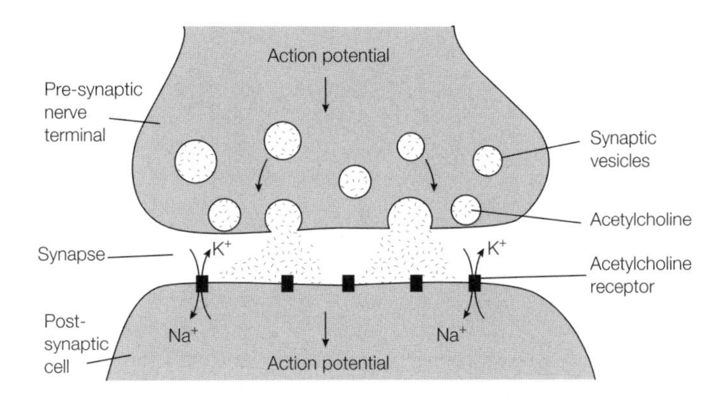


*Fig. 5. Release of a neurotransmitter into the synaptic cleft.*

# FURTHER READING

There are many comprehensive textbooks of biochemistry and molecular biology and no one book that can satisfy all needs. Different readers subjectively prefer different textbooks and hence we do not feel it would be particularly helpful to recommend one book over another. Rather we have listed some of the leading books which we know from experience have served their student readers well.

## General reading

Alberts, B., Bray, D., Lewis, J., Raff, M., Roberts, K. and Watson, J.D. (1994) *Molecular Biology of the Cell*, 3rd Edn. Garland Publishing, New York.

Brown, T.A. (1999) *Genomes*. BIOS Scientific Publishers Ltd., Oxford.

Lodish, H., Baltimore, D., Berk, A., Zipursky, S.L., Matsudaira, P. and Darnell, J. (1995) *Molecular Cell Biology*, 3rd Edn. Scientific American Books, W.H. Freeman, New York.

Mathews, C.K. and van Holde, K.E. (1995) *Biochemistry*, 2nd Edn. Benjamin Cummings, Redwood City, CA.

Stryer, L. (1995) *Biochemistry*, 4th Edn. W.H. Freeman, New York.

Voet, D. and Voet, J.G. (1995) *Biochemistry*, 2nd Edn. John Wiley and Sons, New York.

Watson, J.D., Hopkins, N.H., Roberts, J.W., Steitz, J. and Weiner, A.M. (1998) *Molecular Biology of the Gene*, 5th Edn. Addison Wesley Longman, Harlow.

## More advanced reading

The following selected articles are recommended to readers who wish to know more about specific subjects. In many cases they are too advanced for first year students but are very useful sources of information for subjects that may be studied in later years.

### Section A

Byard, E.H. and Lange, B.M.H. (1991) Tubulin and microtubules. *Essays Biochem.* **26**, 13–25.

de Duve, C. (1996) The birth of complex cells. *Sci. Amer.* **274**(4), 38–45.

Erdmann, R., Veenhuis, M. and Kunau, W.-H. (1997) Peroxisomes: organelles at the crossroads. *Trends Cell Biol.* **7**, 400–407.

Farquhar, M.G. and Palade, G.E. (1998) The Golgi apparatus: 100 years of progress and controversy. *Trends Cell Biol.* **8**, 2–10.

Gupta, R.S. and Golding, G.B. (1996) The origin of the eukaryotic cell. *Trends Biochem. Sci.* **21**, 166–171.

Levy, S.B. (1998) The challenge of antibiotic resistance. *Sci. Amer.* **278**(3), 32–39.

Lichtman, J.W. (1994) Confocal microscopy. *Sci. Amer.* **271**(2), 30–35.

Lopez-Garcia, P.O. and Moreira, D. (1999) Metabolic symbiosis at the origin of eukaryotes. *Trends Biochem. Sci.* **24**, 88–93.

Ludin, B. and Matus, A. (1998) GFP illuminates the cytoskeleton. *Trends Cell Biol.* **8**, 72–77.

### Section B

Netzer, W.J. and Hartl, F.U. (1998) Protein folding in the cytosol: chaperonin-dependent and -independent mechanisms. *Trends Biochem. Sci.* **23**, 68–73.

Nucci, M.L. and Abuckowski, A. (1998) The search for blood substitutes. *Sci. Amer.* **278**(2), 61–65.

Richards, F.M. (1991) The protein folding problem. *Sci. Amer.* **264**(1) 34–41.

Sali, A., Overington, J.P., Johnson, M.S. and Blundell, T.L. (1990) From comparisons of protein sequences and structures to protein modelling and design. *Trends Biochem. Sci.* **15**, 235–240.

Thomas, P.J., Qu, B.H. and Pederson, P.L. (1995) Defective protein folding as a basis of human disease. *Trends Biochem. Sci.* **20**, 456–459.

Wahl, M.C. and Sundaralingam, M. (1997) C–H...O hydrogen bonding in biology. *Trends Biochem. Sci.* **22**, 97–102.

**Section C**

Hampton, R., Dimster-Denk, D. and Rine, J. (1996) The biology of HMG-CoA reductase: the pros of contra-regulation. *Trends Biochem. Sci.* **21**, 140–145.

Kantrowitz, E.R. and Lipscomb, W.N. (1990) *Escherichia coli* aspartate transcarbamoylase: the molecular basis for a concerted allosteric transition. *Trends Biochem. Sci.* **15**, 53–59.

**Section D**

Engelhard, V.H. (1994) How cells process antigens. *Sci. Amer.* **271**(2), 44–51.

Greene, W.C. (1993) AIDS and the immune system. *Sci. Amer.* **269**(3), 66–73.

Janaway, C.A. (1993) How the immune system recognises invaders. *Sci. Amer.* **269**(3), 40–47.

Marrack, P. and Kappler, J.W. (1993) How the immune system recognises the body. *Sci. Amer.* **269**(3), 48–51.

Nossel, G.J.V. (1993) Life, death and the immune system. *Sci. Amer.* **269**(3), 20–31.

Paul, W.E. (1993) Infectious diseases and the immune system. *Sci. Amer.* **269**(3), 56–65.

Puttney, S. (1992) How antibodies block HIV infection: paths to an AIDS vaccine. *Trends Biochem. Sci.* **17**, 191–196.

Schwartz, R.H. (1993) T cells: tolerance. *Sci. Amer.* **269**(2), 48–55.

Van Boehmer, H. and Kisielow, P. (1991) How the immune system learns about self. *Sci. Amer.* **265**(4), 50–59.

Weisman, I.L. and Cooper, M.D. (1993) How the immune system develops. *Sci. Amer.* **269**(3), 32–39.

**Section E**

Barnard, E.A. (1992) Receptor classes and the transmitter-gated ion channels. *Trends Biochem. Sci.* **17**, 368–374.

Bayley, H. (1997) Building doors into cells. *Sci. Amer.* **277**(9), 42–47.

Bhatnagar, R.S. and Gordon, J.I. (1997) Understanding covalent modifications of proteins by lipids: where cell biology and biophysics mingle. *Trends Cell Biol.* **7**, 14–20.

Gahmberg, C.G. and Tolvanen, M. (1996) Why mammalian cell surface proteins are glycoproteins. *Trends Biochem. Sci.* **21**, 308–311.

Gould, G.W. and Bell, G.I. (1990) Facilitative glucose transporters: an expanding family. *Trends Biochem. Sci.* **15**, 18–23.

Hepler, J.R. and Gilman, A.G. (1992) G proteins. *Trends Biochem. Sci.* **17**, 383–387.

Hirschhorn, N. and Greenough, W.B. III (1991) Progress in oral rehydration therapy. *Sci. Amer.* **264**(5), 16–22.

Hsu, S-C., Hazuka, C.D., Foletti, D.L. and Scheller, R.H. (1999) Targeting vesicles to specific sites on the plasma membrane. *Trends Cell Biol.* **9**, 150–153.

Lienhard, G.E., Slot, J.W., James, D.E. and Mueckler, M.H. (1992) How cells absorb glucose. *Sci. Amer.* **266**(1), 34–39.

Pazin, M.J. and Williams, L.T. (1992) Triggering signalling cascades by receptor tyrosine kinases. *Trends Biochem. Sci.* **17**, 374–378.

Pumplin, D.W. and Bloch, R.J. (1993) The membrane skeleton. *Trends Cell Biol.* **3**, 113–117.

Rothman, J.E. and Orci, L. (1996) Budding vesicles in living cells. *Sci. Amer.* **274**(3), 50–55.

Sharon, N. and Lis, H. (1993) Carbohydrates in cell recognition. *Sci. Amer.* **268**(1), 74–81.
Spiegel, A.M., Backlund, P.S., Butrynski, J.E., Jones, T.L.Z. and Simonds, W.F. (1991) The G protein connection: molecular basis of membrane association. *Trends Biochem. Sci.* **16**, 338–341.
Swanson, J.A. and Baer, S.C. (1995) Phagocytosis by zippers and triggers. *Trends Cell Biol.* **5**, 89–93.
Takeda, J. and Kinoshita, T. (1995) GPI anchor biosynthesis. *Trends Biochem. Sci.* **20**, 367–371.

**Section F**

Bridger, J.M. and Bickmore, W.A. (1998) Putting the genome on the map. *Trends Genetics* **14**, 403–409.
Diffley, J.F.X. (1992) Early events in eukaryotic DNA replication. *Trends Cell Biol.* **2**, 298–304.
Diller, J.D. and Raghuraman, M.K. (1994) Eukaryotic replication origins – control in space and time. *Trends Biochem. Sci.* **19**, 320–325.
Earnshaw, W.C. (1994) Mitosis. *Bioessays* **16**, 639–643.
Foiani, M., Lucchini, G. and Plevani, P. (1997) The DNA polymerase α-primase complex couples DNA replication, cell cycle progression and DNA damage response. *Trends Biochem. Sci.* **22**, 424–427.
Greider, C.W. and Blackburn, E.H. (1996) Telomeres, telomerase and cancer. *Sci. Amer.* **274**(2), 80–85.
Gruss, C. and Sogo, J.M. (1992) Chromatin replication. *Bioessays* **14**, 1–9.
Hamlin, J.L. (1992) Mammalian origins of replication. *Bioessays* **14**, 651–660.
Hoheisel, J.D. (1994) Application of hybridization techniques to genome mapping and sequencing. *Trends Genet.* **10**, 879–893.
Hozák, P. and Cook, P.R. (1994) Replication factories. *Trends Cell Biol.* **4**, 48–52.
Lichter, P. (1997) Multicolor FISHing: what's the catch? *Trends Genet.* **13**, 475–479.
Roca, J. (1996) The mechanisms of DNA topoisomerases. *Trends Biochem. Sci.* **20**, 156–160.
Svaren, J. and Chalkley, R. (1990) The structure and assembly of active chromatin. *Trends Genet.* **6**, 52–57.
Travers, A.A. (1994) Chromatin structure and dynamics. *Bioessays* **16**, 657–662.

**Section G**

Adams, M.D., Rudner, D.Z. and Rio, D.C. (1996) Biochemistry and regulation of pre-mRNA splicing. *Curr. Opin. Cell Biol.* **8**, 331–339.
Apirion, D. and Miczak, A. (1993) RNA processing in prokaryotic cells. *Bioessays* **15**, 113–121.
Bachellerie, J.P. and Cavaillé, J. (1997) Guiding ribose methylation of rRNA. *Trends Biochem. Sci.* **22**, 257–262.
Benne, R. (1990) RNA editing in trypanosomes: is there a message? *Trends Genet.* **6**, 177–181.
Burley, S.K. (1996) The TATA box-binding protein. *Curr. Opin. Struct. Biol.* **6**, 69–75.
Chalut, C., Moncollin, V. and Egly, J.M. (1994) Transcription by RNA polymerase II. *Bioessays* **16**, 651–655.
Chan, L. (1993) RNA editing: exploring one mode with apolipoprotein B mRNA. *Bioessays* **15**, 33–43.
Decker, C.J. and Parker, R. (1994) Mechanisms of mRNA degradation in eukaryotes. *Trends Biochem. Sci.* **19**, 336–340.
Draper, E. (1996) Strategies for RNA folding. *Trends Biochem. Sci.* **21**, 145–149.

Fournier, M.J. and Maxwell, E.S. (1993) rRNA processing. *Trends Biochem. Sci.* **18**, 131–135.

Geiduschek, E.P. and Kassavetis, G.A. (1995) Comparing transcriptional initiation by RNA polymerase I and RNA polymerase III. *Curr. Opin. Cell Biol.* **7**, 344–351.

Hodges, P. and Scott, J. (1992) Apolipoprotein B mRNA editing: a new tier for the control of gene expression. *Trends Biochem. Sci.* **17**, 77–81.

Kable, M.L., Heidmann, S. and Stuart, K.D. (1997) RNA editing: getting U into RNA. *Trends Biochem. Sci.* **22**, 162–166.

Lafantavine, D.L.J. and Tollervey, D. (1998) Birth of the snoRNPs: the evolution of the modification guide snoRNAs. *Trends Biochem. Sci.* **23**, 383–386.

Mackay, J.B. and Crossley, M. (1998) Zinc fingers are sticking together. *Trends Biochem. Sci.* **23**, 1–4.

McKnight, S.L. (1991) Molecular zippers in gene regulation. *Sci. Amer.* **264**(4), 32–39.

Morrisey, J.P. and Tollervey, D. (1996) Birth of the snoRNPs: the eukaryotic pre-RNA processing system. *Trends Biochem. Sci.* **20**, 78–82.

Pace, H.C., Kercher, M.A., Lu, P., Markiewicz, P., Muller, J.H., Chang, G. and Lewis, M. (1997) *Lac* repressor genetic map in real space. *Trends Biochem. Sci.* **22**, 334–338.

Pugh, B.F. (1996) Mechanisms of transcription complex assembly. *Curr. Opin. Cell Biol.* **8**, 303–311.

Reeder, R.T. and Lang W.H. (1997) Terminating transcription in eukaryotes: lessons learned from RNA polymerase I. *Trends Biochem. Sci.* **22**(12), 473–477.

Rhodes, D. and Klug, A. (1993) Zinc fingers. *Sci. Amer.* **268**(2), 32–39.

Roeder, R.G. (1996) The role of general initiation factors in transcription by RNA polymerase II. *Trends Biochem. Sci.* **21**, 327–334.

Ross, J. (1996) Control of messenger RNA stability in higher eukaryotes. *Trends Genet.* **12**, 171–175.

Scott, W.G. and Klug, A. (1996) Ribozymes: structure and mechanism of RNA catalysis. *Trends Biochem. Sci.* **21**, 220–224.

Smith, H.C. and Sowden, M.P. (1996) Base modification mRNA editing through deamination – the good, the bad and the unregulated. *Trends Genet.* **12**, 418–424.

Talcott, B. and Shannan-Moore, M. (1999) Getting across the nuclear pore complex. *Trends Cell Biol.* **9**, 312–318.

Tarn, W.Y. and Steitz, J.A. (1997) Pre-mRNA splicing: the discovery of a new spliceosome doubles the challenge. *Trends Biochem. Sci.* **22**, 132–137.

Tijan, R. (1995) Molecular machines that control genes. *Sci. Amer.* **272**(2), 38–45.

*Trends in Biochemical Sciences* (1996) **21**(9). Whole issue devoted to articles on RNA polymerase II and control of transcription.

Valcárcel, J. and Green, M.R. (1996) The SR protein family: pleiotropic functions in pre-mRNA splicing. *Trends Biochem. Sci.* **21**, 296–301.

Verrijzer, C.P. and Tijan, R. (1996) TAFs mediate transcriptional activation and promoter selectivity. *Trends Biochem. Sci.* **21**, 338–342.

Von Hippel, P.H. and Langowski, J. (1995) Action at a distance: DNA looping and initiation of transcription. *Trends Biochem. Sci.* **20**, 500–505.

Wahle, E. (1992) The end of the message: 3′-end processing leading to polyadenylated messenger RNA. *Bioessays* **14**, 113–119.

Wahle, E. and Keller, W. (1996) The biochemistry of polyadenylation. *Trends Biochem. Sci.* **21**, 247–250.

Weiss, K. (1998) Importins and exportins: how to get in and out of the cell nucleus. *Trends Biochem.* **23**, 185–189.

**Section H**

Amara, J.F., Cheng, S.H. and Smith, A.E. (1992) Intracellular protein trafficking defects in human disease. *Trends Cell Biol.* **2**, 145–150.

Bukau, B., Hesterkamp, T. and Luirink, J. (1996) Growing up in a dangerous environment: a network of multiple targeting and folding pathways for nascent polypeptides in the cytosol. *Trends Cell Biol.* **6**, 480–486.

Cedergren, R. and Miramontes, P. (1996) The puzzling origin of the genetic code. *Trends Biochem. Sci.* **21**, 199–200.

Chen, X and Schnell, D.J. (1999) Protein import into chloroplasts. *Trends Cell Biol.* **9**, 222–227.

Ellis, J.R. and Hemmingsen, S.M. (1989) Molecular chaperones: proteins essential for the biogenesis of some macromolecular structures. *Trends Biochem. Sci.* **14**, 339–343.

Hegde, R.S. and Lingappa, V.R. (1999) Regulation of protein biogenesis at the ER membrane. *Trends Cell Biol.* **9**, 132–137.

Helenius, A., Marquardt, T. and Braakman, I. (1992) The endoplasmic reticulum as a protein folding compartment. *Trends Cell Biol.* **2**, 227–232.

Hong, W. and Tang, B.L. (1993) Protein trafficking along the exocytotic pathway. *Bioessays* **15**, 231–239.

Knight, R., Freeland, S.J. and Landweber, L.F. (1999) Selection, history and chemistry: the three faces of the genetic code. *Trends Biochem. Sci.* **24,** 241–247.

Liu, R. and Neupert, W. (1996) Mechanisms of protein import across the outer mitochondrial membrane. *Trends Cell Biol.* **6**, 56–61.

Martin, J. and Hartl, F-U. (1994) Molecular chaperones in cellular protein folding. *Bioessays* **16**, 689–692.

Martoglio, B. and Dobberstein, B. (1996) Snapshots of membrane-translocating proteins. *Trends Cell Biol.* **6**, 142–147.

McCarthy, E.G. and Gualerzi, C. (1990) Translational control of prokaryotic gene expression. *Trends Genet.* **6**, 78–85.

Melefors, O. and Hentze, M.W. (1993) Translational regulation by mRNA-protein interactions in eukaryotic cells. *Bioessays* **15**, 85–91.

Pandey, A and Lewitter, F. (1999) Nucleotide sequence databases: a goldmine for biologists. *Trends Biochem. Sci.* **24**, 276–280.

Peters, J-M. (1994) Proteasomes: protein degradation machines of the cell. *Trends Biochem. Sci.* **19**, 377–382.

Ramakrishnan,V. and White, S.W. (1998) Ribosomal protein structures: insights into the architecture, machinery and evolution of the ribosome. *Trends Biochem. Sci.* **23**, 208–212.

Rapoport, T.A. (1990) Protein transport across the ER membrane. *Trends Biochem. Sci.* **15**, 355–358.

Riis, B., Rattan, S.I.S., Clark, B.F.C. and Merrick, W.C. (1990) Eukaryotic protein elongation factors. *Trends Biochem. Sci.* **15,** 420–424.

Roth, M.G. (1999) Lipid regulators of membrane traffic through the Golgi complex. *Trends Cell Biol.* **9**, 174–179.

Rothman, J.E. and Orci, L. (1996) Budding vesicles in living cells. *Sci. Amer.* **274**(3), 50–55.

Sandoval, I.V. and Bakke, O. (1994) Targeting of membrane proteins in endosomes and lysosomes. *Trends Cell Biol.* **4**, 292–296.

Vitale, A. and Chrispeels, M.J. (1992) Sorting of proteins to the vacuoles of plant cells. *Bioessays* **14**, 151–161.

Wienhues, V. and Neupert, W. (1992) Protein translocation across mitochondrial membranes. *Bioessays* **14**, 17–25.

**Section I**

Brown, T.A. (1995) *Gene Cloning: An Introduction*, 3rd Edn. Chapman and Hall, London.

Capecchi, M.R. (1994) Targeted gene replacement. *Sci. Amer.* **270**(3), 34–41.

French Anderson, W. (1995) Gene therapy. *Sci. Amer.* **273**(3), 96–99.

Gasser, C.S. and Fraley, R.T. (1992) Transgenic crops. *Sci. Amer.* **266**(6), 34–49.

Gerhold, D., Rushmore, T. and Caskey, C.T. (1999) DNA chips: promising toys have become powerful tools. *Trends Biochem. Sci.* **24**, 168–173.

Gilboa, E. and Smyth, C. (1994) Gene therapy for infectious diseases: the AIDS model. *Trends Genet.* **10**, 139–144.

Glick, B.R. and Pasternak, J.J. (1998) *Molecular Biotechnology*, 2nd Edn., ASM Press, Washington.

Mullis, K.B. (1990) The unusual origins of the polymerase chain reaction. *Sci. Amer.* **262**(4), 36–41.

Old, R.W. and Primrose, S.B. (1994) *Principles of Gene Manipulation: An Introduction to Genetic Engineering*, 5th Edn., Blackwell Scientific Publications, Oxford.

Robbins, P.D., Tahara, H. and Ghivizzani, S.C. (1998) Viral vectors for gene therapy. *Trends Biotechnology* **16**, 35–40.

Strachan, T. and Read, A.P. (1996) *Human Molecular Genetics*. BIOS Scientific Publishers, Oxford.

Verma, I.M. (1990) Gene therapy. *Sci. Amer.* **263**(5), 34–41.

Watson, J.D., Gilman, M. Witkowski, J. and Zoller, M. (1992) *Recombinant DNA*. 2nd Edn. W.H. Freeman, New York.

**Section J**

Keller, S.R. and Lienhard, G.E. (1994) Insulin signalling: the role of the insulin receptor. *Trends Cell Biol.* **4**, 115–119.

Sharon, N. (1980) Carbohydrates. *Sci. Amer.* **245**(5), 90–116.

**Section K**

Brown, M.S. and Goldstein, J.L. (1984) How LDL receptors influence cholesterol and atherosclerosis. *Sci. Amer.* **251**(5), 52–60.

Weissmann, G. (1991) Aspirin. *Sci. Amer.* **264**(1), 58–64.

**Section L**

Govindjee, H. and Coleman, W.J. (1990) How plants make oxygen. *Sci. Amer.* **262**(2), 42–45.

Junge, W., Zill, H. and Engelbrecht, S. (1997) ATP synthase: an electrochemical transducer with rotatory mechanics. *Trends Biochem. Sci.* **22**, 420–423.

Youvan, D.C. and Marrs, B.L. (1987) Molecular mechanisms of photosynthesis. *Sci. Amer.* **256**, 42–48.

**Section M**

Smil, V. (1997) Global population and the nitrogen cycle. *Sci. Amer.* **277**(7), 58–63.

Warren, M.J., Cooper, J.B., Wood, S.P. and Shoolingan-Jordan, P.M. (1998) Lead poisoning, haem synthesis and 5-aminolaevulinic acid dehydratase. *Trends Biochem. Sci.* **23**, 217–221.

Warren, M.J. and Scott, A.I. (1990) Tetrapyrrole assembly and modification into the ligands of biologically functional cofactors. *Trends Biochem. Sci.* **15**, 486–491.

**Section N**

Changeux, J-P. (1993) Chemical signalling in the brain. *Sci. Amer.* **269**(5), 30–37.

Rayment, I. and Holden, H.M. (1994) Three dimensional structure of a molecular motor (myosin). *Trends Biochem. Sci.* **19**, 129–134.

Stossel, T.P. (1994) The machinery of cell crawling. *Sci. Amer.* **271**(3), 40–47.

Weber, K. and Osborn, M. (1985) The molecules of the cell matrix. *Sci. Amer.* **253**(4), 92–102.

# INDEX